"A PERFECT REIGN OF TERROR" INSURGENCY IN THE TEXAS HILL COUNTRY 1861 - 1862

By
William Paul Burrier, Sr.
With Archaeology Support From Tim Darby

A *Watercress Press* book
from Geron and Associates
www. watercresspress.com

ISBN-13: 978-09897820-6-7
ISBN-10: 09897820-6-9

Cover image by Wm. Paul Burrier, Sr.
Cover design by 3iii's Graphic Studios

Note regarding the cover phogograph: The background is the
Confederate Battle Flag, with a cutout of Texas showing where
events took place. The flags in the background are that of the state
of Texas and the Black/Red/Gold banner of the 1848 German
Revolution, (the background of the current German flag)
representing German Forty-Eights and their ideals. Between those
flags is a cutout of the area where the Unionists wanted to
establishe the Free State of West Texas. The flags to the left and
right are that of the Confederate States, and of the United States.
Along the bottom of the photo is an 1859 Sharps carbine – which
would have been the weapon carried by the Confederates, a pair
of cap and ball pistols, which both insurgents and Confederate
troops used. Next is a saddle bag of the type the Confederates
used, with a tin cup, an ammo pouch, and a pistol cap holder. To
its left is a cap and ball rifle of the type the insurgents used, and a
pouch of the type the insurgents would have used to carry their
personal items. – WPB

CONTENTS

Maps

Dedication

To the memory of Gregory Krauter; Freethinker extraordinary for dedication, collection, preservation and sharing of cultural heritage.

To the memory of Esther Boerner Weidenfeld; esteemed Germanic Matriarch.

In honor of Anne Seidensticker and Michael Stewart for their openness, acceptance, encouragement, and love.

W.P. Burrier
Leakey, Texas, 2017

When Two Strong Men, Stand Face to Face

*Right: Major "Fritz"
Tegener, leader of the
military wing of the
insurgency in the Hill
Country.*

*Lleft, below:
Lieutenant Colin D.
McRae, leader of the
pursuit force.*

A Perfect Reign of Terror

Introduction

This is the story of the bloody and emotional conflict between Secessionists and Unionists and insurgents in the six-county area of the Texas Hill Country during the War Between the States [Civil War]. The term 'Hill Country' is used to describe the area north and northwest of San Antonio that includes Bandera, Blanco, Gillespie, Kendall, Kerr, and Medina Counties. [Maps A and B, pp 21, 22-23]. The basis of the story is how a group of German intellects attempted to established a new 'German Fatherland' in Texas and how this idea eventuality developed into a Texas German Unionist plan to create the *Free State of West Texas* and the resulting events leading up to the 'Nueces Battle,' or the 'Nueces Massacre' as most current accounts calls it, the battle itself, and the immediate aftermath. Time and space does not allow the telling of the events of late 1863 through 1865 when many more area citizens, especially in Blanco, Gillespie, and Kerr Counties, were killed during the *Haengerbande* [hanging gang] times, also known as the 'Bushwhacker War' times. Groups of vigilantes conducted these killings, some operating from legal units such as the militia of the Third Frontier District and companies of the Frontier Regiment, others from secret associations, both Secessionists and insurgent.

"The Blackest Crime in Texas Warfare" is a good example of the strong emotion raised when anyone reads or talks about the Nueces Battle of August 10, 1862, and the execution of Union insurgents shortly thereafter. There is no doubt that by today's standards the Confederate and state leaders who either ordered, conducted, or did nothing about those executions, should have been tried for war crimes. These events are inexcusable by today's standards of warfare, but it must be remembered the times in which they occurred. Many a Texas Ranger captain earned high praise for deaths of 'outlaws,' 'Mexicans,' and 'Indians' in circumstances not too unlike the insurgents. By the 1860s, 'frontier justice' had become a way of life for many of the men involved, both insurgents and Secessionists. For example, this 'frontier justice' was even formalized in San Antonio in the form of a Vigilantes and Vigilance Committee under the leadership of Asa Mitchell, a prominent and well-respected Bexar citizen. Another example was in Medina County in the form of a Committee of Safety. R. H. Williams, whom many writers often quote to demonize the Confederates, in his book, *With The Border Ruffians*, tells of his taking part in 'frontier justice' when he and a party of eleven ranchers captured 'Mexican horse stealers' in April or May 1862 and hanged four of them. For all of Williams condemnation of the Confederates use of 'frontier justice' when he felt it was proper he turned to Asa Mitchell and his Vigilance Committee for a little 'frontier justice' against ranchers whom had done some of his hired hands wrong.[1]

This book resulted from the author's desire to learn more details about the battle. Surprisingly, when research was started very little detailed data was available. What was found was generally biased and from the insurgent viewpoint. These accounts portrayed the state and Confederate troops deployed to the Hill Country as nothing more than a group of "bullies, bandits, and cut-throats" because "Texas had been drained of her very best men early in the war," and "a man who had been drummed out of the United States regular army in disgrace," led these 'desperados' claimed pro-insurgent descendants and writers. He had a "zeal seemingly inspired more by personal glory and gain than loyalty to the Confederacy … [and] moved into the German settlements and instigated a systematic campaign of terror against those who had not sworn allegiance to the South." He possessed "an arrogance born of sudden authority, [and] burned lonely farm and ranch homes and whole isolated communities." He took livestock and any other property of value and turned it to cash which lined his own pockets. Anyone who resisted was murdered. The Confederates "destroyed crops, burned homes, and lynched over twenty Germans during this period" claims another account. The Confederate units were nothing but a, "roving bands of 'partisan rangers' [who] charged about the countryside and, in the name of the Confederacy, hanged innocent men before the eyes of their wives and children." Indiscriminately the 'rangers' burned the cabins of settlers whose loyalty to the South was thought suspect, and as they hustled the poor souls out into the night, they shouted their own Rebel Yell: "Give 'em a quick look up a tree!" These stories claim that these 'partisan rangers' were not content with hanging, the gang often tortured their victims to death before the eyes of their wives and children. Most writers and historians claim these partisan rangers were "irregular state troops," which operated independent from any Confederate or state control such as Quantrill's Raiders. In fact, the Confederate and state troops deployed into the Texas Hill Country were very much a part of the Confederate Army or Texas State Troops. On April 21, 1862, the Confederate Congress passed a bill to 'organize bands of partisan rangers" and that they shall "be entitled to the same pay, rations, and quarters during their term of service, and be subject to the same regulations as other soldiers." The only major difference between 'partisan rangers' and other cavalry troops were the members of a partisan ranger unit would be paid in full value for any arms and munitions of war captured from the enemy. Beside James Duff's Partisan Rangers other Texas Confederate units organized under this act were: W. P. Lane's 1st Regiment, Texas Partisan Rangers; Gurley's 1st Regiment Texas Partisan Rangers; Isham Chrisum's 2nd Regiment Texas Partisan Rangers; Alexander's 2nd Regiment Texas Partisan Rangers [aka as 34th Regiment, Texas Cavalry]; Leonidas M. Martin's 5th Regiment Texas Partisan Rangers [which was organization by the consolidation of the 9th and 10th Battalions Texas Partisan Rangers].

The insurgents were portrayed as innocent settlers who had organized militia units "for the sole purpose of defending settlements against Indians and outlaws" and were fleeing the area under a proclamation giving them 30 days to leave. Detailed research shows all these accounts are incorrect![2]

A leading Texas historian described the battle as: "A rumble of onrushing cavalry which filled the valley. Horsemen were coming in from the east, and from the south. In no time, it seemed, horses began slashing across the West Nueces ford. Rebel yells split the tense atmosphere, and over the thunder of hoof beats, rifles belched flame into the night. The campers cried out as bullets ripped through their blankets, into their bodies. Those who rose before the onslaught were trampled. Nevertheless, defensive forces were rallied."[3]

These accounts claim after the battle, "Nine of these refugees surrendered as prisoners of war and were brutally murdered." Again, detailed research shows all these supposed 'facts' are untrue or very misleading![4] Respected Texas historians seem content to follow the same general theme. T. R. Fehrenbach's book *Lone Star* uses a very similar description in telling about the event. Even *The Handbook of Texas* follows the same theme. James A. Michener, in his novel *Texas*, has an account of the event. Michener's fictionalize story is closer to the truth than Fehrenbach's *Lone Star* or *The Handbook of Texas*.[5]

As the research delved beyond secondary sources and looked at primary documents, it became clear what had been portrayed was very one sided and biased, and in many cases untrue. The real story has never been completely told, which is the purpose of this book. **It does not present the insurgents side or the Confederate side. It presents the**

factual side! For those who have read a previous account of the battle or had ancestors in the battle and may feel this book is trying to discredit the insurgents or write a revisionist history of the battle, let me address that thought. As an infantry officer who served over three years in combat in three different geographical areas, and who during this time saw many brave young Americans perform outstanding acts of courage, some of which took place during a three-day battle when my command was overrun twice by a large enemy force. Over twenty of my men were killed and another fifty or so wounded, resulting in the award of one Medal of Honor, two Distinguished Service Crosses, and numerous Silver and Bronze Stars earned, so I know a little bit about courage and heroes. As such, I consider the insurgents who stood up for their beliefs as true heroes of the same caliber as the heroes of the Alamo. Both stood up for what they believed and faced almost certain death because of their strong beliefs. Others may have given in and taken a course that was more popular or easier and less dangerous. The insurgents didn't. They took a stand for their beliefs. Like the men of The Alamo, they were surprised at the appearance of a perceived large enemy force and were not fully prepared to fight. Most of the insurgents like the 186 or so Texians at the Alamo on the morning of March 6, 1836, knew the odds were overwhelming chose to stand and fight and not run. Many of the insurgents paid the same price, as did the Alamo defenders. A major difference was there was no escape possible for the Alamo defenders, while there was for the insurgents. Some of the insurgents lived to fight again, not only on the battlefield, but some in state and national political arenas. Like the Alamo, myths have replaced many of the factual facts of the battle.

The myths the Confederates were just a group of bullies, bandits, and cut-throats are not true. They were a normal cross-section of the citizens of Texas at the time. They also believed in their cause and had heroes that day. One was a second lieutenant who rallied his men after they suffered over twenty percent casualties and being repelled after three assaults. Others were a pair of brothers who were killed trying to help each other after receiving serious wounds. For an attacking force to continue after losing over a fourth of their number is a credit to their bravery. Like the insurgents, many of the Confederates lived to fight again and some later served in local and state political arenas.

I have a major problem with the many writers and historians who accept the myths as factual without doing proper research and embrace these myths into the accounts we read over the years and continue to read today. Shame on them! The result is that the factual story of the event is not known. To portray these heroes falsely, both insurgents and Confederates, is totally unfair to these men and their beliefs, actions, and later contribution to Texas and the United States.

This book challenges almost all previous accounts of the battle for truthfulness and historic accuracy. This challenge is against the myths that have been incorrectly presented as facts. Because most of the myths are biased toward the insurgent's view, some descendants, and perhaps even some historians, feel this book is "akin of a Confederate rewriting of events" and is a pro-Confederate version or "trying to justify Confederate actions." Because this book challenges almost everything these descendants and other readers have been lead to believe for over a hundred and fifty-five years it is understood why such a false perception can be reached. It is also understood no matter how much evidence is presented to show the myths are incorrect some are still going to believe this is a pro-Confederate book. The fact is the challenges are against myths that are untrue or greatly biased. Therefore, these challenges are perceived as pro-Confederate. **They are not! They are pro-factual!** Most 'truths' or 'views' or 'beliefs' are not black or white, but gray. The reader is encouraged to look at the documentation presented with an open mind.

A major theme of this book is the insurgents organized and conducted an insurgency. This theme is new. The reader is strongly encouraged to read the prolog, "The Insurgence Model," before beginning the normal chapters. Unless someone has studied how an insurgency is organized and conducted they would not recognize such. Few Texas writers and historians have studied unconventional warfare. One of the major criticism of the research of this book is that the author "has assumed too much from … his Army experience." Like any professional, one gains insight based on academic observation, practical training and experience. Over twenty-four years of studying insurgencies and counter-insurgencies, over two years fighting an insurgency, and a year in Pakistan observing an insurgency, plus an undergraduate degree in history and master's studies in political science qualifies one to examine and conclude the actions of the Union Loyal League meets the definition of an insurgency. Political science professors would never let a student 'assume' something as his final conclusion. A political scientist first makes an assumption or commonsense observation. From there he develops a hypothesis, an educated guess. After testing the hypothesis, the political scientist develops a theory. A theory is when a second or third scientist tests the data and the same results are found. Once a theory is tested enough and accepted, it reaches the final classification, that of being a law or fact. Some political hypotheses and theories regarding the insurgent's organization will be developed in this book. Chapter Two shows how the actions of this insurgents' organization met the definition of an insurgency. Chapter Two also explains why the Unionists conducted the insurgency; to create the *Free State of West Texas*.[6]

The author in the narrative part of this book tried not to get emotional in recounting the events. However, some emotions do show in the endnotes when discussing the myths. These emotions have nothing to do with the events. They show because other writers and historians use the myth as factual events in both fiction and non-fiction articles and books without conducting proper research. I understand the emotions of the descendants of both sides of this tragedy as my family was involved in both insurgents and Secessionists activities.[7]

I was born in Fredericksburg, Texas, the major German settlement in the Texas Hill Country. My paternal grandmother's family, the William Henry Benson family, arrived in Gillespie County by 1855. My great-grandfather, William Thomas Benson, and his brother John Washington Benson, were members of Davis's Company, members of the pursuit force, and took part in the battle. They were later involved in many of the pro-Confederate bushwhacker activities. I am not proud of that, but facts are facts. Their sister, my great-great-aunt, Susan Eveline Benson, was the first wife of Frederick [Fritz] Tegener, the leader of the fleeing insurgent group. William Thomas Benson lived with his sister and brother-in-law, Fritz Tegener, when the 1860 Kerr County Census was enumerated. Here is a case of brothers-in-law fighting against each other in battle. The Benson family's involvement continued after the war. Two of William Thomas' younger sisters and brothers married children of James Billings who, according to a Billings family story, was killed in January 1863 in Gillespie County by insurgents disguised as Indians. James M. Benson married Pricilla Billings and Rebecca Benson married John Billings, who survived the encounter with the Unionists, although seriously wounded.

One of my great-great-grandfathers, George Robert Hollomon, Sr., arrived in Kerr County by 1861. He and his sons served in several home guard and militia units raised for frontier protection and helped pursue insurgents as well as the *Haengerbande.*

My great-grandfather, William Cost Burrier, and his sons, which included William Henry Burrier, my grandfather, moved to Gillespie County in 1882 and knew many of the actual participants. William Cost Burier and another son, Richard Marion Burrier, lived in Fredericksburg and because of their 'Germanic' ancestry made a great effort to become 'Germanized' to include relearning the language. My father, Thomas Watson Burrier, was born in 1894 and grew up in Gillespie County. He served with a Gillespie/Kerr County National Guard company during World War I, during which he saw and experienced the prejudice against the Hill County 'Germans.' He was a great storyteller and I experienced many a happy hour listening to his stories.[8]

My mother's side of the family was living in Blanco, Kendall, and Comal counties by the early 1850s. My great-great-grandfather, Neill Robison, lived near John William Sansom on Curry's Creek and in addition to being a fellow Texas Ranger was a personal friend of Sansoms'. His granddaughter, my grandmother, Jessie Robison Walker, was another great storyteller. Many of my southern beliefs came from her and her stories about the "Yankees." Her daughters, my mother Beatrice Burrier and my aunt Muriel Wagner, retained many of the family stories and passed them on to their children and grandchildren. I pray I do as well and provide them to my daughter, Melissa, and my grandchildren; Tammy, Austin, and Ashley. However, I now understand the stories I heard as a child were biased, based on the storyteller's prejudice. So hopefully, when I pass these stories on the bias will be removed. My great-great-

great-grandfather, Joel Cherry, was one of the first settlers in Blanco County. His daughter, Rebecca, married John Sheppard. Pro-Union bushwhackers raided her home and much of their personal property stolen. They hid their gold and silver in a 'secret desk drawer' that was not found during the raid. That desk had a special place in my Aunt Muriel's home and is now in the home of her son, Don Wagner. Pro-Union bushwhackers disguised as Indians killed one of her sons, my great-great-great uncle, William Irwin [or Erwin] Sheppard, his wife and small child. Pro-Confederate militia units killed Moses Moran Snow, father of Sarah and Moses [Mode] Snow [my mother's aunts] and his nephew, William F. Snow, in early 1864 because of their Unionist activities. A visit to their graves is a very emotional experience.

I consider myself a Texan and a Southerner. As such, the war that took place between 1861 and 1865 was the 'War Between the States' it was between the northern and southern states and was not a true 'Civil War.' However, the war in the Texas Hill Country was truly a 'Civil War.' It was family member against family member and friend against friend. My family is one of many examples.

A major problem in telling the story is the conflicting terms used in the 1860s and their use today: for example, Union, Yankee and Federal. The term Federal is used herein to describe United States government agencies before and after the war. The term Union is used to describe them during the war. Most of the source documents and articles use terms such as 'German,' 'Mexican,' 'Anglo,' and 'American' in telling the stories. Fully understanding that many of the

'German' settlers were pro-Union, this book uses the term 'Unionists' when talking about individuals whose actions resulted in state and Confederate counter-actions. It also includes those individuals who were members of the authorized Medina County Independent Battalion of the 31st Brigade District who supported the Union and members of the unauthorized and secret Bandera County Unionist Company. The term 'Union sympathizer' or 'Unionist sympathizer' is used when talking about known or suspected individuals who supported the Union or the Unionists. In the Texas Hill Country, there were several groups of Unionists. They include several 'Anglos' in far eastern Blanco County and far western Travis County. They also included the Medina and Bandera county Unionists. The term 'insurgents' includes known members of the insurgency and its military arm, the secret and unauthorized military battalion. Therefore, there are many times when these terms are used.

This brings us to another problem regarding the Unionist and insurgents. What was the name they gave to their secret group? Today it is known as the Union Loyal League, but was not the name insurgents used. The term Union Loyal League was first used by John Sansom in his 1905 pamphlet and has been used by everyone since. When researching German-language accounts, I found that at no time did the insurgents use this term. Instead, they called it *Des Organisator* or when translated "the organization." Throughout this book, I will use the term *Des Organisator* instead of Union Loyal League. There will be several times

I will remind the reader of this fact and point out German-language accounts when this term was used.

Three other terms greatly used in the 1860s and not generally used in the same content today are 'Anglos,' 'Germans,' and 'Americans.' These are used to describe individuals based on their, or their parents, location of birth. An 'Anglo,' is someone born in the United States or in the United Kingdom while a 'German' is someone who was born (or whose' parents were born) in one of the 'Germanic' kingdoms of Europe. In many of the 'German' writings they use the term 'American' meaning a non-German who had lived in the United States for an undetermined period of time. The use of these three terms becomes important when describing the Unionists and insurgents because both 'Anglos' and 'Germans' were Unionists and insurgents. It has been generally believed that the majority of the insurgents were Germans. This was not the case. A large part, about a third, of the insurgent movement was Anglo.

Two political types of Unionist are discussed. One is moderate [or 'Grays' as called in the German communities]. They supported the Union, but attempted to avoid a major conflict with state and Confederate authorities. The other is militant [or 'Greens' as called in the German communities]. They were totally opposed to the Confederacy and wanted to take strong action, to include military, against the state and Confederate authorities. The insurgents were part of this group. The term used to describe the Confederacy was harder to define. The term used during early 1861 and 1862 was Secessionists. It means someone who favored withdrawing from the United States and forming a separate country, the Confederate States of America (CSA). At first, the term Confederate or Secessionist was used to describe individuals or units that supported the Texas or Confederate governments during war, but this soon resulted in great confusion, especially when talking about military units. Therefore, the terms state and Confederate are used. State describes the Texas government and its military units. Not all state military units supported the Confederacy. There are times, for the sake of clarification, the term pro-Union or pro-Confederate is used. For example, the Medina County Independent Battalion of the 31st Brigade District was a pro-Union or Unionists unit. Confederate describes the Confederate government and its military units.

This brings up another term problem that being the many types of military units found operating during this period. Even I, a career Army officer, became confused about who these units were and under whose command or authority they operated. This book uses the following terms to describe state and Confederate military units.

1) State, which encompassed three types of units: Home Guard, Militia, and the Frontier Regiment.

2) Confederate - 'regular' Confederate units belonging to the Confederate government, and includes the 'partisan rangers.'

Confederate units were part of the Confederate States Army (CSA). They could be deployed anywhere the Confederate government ordered. Generally, these units were organized by individuals within a state; once organized were 'mustered' into Confederate service. Since an individual organized them from a certain state, normally, their designations included that state's name and often that individual's name. The Confederate government gave the unit a numerical number. There were generally three combat types of such units; artillery, cavalry, and infantry. Within the cavalry there were several different names, such as 'mounted rifles,' 'partisan rangers,' or 'dragoons,' as well as cavalry. Examples of these types of units in this book are Duff's Company of Partisan Dragoons or Rangers, Taylor's 8th Battalion Texas Cavalry, Duff's 14th Battalion Texas Cavalry, 1st Regiment Texas Mounted Rifles, 2nd Regiment Texas Mounted Rifles, 32nd Regiment Texas Cavalry, and 33rd Regiment Texas Cavalry.

Under state control were the three types of units; Home Guard, Militia, and the Frontier Regiment. None of these were under the command of the Confederate military commanders in Texas and could not be ordered to duty outside of the State of Texas. However, there were times when the governor directed these units to cooperate with the local Confederate commander. As a general rule, these units were 'part-time' soldiers. The term 'State Troops' was used to describe these units. As a general rule, membership in one of these units did not exempt a man from the Confederate draft. This resulted in great problems for these units.

There was one state unit that was full-time; the Frontier Regiment. It was designed, as the name implies, for frontier protection. Unofficially, membership did keep men out of Confederate service. Therefore, there was a major desire by many men to enlist in this regiment. Two companies of the Frontier Regiment played a major role in this story, two others a minor role.

The next state unit was the home guard companies organized within a frontier county and designed for protection from Indian attacks. Their term of service was very limited. Sometimes it was only for a certain crisis, sometimes for a certain period. Only a small number of the men was on 'active duty' at any one time, and paid by the state for only those days served. Such units in this book are: Kerr County Minutemen, Blanco County Minutemen, and Pedernales Cavalry Company. Another home guard unit that played a major role was the Gillespie County Company organized by insurgent Philip Braubach. The Gillespie County Rifles, and the Comfort Minuteman Company where home guard type units; but never called to active duty.

The militia units are the most confusing. Texas had authorized a series of militia units before the war. In many cases, they were more of a social club than a functioning military unit. The Gillespie County Rifles was one. When Texas seceded, many of these militia companies were called to active duty. Others were quickly organized. A semi-secret pro-slavery society known as the Knights of the Golden Circle also had military type units, known as castles. These castles quickly went on active duty. The militia that went on active duty or the Knights' military castles played a major role in forcing the surrender of Federal units in Texas. After this was accomplished, some of these units became Confederate units, others disbanded, and some returned to the old militia system. A second series of militia units came into being when Texas prepared itself for war. In late 1861, the state ordered the creation of brigade districts. Every county was in one of these districts. Depending on the population of the county, companies were organized within precincts, within a county, or by combining counties. The purpose of these brigade districts was to repel invasion of the state. Every eligible male within an area was required to join one of these militia companies. These companies were organized into battalions and regiments. This was well underway when the Confederate government passed its draft law in April, 1862. A large number of men in these militia units were now subject to the draft. This resulted in almost complete disintegration of those militia units. Therefore, the militia units were of little overall importance. But this was not known until after the Confederate draft. Until then there was a major effort by the Unionists to control the militia units. There were a few times these units were called upon for service. The 31st Brigade was the brigade district in which the Hill Country counties belonged.

In January, 1864, another type of militia unit was forced. This was a hybrid between a home guard and a militia unit. Because the Frontier Regiment was transferred to Confederate service, a new frontier organization was organized. The Texas frontier was divided into three frontier districts. Each frontier district was commanded by a state major, later a state brigadier general. The frontier counties were authorized these new militia companies. They were similar to the old home guard companies. The State of Texas passed a law, which exempted men in these companies from the Confederate draft. These companies became a popular way to avoid the draft. Generally, only a third to a fourth of the men in these units were on active duty at any one time. Many of the companies from the Hill Country were made up of mostly pro-Union men. The Hill Country counties were in the Third Frontier District.

Like military units, there were two types of officers, state and Confederate. This book describes which type of officer is being discussed.

There were also the military units organized by the Unionist. There were two in the Hill Country. One was the insurgent battalion of *Des Organisator* that had three companies. Gillespie, Kendall, and Kerr Counties each had a company. The other Unionist unit was in Bandera County. Very little is known of this company. These four companies were secret organizations. Their exact size and organization are not known. Several reports claim the insurgent battalion had at least five hundred members, but it likely only had about two hundred and fifty. A major commanded the battalion. A captain and a lieutenant commanded the Bandera Company and its strength was somewhere between thirty and fifty men.[9]

The existence of so many different types of military units has caused great confusion for many researchers, writers, historians, and thus readers. For example, Jacob Kuechler commanded three different companies about the same time. First was the unit he raised to be a company for the Frontier Regiment. Second, was Company E of the 2nd

Regiment of the 31st Brigade District, which was an authorized militia company. Third was the Gillespie insurgent company of the *Des Organisator* military battalion, which was a secret and unauthorized unit. Many researchers, writers, and historians have failed to realize such and have tended to lump all three into one unit, which has caused greater confusion for the reader. The most recent example of not only confusing Kuechler's companies is the book *The German Settlement of the Texas Hill Country.* [10]

As stated, the narrative tells the story about the Nueces Battle. The story goes into great detail in order to give the reader a sense of identifying with the individuals, events, and locations where these events took place. When recounting the events, the actual quotes of participants are used whenever possible. In those cases, where there is no first-hand account, secondary sources are used from those who had ties to the events or did early research into the events. Oral interviews and written family accounts that had ancestors in the battle or related events were also a major source. The axiom "No family story is 100% true, but every family story is based on something that did happen" was fully used. These family stories were checked against primary sources, to the maximum extent possible. There were times when the family story could not be verified and when that was the case the account is identified as a family story.

The narrative and the endnotes include detailed accounts of the events. They also include accounts or myths that have been told so often they are generally accepted as facts.

Map A – Texas in 1860

Map B – The Texas Hill Country

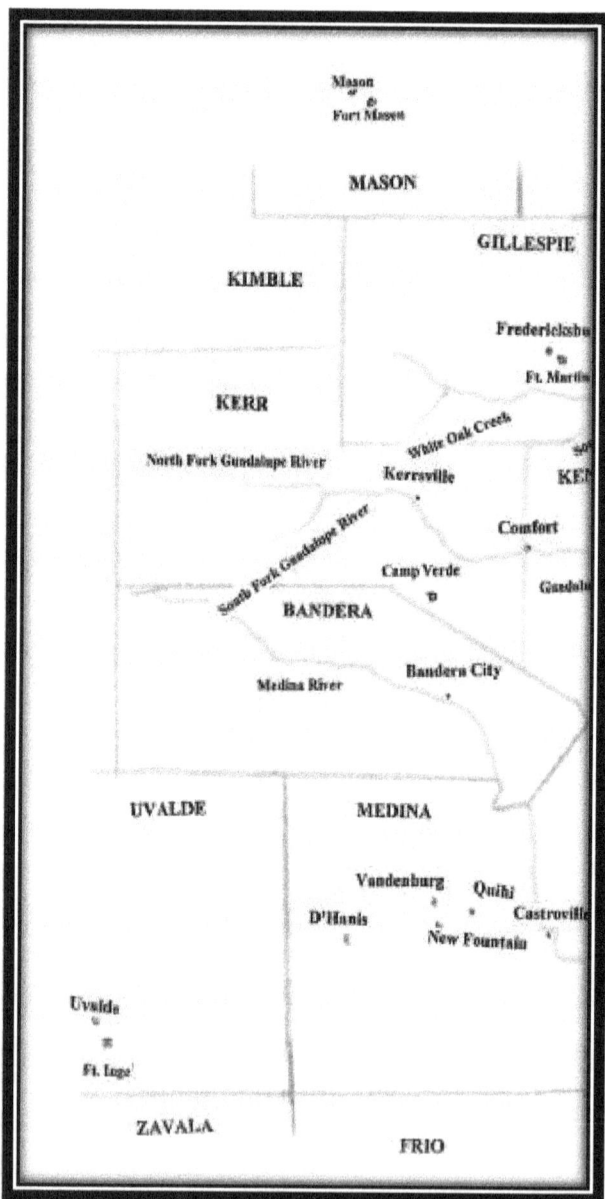

Mason
or
Fort Mason

MASON

GILLESPIE

KIMBLE

Fredericksbu

Ft. Martin

KERR

White Oak Creek

North Fork Guadalupe River

Kerrsville

KE

South Fork Guadalupe River

Comfort

Camp Verde

Guadal

BANDERA

Bandera City

Medina River

UVALDE

MEDINA

Vandenburg

Quihi

D'Hanis

Castrovill

New Fountain

Uvalde

Ft. Inge

ZAVALA

FRIO

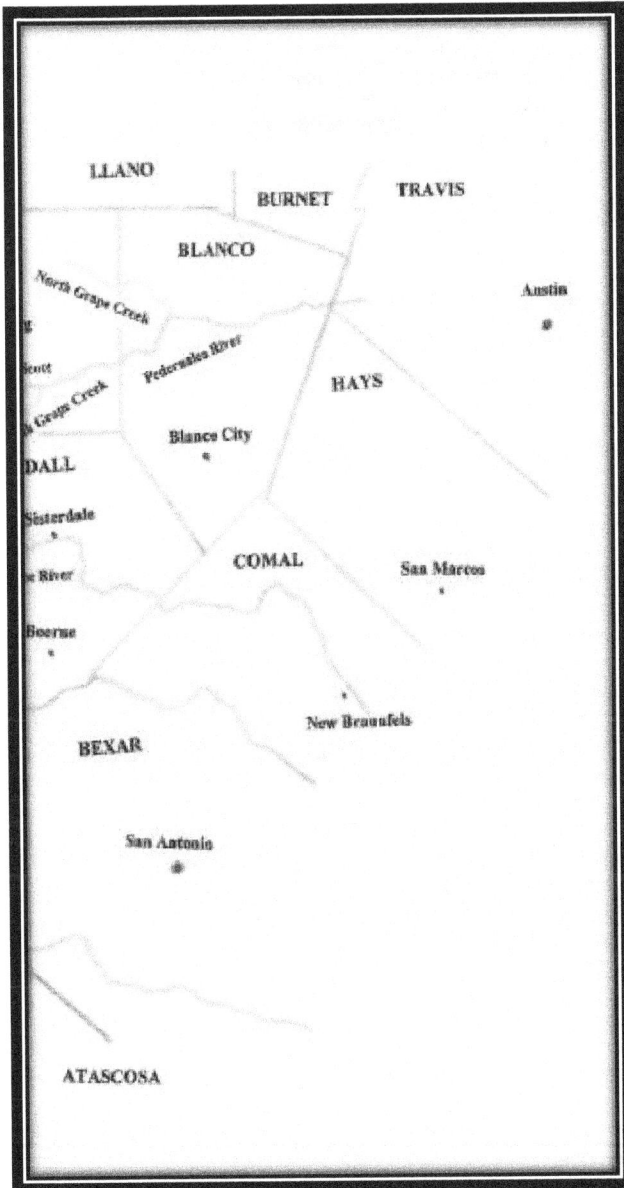

Endnotes - Introduction

1. Interview with August Hoffmann contained in newspaper article entitled
"The Blackest Crime in Texas Warfare" by Helen Raley in the *Dallas
Morning News*, May 5, 1929. This headline is the most often quoted source
to show what a hideous crime the battle (or massacre, as many writers claim)
was. This newspaper account is only one of many articles designed to
'demonize' the state and Confederate troops who took part in the battle. For
example, see the Handout from the Comfort Heritage Foundation, which
maintains the *Treue der Union* Monument in Comfort. Others are *The
Handbook of Texas* in Two Volumes, Walter Prescott Webb, editor,
(Published by Texas State Historical Association, Austin, Texas), 1952 and
The Handbook of Texas of Supplement, Eldon Stephen Branda, editor,
(Published by Texas State Historical Association, Austin, Texas), 1976,
considered Volume III of *The Handbook of Texas*, Volume II, p 18. Also
see *The New Handbook of Texas* in Six Volumes, Ron Tyler, et al eds.,
(Published by Texas State Historical Association), 1997, Volume 4, p 778;
Constitution of the Committee of Safety of Medina County, copy located in
the Medina County Library, Castroville, Texas. Williams' account is in *The
Border Ruffians*, by R. H. Williams, (University of Nebraska Press, Lincoln,
Nebraska), 1982, pp 199200 and 273-274.

2. *The Story of Fredericksburg, its past, present, points of interest and annual
events*, compiled and edited by Walter F. Edwards, (Fredericksburg
Chamber of Commerce, Fredericksburg, Texas), n.d., p 45. Also see *Tales
of Old Fredericksburg* by Walter F. Edwards, (Published by Walter F.
Edwards, Fredericksburg, Texas), 1975, p 2; Raley, *Blackest Crime in Texas
Warfare*; and *A Hundred Years of Comfort in Texas* by Guido E. Ransleben,
(The Naylor Company, San Antonio, Texas), 1974, p 121. This book has
been referred to as the best overall source on the Nueces Battle, because it
contains so many eyewitness accounts, to include one from "the
Confederate side". However, when checking the versions contained in this
book with the actual sources quoted many discrepancies were found, all
slanting the book toward the insurgent view. While the author should be
commended for locating and putting so many difference accounts in one
source, the reader needs to fully understand the author was a residence of
Comfort, Texas, and was biased in what he chose to include in the book.
This book was not written by a neutral 'historian' but by an insurgent
descendant with a very biased view. The major source of the myth the
insurgent's military units was just the local militia designed for protection
against Indians and other outlaws is *Battle of Nueces River in Kinney
County, Texas August 10, 1862* by John W. Sansom, (Published by John W.
Sansom, San Antonio, Texas), 1905, p 2. An example of this myth is 'They
Had Enough of Duff' by Louis B. Engelke *San Antonio Express Magazine*,

day and month not known, 1952. An example of the myth regarding the insurgents were traveling under some sort of proclamation is *Germany Pioneers in Texas*, by Don H. Biggers, (Published by Fredericksburg Publishing Company, Fredericksburg, Texas), 1925, p 58. Research has established no such proclamation was issued. What was issued a year before the Confederate Congress passed a law requiring everyone to either take the oath of allegiance to the CSA (CSA) or leave within forty days. Other accounts that are biased and inflammable included "Massacre by White Men" by Ruel McDaniel 'Frontier Times', Volume 43, No. 5 (August-September 1969); "Mainstreams ... The 'Last Gentlemen's War' Also Had Its My Lai" by Edwin a Roberts, Jr., 'National Observer', March 31, 1973; "Defying The State Of Texas: German Immigrants Died At the Battle of The Nueces" by Phillip Rutherford 'Civil War Times Illustrated', Volume 18, No. 1 (April 1979); *Confederate irregular Warfare 1861 – 1865: Partisan Rangers Units and Guerrilla Commands* by Bertil Haggmann, LL.M. http:hem.passaagen.se/csa01 and "Bloody Ground" by Mary Clare 'Civil War', Volume 70, (October 1998).

3. Engelke *Enough of Duff.*

4. Ibid.

5. *Lone Star – History of Texas and the Texans,* by T. R. Fehrenbach, (Wings Books, New York), 1968, p 363–364; Prescott Webb, et al, ed, *Handbook of Texas*, Volume II, p 290; Roy Tyler, et al, ed, *New Handbook of Texas*, Volume 4, p 1054; and *TEXAS*, by James A. Michener, (Random House, New York, New York), 1985, pp 625-626.

6. 'Historical Friction' by Helen Thorpe, '*Texas Monthly*', October 1997, pp 74-81 and *Scope and Methods of Political Science* by Alan C. Isaak, (The Dorsey Press, Homewood, Illinois), 1969, various pages.

7. One of the best examples of fiction is A *House Divided* by Marj Gurasich, (Texas Christian University Press, Fort Worth, Texas), 1994, pp 47 57 and pp 75-78. In this account the author has Duff remaining in the Hill Country for an extended period hunting down and murdering "Yankee-lovin' Germans." James Duff was only in the Hill Country a total of six to eight weeks on two separate trips. The account of the battle is the normal myth account, which is almost all incorrect. Another recent work that contains the age-old myths is *Sophie's War: The Journal of Anna Sophie Franziska Guenther* by Janice Shefelman, (Eakin Press, Austin Texas), 2006 pp 70, 105-107, 118, 123-124, 132-149, and 170-171.

Introduction

8. Family stories told to the author by his father and other Benson family members.

9. The term applied to these insurgent units in almost every other account is militia. These accounts claim, or imply, the Unionist units were part of the state militia system. A recent example is *Death on the Nueces: German Texan Treue der Union* by Rodman L., Underwood (Eakin Press, Austin, Texas), 2000, p 27-28. This is not correct. They were secret and purely insurgent units. They were secret and unauthorized military units. These insurgent units were made up of insurgents and organized outside of the state militia system. All the Hill Country counties had authorized militia units in which the insurgents were also members. These authorized militia and home guard companies had the mission to protect citizens from Indian and other attacks. The existence of these secret and purely insurgent units is what makes the insurgent movement in the Gillespie, Kendall, and Kerr area totally different from all other Texas Unionist movements. Therefore, the term used for these is military, not militia.

10. *The German Settlement of the Texas Hill County* by Jefferson Morgenthaler (Mockingbird Books, Boerne, Texas) 2007, various pages.

PROLOG: The Insurgency Model

To fully understand the 'Perfect Reign of Terror' that existed in the Texas Hill Country in the early years of the War Between the States it is necessary to examine the individuals involved, their experiences, and methodology they followed.

The first thing to understand is the methodology the insurgents used. Many descendants of the Texas Civil War Unionists and Texas historians deny the idea that the Texas Unionists conducted an 'Insurgency', which is a localized *Internal War* between a constituted government and rival elements originating in the same territory. United States Army Colonel Frank King argues conversely, "If it looks like a duck, walks like a duck and quacks like a duck, there's a very good chance it is a duck." When one understands how an insurgency is organized, planned, and conducted and compares this to the Texas insurgent organization known today as the Union Loyal League, one realizes it meets all the requirements of an insurgency. The activities of the Union Loyal League in 1861 and 1862 looked like, acted like, and sounded like an insurgency. It was an insurgency.[1]

A leading French military scientist defines warfare as, "An interlocking system of actions-political, economic, psychological, military-that aims at the overthrow of the established authority in a country and the replacement by another regime." Carl von Clausewitz, the father of modern military science, defines war as "an act of violence intended to compel our opponent to fulfill our will", with three major principles: (a) to conquer and destroy the armed power of the enemy; (b) to take possession of his material and other sources of strength, and (c) to gain public opinion. War, he says, "Is a continuation of politics by other means." Clausewitz argues that war is for political objectives, but once war begins military objectives become paramount.[2]

Normally, when one speaks of war it is one nation's military fighting another. As part of warfare, there are types or sub-sets, one of which is rebellion. This is where one part of a country rebels against another. The American War Between the States, was not a true rebellion. During the war the Confederate States was never recognized by any major nation as a separate nation. It was recognized by many European nations as a belligerent, which gave it some legitimacy.

The North's goal was the destruction of the rebellion and to reunite the states. Therefore, it met the definition of warfare described above. The South's goal was never to destroy the North. For both the North and the South it became 'total war.' All the resources of each 'nation' were applied in the struggle. Another type of warfare is revolution: a sudden, radical, and complete change in government. It is an overthrow or renunciation of one government, or the substitution of one government or leader for another. The South did not want to replace the Northern government; therefore, it was not a revolution.

An irregular force fighting small-scale, limited military actions against conventional military forces, characterizes guerrilla war. It is often confused with and used interchangeably with revolution or insurgency. It is not the same. Guerrilla war is part of a larger war, which can be a conventional war, a revolution or an insurgency. During the War Between the States, there were many examples where guerrilla war supplemented the two armies.

A third type of warfare is insurgency. An insurgency, involves more than just military actions. It brings all of the military components, as well as the political concepts, strategy, and tactics, into a single coherent operation. It is designed to force a change in government not necessarily the destruction of the enemy's military force. The political objective is paramount in an insurgency and is supported by military operations, unlike other forms of warfare. It is "a condition of revolt against a government that is less than an organized revolution and that is not recognized as a belligerent." As stated before an insurgency is, "A localized *Internal War* between a constituted government and rival elements originating in the same national territory, which may be guerrilla, civilian-insurrectional, or terrorist in nature." *Revolutionary War* may begin as an insurgency, but one need not develop into the other. The term insurgency correctly applies to localized conflicts, often caused by ethnic or regional demands for autonomy or secession. To better understand an insurgency, it is necessary to understand the Insurgency Model and how an insurgency fits into War or Warfare.[4]

Insurgencies can be divided into three phases. Phase I – a cellular development of resistance against an incumbent political regime by organizing a secret political cell, or

Hard Core Cell

The Brain Trust

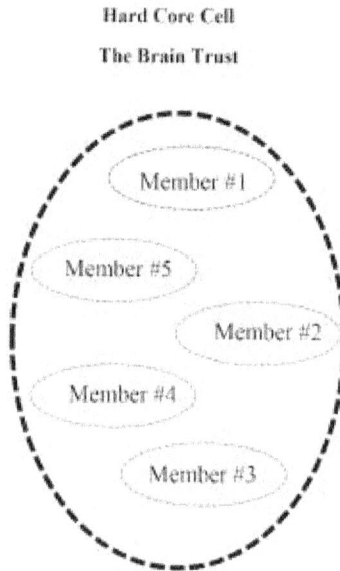

Develops specific long term goals
Develops plans for the insurgency
Recruits members for political front
◯Hard Core – Brain trust individuals and cell

Figure A

group, where often a blood oath, or an agreement to kill any member who exposes the cell, is taken. This is called the 'hard-core cell' or the brain trust. [Figure A.] The hard-core cell develops specific long-term goals and plans for the insurgency. Once organized this hard-core cell begins recruiting supporters to form a second secret political organization, or a 'political front'.[5]

This political front claims to represent several groups opposed to the 'threat'. This second cell or political front is actually controlled by the first cell. The political front uses the façade of being democratically elected by the areas or groups, which it claims to represent. Each member of the hard-core cell is 'elected' to the political front, or political arm of the organization. The hard-core cell carefully chooses the other members of the political front. [Figure B]. Often this political front also has a blood oath and a grand unifying slogan, such as 'United We Stand.'

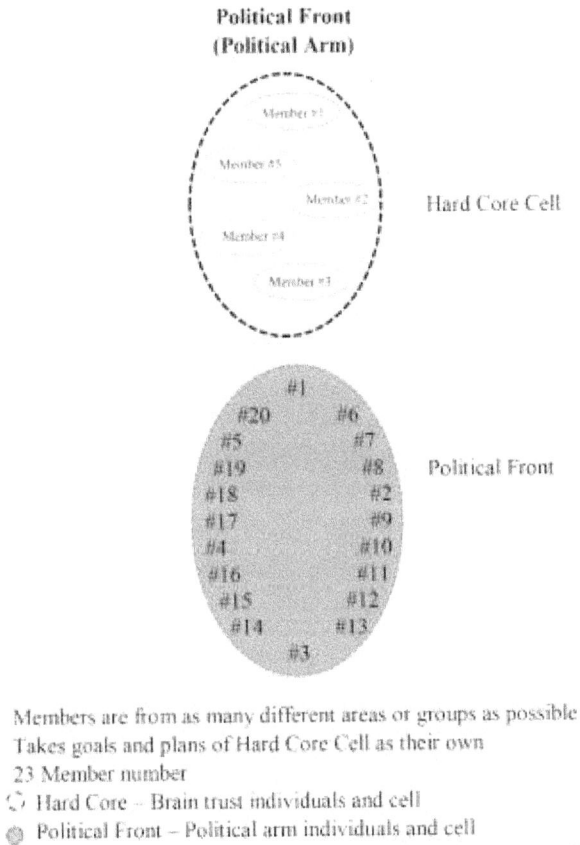

Political Front
(Political Arm)

Member #1
Member #5
Member #2
Member #4
Member #3

Hard Core Cell

#1
#20   #6
#5   #7
#19   #8
#18   #2
#17   #9
#4   #10
#16   #11
#15   #12
#14   #13
#3

Political Front

Members are from as many different areas or groups as possible
Takes goals and plans of Hard Core Cell as their own
23 Member number
◌ Hard Core – Brain trust individuals and cell
◉ Political Front – Political arm individuals and cell

Figure B

31

The political front develops long-term goals, plans and programs, based upon the long-term goals and plans developed by the hard-core cell, however, the political front normally does not know they are 'rubber-stamping' goals, plans, and programs already developed. The organization of these two cells is the most important of all the activities of the insurgency.[6]

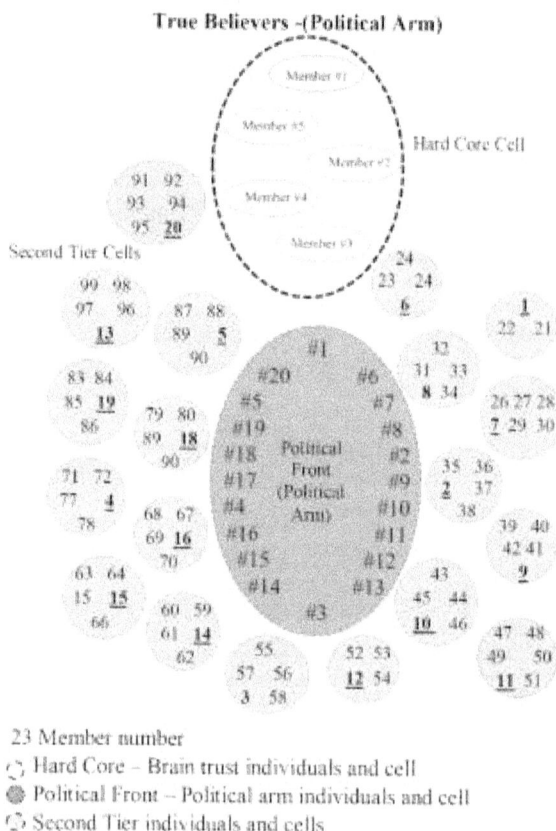

Figure C

Each member of the political front is responsible for organizing a cell of 'true believers', normally about five or six members. [Figure C]. At this point, there are three secret organizations. First, is the hard-core cell, or brain trust, which tries to remain secretive. Next is the political front, which usually consists of no more than twenty members. Third are the cells each member of the political front organizes, called 'second tier' cells. Normally, there are about five or six members in each second-tier cell. At this early stage, there are about 100 members of the insurgency.

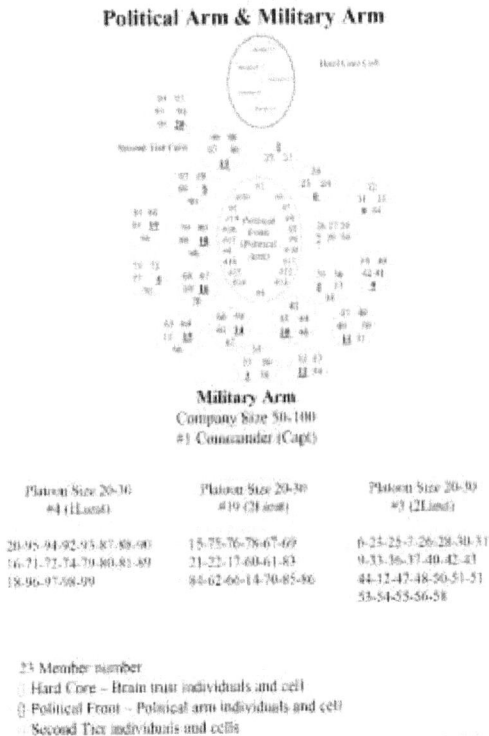

Figure D

Once this political front completes its initial organization, it organizes a secret military arm. This is a paramilitary, or guerrilla unit. The commander of the military arm is a member of the hard-core cell, as well as the political front. The military arm is under the control of the political front. [Figure D]. These part-time, paramilitary units are normally called 'self-defense forces'. They claim to be only for the protection of the local area or population against an outside threat, such as the unpopular government or another common enemy. One of the actual uses of this secret paramilitary unit is the protection of the political leaders, the hard-core cell. A second use is to conduct limited military or guerrilla actions against the 'threat' or disliked government. Another major purpose is to impress the local citizens, intimidate or threaten any opposition, which likely results in injury or death of any major adversary. Most members of this paramilitary unit are not fully aware of its designed use and believe it is a true self-defense force. They are unaware the paramilitary has a larger role until later.[7]

Simultaneously with the organization of the military arm, each second-tier cell member organizes cells of his own, with again only about five or six members. [Figure E]. If the members of the true believer cells, or the second-tier cells, are of military age they become part of the military element as well. [Figure E]. This process continues until there are numerous cells; 'third tier', 'fourth tier', and so on, throughout the target area. These third and fourth tier cells are mainly individuals of military age who become members of the military arm. The identity of each cell is kept secret

Second & Third Tier Cells

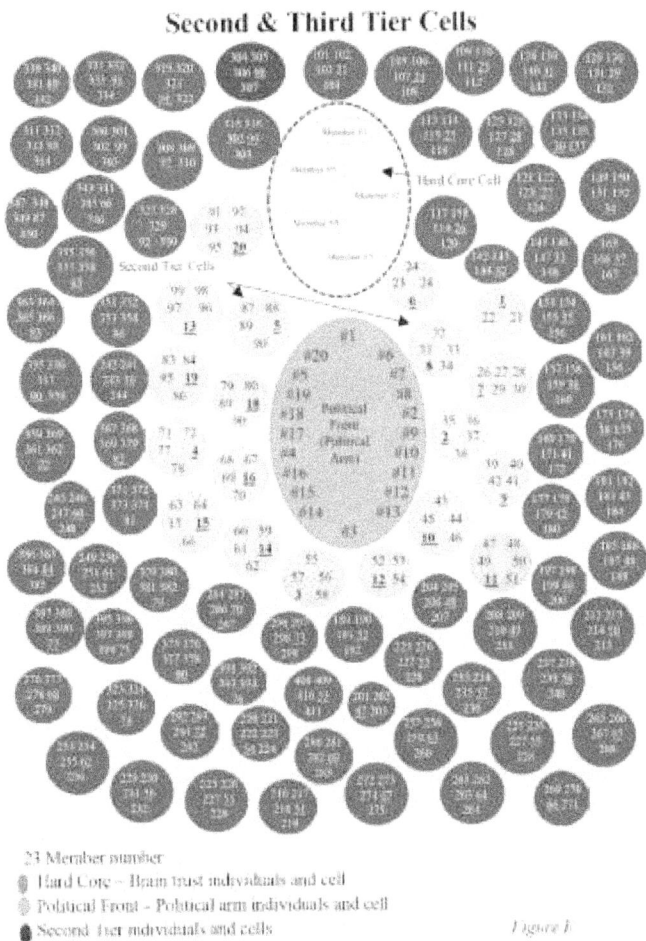

23 Member number
- Hard Core – Brain trust individuals and cell
- Political Front – Political arm individuals and cell
- Second Tier individuals and cells

Figure F

from other cells. Secret signs, handshakes, and code words
are used for recognition.

The military arm is increased to company size [50-100
men] and to battalion size [200-500 men] toward the end of
Phase I. [Figure F]. At this point, it is not designed to take
on a 'regular' military force in open combat. The members

of these paramilitary units are not gathered together, or enmasse, at this point as is the case of a conventional military unit. They meet and operate in small cells of four or five individuals. In the later part of Phase I, the unit may come together in a few cases to conduct military operations, but quickly disperses back into the small cells.[8]

One technique the political front utilizes is to obtain positions in the government and use these positions to control the government, or a part of the government. They have members join authorized government military units and other governmental agencies. They receive military training and equipment, as well as, other governmental training.

Military Arm
Battalion Size 150-400
#4 Deputy Commander (Major)

Company Size 50-100 #20 (Capt.)	Company Size 50-100 #19 (Capt.)	Company Size 50-100 #3 (Capt.)	Company Size 50-100 #6 (Capt.)
Plt. Plt. Plt. 71(Lt.) 18 (Lt.) 79 (Lt.)	Plt. Plt. Plt. 14(Lt.)15(Lt.) 17 (Lt.)	Plt. Plt. Plt. 9(Lt.)12 (Lt.) 40 (Lt.)	Plt. Plt. Plt. 7 (Lt.) 26 (Lt.) 31 (Lt.)
290-291-292-293-294	245-246-247-248-249-	55-225-226-227-228-	101-102-103-104-96-
295-296-95-94-92-93-	250-251-16-17-75-76-	233-234-56-229-230-	97-98-99-23-109-110-
87-88-90-288-290-72-	78-67-69-252-255-	231-232-235-236-56-	111-112-105 106-
74-70-80-85-86-89-	277-21-60-61-83-84-	42-43-44-237-238-	107-108-25-113-114-
297-298-81-297-298-	62-66-253-254-22-	239-240-47-48-50-51-	115-116-28-117-118-
81-90-299-327-328-	256-271-273-274-275-	52-161-162-163-52-	119-120-121-122-123-
329-330-331-332-333-	276-278-279-280-282-	53-54-164-404-405-	124-125-126-127-128-
334-323-324-325-326-	283-284-285-286-287-	406-407-169-170-171-	129-130-131-132-30-
347-348-349-350-343-	269-170-271-265-267-	172-177-178-179-180-	133-134-135-136-137-
344-345-346-338-340-	268-363-368-355-366-	189-190-191-192-408-	31-138-139-140-141-
341-242-353-336-337-	359-360-361-362-355-	409-37-168-220-221-	142-143-144-33-145-
338-379-380-381-382-	356-357-358-351-352-	222-223-229-85-230-	146-247-148-149-150-
271-372-373-374-375-	353-254-395-496-597-	231-232-234-235-86-	151-152-153-154-155-
376-377-378-367-368-	398-399-391-392-393-	70-237-238-239-240-	156-36-157-165-166-
369-370-382-383-379-	394-387-388-389-390-	236-158-159-160-165-	167-173-174-175-176
380	383-384-385-277	166	

Figure F

These clandestine placed individuals provide vital intelligence to the political front. A second objective is to bring the government military unit or governmental agency under the control of the insurgency, or at the very least neutralize its effectiveness, if called upon by the government to fight the insurgency.[9]

A second technique is the political front establishes an overt organization, such as a trade union or a legal political party, which meets the needs of the citizens. This overt organization invites other such organizations to meet and form some type of 'united front', which is controlled by the secret political front. The other organizations are not aware they are being manipulated.

At the latter part of Phase I, portions of the political front and part of its military arm may move into a 'safe area' such as mountains, jungles, swamps, or across an international border, especially if there is a strong outside military threat. The political front starts functioning as a shadow government. The paramilitary unit becomes a full-time military unit and begins extensive training. The absence of these individuals is explained with some simple explanation such as they are gone to take care of some need such as hunting or harvesting.[10] Phase II is the actual conduct of hostilities. This starts with small raids, assaults, assassination, or demonstrations against any individual or group who can be easily identified with the government, or who can be identified as a threat to the citizens who support the insurgency. At first, only the threat of overt action is sufficient to intimidate the opposition. These actions gain popular support because it seems they are protecting the local population from danger. In most cases, the government authorities are not aware of an insurgency until it reaches this second phase. To provide a cover for their actions the political front explains these threats or raids as just some 'radical' or 'outlaw' element. This disguises the fact the actions are planned by a secret political organization and conducted by a secret paramilitary unit.

At this stage, it may be possible for the insurgency to go for a quick final victory in the form of a coup d' etat or *putsch*. A coup d' etat or *putsch* has many of the outward appearances of a revolution. It is a very dangerous action unless the certainty of success is great. If it fails, which happens in most cases, the entire structure of the insurgency is revealed. Reaction comes quickly, resulting in the insurgency leaders being arrested, jailed, and likely executed. The fortunate ones are able to flee to 'safe areas' or perhaps another country.[11]

Phase II normally lasts a long time, unless an outside force intervenes in support of the insurgency. This phase wears down the enemy's will and desire to fight. By the end of Phase II the insurgency has large full-time military units and control of large areas.[12]

Phase III is the violent climactic offensive by large military units, often from an outside power, and the replacement of the government by that of the insurgency. A characteristic of an insurgency is the ability to move back to a previous stage if there is a major reversal, or move forward if there is a sudden major gain or victory.[13]

Endnotes – Prolog: The Insurgency Model

1. Numerous briefing by Colonel Frank King, Commander, 2nd Psychological Group during period February 1980 to July 31, 1985.

2. *Modern Warfare- A French View of Counter-insurgency* by Roger Trinquier (Published by Frederick A. Praeger, New York and London), 1961; *Principles of War* by General Carl von Clausewitz (The Military Service Publishing Company, Harrisburg, Pennsylvania) 1942, p 45; and *Carl von Clausewitz On War* Edited by Anatol Rapoport (Penguin Books, Middlesex, England), 1968, pp 67, 101, and 119.

3. Webster's New Collegiate Dictionary (G. and C. Merriam Company, Springfield, Massachusetts), 1979, pp 594 and 984.

4. *The Counter-Insurgency Era – U. S. Doctrine and Performance, 1950 to the Present* by Douglas S. Blaufarb (The Free Press—A Division of Macmillan Publishing Co., Inc., New York, New York), 1977, p 3 and Webster, p 984.

5. *The Dictionary of Modern War,* by Edward Luttwak and Stuart L, Koehl (Gramercy Book, New York, New York) 1991, p 307; *War in the Shadows – The Guerrilla in History* in two volumes by Robert B. Asprey (Doubleday and Company, Inc., Garden City, New York), 1975, Volume I, xi; and *Counter-Insurgency Warfare* by John S. Pustay (The Free Press—A Division of Macmillan Publishing Co., Inc., New York, New York), 1965, p 5.

6. Ibid.

7. Blaufarb *U. S. Doctrine on Counter Insurgency*, pp 3–4. Also see *Strange Vigour – A Biography of SUN YAT— SEN* by Bernard Martin (Kenikat Press, Post Washington, New York and London), 1944; *MAO TSE-TUNG on Guerrilla Warfare* by Brigadier General Samuel B. Griffith, USMC (Ret) (Published by Praeger Publishers, Inc., New York, New York), 1961; *MAO TSE*-TUNG *Basic Tactics* Translated by Stuart R. Schram (Frederick A. Praeger Publishers, New York, New York, Washington, D. C. and London, England), 1966; Mao's *Way* by Edward E. Rice (University of California Press, Berkeley, Los Angles, and London), 1972; *Banner of People's War, The Party's Military Line* by General Vo Nguyen Giap (Praeger Publishers, New York, Washington, and London), 1970; *The Counter Insurgency Manual,* by Leroy Thompson (Stackpoles Book, Mechanicsburg, Pennsylvania), 2002; and *Sun Tzu: The Art of War*, Edited by James Clavell (Delacorte, New York, New York), 1983.

8. Griffith *Mao Tse-Tung on Guerrilla Warfare*, Chapter 5.

9. Blaufarb *U. S. Doctrine on Counter Insurgency*, pp 3–4; Martin Sun *Yat-Sen*; Griffith Mao *Tse-Tung on Guerrilla Warfare*; Schram *Mao's Basic Tactics*; Rice's *Mao's Way*; Giap *Banner of People War*; and Thompson, *Counter Insurgency Manual*.

10. Ibid.

11. Blaufarb *U. S. Doctrine on Counter Insurgency*, pp 3–4.

12. Ibid.

13. Ibid.

CHAPTER 1 – The Setting

The barking of dogs awakened fifty-year-old Tennessee native Robert A. Gibson at his Gillespie County Texas farm early the morning of May 18, 1862. Something was wrong! He first thought it was an Indian raid. He looked out his widow and saw his barn and wooden fences ablaze. It wasn't Indians. Local insurgents had warned him that if his sons, eighteen-year-old Samuel and sixteen-year-old Robert, joined either state or Confederate military forces, he would pay the price. They had made good on their threat.[1]

Later that day nineteen-year-old Oscar Basse, the son of Pastor Henry S. Basse, walked into the Doebbler Beer Hall in Fredericksburg, Texas. He had just ridden into town from his Ranger camp. As he entered, he was greeted by jeers from insurgents drinking beer, jeers which quickly turned to violence. An insurgent rose from his chair and walked toward Basse. He struck him in the head. Other insurgents joined in. Basse was severely beaten in retaliation for joining the Texas military force and because his twenty-two year-old brother, Carl, had joined a Confederate company. A friend of Basse's came to his aid, only to receive his own beating. The Gillespie County sheriff, Philip Braubach, was present but only watched, as he, too, was an insurgent. Leaving the two Confederate supporters bleeding and lying on the floor, the insurgents walked outside and pulled a third man from his horse. He received the same kind of savage beating about the head. A fourth man walked by whom the insurgents also believed was a Confederate sympathizer and he also was beaten. Braubach and the insurgents next turned their attention to twenty-two-year-old Joseph Poetsch and warned him if he did not join them something worse would happen to him and his father, Peter.[2]

The insurgent sheriff learned that one of his former home guard members, twenty-two-year-old Charles Schwartz, was about to join the Confederate Army. Sheriff Braubach warned Schwartz if he joined he would be arrested on an outstanding warrant. Schwartz joined up anyway, whereupon Braubach immediately arrested him and placed him in the local jail. Braubach threatened Schwartz that he would not live to be released. When Schwartz's new company commander learned of the situation he quickly got him released on bond.[3] Leaders of the militant insurgent organization had just voted to execute a fellow insurgent whom they believed to be a traitor. They now looked at a hand in the flickering candlelight that contained several straws. After each had drawn their straw, twenty-year-old Ernst Beseler realized he had drawn the death straw. His opportunity came on the morning of July 5, 1862. The target, twenty-seven-year-old Scot, Basil Stewart, and a twenty-two-year-old Negro slave belonging to Henry Attrill were driving cattle from Comfort, Texas, north to the Attrill farm. Beseler hid along a mountainside as the two men entered a canyon. He took careful aim and fired, killing Stewart instantaneously. Beseler triumphantly swung his rifle above his head and returned to his home.[4]

Frontier justice was making its way into both the Anglo communities and the German-Texas communities of

Fredericksburg and Comfort. Seeds of hate were becoming embedded into the lifestyle of Secessionists and insurgents. A new method of settling disputes was being established: taking the law into their hands.[5] Texas was a Southern state and it didn't take long for the Confederates to retaliate for the killing of Stewart. Within days many of the Texas Unionists fled toward Mexico. The United States Consul at Matamoros, Mexico described the Texas situation as, "a perfect reign of terror."[6]

As stated in the Introduction to fully understand the 'Perfect Reign of Terror' that existed in the Texas Hill Country in the early years of the War Between the States it is necessary to examine the individuals involved, their experiences, and methodology they followed. The methodology is discussed in the Prolog, The Insurgency Model.

The Hill Country insurgency of 1861 – 62 was led by German radical elements known as Freethinkers (someone who did not believe in an organized church) and Forty-Eighters (one who had taken part in the German Revolution of 1848). Their background is an important element in understanding that insurgency.

Germans had been interested in Texas for some time. Even before Moses Austin received permission from Spanish and Mexican authorities to bring American colonists into Texas in 1821, Germans looked at Texas as a location for colonization. In 1818, J. Valentine Hecke, a former Prussian army officer, traveled extensively through the western parts of the United States. He visited Texas in 1819 and early 1820. In 1821, upon his return to Prussia, he published a book *Reise durch die Vereinigten Staaten* [Travels Through the United_States]. Hecke saw Texas as a place where a 'New Germany' could be founded. He advised that Prussia purchase Texas from Spain and colonize it. Prussia would organize a commercial company along the lines of the British East India Company. This company would defray all expenses of administration and would derive all the profits. Hecke felt Prussia could induce German immigrants who were already in the United States, and fallen on hard times, to move to Texas. Prussia would furnish them free transportation and give them land either by outright grants or credit on very easy terms. Each emigrant would be granted 50 acres of fertile land, which would be sufficient to support him and his family.

Hecke recommended Prussia send over 10,000 former soldiers, who would also be given land as a gift or on very east credit terms. Between the individual immigrants and the former soldiers, an effective militia would be available. Prussia would also furnish regular troops for the protection of the colony against Indian depredations, or any other hostile aggression. Prussia's navy would be built up through this colonial possession and Prussia would become rich and powerful with its 'New Prussia.' Nothing immediately came of Hecke's recommendations, but the idea of a 'New Germany' in the immense territory of North America west of the Mississippi River took root in the minds of many Germans.[7]

One such German was Gottfried Duden from the Duchy of Berg. Duden published a book *Report on a Journey to the West States of North America and a Stay of Several Years Along the Missouri (During the Years 1824, 25, 26, 1827)*. In it Duden urged German emigrant to settled in Missouri. Duden addressed slavery. He focused on three major topics: recognition of slavery, minimization of undesirable ramifications, and emphasis on its potential benefits for German emigrants. He reminded his readers that Persia, Egypt, and Rome managed to thrive under the burden of slavery. Duden urged Germans to emigrate and established communities which would become centers of German culture: a second *Vaterland*. If enough Germans came, they might be able to establish an all-German state.[8]

By the early to mid-1830s, many of the German kingdoms were in political upheaval. Many young German patriots, particularly among university students and intellectuals, desired political and religious freedom and a more liberal society. The political leadership of the various German royal families, suppressed this movement. Leading this persecution were Austria and Prussia. The result was political revolts in Rhenish Palatinate in May, 1832, and in Frankfurt and Rhenish Bavaria in 1833. These revolts were severely crushed, resulting in many with liberal ideas fleeing. Many of these liberal leaders looked to America, and particularly the new Republic of Texas, as a place where German colonies might be established to preserve the German 'way of life.' They looked for a way to emigrate to the United States and to the Republic of Texas. At the same time, the German petty princes looked for a way not only to get rid of the 'political radicals' or *Dreissiger* [political refugees of the 1830s] but of paupers and petty offenders. Many German states eager to get rid of these 'troublemakers' offered them pardons on condition that they agree to leave and never return. German societies were formed to assist in this emigration.[9]

Many of these *Dreissigers* were *Freidenker* [Freethinkers], someone who did not believe in an organized church; a vague term with more emotional meaning than clear definition. The dictionary identifies a Freethinker as a "Person who forms his religious opinions independently of authority or tradition." Other definitions suggest an attitude of liberalism unencumbered by dogma and the status quo. The term also connotes agnosticism, if not outright atheism. Many of the early Freethinkers were neither true agnostics nor true atheists. These early Texas Freethinkers considered the notion of Deity irrelevant and opposed clerics and churches. If they acknowledged the existence of a traditional Judeo-Christian God, they did not do so with friendliness or affection.[10]

Others describe the Freethinkers as either moderate reformers or being militantly anticlerical; the forerunners of some of the more liberal Protestant faiths of Unitarianism or Universalism and liberal Congregationalism. Some claim a Freethinker believes in the potentialities of man. They ask for no intervention from a supernatural force. They accept all the implications of the concept of evolution, the perfectibility of man, and 'enlightened progress', along the road of philosophy and reason. The more extreme hoped to build a 'new order on the ruins of Christianity.' Freethinkers follow the 'practice of truth'. Some advocate a religion of humanism to transcend the divisive forces of nationalities, states, churches, and social groups, and make men brothers of one human family, living in a social order of liberty, equality, and fraternity.[11]

The early Freethinkers desired free schools rather than parochial ones and wanted to be free of any religious influence. One prominent disciple of the movement said, "The Freethinker has a book of revelation older than the Bible [or] Koran … It is the book of Nature and of World History." A Freethinker belongs to no sect, but is free from all. In his book mankind learns to read; from it man draws all knowledge that leads him to happiness in the heaven on this earth. All that is required is to make a proper use of his five senses and to use his human reason to study nature and its laws and to serve it; to analyze everything — all events — in their natural sources and origin, and in their natural activity. This Freethinker belief caused one traveler to Germany in the 1830s to state, "People of intelligence here turn away from the church in general." Another stated, "Perhaps more than half of the late emigrants [to the United States and Texas] are almost entire strangers to Christianity."[11]

Anne Stewart, a Comfort, Texas, historian and a student of the Texas Freethinkers, provides one of the best explanations, defining "Freethinker" as an umbrella term used to describe a variety of people who believe in rational thought rather than a supernatural religious being, university radicals who wished to practice a Utopian or communitarian way of life, intellectuals who have no interest in conventional worship, and some individuals who sought a way of life free from governmental and theological interference.[12]

One of the main characteristics of the Texas Freethinkers was their non-participation in church activities. One Hill Country German scholar had this to say about their lack of church support, "These pioneers said frankly that they left Germany to escape not only political persecution but also religious oppression." In very few homes was there a Bible or any religious literature. This remained true for decades. In the Sisterdale and Comfort area, "it is not uncommon to find German-American livestock ranchers who boast that the Bible has never been allowed in their homes." As late as the 1980s, the area was the only rural district in Texas in which the majority of the inhabitants were known to be agnostics. There were no prayers. The attitude toward organized religion ran the gamut from mild anticlericalism to bitter denounciation.[13] Funerals were, and remain, extremely important for the Freethinkers. Their attitude about funerals played a major role in the way the insurgents recounted the Nueces Battle and the fact the Confederates did not bury the bodies of the killed insurgents. "At funerals, sentimental German ballads were sung ... When occasionally a minister conducted a funeral and asked for the Lord's Prayer, they [the Freethinkers] did not know it." Funerals were large because of so many relatives attending. A German lodge conducted the service and the message was 'Rest in Peace.' The life of the deceased was told, often a man skilled in public speaking read a eulogy. There was no mention of immortality because no one believed in it. "We live in our children. That is our only immortally," was their creed. This attitude was followed by several prominent Hill County families which, "For four or five generations maintained Freethinking practices, like secular funerals." Again, as late as the 1980s, the Sisterdale Cemetery was almost totally devoid of Christian symbolism of any kind.[14]

One major result of the failed 1830s revolutions was reawaking the idea of a 'New Germany' in America.' This was a period among Freethinkers and other 'higher classes' of a feeling of *Europamuede* [tired of Europe]. "The time for emigration was ripe," they said. Several German newspapers printed articles calling for emigration under the theme *ausdauerndst Patriot* or that the conditions were so bad it was alright for even the most patriotic individuals to leave his homeland with a clear conscience.[15]

Under these conditions German emigration societies were formed to create the 'New Germany'. The most important of these German emigrations societies was the *Giessener Auswanderungsgesellschaft* [Giessen Emigration Society]. The societies' basic concept was for German emigrants to occupy an unsettled and unorganized territory in North America so a new German republic would arise. There were two schools of thought on how this 'New Germany' should be created. The colonies "might or might not be members of the ... [United States]." In any case, it would be predominantly German in character. If it was a member of the United States, it would "Maintenance ... a form of Government, which ... assured the continuance of German custom, German language, and create a genuine free and popular life." If the United States would not grant this, "why then the German state would secede and set up a National Government of their own."[16]

An 1834 Stuttgart newspaper outlined some of the details. It pointed out that in order for a state to be admitted into the United States, the law required 60,000 free inhabitants above 25 years old. Many of the 60,000 required inhabitants would be Germans already living in North America, the remainder would be new immigrants. "The plan is so popular that scarcely any doubts are entertained of its [not] being successful," wrote the paper.[17]

To many, this plan sounded like nothing but a 'greenhorn scheme,' but when Texas gained its independence in 1836 it seemed more feasible. Conditions in Texas were now very favorable. It was an independent country with vast amounts of territory. The population was small so a large German immigration in a short time would make the German population predominant. The idea of bringing Germans already living in North America to Texas gained new supporters, both in Texas and in the United States. A society had already been formed in Philadelphia in 1836 with the objective of "the unification of the Germans in North America and by this means the founding of a new German Fatherland." In early 1839, the *Germania Gesellschaft* was organized in New York with the purpose of establishing a

German state in Texas. In late 1839, the society sent one hundred thirty colonists of "all classes and stations in life" to Texas. The colonists reached Galveston on Christmas 1839 where they learned of a yellow fever epidemic at Houston. The news deterred most of the colonists from staying, and many returned to New York. A few went on to Houston and eventually settled at Houston or Cat Springs in Austin County.[18] Within Texas two societies were formed to support new German immigrates whether from the United States or aboard.

The first of these was *Deutscher Verein fuer Texas* [German Society of Texas] founded on November 29, 1840 in Houston. The second was the *Teutonia Orden* [Teutonic Order] founded in early 1841 at the German communities of Cat Springs and Industry. These were designed to, "Assist [German] newcomers with advice and necessary aid and succor" and "making the Germans a strong political factor in Texas."[19]

Thus, far the German Liberals and Freethinkers' efforts to establish a New Germany in Texas had showed little results. The next major German migration revived the New Germany hopes. These were the Castro Colonies and the *Adelsverein,* which resulted in three principal towns and counties and began to look like a New Germany. Most of these colonists were farmers, craftsmen, and tradesmen. Instead of Freethinkers, they were mainly Catholic or Lutheran. The Castro Colonies began to arrive in early 1842 when Henry Castro received contracts from the Republic of Texas to settle about 600 families along and near the Medina

River west of San Antonio. Castro formed the *Societe de Colonisation Europee-Americain au Texas* [European-American Colonization Society to Texas] in February 1843. The first immigrants arrived at Galveston on New Year's Day 1843. They consisted mainly of Swiss, French, particularly from Alsace, and Germans from the Rhine River area, particularly Baden, Hessen, Nassau, and seventeen from Prussia. It was early September, 1844, before the first group reached their land grants on the west bank of the Medina where they established Castroville.[20]

Once Castroville was established, the colonists moved west and established the towns of Vandenburg and Quihi in early 1846. In the spring of 1847, the town of D'Hanis was founded, and later that year the town of New Fountain. On February 12, 1848, Medina County, which took in the five settlements was created. The 1850 Medina Census shows 177 heads of households or about 909 citizens. This grew to 363 heads of household or just over 1,300 citizens by 1860. Medina County was predominantly Catholic with its first Catholic church completed in November 1846. In 1852, a Lutheran church was organized.[21]

The majority of the Castro colonists came to Texas for economic, not political, reasons. However, "Many … became politically active as a result of the anti-immigrant Know-Nothing movement of the early 1850s." They generally opposed secession and were Union supporters during the Civil War.[22]

The *Verein zum Schutze deutscher Einwander nach Texas* [Society for the Protection of German Immigrants in

Texas], more commonly known as the *Adelsverein* [League of the Nobility], was organized on April 20, 1842 by twenty-one German noblemen at Biebrich on the Rhine. Its goal was the establishment of a New Germany in Texas by an organized mass emigration. Prince Carl of Solms-Braunfels, the first commissioner of the *Adelsverein* outlined the plan in a December 6, 1845 letter to William Kennedy, the British Consulate at Galveston. He stated in part "Should the enlightened Government of Her Majesty the Queen approve the idea [of a Germany State in Texas], it would be easy to make some arrangement with the direction of the said Association [*Adelsverein*], and send, even by the end of next Spring, some twenty, or thirty, thousand individuals, well-armed and equipped, to this Country." Prince Solms continued explaining, "This number of men could be got in Germany, which suffers from a redundant population that causes three times the amount to leave their native shore annually for the United States. English and Germany ships could carry them to this Country — able and active young officers, of every arm, would accompany them. English arms — by the testimony of every English officer, who served in the wars of the Peninsula — were effective weapons in the hands of the German soldiers of the Legion." Solms clearly stated the purpose of this force, "They would do as well to stop American [United States] encroachment towards the South. In fact, this force once established, I may pledge my word for the safety of the future, on this side." Prince Solms continued, "At the present moment, the German Association has a right to introduce as many emigrants into Texas as they are able to transport, and so large a number, promptly established, who will dare drive them out?"[23]

Despite later writers claim the *Adelsverein* failed because it was not properly financed, Prince Solms did not share that idea. He wrote to the Texas Congress that while he had already spent a lot of money of Texas, he was prepared to expend a great deal more, but "only for a free Texas; for the state of Texas, not a cent."[24]

The *Adelsverein* was reorganized on June 18, 1843 and formally constituted on March 25, 1844. Its official new goal was both philanthropic [to promote human welfare] and commercial. It planned to do this by enabling mass German emigration to Texas and establishing German settlements, which would supply markets for German industry and promote the development of German maritime commerce. However, recent studies show the *Adelsverein* leaders had not given up the idea of establishing a New Germany and that this was their real goal. This plan was aborted when Texas became a state in 1845, but the idea of a free state in western Texas remained a goal of the liberal Freethinkers and later the Forty-Eighters.[25]

The first *Adelsverein* officials, Count Joseph of Boos-Waldeck and Count Vikter of Alt-Leiningen-Westerburg, arrived in Texas in the summer of 1842. They spent "the remainder of 1842 familiarizing themselves with the social, political, and economic realities of the young republic." The *Adelsverein* wanted to establish "a slave plantation, or possibly, even in a string of them." *Adelsverein* approved a plan of Count Boos-Waldeck to create the plantation. Boos-

Waldeck and the *Adelsverein* felt the plantation would replace the castle, The German settlers to take the place of the yeomen and tradesmen ... [and] The Negro slaves were to take the place of medieval serfs and bondsmen, while the German noblemen at the top of this hierarchy bask in a life of leisure and privilege. By July 1844 the plantation, now named Nassau, was up and operating with at least 15 slaves. By December 1847 Nassau had at least 24 slaves, making the *Adelsverein* one the largest slave owners in Texas.[26]

The first *Adelsverein* settlers arrived in Texas in December 1844 at Carlshafen, later known as Indianola. They settled at Comal Springs, on the road between San Antonio and Austin. They established the town of New Braunfels on March 12. Comal County was established a year later in March, 1844. By 1850, New Braunfels was the fourth largest town in Texas. The 1850 Comal Census shows 1,723 citizens, of which 1,298 lived in New Braunfels. The 1860 Comal Census shows 3,837 citizens.[27]

In November 1845, the *Adelsverein* completed plans to establish a second colony. This one was further inland, just north of the Pedernales River. A survey party laid out a wagon route between December 1845 and February 1846. The new settlers left New Braunfels about April 23, 1846. They arrived on May 8, 1846, and established the town of Fredericksburg. On February 23, 1848, Gillespie County was established. The U. S. Army established Fort Martin Scott, about three miles southeast of Fredericksburg in February 1848. Other communities were established. By 1850 the population of Fredericksburg was 754. The 1850

Gillespie Census showed 1,235 white citizens. By 1860, Gillespie County had 2,703 citizens.[28]

The vast majority of the *Adelsverein* settlers, like the Castro colonists, came to Texas for economic, not political reasons. They were also farmers, craftsmen, artisans, and laborers. As pointed out by a modern Texas German scholar, "The German peasant of the nineteenth century was not a politically oriented being." Yet another points out the German settler's "High level of literacy and strong concentrations in a number of counties made them much more of a political factor." The *Adelsverein* immigrants were further down the social scale than the literate Freethinkers and their attitudes toward slavery "ranged from indifference to hostility." The problem was that the more highly-educated Freethinkers and later the Forty-Fighters easily influenced them. The result was sons of several of the Gillespie *Adelsverein* settlers would shed their blood at the Nueces River in August, 1862, and on the banks of the Rio Grande in October, 1862.[29]

The story of *Die Vierziger* (the Forty) is critical to understanding the later *Des Organisation* [Union Loyal League]. Several members of *Die Vierziger* would play major roles in the *Des Organisator* and its insurgency. The most prominent of these was Jacob Kuechler, who later commanded the *Des Organisator* Gillespie Company. The Forty was an 1847 fraternity of German students with chapters at the universities of Heidelberg, Giessen, and Darmstadt. It was patterned in part on the dreams of a communistic utopia. Its' slogan was "friendship, freedom, and equality." The members were mostly young Freethinkers. Most were from families in business, sciences, and government in the Duchy of Hesse-Darmstadt. The *Adelsverein* leaders recruited them to boost the emigration society's reputation and to encourage young professionals to find new markets for their talents in Texas. *Die Vierziger* agreed to establish a colony of professional men who would attract additional individuals and families.[30]

As best as can be determined, only thirty-one of the original forty actually made the trip. On July 4, 1847, most arrived at Galveston on the *St. Pauli*. Others, previously arrived, joined them, making the total thirty-five. The *Adelsverein* gave them $10,000 in gold for their expenses to establish the colony. The Forty reached the Llano River on September 30, 1847 and established the *Lateinische Ansiedlung* [Latin Settlement] of Bettina, located where Elm Creek flows into the Llano River on October 1, 1847. It was the first of five or six Latin Settlements in Texas.

They were called Latin Settlements because at that period in German culture a knowledge of Latin was considered to be both a prerequisite for higher learning and a sign of education The individuals were called *Luteiner* (Latin Ones). These Latin Settlements were "the centers of light, from which higher ideas of life were customary among the ordinary settlers spread among wide portions of the country. Especially in educational matters, these men set the standard, not only for their German countrymen, but for their Anglo neighbors as well."[31] The Bettina communist experiment failed by the summer of 1848, because while some of the Forty worked hard, others idled away their time in the shade of live oak trees, where they dreamed of bygone student days and lived to the convenient principles expressed in the words *Ede, bibe, post mortem nulla voluptas* [Eat, drink for after death there is no joy].[32] After the Bettina failure, most of the Forty moved into urban areas such as San Antonio, New Braunfels, and Fredericksburg, where their skills could be put to use. Some of the Forty, now known as *Darmstaedter Kolonie* [Darmstadt Colony], attempted to establish Latin Settlements at Darmstaedter Farm, near New Braunfels, and at Tusculum, near Boerne. Some moved to Sisterdale, a settlement founded in 1847 by another Freethinker, Nicholaus Zink. By 1850, the remnants of *Die Vierziger* was a small group of five living together north of New Braunfels 'in the mountains' near Sisterdale.[33]

Der Achtundvierziger [The Forty-Eighters] is a collective term for supporters of the European revolution of 1848-49, which in Germany culminated in the calling for a constitutional parliament in March 1848, resulting in the Frankfurt National Assembly from May 1848 through June 1849. It favored unification, constitutional government, and guarantees of human rights. After the failure of their revolution many were arrested, imprisoned, and executed. Others fled. About 4,000 came to the United States, of whom about 100 came to Texas and settled mainly in San Antonio, Comfort, and Sisterdale. Terry G. Jordan, a German-American historian, claims that about 5-10 percent of the Hill County German immigrates were a "vociferous and influential minor" of Freethinkers [and Forty-Eighters] most of whom was university-educated political refugees.[34]

An expert on the Forty-Eighters identifies two major groups: 'Liberals' and 'Democrats.' The Democrats were referred to as 'radicals' using the term common in the German political vocabulary of the time. It was mainly the 'Democrats/radicals' who left Germany following the 1848 Revolution.[35] The Liberals were the moderate wing of the reform movement. They were much more numerous than the Democrats/radicals. The Liberals wanted to transform the German Confederation on the model of Great Britain; that is, a limited monarchy, representative government, restricted suffrage, and free enterprise. Their ideal was a state in which every citizen enjoyed personal freedom enshrined in the fundamental law, but an elite of wealth and talent that dominated political life. Theirs was the ideology of the upper bourgeoisie of well-to-do industrialist, merchants, finances, and entrepreneurs.[36]

The radicals were mainly the lower middle classes of publicists, lawyers, doctors, teachers, tradesmen, and shopkeepers. They felt the program of the Liberals was timid and calculating, designed to disguise oligarchic aims with highsounding phrases. It was not much better than the existing system, which was trying to perpetuate an authoritarian and hierarchical form of government. Their own convictions were altogether different. They believed in popular sovereignty, in the supreme authority of a majority of the people, regardless of class, wealth, or education. They advocated the establishment of a national parliament elected by equal manhood suffrage. They were opposed to socialism, but did agree the state should intervene in economic life to alleviate the worst excesses of unrestrained competition. They demanded a total reform of the political system. They believed this was the way to invincible progress, of the idea of freedom, that society stood on the threshold of a new golden age, and that their generation had been chosen to witness the triumph of liberty in public life. Legislation must reflect the needs of society as a whole, and only the people could decide what those needs were. All forms of government rejecting the principle of popular sovereignty should be swept away. Popular sovereignty was the only sound foundation of government.[37]

By the year 1848, a very large portion of those classes which took an interest in public matters had become imbued with ultra-democratic notions. They believed all monarchies, no matter how strictly limited, were merely forms of oppression. All kings and princes were enemies of mankind. An enthusiastic belief in 'Liberty' was, with most of them, coupled *with fanatical intolerance of all who disagreed with them*. The great majority of the radical leaders were literary men, but their following came almost exclusively from the small trade and workingmen of the cities – the same source of the Texas Forty-Eighters in 1854 and 1862. Most of the radicals were young men. The more experienced and seasoned politicians were in the Liberal Wing of the reform movement. Another ingredient of the radical was that he was typically a Freethinker, if not an outright atheist. He was a humanist in the more militant sense in that he was opposed to conventional religious views concerning man's other worldly commitments.[38]

Der Deutsche Turnverein [The German Turner Club], or athletic or gymnastic club, played a major role in the German 1848-49 Revolution and in turn the early activities of the Forty-Eighters in Texas. One German historian states, "The *Turnverein* were one of the most influential groups in the Revolution of 1848." They provided the nucleus of the revolutionaries' military arm. Friedrich Ludwig Jahn founded the *Turnverein* in 1811. It became an effective vehicle for German patriotism and nationalism. The Turner goal was to develop a sound, well-coordinated body, a mind sensitive to liberty, a love of country, and an intellect disciplined for the political struggle to make the Germans united and free. Turner groups spread all over Germany with memberships of energetic and politically conscious young men, many of who were university students. The Turners were one of the liberal groups the royal families declared unlawful. The Turners went underground. When the 1848 Revolution broke out, Turners flocked to the black, red, and gold banner. They fought on the barricades and in the streets, and when the revolution collapsed, paid a heavy price for their rebellion. In American and Texas, it would be the *Turnverein* who the ultra-radicals of 1854 and the *Des Organisator* in 1861 – 1862 would first turn to for their military arm. Like in Germany, they quickly responded and again paid with their blood.[39]

The spark that initiated the 1848 revolutions occurred in Paris in February when Louis Phillipp of Orleans was forced to abdicate and a republic established. The revolution quickly spread to Germany. A force of six thousand people, supported by a five hundred-man *Turnverein* unit expelled the Duke of Baden. A provisional government was established. The revolution spread to Austria, particularly Vienna, and to Prussia, particularly Berlin. On March 13th Prince von Metternich, the Austria Prime Minister, resigned. A parliamentary in situation, the *Reichstag*, was formed. Next came Prussia. On March 18th barricades appeared in the streets of Berlin. The Prussian king, Friedrich Wilhelm IV, ordered his army to crush the revolutionaries. Bloody street fighting resultee, and many revolutionaries, including *Turnverein,* died on the barricades. The next day, King Friedrich capitulated and withdrew his troops. He acknowledged the legitimacy of the revolution. He promised a constitution and a parliament. On May 18, 1848, the Frankfurt Parliament convened. With both Austria and Prussia in the hand of revolutionaries, it appeared they had won.[40]

The Forty-Eighters had followed the Insurgency Model as explained in the Prolog. One of the first things they did was to establish small secret political groups, or cells. Next, they developed a program called the *Offenburg Program.* They next recruited additional members from the tradesmen, craftsmen, and laborers of the general German population and created a political front. The next step was the creation of secret paramilitary units. The nucleus of these was the *Turnverein.* The secret political front developed into the *Reichstag* and the Frankfurt Parliament. These actions were almost a perfect example of an insurgency's Phase I and II. Unfortunately, for the revolutionaries, the movement became a full revolution, or Phase III of an insurgency, before the leaders had firm political plans on how to run a government.[41]

The Frankfurt Parliament had no program on which all could unite. The major debate was between the Liberals and the Radicals. It centered on whether the new government would be a republic or a constitutional monarchy. The Radicals wanted nothing to do with any kind of monarch. The Frankfurt Parliament only had a short window of opportunity to establish a constitution and put a government in place, before the crown princes reacted. The Radicals' trait of *fanatical intolerance of all who disagreed with them* prevented them from compromising with the Liberals. King Friedrich regained control of the army. The army and the military forces of Hesse-Darmstadt and Nassau, united and regained control of Berlin and Vienna. By the summer of 1849, the revolution was over. Many of the leaders were arrested. Some were executed, some given prison sentences, while others fled.[42]

When the Radical Forty-Eighters reached Texas, their goals had not changed. They saw in America basically the type of government they envisioned in Germany. They "found in the United States a political peace and unity long desired." However, their first contact with Texas and American political life left many of them tremendously disillusioned. The American idea of a democratic republic was very different from that of the radical Forty-Eighter.

Representative institutions seemed to them hardly more than makeshift, which to them was a miserable compromise between aristocracy and democracy. They still dreamed of a pure democracy, in which the people should govern directly. "Many of the idealistic saw it as the destiny of the Germans and Americans to reunite in the common struggle to extend the frontiers of human liberty."[43]

There is an anecdote told by the Germans that illustrates their naïve, even arrogant nature. Shortly after arriving, one met another on the streets of a major American city. They had both been members of the Frankfurt Parliament. "What, are you here too?" one cried. "When did you arrive?" "Last week" replied his friend, and continued, "But listen, they manage things horribly in this country. And this is what they call a republic? Well, that must be changed!" Another incident occurred in San Antonio on July 4, 1853, during a German celebration, which the mayor was present. One speaker included the following remark in his speak, "We Germans who invented freedom don't need to learn it from Americans." Fortunately, the San Antonio mayor didn't understand the comment.[44]

When the Radical Forty-Eighters first arrived, they did not intend to make Texas and North America their home. They were looking for a safe harbor where they could remain until, as they said, *"es wieder losgeht"* [until it breaks out again]. But as time elapsed, they ceased their attempt at revolutionizing Germany and started taking part in Texas affairs. They soon became embroiled in local and Texas politics. Radical Forty-Eighters supported by the

Freethinkers began to push for a more liberal Texas society and for political leadership among other German-Americans. The issue of slavery was a lightning rod for them. The editor of the *Neu Braunfelser Zeitung,* Ferdinand Lindheimer, a *Dreissiger*, cautioned the Forty-Eighters not to antagonize the Anglo-American settlers by meddling in their affairs. He warned them that the problem was an old one and new arrivals [the Greens] did not understand all the issues. "When in Texas, do as the Texans do," he wrote. Anything else is suicide and brings tragedy to all of our Texas Germans. Others pointed out "In the North that ideal [of opposition to slavery] was not at all out of place," but in the South and Texas in 1851 it was extremely dangerous." Another said, "For all their avowed liberalism, the Forty-Eighters had difficulty understanding democratic processes and government." Above all, they could not tolerate compromise. "Had more Germans heeded Lindheimer's counsel" says one German scholar, "the tragedy that followed might have been averted." It was not averted. The Forty-Eighters and Freethinkers challenged Texas' 'peculiar institutions.' In the Texas Hill Country, they and their sons would be the nucleus of the *Des Organisation* [Union Loyal League]. They would also pay with their blood for these ideas.[45]

An interesting phenomenon developed among the German immigrant leadership who arrived before 1848 and were generally *Die Dreissiger* and those who arrived after 1848 that were generally the Radical Forty-Eighters. This was true, not only in Texas, but across the sphere of German immigrants in North America. This struggle became known as the fight between the Grays [older Germans] and the Greens [Greenhorns]. The Grays were more conservative and had begun the process of assimilation into Texas and American society, while still maintaining their German traits. The Greens were still full of revolutionary ideas and a desire "to infuse Anglo-American civilization with a 'Germanizing process' and to revive the cultural intellectual interest of their older countrymen." [46]

The Grays had the advantage of a longer residence in the country, greater familiarity with conditions, greater wealth, and established connections. The Greens advantage was in superior numbers and enthusiasm. They had no personal reasons to attach themselves to any particular party organization, while the Grays, by reason of habit and the manifold personal interest which party affiliations creates, found it difficult to sever their connection with the Democratic Party, even when they became increasingly disgusted with the growing pro-slavery leaning of the party. The inexperience of the Greens led them to favor all sorts of Utopian schemes, including the German state idea, which most of the Grays had outgrown, and in Texas the idea of a free state in western Texas. In return for the cold water which the the Grays poured over these exuberances, the Greens attacked their opponents in bitter tirades, charging them with being traitors to the German nationality, with having no love for anything except their own pecuniary interest. [47]

The Forty-Eighters had a tendency to suspect others of being spies in the pay of the German governments. When they arrived in the United States and Texas, this distrust of their fellow refugees remained. "How many had the German governments hired to come to America with them to report back on their activities?" asked the Greens. "Many of the exiles in [America] were but too ready on the slightest evidence to charge one of their colleagues with being a spy," writes a German-American historian. In Texas, this tendency was carried forward to their organization of an insurgency and the *Des Organisator*. There was always fear that the Confederates had planted spies among them. When someone was discovered who had given information to the Confederate authorities he was banded a spy that had been in the employment of the Confederate government all along.[48]

When the Radical Forty-Eighters reached Texas, they found other German liberal immigrants, the Freethinkers, already settled into the areas of San Antonio, Sisterdale, and the new community of Comfort about fifteen miles east of Sisterdale. The radicals quickly merged with the Freethinkers and the distinction between the two became blurred. While not all Forty-Eighters were Freethinkers and not all Freethinkers were Forty-Eighters they shared many common values such as being well-educated and ultra-liberal views. Many of the Radical Forty-Eighters joined the Freethinkers at Sisterdale and established the only Latin Settlement that succeeded. By the mid-1850s, Sisterdale had among its population six barons, nineteen settlers with university educations – including three Ph.D.s – a medical doctor, a professor, an architect, a vintner, a mining engineer, an ex-customs official, a lawyer, a merchant, a squire-politician, a foundry manager, a family of industrial chemists, and a former military officer. It consisted of some thirty farms, distributed with a radius of about six miles. Not even San Antonio could match Sisterdale's cultured assembly. A few of Sisterdale's residents moved to the new town of Comfort in the mid-1850s. Comfort also attracted large numbers of Freethinkers and Forty-Eighters. "Nearly all Germans in Comfort were Freethinkers as were their brilliant ancestors in Sisterdale" wrote a new arrival.[49]

Practically every Sisterdale farm had a good library. The latest in literature was on hand, and a part of each Sunday was set aside for discussions and lectures. The schoolhouse served as a meeting place. It was there when one of the Freethinkers was discussing the doctrine of socialism when a dozen of more Comanches appeared at the front door and with a perplexed expression on their faces, watched the speaker who was so engrossed in his text that he was completely unaware of their presence. Another Indian visit almost resulted in bloodshed. About forty men and twelve women were discussing Feuerbach, a German Freethinker, when about fifty Lipan Apaches arrived and demanded an ox or they would attack. The startled Germans quickly agreed and the Indians left with one of the best oxen in the settlement. Each family contributed funds to repay the unlucky farmer. After the Indians departed the discussion continued. The people of Sisterdale were "as choice as one could find in the best circles of the civilized world," said one visitor. The discussions were on principles of government, science, philosophy, and world affairs. It was "a center of light from which high ideals of life spread through the country." There was talk of ideas, of history, government, the arts, books, and music. One subject was never discussed: religion.[50]

In Texas, the Forty-Eighters found a large German population that was politically unorganized and without political direction. They quickly moved to fill this void. The method was the same used in the 1848 Revolution. Three instruments helped them: German-language newspapers, *Verein* [clubs or societies], and the local *Turnverein*. Sisterdale Forty-Eighter, Adolph Douai, who received his Ph.D. from the University of Leipzig in 1841 and had spent over a year in prison for his activities in the 1848 Revolution, established the *San Antonio Zeitung* in July, 1853. It didn't take long for him to announce he was a radical and it was "his determination to regard every political question from the point of social progress." Slavery was his major topic. Since the newspaper was circulated only among Germans, the Anglos took little initial notice. However, by 1855, he was publishing articles in English.[51]

The *Verein* developed in every aspect of the German immigrant's life. There were four main types. First, the social club, which promoted the spirit of good fellowship. Second, the trade and agricultural societies, designed to improve knowledge and methods of trade, farming, and stock rising. Third, the literary, dramatic, and singing societies, which promoted the cultural development of their members. Fourth, was the hotbed, the political clubs, designed to enable Texas Germans to act in unison in political affairs and a forum for discussing political questions. The Forty-Eighters, supported by the Freethinkers, were responsible for organizing many of the *Verein.* Almost every Texas German community had some type of *Verein.* Most began discussing political issues.[52]

The first large numbers of *Turnverein* appeared in Texas with the arrival of the Forty-Eighters in 1848. The first *Turnverein* in Texas was established in 1851 in Galveston. *Turnverein* in Houston followed in 1854, San Antonio in 1854, New Braunfels in 1855, and Comfort by 1860. These Turnverein were generally military-type companies which manufactured their uniforms and weapons. In October, 1850, the *Turnverein* formed a national organization, and in 1851, named it the *Sozialistischer Turnerbund,* which in "general subscribed to the principles underlying the radical Free Soil Party and urged all members to support that party in every way possible."[53] The Radical Forty-Eighters, now known in the Texas press as 'Ultra-Radicals,' took the old *Adelsverein* idea of a 'New Germany' in western Texas and modified it to the idea of establishing a free state in western Texas. They began to organize for this goal. There was a period of seeking the right combination of organizations. In July, 1851, they formed 'The German Political Association of West Texas' made up of "liberal thinkers, including several doctors and merchants, and at least two former members of the Bettina Colony," *Die Vierziger.* Little came of this association.[54]

In October, 1853, the ultra-radicals found a vehicle to bring the many *Verein* together under their leadership. The various *Saengerbunde* [singing clubs] decided to hold a *Staats Saengerfest* [State Singing Festival]. All German *Saengerbunde* were invited to this *Saengerfest*. It was held on October 16, 1853, in New Braunfels. It was a huge success with singing clubs from across the state attending. Immediately the singing clubs made plans for future *Saengerfest*; a second to be held at San Antonio on May 14-15, 1854, with a third in New Braunfels in May 1855. The singing clubs created a *DeutschTexanischer Staats-Saengerbung* [German-Texas State Singers' League] to sponsor these events. This was the first statewide German organization.

After the first *Saengerfest,* August Siemering, Ottomar von Behr, Louis von Donop, Christian Wilhelm Rhodius, all of Sisterdale, and Adolph Douai, the editor of the *San Antonio Zeitung,* realized this was the opportunity they had been looking for, while having a few drinks in a New Braunfels tavern. That night they formed an umbrella organization, the Committee of Correspondence, to coordinate the activities of German cultural organizations, especially the *Saengerhund.* While this committee was not organized in complete secrecy, it fits the Insurgency Model as a 'hard-core cell' containing sub-secret organizations. Later the committee would be split. The part planning the statewide convention became known as 'The Bureau' while the other remained in the shadows. Siemering, von Donop, Rhodius, and von Behr returned to Sisterdale, on the morning of October 17th. That afternoon they met with fellow Sisterdale ultra-radicals and organized *Der freie*

Verein [The Free Union] with sixteen members and Dr. Ernst Kapp, who was both a Freethinker and Forty-Eighter, as president. August Siemering was elected secretary. The stated purpose of *Der freie Verein* was, "to strive for the greatest possible *Freihit des Geistes* [intellectual freedom] and to spread it nach *allen Richtungen* [in all directions]." According to one Gillespie German historian, "This was the beginning of the Union Loyal League." *Der freie Verein* and later *Des Orzanisator* were the 'political front' described in the Insurgency Model. *Der freie Verein* immediately took major action. One was to align itself with a national German organization of Freethinkers and Forty-Eighters named *Bund freier Maenner* or sometimes known as *Freimaennerverein* [League of Free Men]. In November, 1853, Der *freie Verein* hosted the Texas convention of the *Bund freie Maenner* at Sisterdale. Records of this convention have not been located, but it would seem safe to believe other like-minded Texas ultra-radicals were invited. Three months later, in February 1854, the national *Bund freier Maenner* held its second convention in Louisville, Kentucky. It adopted a set of resolutions that became known as the 'Louisville Platform'.[55]

Armed with a 'Platform' on March 15, 1854, the Bureau, now increased to about ten members (who included Heinrich Guenther, Hugo Frederick Osswald, Dr. Wilhelm Keidel, Julius Schlickum, Theodor Rudolph Hertzberg, August Siemering, and Louis Schuetze) sent, under *Der freie Verein,* invitations to all German-Texas *Verein* to attend a statewide convention on May 15, 1854, immediately following the second annual *Saengerfest*. Since most of the *Saengerbunde* were going to be in San Antonio anyway, they agreed to attend and send delegates. It appeared the ultra-radicals were about to be successful in their efforts to unite the various German communities into a single voice, controlled by them. Their plans were, after the San Antonio Convention, to organize the German Texans into a political party with the ultimate goal of a free state of West Texas. This planned German political party was the overt organization that invites other groups and forms a 'united front' diagrammed in Figures C,D, and E of the Insurgency Model. At this point, the only thing missing for the ultra-radical actions to meet the definition of an insurgency, except for secrecy, was the absence of a military arm. But, as will be seen, they had that in the form of the San Antonio *Turnverein.* The Texas ultra-radicals had learned well from their 1848 Revolution.[56]

At just this time, another player entered the ultra-radicals inner sphere, Frederick Law Olmsted, who would later design New York's Central Park. Olmsted was an Abolitionist who was traveling through Texas to gather material for a book. He met Adolph Douai, the *San Antonio Zeitung* editor and member of the Committee of Correspondence, the intermost group of the German-Texas ultra-radicals. They conceived a plan, sometimes called the 'Deep Water Plan' because it ran through deep water and demanded all their navigation skill. At this time, pro-slavery and antislavery forces were contending for Kansas. Olmsted had contacts with abolitionist groups, such as the New

England Emigrant Aid Company of Boston, who sent immigrants and supplies into Kansas. "Why not do something similar in Texas?" Some of the more zealous Kansas freesoilers were already saying they would "take Western Texas next"; pointing out the method by which Maine had gained her independence from Massachusetts and became a free state in 1820.

The western Texas area included large numbers of Germans and Mexicans, and very few slaveholders. Friedrich Kapp, a Forty-Eighter and northern abolitionist, who was also a nephew of Sisterdale's Ernst Kapp, urged his northern readers to consider making their homes in western Texas. Based on voting records of August, 1853, Douai informed the northern abolitionists that there were sufficient numbers in western Texas to vote for a free state. He pointed out that of 50,000 Anglos living in the area, some 10,000 were born north of the Mason-Dixon Line. Added to these 10,000 there were another 11,000 Germans and other Europeans, and 25,000 Mexicans who could be persuaded to vote for a free West Texas. The western part of Texas could be divided from the eastern part as provided in the 1845 terms of annexation agreed upon between the United States and Texas. The German-Texas settlements enjoyed a strategic position stretching from the Gulf of Mexico to the Plains. To the south lay the Gulf of Mexico, to the north the desert, and to the west the antislavery Republic of Mexico. They believed they could raise ten to fifteen thousand former German soldiers, ready to take to the field on short notice. "We were not only impregnable behind the Colorado River, but we also could train and arm the fleeing negroes [and Mexicans] and advance from one river defense line to another," Douai wrote years later. "We would, with the support of the North, roll back slavery to Louisiana and ever further to the East." The ultra-radicals inner circle, The Committee of Correspondence, pledged themselves to secrecy until the time was ripe. Meanwhile, Olmsted and Douai would play key roles. Olmsted would return to New York to organize support in the North, and plant stories designed to attract free-soil settlers. Douai would hold his press ready until the moment came.[57]

The American Forty-Eighters felt they could draw on those 215,000 Germans who arrived in the United States in 1854 to form that base. The German revolutionaries quickly embraced the challenge and opportunity to revive their frustrated idealism. The fight for freedom in Europe became the struggle against slavery and slavery's extension. While using the vote, they also were aware that one of the reasons for the 1848 failure was a lack of armed forces. They developed the idea that not only should they use the ballots, but use the bayonet when "counting up voters." They now believed that a revolutionary crisis justified "extraordinary military initiatives." These ultra-radicals threw their support behind the Boston New England Emigrant Aid Company. Kansas became their first focal point, with Texas second. The need for armed resistance in Kansas became part of the record in the financial statements of the company. The Forty-Eighters advocated Free State principles and direct immigration to Kansas. These emigrants should only be armed individuals or groups. One leading Forty-Eighter of the time summed it up as, "The more we succeed in resisting slavery and driving it back, the closer we come to our final goal — its total abolition. Each battle against slavery is at the same time a service to humanity and freedom."[58]

As in time after time the first group the Forty-Eighters turned to where the Kansas *Turnverein*. They quickly came together and formed a battalion. Their first action was to prevent Missouri citizens from taking part in the upcoming Kansas elections. They then became the basic force battling pro-slavery border ruffians. Kansas became a free state.[59] Unlike Kansas, the Texas Forty-Eighters before the start of the May 1854 San Antonio Convention had problems. First, it became general knowledge that "something was up in the German communities." Several Anglo newspapers reported, "An organization has been formed among the Germans of Texas, and that resolutions had been adopted demanding, among other things, the gradual emancipation of slaves; that the same rights be given to the free colored men as to the whites," and that these resolutions would be discussed at the *Saengerfest* to be held in May 1854. Second, some of the ultra-radical leaders had second thoughts about the events and withdrew. Ernst Kapp, the President of *Der freie Verein* was one; he did not attend the convention.[60]

Undaunted, the ultra-radical pushed ahead. Just before the convention, several of August Siemering's letters were published in German newspapers briefly outlining the agenda of the upcoming convention. The Bureau developed a slate of resolutions for the convention to pass, which was the Louisville Platform. They also developed a slate of officers to be nominated, which were their names.[61]

The San Antonio Convention met immediately after the *Saengerfest* on May 15, 1854, in Vauxhall Garden on Alamo Street. It had delegates and representatives attending from San Antonio, Sisterdale, New Braunfels, Fredericksburg, Coleto, Grape Creek, Pedernales, Victoria, La Grange, Seguin, Indianola, and Castroville – up to 200 individuals. The first order of business was the election of officers. The convention approved the Bureau's slate and appointed them to the resolution committee. Next, as recommended by the Bureau, the convention approved the plan of forming a political party. The Bureau, now the resolution committee, presented the Louisville Platform disguised as their work. After a great deal of discussion, the resolution – now called a platform – was approved. In part, the platform called for the abolishment of slavery. This resulted in a 'fire storm' among the Anglo-Texans. Despite efforts by the Grays, the German Texans were branded as abolitionists with Sisterdale the center of absolutist activates.[62]

The backlash was so great that a mob threatened to destroy Douai's *San Antonio Zeitung*. The ultra-radicals had to call on its military element, the San Antonio *Turnverein,* for protection. The Turners armed themselves and took to the streets, saving Douai's newspaper. But Douai was finally forced to sell his newspaper and move North. The ultra-radicals had misjudged their support from the Texas German community. It would not be the last time.[63]

From 1854 until secession, many of the key ultra-radicals moved into the settlements of Gillespie County. Because of its isolation from the remainder of the state, they were able to greatly influence the citizens with their radical ideas and beliefs. These key individuals included Louis Schuetze, Jacob Kuechler J. Rudolf Radeleff, Phillip Braubach Julius Schlickum, and August Siemering. *Der freie Verein*, through its Committee of Correspondence, continued to provide leadership for the Hill Country ultra-radicals. In the summer of 1861 *Der freie Verein* gave birth to a new organization; *Des Organisator*. With the Union Army's help, a free state in western Texas could become a reality. This time their actions would be done in complete secrecy. But every so slightly, the word that the insurgents wanted to create the free state of West Texas began to creep out.[64] They now began to understand how to conduct an insurgency.

A major fear of many Texas Anglo leaders, as the sectional crisis heated up, was an incident in the Mexico War, less then fifteen years before, where a group of recent immigrants joined an enemy force and fought against the United States and Texas. Many Texans had fought in the Mexican War and remembered the treachery of the San Patricio Battalion. This unit was filled with Irish and German Catholics who had deserted the American side and joined the Mexican army. The Mexican government, aware of prejudice against immigrants, had a campaign to win the foreigners, especially Catholics, to its cause. It urged the immigrant soldiers to throw off the burden of fighting for the

'Protestant tyrants' and join them in driving the 'Yankees' out of Mexico. Mexican propaganda insinuated the U. S. intended to destroy Catholicism in Mexico, and if Catholic soldiers fought on the side of the Americans, they would be warring against their religion. By November, 1846, General Santa Anna organized these immigrant deserters, and other foreigners, into the San Patricio Battalion. The unit saw action at Monterrey, Saltillo, Buena Vista, and Churubusco. The battalion almost single-handedly halted the American advance at Churubusco in August, 1847. Therefore, instead of looking at the German settlers as friends and neighbors, Texans now considered them as foreigners and outsiders. Actions which might have been tolerated by friends and neighbors, were considered disloyal and traitorous when committed by foreigners and outsiders. Texans wanted no repeat of a group of immigrants fighting against them in the future "killing Texas boys."[65]

This fear was brought home to Texas Confederate leaders early in the war. They needed to look no further than the events of Kansas in late 1850 and to Missouri early in 1861. In both cases a large German immigrant group controlled the states for the Union. The conditions were similar in Texas. Both states had a large German population. An influential part of that population were Forty-Eighters and Freethinkers. The Forty-Eighters and Freethinkers settled near a metropolitan area. In Missouri, it was near St. Louis; in Texas, it was near San Antonio. In both states, they established 'Latin Settlements.' In Missouri, it was Belleville, just southeast of St. Louis. In Texas, it was

Sisterdale, just northwest of San Antonio. In both states, the Forty-Eighters organized youth clubs, or *Turnverein*. In both states, the *Turnverein* had shown they were not afraid to arm themselves to protect German organizations. In 1854, it was to prevent Douai's *San Antonio Zeitung* from being destroyed. In both states, the Turnverein were very visible. They wore uniforms and marched in local parades. The Kansas and Missouri Forty-Eighters and their *Turnverein* openly spoke about extending "our hands to our brothers in Texas."[66]

Texas Confederate leaders saw the impact of these German *Turnverein*. The head of the *Turnverein* Society in St. Louis organized an armed Turner company even before Lincoln's call for 75,000 troops in April 1861. These Turners, in a repeat of what had taken place in Texas in 1854, prevented the destruction of the St. Louis Unionist newspaper *The St. Louis Democrat*. This Turner company quickly grew into a regiment, the 17th (Turner) Regiment Missouri Infantry. A sister Turner unit, the 1st Regiment Missouri Infantry, joined it. These two Turner regiments were the core of a German five-regiment force that saved the St. Louis arsenal from Secessionists' hands and eventually kept Missouri in the Union.[67]

Of the 85,000 Union soldiers from Missouri, 30,000 were Germans. One out of every three German males in Missouri joined the Union Army. The leaders of the Missouri German units were Major General Peter Osterhaus and Major General Franz Sigel, both of whom had been significant leaders in the 1848 German Revolution.

Osterhaus and Sigel led Union divisions and later corps. Many Missouri Germans who did not serve as Union soldiers joined a pro-Union militia, called the Home Guard. The Home Guard was at open war with a pro-Confederate militia named the State Guards. Like the San Patricio Battalion in the Mexican War, foreigners were killing Southern boys in Missouri and on other battlefields. The Texas Confederates were determined this was not going to happen in Texas.[68]

Endnotes – The Setting

1. Report, James Duff, Captain, Commanding Company of Partisan Ranger, Headquarters Camp Bee, San Antonio, Texas, June 23, 1862, 'War of the Rebellion: A Compilation of the Official Records of the Union and Confederate Armies', (OR), Series I, Volume 9, pp 785-787 and Letter, R. A. Gibson, Camp Davis, March 31, 1864, Adjutant Generals Correspondence, (A.G.C.) Texas State Archives, (TSA), Austin, Texas.

2. "Records of the Confederate Military Commission in San Antonio July 2 — October 10, 1862" (CMC). Edited by Alwyn Barr, 'Southwestern Historical Quarterly', Volumes LXX (July 1966), pp 93-109; LXX (October 1966), pp 289-313; LXX (April 1967), pp 623-644; LXXI (October 1967), pp 258-263, 267-268 and 272-277; LXXIII (July 1969), 83-90; LXXIII, pp 243-274.

3. Ibid.

4. Letter, Ernest Cramer, dated October 31, 1862, Monterrey, Mexico to 'My Dear Parents'. A copy in Germany is located in *Deutsche im Amerikanischen Buergerkrig: Briefe von Front und farm 1861 – 1865* by Wolfgang Helbich and Walter D. Kamphoefiner, Published by Ferdinand Schoeningh Muenchen, Germany) 2002, pp 473-474; "Fredericksburg, Texas During The Civil War and Reconstruction" by Frank W. Heintzen, Master of Arts Thesis, St. Mary's University, 1944, p 44: *Die Deutschen in Texas Waehrend Des Buergerkrieges* [The Germans In Texas During The Civil War] Taken From Notes of Judge A. Siemering , San Antonio, Texas 1875 in *San Antonio Freie Press fuer Texas* May 4, 1923 to June 12, 1923, Translated by Ronni Pue and Helen Dietert, June 1, 1923; *Deutsche Pionier*

– *Zur Geschichte des Deutschthums in Texas* by Adolf Paul Weber (Published by Adolf Paul Weber, San Antonio, Texas), 1894, pp 112-113; Newspaper article, "True to Union, Texas Germans Fell on Nueces" by Elmer Kelton in *San Antonio Light*, August 12, 1962. Copy provided by Gregory J. Krauter, Comfort, Texas; *Texas in The Civil War, Death On the Nueces, The Minna Stieler Stories* by Anne and Mike Stewart, (Published by Anne and Mike Stewart, Comfort, Texas), 1997, p 21; "The Fleeing Sixty—A True Story" by Elmer Kelton in 'Ranch Romances Magazine', January 18, 1952, pp 88-90; *San Antonio Herald*, July 26, 1862. The author is very grateful to Anne Stewart of Comfort for a copy of this article, and Eighth U. S. Census, 1860 Gillespie County Texas Census, Schedule 2, Slave Inhabitants in Gillespie County Texas, p 405.

5.   *San Antonio Herald*, February 2, 1861 and February 7, 1861

6.   Letter, Leonard Pierce, U. S. Consul, Matamoros, Mexico to Hon. W. H. Seward, Secretary of State, Washington, D. C., May 5, 1862 and Letter, C. B. H. Blood, U. S. Consul, Monterrey, Mexico to Hon. W. H. Seward, Secretary of State, Washington, D. C., dated May 28, 1862, OR, Series I, Volume 9, 684-686 and Dispatches from United States Consuls in Matamoros and Monterrey, Mexico, 1826-1906, Volumes 7-9, January 1, 1858 – December 28, 1869.

7.   *Reise durch die Vereinigten Staaten von Nord-Amerika in den jahren 1818 und 1819*, by J. Valentine Hecke (Published by J. Valentine Hecke, Berlin), 1820, pp 199-200 as quoted in German Element in Texas by Moritz Tiling, Published by Moritz Tiling, Houston, Texas 1913 – 7- p 11; *Germans In Texas: A Study In Immigration*, by Gilbert Giddings Benjamin (Published by Jenkins Publishing Company, Austin, Texas), 1974, pp 3-5; and *The History of the German Settlements in Texas 1831 – 1861*, by Rudolph Leopold Biesele (Eakin Press, Austin, Texas 1987, pp 23-24. Another travel book that had great influence upon Germans was *Bericht ber eine Reise nach westlichen Staaten Nord Amerika's und einen mehrjahrigen Aufenthalt am Missouri*, by Gottfried Duden, (Published by Gottfried Duden, Switzerland) 1829. This book described Texas as a 'New Canaan' where the oppressed would fine peaceful homes and a comfortable living. It had great influence in southwest Germany along the Rhine River, especially in Baden, Wurttemberg, Hessen Rhenish Prussia, Hanover, and Oldenburg.

8.   *Report on a Journey to the Western States of North America and a Stay of Several Years Along the Missouri (During the Years 1824,25,26, 1827)* by Gottfried Duden, translated by James W. Goodrich, (The State Historical Society of Missouri and the University of Missouri Press) 1980 pp xv –xvi.

9. Benjamin Germans *In Texas*, pp 3-5; Tiling *German Element in Texas*, pp 1213; and *Refugees of Revolution: The German Forty-Eighters in America* by Carl Wittke (Published by University of Pennsylvania Press, Philadelphia, Pennsylvania), 1952, pp 9-11.

10. Webster, p 454 and Ron Tyler, et al., eds. *The New Handbook of Texas*, Volume 2, p 1173.

11. Wittke *Refugees of Revolution*, pp 122-124.

12. *The Forty-Eighter Political Refugees of the German Revolution of 1848 by* A. E. Zucker (Published by Columbia University Press, New York, New York), 1950, p 55, and *The Spirit of 1848: Germans Immigrants, Labor Conflict, and the Coming of the Civil War* by Bruce Levine (Published by *University* of Illinois Press, Chicago, Illinois), *1992*, p 55.

13. "The Town with No Sunday Houses—Comfort, Texas" by Anne Stewart 'The Journal', Volume XXI (Fall 1999), p 210.

14. *The German Texans* by Glen E. Lich (The University of Texas Institute of Texan Cultures at San Antonio), 1996, p 135 and "A Religious Geography of the Hill Country Germans of Texas" by Terry G. Jordan "Ethnicity on the Great Plains" University of Nebraska Press, Lincoln, Nebraska, 1980, pp 113, 116.

15. *A Yankee in German America: The Texas Hill Country* by Vera Flack (The Naylor Company, San Antonio, Texas), 1973, p 30; Jordan Religious Geography, pp 113, 122; and Roy Tyler, et al, eds., *New Handbook of Texas*, Volume 2, p 1173. One example illustrates the Freethinker practice very well. In a 1999 article, Comfort Historian Anne Stewart explains in the late 1980s the Comfort Heritage Foundation Archives received a gift of over 700 eulogies given by one man during his lifetime, *not one in a church.* Stewart further explains because of the lack of Freethinkers membership in churches, forced local Freethinkers and Forty Eighters to conduct their own funerals. They gave the eulogy for the departed and directed the local *Liedertafel* [Singing Society] in appropriate songs, not found in hymnals. She points out that the first church was built in 1902, forty-eight years after the founding of Comfort and only because the railroad was coming west and they would not build a line through a town with no church. Stewart *Town with No Sunday Houses*, p 2. There remains in Comfort a small group of Freethinkers, most of whom are descendants of the original settlers. For a study of other socialist experiments see *Marxists And Utopias In Texas: The Lone Star State's Pioneer Flirtation With Socialism-Communism* by Ernest G. Fischer (Eakin Press, Austin, Texas), 1980 and an attempt by the

Mormons to establish a colony in the Texas Hill Country see *The Texas Republic and the Mormon Kingdom of God* by Michael Scott Van Wagenen (Texas A and M University Press, College Station, Texas), 2002; *Lyman Wight: The Wild Ram of the Mountain* by Jermy Benton Wight, (Afton Thrifty Print, Bedford, Wyoming), 1996 and *The Lyman Wight Colony in Texas: Came to Bandera in 1854* by J. Marvin Hunter, (The Bandera Bulletin, Bandera, Texas), n.d. For a more recent study on the Freethinkers in the Texas Hill County see *Frontier Freethinkers in the Texas Hill County* by Edwin E. Scharf, (Reality Publications, Helotes, Texas), 1998.

16. Biesele *German Settlements*, pp 13-14 and Benjamin *Germans In Texas*, pp 3-4.

17. Tiling *German Element in Texas*, p 15; Biesele *German Settlements*, pp 18-19; and 'German Political Refugees in the U. S. (1815 – 1860)'', by Ernest Bruncken located in *The German-American Forty-Eighters 1848-1998* Edited by Dan Heinrich Tolzmann (Published by Max Kade German-American Center, Indiana University-Purdue University at Indianapolis), 1999, p 17.

18. Biesele *German Settlements*, p 18.

19. The families that settled at Houston included Usener, Schweickart, Habermehl, Bottler, Karcher, and a single man by the name of Schnell. Most of the single men settled at Cat Springs. Biesele *German Settlements*, pp 19-20 and 37-39; Benjamin *Germans In Texas*, p 7; Tiling *German Element in Texas*, pp 44-46; *Germans and Texans: Commerce, Migration, and Culture in the Days of the Lone Star Republic* by Walter Struve (University of Texas Press, Austin, Texas), 1996, pp 43-48.

The original members of *Deutscher Verein fuer Texas* included: Theodor Miller; Henry F. Fisher; Charles Gerlach; Conrad Franks; Robert H. Levenhagen; Henry Levenhagen; Jacob Schroeder; Joseph Sandman; Gottlieb Gasche; Martin Rumpff; Wilhelm Schroeder; I. Hermann; Gustav Erichson; Jacob Buchmann; I. I. Knoll; A. Jung; Emil Simmler; Friedrich Otto; Christian Rientiz; Charles Baumann; Wendelin Bock; Henry A. Kuykendall; Ulrich Fischer; Karl Fischer; John H. Mueller; Friedrich Schiermann; John Koop; Daniel Super; Peter Bohl; Joseph Ehlinger; Johann Buhn; Anton Brueggemann; Wilhelm Ewald; Casper Gerlach; Friedrich Lemsky; Friedrich Barthold; Johann Grunder; Christian A. Kasting; Peter Dickmann; Wilhelm Weigand; Anton E. Spellenberg; Johan Wilhelm Schrimpf; Dr. I. Anton Fischer; a Dr. Witt; A. Schanten; and Johann Schweikart. Tiling *German Element in Texas*, pp 51-54; Benjamin *Germans In Texas*, pp 7-8; and Biesele *German Settlements*, pp 220-221.

20. *Castro's Colony: Impresario Development in Texas 1842-1865* by Bobby D. Weaver (Texas A and M University Press, College Station, Texas), 1986, pp 14-50 and *Henry Castro: A Study of Early Colonization In Texas* by Cornelia English Crook (Printed by D. Armstrong Co., Inc., Houston, Texas), 1988, pp 73-74.

21. Roy Tyler, et al, eds., New *Handbook of Texas*, Volume 2, pp 624-625, Volume 4, pp 602-605 and 992, Volume 5, p 386, and Volume 6, p 699.

22. Weaver *Castro's Colony*, p 124.

23. Roy Tyler, et al, eds., New *Handbook of Texas*, Volume 1, pp 30 - 31; *Reise durch die mexikanischen Provinzen Tumalipas, Cohahuila und Texas im Jahre 1834* by Eduard Ledecus (Leipzig Germany), 1837; Bruncken, German *Political Refugees*; *Geschichte und Zustaende der Deutschen in Amerika* by Franz Loeher (Cincinnati, Ohio), 1847; *Das Deutsche Element in den Vereinigten Staaten von Nord-Amerika 1818-1848* by Gustav Koerner (Cincinnati, Ohio), 1880, all quoted in Biesele's *History of German Settlements*, pp 21-41. Two advance members of the *Adelsverein* went so far as to explain to Freidrich Ernst, the first German settler to bring his family to Texas that they "had the idea to establish Texas as [a] German colony and to organize a monarchy." "Along the Indianola Trails To New Braunfels and San Antonio" by Linda Wolff 'The Journal', (Volume XXVIII), Summer 2006, pp 129-130. "The Situation in Germany and in Texas 1840—1860: Julius Theodor Splittgerber (1819 – 1897)" by Kenn Knopp, 'The 'Journal', (Volume XMVII), Winter 2006, p 337; *British Diplomatic Correspondence Concerning the Republic of Texas—1838 – 1846* Edited by Ephraim Douglass Adams, (The Texas State Historical Association, Austin, Texas), n.d, pp 389-390. Prince Johannes von Sachsen-Altenburg, Duke of Saxony presented the clearest source for the *Adelsverein* having an overall master plan to establish a German state in Texas on February 21, 2003 at New Braunfels and on April 5, 2003 at Fredericksburg; *San Antonio Express-News*, February 23, 2003; and "Historian Tells of Area Settlers Breakaway Plan" by Roger Croteau 'The Journal', (Volume XXV), Spring 2003, pp 84-54; *The Tragedy of German-America: The Germans in the United States of America during the Nineteenth Century—and After* by John A. Hawgood (G. P. Putnam's Sons) 1940, pp 93-107 and 137-200. In the geography of Christoph Feuge, one of the *Adelsverein* settlers, says the plan was to establish a *feudal* (feudal estate). *Christoph Feuge: A German Pioneer's Story* by Robert Lamar Feuge Ph.D., Llumina Press (Coral Springs, Florida) 2009, p 17. The source that shows the later Commissioner-General of the *Adelsverein*, John O. Meusebach, knew of the plan to establish a German state. *John O. Meusebach: German Colonizer in Texas* by Irene Marschall King (University of Texas Press, Austin, Texas), pp 51

and 60-61. Sources showing the German liberals wanted a free state in western Texas are: *Lone Star and State Gazettes: Texas Newspapers before the Civil War* by Marilyn McAdams Sibley (Texas A and M University Press, College Station, Texas), 1983, pp 232-237; *FLO: A Biography of Frederick Law Olmsted* by Laura Wood Roper (The John Hopkins University Press, Baltimore, Maryland), 1973, pp 95-98; *A Clearing in the Distance: Frederick Law Olmsted and America in the 19ᵗʰ Century* by Witold Rybczynski (Simon and Schuster, New York, New York), 1999, pp 131-138; and *A Journey Through Texas: Or, a Saddle Trip on the Southwestern Frontier* by Frederick Law Olmsted (University of Texas Press, Austin, Texas), 1978, pp 138-146 and 169-201. The idea of a free state in West Texas was even printed in Northern newspapers. The Keosauqua, Iowa *Western American* of February 28, 1852, stated "The Free State of West Texas. Western Texas, it is said, is preparing to ask admission into the Union as a separate State." A source that tells about one of the Prussia officers' involvement in the Adelsverein see *Julius Theodore Splittgerber (1819-1897), Volume One: His Life and Times* by Mary Lewis Turner (Watercress Press, San Antonio, Texas) 2003, Part II, pp 53-120.

24. Ibid.

25. Prince Sachsen-Altenburg's Presentation, February 21, 2003 and April 5, 2003.

26. *Nassau Plantation: The evolution of a Texas German Slave Plantation* by James C. Kearney (University of North Texas Press, Denton, Texas), 2010, pp 36, 245-246.

27. Biesele German *Settlements*, Chapter V; Ron Tyler, et al, eds., New *Handbook of Texas*, Volume 2, pp 237- 238 and Volume 4, p 87; and *History of New Braunfels and Comal County, Texas 1844-1946* by Oscar Haas (Office Centers, Inc., Austin, Texas), 1976, p 48.

28. *Fest Ausgabe fuenfzigjaehrigen Jubilaeum der Deutfchen Kolonie Friedrichsburg, by* Robert Penniger (Herlag von Robert Penniger, Fredericksburg, Texas) 1896, pp 47-48; *Fredericksburg, Texas ... The First Fifty Years,* A Translation of Penniger's 50ᵗʰ Anniversary Festival Edition by Dr. Charles L. Wisseman, Sr. (Fredericksburg Publishing Co., Inc.), 1971, pp 28-29, 47; Biesele *German Settlements*, Chapter V and pp 139-141; Ron Tyler, et al, eds., *New Handbook of Texas*, Volume 2, pp 140-141, 237-239, and 1161; *Pioneers in God's Hills: A History of Fredericksburg and Gillespie County: People and Events* Compiled by Gillespie County Historical Society, Volume I (Eakin Press, Austin, Texas), 1960, XIX. In September 1850, the U. S. and Mexican Boundary Commission stopped in

Fredericksburg. They reported there were about 500 people living in the town. *Personal narrative of Explorations and Incidents in Texas, New Mexico, California, Sonora, and Chihuahua 1850 – 1853* by John Russell Bartlett, Two Volumes (The Rio Grande Press, Inc., Chicago, Illinois) 1965, Volume I, pp 59-60.

29. *German Seed in Texas Soil: Immigrant Farmers in Nineteenth Century Texas* by Terry G. Jordan (University of Texas Press, Austin, Texas), 1966, pp 182-183 and "New Perspectives on Texas Germans and the Confederacy" by Walter D. Kamphoefner, 'Southwestern Historical Quarterly', Volume CII, (April 1999), pp 166-179). For a fuller discussion of what life was like for Germans in the early days of the Republic and for the *Adelsverein* leaders and settlers see: *Sketches of Life in The United States of North America and Texas as Observed by Friedrich W. von Wrede* Translated by Chester W. Geue, (Texian Press, Waco, Texas), 1970; *Hurrah for Texas: The Diary of Adolphus Sterne 1838 – 1851* Edited by Archie P. McDonald, (Eakin Press, Austin, Texas), 1986; *Gustav Dresel's Houston Journal: Adventures in North America and Texas, 1837 – 1841* Translated by Max Freund, (University of Texas Press, Austin, Texas), 1954; *Roemer's Texas 1845 to 1847 by Dr. Ferdinand Roemer* Translated by Oswald Mueller, (Eakin Press, Austin, Texas), 1995; *Texas in 1848 by Viktor Bracht* Translated by Charles Frank Schmidt, (German-Texas Heritage Society, Austin, Texas), 1991; *The Cypress and Other Writings of a German Pioneer in Texas by Hermann Seele* Translated by Edward C. Breitenkamp, (University of Texas Press, Austin, Texas), 1979; *The Cabin Book or National Characteristics* by Charles Sealsfield Translated by Sarah Powell, (Eakin Press, Austin, Texas), 1985; *A Life Among the Texas Flora: Ferdinand Lindheimer's Letters to George Engelmann by* Minetta Altgelt Goyne, (Texas A and M University Press, College Station, Texas), 1991; *Letters From A Texas Sheep Ranch* by Harry James Brown, Illinois Press, Urbana, Illinois), 1959 *Kendall of the Picayune: Being His Adventures In New Orleans, On The Texan Santa Fee Expedition, In The Mexican War, And The Colonization Of The Texas Frontier* by Fayette Copeland, (University of Oklahoma Press, Norman, Oklahoma), 1943, pp 273-320; *The Diary of Hermann Seele and Seele's Sketches from Texas: Pioneer, Civil and Cultural Leader, German-Texan writer* Translated by Theodore Gish, (German-Texan Heritage Society, Austin, Texas), 1995; *Eagle in the New World: German Immigration to Texas and America* by Theodore Gish, (Texas A and M University Press, College Station, Texas), 1986; *A Sojourn in Texas, 1846-47: Alwin H. Soergel's Texas Writings* Translated by W. M. VonMaszewski, (German-Texan Heritage Society, Austin, Texas), 1992; and *Voyage to North America 1844-1845: Prince Carl Solms's Texas Diary of People, Places, and Events* Translated by Wolfram M. Von Maszewski, (University of North Texas Press, Denton, Texas), 2000.

30. Ron Tyler, et al, eds., *New Handbook of Texas*, Volume 2, p 1131.

31. The 27 who arrived on the *St. Pauli* included: Ernst Otto Amelung; Heinrich Backhofen; Peter Bub; Adam Deichert;Christoph Flack; Wilhelm Friedrich; Adolph Hahn; John Hoerner; Heinrich Kattmann; Adam Koeppel [Kappel]; Jacob Kuechler; August Lerch; Frederick Louis; Friedrich Michel; Eduard Mueller; Jacob Obert; Louis Reinhardt; Friedrich Schenk; Gustav Schleicher; Theodor Schleuning; Leopold Schultz; August Strauss; Adam Vogt; Julius Wagner; Carl Wendt [Wundt]; Franz Zentner; and Philip Zoeller. Two others who arrived in April 1847 were Hermann Spiess and Dr. Ferdinand von Herff. Christian von Hess and August Vogelsang who had previously been members of the Forty joined them at Carlshafen. A Canadian, named Rock or Rockan also joined. Others who 'may have' joined them included men with the surnames of: Fuch; Hermann; Kappelhof; Mertins; Neff; Ottmer; and a Wilhelm Zoeller. The Forty hired a female cook who could speak English, Julie Herff from Baden. Tolzmann's *German-American 48ers 1848-1998*, p 23; Ron Tyler, et al, eds., *New Handbook of Texas*, Volume 2, p 1131 and Volume 4, p 103; Beseler's *German Settlements*, pp 156-157; *Texas Indian Fighters* by A. J. Sowell, (State House Press, Austin, Texas), 1986, pp 663-667; *Historical Images of Boerne, Texas* by Garland Perry, (ECON-O-PRINT, San Antonio, Texas), 1982, p 179; and *German Pioneers on the American Frontier: The Wagners in Texas and Illinois* by Andreas Reichstein (University of North Texas Press, Denton, Texas), pp 2001,42-49.

32. Biesele *German Settlements*, p 157 and "Texas German Grave markers: Lateiner, Freethinkers, and Other Intellectuals" by Scott Baird <u>The Journal</u>, Volume XXX11, Spring, 2010.

33. While at Tusculum, the *Darmstaedter Kolonie* were joined by six other Freethinkers from Darmstadt: Ernst Dosch; Ernst von Lichtenberg; Ludwig von Lichtenberg; and Hermann Schenck. The *Darmstaedter Kolonie* were likely: Fritz Kramer; Wilhelm Friedrich; Christian Flack; Gustav Theissen; Adam Vogel; Jacob Kuechler, Leopold Schultz, and Rudolph Carstanjen. Andreas Reichstein, in his book *German Pioneers*, claims the Forty. on their way to the Llano River, established the *Darmstaedter Farm*. Nicholaus Zink and his wife, Louise von Khueser, came to Texas as a surveyor with the *Adelsverein* in 1844. He was born on February 4, 1812 in Bamberg, Bavaria. He and Louise were divorced in 1847 and he moved to land on Sister Creek, where he built a large log house with an upper story, the first building in Sisterdale. Zink married Elisabeth Mangold on January 19, 1848 in Bexar County. In 1850 they sold their home to Eduard Degener, the leader of the Union Loyal League. Zink died on November 3, 1887 and is buried at Welfare, Texas. The remnants were: Wilhelm Friedrich; Jacob Kuechler;

Christoph Flack; Leopold Schultz; Frederick Kramer, and the cook, Julia Herff. Several sources claim a man named Keller was a member of *Die Vierziger* remnants. This is incorrect. Misreading the 1850 Comal Census causes the error. The name appears to be Keller; it is actually Jacob Kuechler. Seventh U. S. Census, 1850 Comal County Texas Census, p 59; Beseler's *German Settlements*, p 157; Ron Tyler, et al, eds., *New Handbook of Texas*, Volume 6, p 1154; Reichstein German Pioneers, 48; and Bexar County Marriage Records, Volume D2, p 141.

34. Tolzmann German-*American 48ers 1848-1998*, 31 and Ron Tyler, et al, eds., New *Handbook of Texas*, Volume 2, P 1131.

35. "The Two Worlds of the Forty-Eighters" by Theodore S. Hamerow, in *The German Forty-Eighters in the United States* Edited by Charlotte L. Brancaforte, (Peter Long Publishers, New York, New York), 1989, pp 27-29.

36. Ibid.

37. Ibid.

38. Tolzmann *German-American 48ers 1848-1998*, pp 31-32 and Wittke *Refugees of Revolution*, p 4.

39. Wittke Refugees *of Revolution*, pp 147 - 148; Zucker Forty-*Eighters*, p 80 and *A Brief History of the America Turnerbund* by Henry Metzner, (National Executive Committee of the American Turnerbund, Pittsburgh, Pennsylvania), 1924.

40. "The Mathilda Doebbler Gruen Wagner Story" by Winifred S. Cade, in *The Golden Free Land,* Compiled by Crystal Sasse Ragsdale, (Landmark Press, Austin, Texas), Tolzmann *German-American 48ers 1848-1998*, 1976, p 157; Levine's *Spirit of 48*, pp 38 and 46, *The Mainstream of Civilization* by Joseph R. Stayer, et al, (Haracourt, Brace and World, Inc., New York, New York), 1969, pp 586-587; Hamerow *Two Worlds of Forty-Eighters*, pp 29-30; Tiling *German Element in Texas,* p 121; Zucker *Forty-Eighters*, pp 4-6; Tolzmann *German-American 48ers 1848-1998*, p 76; Levine *The Spirit of 48*, pp 35-38 and 46 and Wittke *Refugees of Revolution*, pp 20-21.

41. Interview with Dr. Ansgar Reiss, Regensburg University, Germany, Regensburg, Germany, October 6, 1998 at University of Dallas and Wittke *Refugees of Revolution*, pp 21 and 24.

42. Wittke Refugees *of Revolution*, pp 21 and 24; Levine *The Spirit of 48*, p 36; Hamerow *Two Worlds of Forty-Eighters* pp, 29-31; Zucker *Forty-Eighters*, pp 21-27 and Tolzmann *German-American 48ers 1848-1998*, pp 76-77.

43. "The Battle of the Nueces, August 10, 1862" by Robert W. Shook in 'Southwestern Historical Quarterly', Volume LXI, (July 1862), p 31.

44. Lich *German Texas*, pp 92-94; Tolzmann *Germany-American 48ers 1848-1998*, pp 35-36; Jordan *German Seed*, pp 106-111 and 180-185; Kamphoefner *New German Perspectives*, p 171; and *Adolf Douai, 1819 – 1888: The Turbulent Life of a German Forty-Eighter in the Homeland and in the United States* by Justine Davis Randers-Pehrson (Peter Lang Publishers, New York, New York) 2000, pp 189-190.

45. Ibid.

46. Tolzmann *German-American 48ers 1848-1998*, p 35.

47. *Secession and the Union in Texas* by Walter L. Buenger, (University of Texas Press, Austin, Texas) 1984, pp 80- 83, 91-95, 98-100 and 133-134 and Tolzmann *Germany-American 48ers 1848-1998*, pp 43-46.

48. Tolzmann *Germany-American 48ers 1848-1998*, p 45.

49. "The Clash of Utopias: Sisterdale and the Six-Sided Struggle for the Texas Hill County" by Michael P. Conzen in *Cultural Encounters with the Environment: Enduring and Evolving Geographic Themes,* Edited by Alexander B. Murphy and Douglas L. Johnson, (Rowman and Littlefield Publishers, Inc., New York, New York), 2002, pp 45-52; Flack *A Yankee in Texas Hill Country*, pp 29-31 and Siemering's Germans During Civil War, May 21, 1924.

50. *Die Lateinische Ansiedlung in Texas* [The Latin Settlement in Texas] by August Siemering, Translated by C. W. Geue, 'Texana' (Volume V, Summer 1967); Flack *A Yankee in Texas Hill Country*, p 29; Randers-Pehrson *Adolf Douai*, p 196.

51. Biesele German *Settlements*, pp 202 and 225; Benjamin Germans *In Texas*, pp 9697; Fischer *Marxists And Utopias in Texas*, pp 80-86 and "Goethe on the Guadalupe" by Glen E. Lich in *German Culture in Texas: A Free Earth: Essays from the 1978 Southwest Symposium* Edited by Glen E. Lich and Donna B. Reeves (Twayne Publishers, Boston, Massachusetts), 1980, pp 1129-1171.

52. Biesele *German Settlements*, pp 50-60, 196-198, 208-210 and 220-227 and Benjamin *Germans In Texas*, pp 116-119.

53. Zucker Forty-Eighters, pp 97-99; Roy Tyler, et al, eds., New *Handbook of Texas*, Volume 6, pp 597-598; Rifles *and Blades of the German-American Militia and the Civil War: A Comprehensive Illustrated History of the Turners, a Unique German-American Gymnastic Society, and their Role in the Events Before and During the American Civil War* by Thomas B. Rentschler, (Blue Hills Press, Hamilton, Ohio) 2003 and Metzner *Brief History of Turners*.

54. Conzen *Clash of Utopias*, pp 52-53.

55. Biesele German *Settlements*, pp 196-198; Wittke Refugees *of Revolution*, pp 122-132 and 163-165; Zucker *Forty-Eighters*, pp 56 and 77; "Sam Houston and "The Texas Hill Country Germans" by Kenn Knopp 'The Journal' (Summer 2006), Volume XXVIII, p150; "The Texas State Convention of Germans in 1854" by R. L. Biesele 'Southwestern Historical Quarterly' Volume XXXIII, (April 1930), pp 247-251; Levine *Spirit of 1848*, pp 96-99, 102-108 and 149-150; Tolzmann' *48ers 1848-1998*, pp 96-105; Tiling's *German Element in Texas*, 138; Randers-Pehrson's *Adolf Douai*, p 199.

56. Biesele German *Settlements*, pp 196-198; Biesele German *Convention*, pp 257-261 and Sibley *Lone Stars and State Gazettes*, pp 232-233.

57. *The Howling of the Coyotes: Reconstruction Efforts to Divide Texas* by Ernest Wallace, (Texas A and M, College Station, Texas) 1979, p 14; Sibley Lone *Star and State Gazettes*, pp 232-233 and Roper *Biography of Olmsted*, pp 101-104. For a more detail article on Olmsted's activities to support the Free-Soil Movement in Texas see "Frederick Law Olmsted and the Western Texas Free-Soil Movement" by Laura Wood Roper 'The American Historical Review', Volume LVI, No. 1, (October 1950), pp 58-64 and Randers-Pehrson's *Adolf Douai*, Chapters 9 and 10.

58. "German Republicans and radicals in the Struggle for a Slave-Free Kansas" by Frank Baron Yearbook of German- American Studies, Volume 40, (2005), pp 3-22.

59. Ibid.

60. Benjamin *Germans In Texas*, p 97.

61. Biesele *German Settlements*, pp 197-198 and Biesele *German Convention*, pp 250-251.

62. Biesele *German Settlement*, pp 197-203; Tiling, *German, Element in Texas*, p 141; Benjamin *Germans In Texas*, pp 98-101; Biesele *German Convention*, pp 255-261; Conzen *Clash of Utopias*, p 53; "The Texas Germans in State and National Politics, 1850-1865" by Ada Maria Hall, (M. A. Thesis, University of Texas, Austin, Texas), 1938, pp 7-13; Buenger *Secession in Texas*, p 92; "A History of Gillespie County, Texas, 1846-1900" by Sara Kay Curtis, (M. A. Thesis, University of Texas, Austin, Texas), 1943, p 54; "A New Perspective for the Antebellum and Civil War Texas German Community" by Melvin C. Johnson, (M. A., Stephen F. Austin State University, Nacogdoches, Texas) 1993, pp 59-68; Marten *Texas Divided*, pp 27-28; Siemering's Latin Settlements, pp 129; and Olmsted *Journey Through Texas*, pp 435-439.

63. Siemering's Latin Settlements, pp 127-131.

64. *San Antonio Weekly Herald*, June 7, 1862 and *San Antonio Weekly Herald*, June 21, 1862.

65. Ron Tyler, et al, eds., New *Handbook of Texas*, Volume 5, pp 871-872 and *Fourteen Presidents Before Washington* by Herman D. Hover, (Dodd, Mead and Company, New York, New York), 1985.

66. "The Kansas Campaign" by Frank Baron Yearbook of German-American Studies, Supplemental Issue, Volume 1 40.

67. *The Germans in the American Civil War* by Wilhelm Kaufmann, translated by Steven Rowan and Edited by Don Tolzmann with Werner D. Mueller and Robert E. War, (John Kallmann Publishers, Carlisle, Pennsylvania), 1999, p 63, 72, 112, 114, and 277-278.

68. Peter Joseph Osterhaus was born on January 4, 1823, in Koblenz, Prussia. He attended the Berlin Military Academy and served as a Prussian Army officer. When the 1848 Revolution began, Osterhaus joined the revolution. Osterhaus fled Germany after the revolution failed and first settled in Illinois, where he became friends with a young attorney named Abraham Lincoln. In 1860 he moved to Saint Louis where he established himself as a merchant and bookkeeper. He helped organized the "Home Guards," a pro-Union military unit. He commanded the largely German First Division and later the XV Corps. He died on January 2, 1917, in Duisburg, Germany.

Franz Sigel was born on November 18, 1824, in Sinsheim, Baden. He graduated from Karlsruhe Military Academy in 1843 and entered the army of the Grand Duke of Baden. When the 1848 German Revolution broke out, he resigned his commission and joined the insurgents where he commanded one of their armies. When they organized a provisional government in 1849 Sigel became the minister of War. After the failed 1848 Revolution, he fled to Switzerland, then to England. Sigel fled to New York in 1852 and moved to St. Louis in 1857 and taught school. Sigel died in New York in 1912.

The Germans of Missouri also had their 'Nueces Massacre'. On June 19, 1861, about 200 German home guard militiamen were surprised by a force of Anglo State Guards at Cole Camp, Benton County. The result was about 35 German home guards were killed, or as Union sympathizers said, "massacred." *Yankee Warhorse: A Biography of Major General Peter Osterhaus* by Mary Bobbitt Townsend, University of Missouri Press, Columbia, Missouri, 2010; Yankee *Dutchman: The Life of Franz Sigel* by Stephen D. Engle, (Louisiana State University Press, Baton Rouge, Louisiana), 1993, pp xiv-2; Kaufmann *Germans in Civil War*, pp 63, 72, 112, 114, 200 and 277-278; "Benton County Lutherans and the battle of Cole Camp June 19, 1861" by Roger Moidenhauer, 'Concordia Historical Institute Quarterly', Number 61 (Winter 1988), pp 15-16 and "Killed by Rebels: A Civil War Massacre And Its Aftermath" by Robert W. Frizzell, 'Missouri Historical Review', Volume 71, Number 4, 1977.

CHAPTER 2 – The Killing Starts
Unionists Prepare to Oppose the Confederacy

April[1] 1860:

Twenty-seven-year-old Dennis Kingston, a native of Cork County, Ireland rode into the German-Texas town of Fredericksburg in the Texas Hill Country County of Gillespie. The five-foot-six-inch, gray-eyed, sallow-complexioned former U. S. soldier left his eighteen-year-old wife, Martha, and young son, William, at their small farm near present-day Loyal Valley in southern Mason County. Kingston was a long way from his native Ireland. In 1854, he stowed away on a ship to New York. On September 6, 1854, he enlisted in the First U. S. Infantry. After serving his five-year enlistment, most of it on the Texas frontier, he was discharged at Camp Cooper in September, 1859. He grew accustomed to the rugged ways of army life and the hardships of the frontier. He stopped at John M. Hunter's store and saloon in downtown Fredericksburg for a few drinks.[2] Hunter's saloon was a popular gathering place for former soldiers, frontiersmen, and German settlers. Hunter was experienced with their sometimes-drunken sprees and the violent result. Ten years before Hunter had killed a Fort Martin Scott soldier during a fight at his store. The soldier's friends returned and burned down his establishment. Hunter escaped death only by hiding in a friend's house on Live Oak Creek. He was tried and acquitted of the killing, returned to Fredericksburg, and rebuilt his store.[3]

It is not clear if Kingston was already drinking when he entered the saloon. According to one story, told by a member of the Biberstein family, Kingston had "looked too deep into the bottle" and was so obnoxious Hunter ended up hitting Kingston over the head with an ax handle and threw him out of the store.[4]

When Kingston regained consciousness, he got to his feet, stumbled away, and located a revolver. He returned to the saloon and fired a single shot into the veranda, killing a "well-beloved German from Boerne named Louis, an energetic and successful hunter and head forester." Louis was thirty-seven-year-old Frederick Louis, a Freethinker, a former member of *Die Vierziger,* one of the founders of Boerne and an 1854 ultra-radical. Kingston attempted to escape but the citizens of Fredericksburg quickly captured him and dragged him to the county jail on the market place. Here Edward Maier, the local sheriff, placed him in jail. The next morning Kingston was found "stiff dangling from a rafter." The rumor quickly spread that Louis' "German friends" were responsible for Kingston's death. These 'German friends' were fellow Freethinkers, Forty-Eighter, and *Die Vierziger* members.[5]

Kingston's death was the first lynching in Fredericksburg. It would not be the last. Within the next five years, over fifty area men would be hanged or otherwise killed. The origin of most would be the result of Kingston's murder. His death and that of his father-in-law, Samuel Shallenberger, began the polarization of several families into various groups. After the start of the War Between the States, two more deaths – that of Basil Stewart and James Billings – accelerated this polarization. This led to many families/groups using the war to settle personal grudges, which in turn led to counter-reprisals and more deaths. This was especially true during the 'Bushwhacker War' during the period of late 1863 to mid-1864.

Martha Kingston was very worried when her husband did not return. She previously had several bad experiences with area Germans. Her father, Samuel Shallenberger [Schellenberger], was killed in 1856 when he, "was shot in the back by a German buffalo hunter in a dispute over a house in Fort McKavett." She and Dennis were having problems with other area Germans. One was a family named Biberstein. The previous June when Dennis was away on business he wrote her a letter. It cautioned her, "If you think that there is any danger where you are with that Beberstein [sic], do leave. You are better to go nearer to Fredericksburg to live until I come home."[6]

Now her husband had not returned. She hitched up the buggy and rode into Fredericksburg. When she learned of his death, according to the Biberstein account, she "swore that she would send at least ten Fredericksburg Dutchmen into the beyond as sacrifices of atonement." This atonement came on August 10, 1862 when seven wounded Germany insurgents were killed after the Nueces Battle.[7]

March, 1861:

It was a hot summer day. Several men stood on top of a bluff overlooking Johnson Creek in western Kerr County. A closer look revealed three of them had their hands tied with ropes around their necks. One of these was William Tegener, a native of Prussia. He and two brothers, Gustav and Frederick, had left Prussia after the failed 1848 Revolution. They arrived in Texas about 1854 and purchased land near Comfort. Other men stood a short distance away from Tegener and the other two "inoffensive Germans." The three men who stood about to be hanged had expressed their support for the Union. This cost them their lives. The three were soon dangling from the ropes. After they were hanged their bodies were thrown from the 75-foot bluff into Johnson Creek below. Unionist swore revenge.[8]

Which course of revenge should mankind follow?
Eye for eye, tooth for tooth, hand for hand, foot for foot. –
Exodus 21:24
Or
To me [The Lord] belongeth vengeance, and recompence. –
Deuteronomy 32:35

The election of Abraham Lincoln as President of the United States on November 6, 1860, resulted in the Southern States withdrawing from the Union and forming the Confederate States of America. Texas was one of the states to join the Confederacy. Texas called a Secession Convention on January 28, 1861. On February 1st, by a vote of 166 to 8, the convention voted to withdraw from the Union on March 2, 1861, unless secession was rejected by a statewide election on February 23rd. In the February 23rd election Texas voters approved secession 46,188 to 15,149.

One of the myths clamed by insurgent descendants and some writers is the ordinance of secession passed because: "A light vote was polled, and the proposition carried by a small margin," or "Not only was there a very light vote polled on the question whether or not Texas should secede, but only a bare majority of the votes cast spoke in favor of secession" or [many] "of the eligible voters preferred [not to vote] fearing retaliation regardless of the results. It is also said that a number of those voting for secession did so because they feared retaliation from mob-inspired secessionist from neighboring counties." It is not certain how many voters there were in Texas in February, 1861.

In the presidential election of 1860, 63,733 votes were cast. A total vote of 61,337 was cast in the special election of secession. This was only 2,436 less than in the presidential election, hardly a light vote. The ordinance of secession passed by over a 75% majority, hardly a bare majority. Had the other 2,436 all voted against secession the ordinance still would have passed by a majority of almost 73%; a 'landslide' in today's terms. Dale Baum, in his excellent study of voting patterns in Texas during the war, concluded, "Voting irregularities [in the ordinance of secession] occurred in far fewer Texas counties than allegations made by contemporaries of the period would suggest."[9]

Nineteen Texas counties voted against secession. Of these nineteen, ten were in the Austin and the Hill Country area. The anti-secession vote here was attributed "to the large German population [in the Austin {area} and Hill Country]", the effort of Austin Unionists, and to several Unionist newspapers. These Hill Country counties included Blanco, Gillespie, Medina, and the parts of Blanco, Comal, and Kerr counties which contained large numbers of German settlers and became Kendall County in early 1862. The vote in Blanco County was 86 for secession and 170 against. The vote in Gillespie County was 16 for and 398 against. Medina County voted against secession by a vote of 140 for and 207 against. Kerr County was the only Hill Country County, which had a large percent for secession. In Kerr County, the vote was 76 for and 57 against. Of the 57 opposed, 42 came from the Comfort area. Fifteen citizens in the Comfort area voted for secession.

It is interesting to note that almost as many citizens of the Comfort area voted for secession than did so in the entire area of Gillespie County. Comal County, which contained a large German population, voted in favor of secession by a vote of 239 for and only 86 against. Many of the German leaders in Comal County urged the citizens not to oppose the majority of Texans and vote for secession as a states rights issue and not an issue of slavery. However, unlike the majority of Comal County, in Boerne, which was in Comal County at the time, only six voted for secession while 85 voted against. Thus, in the basic area which became Kendall County in early 1862, the vote was 21 for and 127 against. The other Hill Country county, Bandera, supported secession by one vote. There the vote was 33 for and 32 against. Five Austin-area counties voted against secession; Travis, Burnet, Williamson, Bastrop, and Fayette. Travis voted 450 for and 704 against. In Burnet, the vote was 157 for and 248 against.

Williamson voted 349 for and 480 against. In Bastrop, the vote was much closer; 335 for and 352 against. Fayette voted 580 for and 626 against.[10]

Uvalde County voted against secession by a vote of 16 for and 76 against. Professor Baum's study of voting patterns concludes that Uvalde County had the largest percent of inflated Unionists strength. He concluded, "If all Uvalde County voters who participated in the 1860 election returned to the polls three months later to vote in the secession referendum, then over half of them would have had to 'switch' ... to the Unionist camp to achieve the over-whelming anti-secessionist vote registered in their county." He points out that, "three of the county's four precincts mysteriously failed to report returns [helps explain] this large conversion of ... men into Unionists." Professor Baum concludes that Reading W. Black, a Unionist and the chief justice [county judge], "shaped the lopsided Unionist majority by allowing the U. S. soldiers at Ft. Inge to vote and not opening the polls in the other three precincts."[11]

Eight counties in northern Texas also voted against secession. This was attributed to the effort of such leaders as James W. Throckmorton. These counties included: Jack, Montague, Cooke, Grayson, Fannin, Lamar and Collin. Jack voted 14 for and 76 against secession. In nearby Montague, the vote was 50 for and 86, against. Cooke opposed secession by 137 for and 221 against. Grayson voted 463 for and 901 against. Fannin likewise voted against by 471 for and 656 against. In Lamar, the vote was 553 for and 663 against. Collin opposed secession by a vote of 405 for and 948 against. Many claim this vote showed opposition against slavery. This was not necessarily correct. For many, the vote was not opposition to slavery but a desire to preserve the Union.[12]

There were many in major leadership positions in Texas who supported the state remaining in the Union, or if that failed for Texas to go it alone as a separate nation, the Republic of Texas. These included Governor Sam Houston. Governor Houston was under pressure to use force to prevent the state from seceding. At first, it appeared he seriously considered using the U. S. Army and Texas Unionists to keep Texas in the Union. Houston received visits from several Hill Country Unionists who offered to provide troops to oppose the Secessionists. Among those Unionists were John W. Sansom of Curry's Creek and Ferdinand Simon of Leon Springs. At that point Houston was still undecided, but told Sansom he was "opposed to war and by my acts shall not bring it about ... I hope we will have no blood shed." Simon told Houston he could bring 1,000-armed men to keep him in power.

Another Hill Country Unionist who offered to raise troops was Noah Smithwick of nearby Burnet County. Houston's reply was, "I have seen Texas pass through one long, bloody war. I do not wish to involve her in civil strife. I have done all I could to keep her from seceding, and now if she won't go with me I'll have to turn and go with her." Houston declined all such offers. By the end of March, 1861, Houston informed U. S. military officials that he declined such help and requested any troops being assembled for such purpose to withdraw from Texas.[13]

Other Unionists appeared ready to fight in support of Federal troops. One of the more serious incidents took place in San Antonio as the Committee of Public Safety moved to force Federal Major General David Twiggs, the commander of the Department of Texas, to surrender. Unionists – mainly Germans – formed military companies. As Texas troops under the command of Ben McCulloch marched into San Antonio on the night and early morning of February 16, 1862, "Companies of Union citizens, [very likely *Turnverein*] well-drilled and well-armed, were marching and counter-marching, presenting an imposing contrast to the other party, [McCulloch's forces] and a conflict seemed inevitable." A member of McCulloch's force commented on what he expected; "a sharp tussle, not only with the U. S. troops but with the Germans, who made up quite half the population of the town." General Twiggs surrendered his department and the threat of armed conflict diminished.[14]

After Texas seceded and joined the Confederacy, Houston refused to take the oath of allegiance to the Confederacy and was forced from office. After Lincoln called for 75,000 volunteers, sent Union troops into Virginia, and approved the proclamation of martial law in Missouri, Houston concluded by these actions, which in his eyes he considered usurpation, that Lincoln had dissolved the Union and absolved the states and the people of the states from their allegiance to the Federal government. Houston now threw his support with the Confederacy saying, "The time has come when a man's section is his county ... I am now a conservative citizen of the Southern Confederacy: and [give] to the constituted authorities of the country, civil and military, and to the Government which a majority of the people have approved and acquiesced in, an honest obedience." In a speech in May, 1861, Houston reaffirmed his support of the South, "All my hopes, my fortunes, are centered in the South, the people whose fortunes have been mine through a quarter of a century of toil, threatened with invasion, I can but cast my lot with theirs and await the issue." He retired to his home in Huntsville where he died on July 26, 1863.[15]

Others such as former Governor E. M. Pease and George W. Paschal, the leader of the Texas Union Party, left their law practices and attempted to withdraw into private lives. Some early Unionists like James W. Throckmorton, Eber W. Cave, and B. H. Epperson, gave their support to the state and actively supported the Confederate government. Others, like James P. Newcomb, the Unionist editor of the *Alamo Express*, fled to Mexico and waited out the war. But some such as E. J. Davis, A. J. Hamilton and John L. Haynes, gave active support to the Unionist movement in Texas. Davis and Haynes later organized and commanded Union Army units made up of Texans.[16]

With the secession of Texas, state and Confederate authorities faced four major problems. First; the presence of two to three thousand Union troops. This was about ten to fifteen percent of the entire Union Army. The major purpose of this force had been the protection of the Texas frontier against Indian attack. [Map C, p 130] Second; the replacement of this force by either state or Confederate units. Third; organizing and preparing the state for war, which everyone saw coming. Fourth; controlling the area within Texas where there was strong Union sympathy and in some cases outright hostility to Confederate authority.[17]

The problem of Union troops on Texas soil was quickly resolved when by late May, 1861, all Union troops had either fled or had surrendered to state militia units. The problem of frontier protection was not as quickly resolved. Texas had never been completely satisfied with the frontier protection the Federal government provided; a force largely composed of infantry units. In the Texas view, this frontier force should be mainly cavalry, and strike against Indian encampments – not just react to Indian attacks. Therefore, since statehood Texas had wanted a force of Rangers or mounted Texas units posted on the frontier paid, supplied by the Federal government but under the command of the State of Texas. This failure by the Federal government to provide frontier protection was one of the reasons Texas gave when it seceded. Several solutions for frontier defense were attempted. Even before secession, the Texas Legislature took action on this problem. On February 6, 1861, Governor Houston requested laws and funding for frontier defense. The legislature responded and directed all frontier counties to organize ranger or minutemen companies for home defense.

Each of the Hill Country counties, except Bandera and Medina, organized home guard companies. The reason

Bandera and Medina Counties did not raise home guard companies was because their citizens refused to pledge allegiance to the Confederacy, which was a requirement. They did organize 'unauthorized' units, which were Unionist in nature. A few months later Bandera and Medina Counties had authorized militia units for home defense. These home guard companies were to have between twenty and forty members. They were not very effective for several reasons. The funding only allowed about twenty-five percent of the unit to be in the field at one time. Since these units were made up of citizens from one county, there was poor communication and lack of cooperation between various units. One example of good coordination between the home guard units was in late March and early April 1861. An Indian raid killed three citizens of Uvalde County; Henry Adams, Henry Robinson, and Julius Sanders. They raided C. C. Quinlan's Kerr County home and Lemuel Haywood's ranch and stole over two-dozen horses from a Mr. Rossey in Bandera County. They continued and stole several horses from D'Hanis and in Sabinal County, from a Mr. Crabb, James Riley on Grape Creek, and William Miller living near the head of the Pedernales River, and attacked and shot one of John Doss' Negroes in Gillespie County. Several home guard units went after them. A group of Gillespie citizens went after one group of Indians. In the ensuring fight, John Cadwell and James Wilcox were wounded. The militia company of Lieutenant James Paul at Camp Verde was on a scout in far west Kerr County when they came across the home guard company from Kerr County, also on a scout.

They joined forces and followed the Indians. The Indians escaped but the units recovered horses and other items belonging to Adams and Robinson.[18]

R. H. Williams' book, *With The Borden Ruffians* is a source almost every writer or historian uses to demonize James Duff, later the Hill Country provost marshal. No writer has ever checked the accuracy of Williams' stories. A detailed comparison of various accounts of Williams' compared to official records shows Williams' accounts are extremely exaggerated and, in many cases, untrue. This account is one.

Both Paul and Harbour, the commander of the Kerr County home guard unit, wrote accounts of this incident. They match in every detail. Williams also tells of this incident, but his account is almost totally incorrect. For example, he says he was the orderly sergeant [first sergeant]. Paul's Muster Roll shows Williams was a private. Williams says Paul's detail consisted of 25 men. Paul's report says 10 men. Williams says they came across a farm where two white men had been killed and goes into great detail about the grieving women. Paul's and Harbour's reports say nothing about such an event. Williams says there were three other ranches raided and all the white men and women were killed. Paul's and Harbour's reports say nothing about other ranches being raided. A check of newspapers of the time says nothing about any other Indian raids during this time, except in the above-mentioned accounts. Williams says at least three Indians were killed. Both Paul and Harbour say no Indians were killed. Williams says they captured 15 ponies.

Both Paul and Harbour say they captured 2 horses and one mule. This is an example that shows Williams' various stories must be very carefully compared to other sources to find the actual facts. Williams is not a very dependable source.[19]

One, if not the first, of these home guard companies organized was the Minuteman Company of William T. Harbour in Kerr County, as referenced above. It was organized on February 27, 1861 and consisted of 40 members. It is not known how long it was in service, but likely until early 1862. In Blanco County, William A. Blackwell organized a 40-man company of rangers or Minutemen in May. On August 24, 1861, George Freeman organized a 61-man Blanco company named the Pedernales Cavalry Company. Another early home guard company was that of Charles H. Nimitz of Gillespie County. It was organized by June, 1861. It is not known how long it remained as a unit. The last located muster roll is dated July 30, 1861; but other records show it was still in existence as late as March, 1862. One home guard company which became very controversial was that of Philip Braubach of Gillespie County. It was organized on February 24, 1861 and served for at least a year.[20] [Appendices A-E, pp 604-614]

In addition to the home guard units, two Confederate units were raised for frontier protection. These were the 1st and 2nd Texas Mounted Rifles. It was hoped these two regiments would provide the type frontier force wanted by Texas. It was not to be. A major dispute arose between Texas and Confederate authorities over which government had jurisdiction over or command of the Mounted Rifles. The issue was resolved with the Confederate government in Richmond, Virginia retaining control. This meant that at any time these troops could be withdrawn from frontier protection and deployed to meet Union Army advances or attacks.

For the first year of the war, the 1st and 2nd Regiments were deployed along Texas' western frontier. The 1st Regiment was deployed on the Texas frontier from the Red River to Fort Mason, near the present town of Mason. The 1st Regiment also garrisoned Fort McKavett, about eighty miles northwest of Fredericksburg in present Menard County.

The 2nd Regiment was deployed along the Rio Grande from Fort Brown to Fort Duncan, at Eagle Pass. It had companies and detachments in many of the abandoned Federal posts. Forts Inge, in Uvalde County, and Clark, in present Kinney County, Camps Wood, in present Real County, Hudson in present Val Verde County, and Verde, in Kerr County were among those garrisoned.[22]

For several reasons, Indian raids against the frontier settlers fell dramatically in 1861 and early 1862. Historians have never completely explained why there were fewer raids during this period from early 1861 until mid-January 1862. David Paul Smith in his book *Frontier Defense in the Civil War* gives two of the most logical reasons. He says, "the Indians overawed by the magnitude of the great war, drew back from the settlements, hoping that the white men would kill one another to the last man." Or "a more practical explanation may be that the constant warfare of the previous two years, where punitive expeditions severely punished the raiding tribes, along with earnest attempts by authorities at treaty negotiations, somewhat abated the Indian incursion."[23]

This temporary decline of Indian attacks and the demand for more Confederate units in the east, made it clear to Texas these two regiments were going to be withdrawn and not replaced by the Confederate government. Therefore, Texas attempted a third solution for its frontier protection: its own full-time regiment force under state control. The primary unit raised for this purpose was known as the Frontier Regiment. On December 21, 1861, the Texas Legislature approved a bill creating this state regiment. It was to have a total of ten companies with one hundred men each. Nine companies were to be recruited from the area it was to protect, and one company from the remainder of the state. The field grade officers of the Frontier Regiment were the same as that of a Confederate regiment. It had a colonel, a lieutenant colonel, and a major.[24]

The law that established the Frontier Regiment required the governor to appoint a 'competent person' for each designated area as enrollment officer. Once at least sixty-four men were enrolled, elections could be held for company officers. This portion of the law resulted in competition between individuals wanting to be the enrollment officer which would likely result in that man being elected captain of the company. The governor appointed the field grade officers. They were: James M.

Norris, colonel; Alfred J. Obenchain, lieutenant colonel; and James Ebenezer McCord, major. Obenchain was killed after a quarrel with one of the regiment company commanders in the summer of 1862. McCord was promoted to lieutenant colonel and Buck Barry was appointed major.[25] From the Texas Hill Country, two companies were authorized. One company was made up of men from Gillespie, Hays, and Kerr Counties. The second company was made up of men from Bandera, Blanco, Medina, and Uvalde counties.[26] The area just north of the Hill Country was authorized one company made up of men from San Saba, Mason, Llano, and Burnet counties. The area generally south of the Hill Country was authorized one company made up of men from Frio, Atascosa, Live Oak, Karnes, and Bee counties.[27]

The Frontier Regiment began organizing in December, 1861. By April, 1862, it was stationed along the Texas frontier. Each company was split between two locations or camps. A total of eighteen camps were established, running generally south-southwest from near the Red River in Wichita County to the Rio Grande River at Eagle Pass. [Map C, p 130]. The Texas government now had what it had long desired; a full-time mounted regiment and numerous 'ranger' companies.

Another myth which insurgent descendants, Texas writers, and historians all claim is that after secession and withdrawal of Federal forces, the Texas frontier had "no protection at all. In a word, the settlements were left wide open to thieves, desperados, and all kinds of Indian depredation." The fact is that after secession, the Texas frontier had just as good, if not better, protection then it had before. At the beginning of the war, the Federal government had most of four regiments stationed in Texas – somewhere between ten and fifteen percent of the army; thirty-seven companies or about 2,166 troops. But the the majority of this force were infantry and unsuited for mounted warfare against the Indians. Of these thirty-seven companies, only ten were cavalry and suitable for offensive operations. Frontier settlers continued to complain about the inadequacy of the Federal force. It was only when Federal forces were supplemented by Texas forces [Rangers], were Indian attacks reduced. In the first year of the war, the Confederates deployed two regiments of cavalry or twenty companies for frontier protection. Even when these regiments were withdrawn, the state deployed the Frontier Regiment of ten companies. A national force made up of parts of three cavalry regiments, plus local militia units, supplemented this national force. According to David Paul Smith, author of the best-known study of frontier defense, *Frontier Defense in the Civil War*, these forces were not only better than provided by the U. S. Army, they "more than held their own against the Indians." [28]

One of the methods Texas used to solve the third problem – that of organizing for war – was to create brigade districts. On December 25, 1861, Governor Lubbock signed a bill whereby each county in the state was placed in one of these districts. All free white males between the ages of eighteen and fifty were required to enroll in one of these state units or in the Confederate Army. Companies were organized into battalions and regiments within each district. The bill required each brigade district to maintain a strength of at least 400 men, with up to as many as 1,200. The Hill Country counties were in the 31st Brigade District, headquartered at New Braunfels in Comal County and commanded by State Brigadier General Robert Bechem, a 42year-old New Braunfels merchant from Prussia. The counties of the 31st Brigade were Atascosa, Bandera, Blanco, Comal, Concho, Dawson, Gillespie, Llano, Mason, Maverick, McCulloch, Medina, Menard, San Saba, Uvalde, and Zavala.[29] [Appendix F, p 616]. The law which created the brigade districts was passed just four days after the law authorizing the Frontier Regiment. Texas now had three types of State Troops: home guard companies, the Frontier Regiment, and the militia of the brigade districts.[30]

Gillespie County had an entire militia regiment, the 2nd composed of two battalions of three companies each. Bandera, Blanco, Kendall, and Kerr Counties were in the 3rd Regiment of the Brigade. Kendall County had three companies, while Bandera, Blanco, and Kerr Counties each had one company. Medina County had an independent battalion of four companies.[31] [Appendices G, H, and I, pp 617-619].

A major problem for the Confederates existed with the Hill Country militia units of the 31st Brigade District. Many, if not most, of the leaders of these sixteen local companies were insurgents, Unionists, or Union sympathizers. Most of the local citizens of the six-county area were under arms and undergoing military training. These sixteen local militia companies were designed to protect the area in case of Indian or Union invasion. The key question was, "Would these Hill Country citizens, who had voted against secession, fight for or against the Confederacy in the event of a Union invasion, or would they rise up in arms and fight with the Union Army when it came?"

Thus, a year after secession, Confederate authorities in Texas had taken major action to solve three of their four major problems. They had removed all Federal troops from Texas. Frontier protection had been provided with a combination of home guard, the Frontier Regiment, and Confederate troops. The brigade districts gave Texas a force to help repeal a Union invasion. Most Texas males were either in Confederate service, in state units of the Frontier Regiment, or the militia of the brigade districts.

The fourth problem – that of strong Unionist sympathy – was never solved. In the Secession Referendum of February, 1861, about one-fourth of the citizens voted against secession. This number gave either active or passive support to the Texas Unionist cause, especially from early 1861 to mid-1862. This number was much higher in the Texas Hill Country. Some studies show about one-third of the total population remained neutral, one-third actively supported the Confederacy, and the remainder supported the Union, especially by 1864. Some 2,132 whites and 47 blacks from Texas served in the Union Army.[32]

Many areas in Texas had different methods of supporting the Union cause and Texas Unionists. One of the early Unionist reactions was in the northern counties of

Denton, Wise, Cooke, and Collin, which had voted against
secession. This Unionist reaction was a dismemberment
plan, whereby these counties would form a separate state and
apply to join the Union. This quickly failed, but
demonstrated that the idea of a free state was on Unionists
minds.[33] Unionists in many parts of the state organized
'Peace Parties,' sometimes known as 'Loyal Leagues.' Most
were organized sometime after September, 1861, but many
areas, including the Texas Hill Country, had such an
organization before that date. The members of these 'Loyal
Leagues' came from a cross section of the communities
including professional men clergymen, merchants, and
farmers.

Texas Confederate authorities received information
that these secret organizations had secret signs, grips, and
passwords. The Loyal Leagues had, at least among the Anglo
Unionist, three degrees. The first degree bound the member
to secrecy and obligated him to avenge an attack on a fellow
member. The second degree tested the candidate on robbery
and 'Jayhawking' [or to make war on a southern
sympathizer]. The third degree pledged him to support the
movement to reestablish the Union. The pledge of each
member initiated into the League was do all he could for the
North, and all he could against the South, and to reinstate the
United States Constitution by any means necessary. He was
to go to the aid of any member, who was arrested, to support
their families if they were lost, and to kill anyone who
betrayed the organization.[34]

The sign of recognition was for one member to pass the fingers of the right hand slowly over the right ear. The correct response being to pass the fingers of the left hand slowly over the left ear. To confirm recognition, the leagues had a password: 'Arizona,' and a handshake. The handshake was the common grip with the forefinger extended and pressed on the inside of the other man's wrist.[35] The leagues were strongest in the northern counties of Denton, Cooke, Wise, and Collin. In central east Texas, Leagues existed in Austin, Colorado, and Fayette Counties.

Leagues also existed in South Texas from Zapata County to Fort Brown, in Cameron County. Leagues were also strong in Travis and Bexar counties. The League in Travis County first met on February 9, 1861. It remained strong until public opinion changed in early 1862, at which time many of its members fled into the hills above Austin. Others fled to Mexico and many entered the Union Army. One Austin Unionist gave his reason for fleeing when he said, "Every man that is not willing to support the Southern Congress is to be beheaded."[35]

Three-quarters of the residents of Austin were Unionists or Union sympathizers and provided many of the state Unionists leaders. By the spring of 1861, they had organized a military company. This Unionist company met in the second story of a dry goods store on the corner of Pecan Street and Congress Avenue, which was owned by Unionists George Hancock and Morgan Hamilton. This military company, 'The Capitol Guards,' did not submit a muster roll to the governor until February, 1862, and marched "under no flag." The author found no muster roll for this early company. Martin in *Texas Divided* says the captain was George Hancock. The Saturday, April 27, 1861, *Austin State Gazette* refers to this company as 'The Capitol Guards.' Other members of the company included Thomas Duval, A. J. Hamilton, John Hancock, John T. Allan, William P. de Normandie, E. M. Pease, Morgan Hamilton, George W. Paschal, and James Bell. In February, 1862, this company was organized as part of the state troops belonging to the 26th Brigade District. A muster roll for that 89-man company was located.[36]

The aims, or objectives, of the Leagues were to avoid the draft, provide a spy system for the Northern army, desert, during battle if drafted, and prepare the way for an invasion by Union troops. The Unionists believed the North planned to invade Texas with two armies; one from Kansas and the second by way of Galveston. With the cooperation of the Texas Unionists, the two armies would meet in Austin. [Map D, p 132]

Early in the War, Union officials discussed a plan for such an invasion. Union General George B. McClellan planned such an operation as described. It called for an overland drive from Kansas in coordination with a landing on the Texas coast. James H. Lane, an Abolitionist senator from Kansas, was to command 20,000 troops for the overland route. Nathaniel P. Banks, a major general and former governor of Massachusetts, offered to enlist 15,000 troops for the seaborne part of the operation. The plan failed when McClellan did not get early support of military officials in Kansas. According to McCaslin, Union Secretary of War Edwin M. Stanton continued to insist he supported such an expedition. The plan was cancelled in favor of an attack on New Orleans in the spring of 1862.[37] Like Austin, San Antonio and Bexar County had a large and active Loyal League. A San Antonio German expressed his views in a letter of February 20, 1861, to the New York German-language newspaper, *New York Democrat* saying if the United Stated would only send an army, nothing could prevent it from retaking Texas. James P. Newcomb, editor of the *San Antonio Alamo Express,* published several pro-Union articles. On February 20, 1861, he reported on a parade of "Young Americans" with wooden guns and carrying the Union flag. On February 23, he ran an article about Washington's Birthday Parade the previous day. In it he spoke highly of Ward Company Number 2 who carried a "large blue flag, upon it a rattle snake in a striking attitude and the motto, 'Don't Trend On Me,' the old flag of the Republic under which Washington fought the early battles of the Revolution." Ward Company Number 3, "carried a new stars and stripes presented to them the evening before." On February 25, 1861, Newcomb ran an article about, "Some enthusiastic Union men got up a torchlight procession." The Alamo Rifle Band, a local militia company, accompanied the procession. Newcomb claimed that at least five hundred persons were in the procession and serenaded several of San Antonio's prominent citizens who returned the compliment in speeches, "glowing with patriotism." In the same issue, Newcomb reminded the naturalized citizens of the oath of citizenship they took to, "Support the constitution of the United States, so help you God."[38]

Newcomb continued with his pro-Union views. On April 9, 1861, he wrote the Confederacy was "conceived in sin, shaped in iniquity, and born out of due time, because it was rushed into the world with indecent haste expressly to prevent the people from beholding its deformities." He went on and stated, "President Davis was vain, proud, weak, imprudent, ambitious, and unprincipled, a vile traitor, a trained rebel, and an inflated bigot." Newcomb also criticized the capture of 300 Federal troops on May 13th as they were evacuating Texas. That was too much for the local Knights of the Golden Circle, a semi-secret pro-slavery organization. That night they burned his paper and Newcomb left Texas.[39]

There were others in San Antonio who openly expressed support for the Union. After the evacuating Federal/Union troops surrendered to Texas militia units in April and May 1861, most of the enlisted men remained in San Antonio as prisoners of war. One of these men, Trooper Stephen Schwartz, told about a San Antonio saloonkeeper named Saddour whom he described as "a Union man to the backbone." Schwartz and his fellow soldiers were welcomed at Saddour's saloon where they "drank, sang patriotic songs, and cheered one another and Abraham Lincoln." Saddour sold them three pints of beer for the same price as he charged Confederates for a much smaller quantity. When called upon to justify this, he "explained his prices by tossing a large 'Union' and a small 'Secesh' glass onto the floor." The former bounced off the floor unharmed, while the latter burst into pieces, proving what Saddour believed to be an obvious point about the nature of the Union and that of the Confederacy, respectively.[40]

By the summer of 1861, Unionists in Texas were such a problem Governor Clark issued a proclamation regarding the actions of the citizens [Figure I, p 120]. The proclamation ordered all communication between Texas and the United States discontinued. It also ordered that Texans and the United States conduct no trade, and made it treason for any Texan to pay any debt owed to the United States or any United States citizen.

The proclamation further directed tht no United States citizen be allowed to visit Texas. Finally, it ordered all citizens of the United States within the limits of Texas to leave within twenty days. The proclamation stated anyone not complying be arrested and treated as spies.[41] The problem of Unionists and anti-Confederate activity was not limited only to Texas. Other Confederate states had problems; Jones County, Mississippi; North Georgia; Kinston County, North Carolina; East Tennessee; Kentucky; Louisiana; and Arkansas. The most serious was in western Virginia. This eventually led to the creation of the state of West Virginia.[42]

To deal with these activities, the Confederate Congress passed a series of laws dealing with 'Alien Enemies.' The first was passed on August 8, 1861, entitled 'An Act Respecting Alien Enemies.' The second was entitled 'An Act For The Sequestration Of The Estates Property and

Effect Of Alien Enemies, And For Indemnity Of Citizens Of The Confederate States, and Person Aiding The Same In The Existing War With The United States.' It was also passed on August 13, 1861.[43] The first law required any citizen of the United States residing within the Confederate States to either become a citizen of the Confederate States or make a declaration of such intention. If he did not, then he had forty days to leave the Confederate States. The citizens of Delaware, Maryland, Kentucky, Missouri, and the District of Columbia and the territories of Arizona and New Mexico as well as the Indian Territory south of Kansas were exempt.

PROCLAMATION BY THE GOVERNOR OF THE STATE OF TEXAS

WHEREAS, There is now a condition of hostilities between Governments of the United States and the Confederate States of America; and whereas, the Congress of the latter Government has recognized the existence of war with all of the United States except the States of Tennessee, Missouri, Kentucky, Maryland and Delaware, and the Territories of Arizona, New Mexico, and the Indian Territories situated between Kansas and the State of Texas; and, whereas, the late intimate commercial and political association of the people of the State of Texas, as a member of the Confederate States of America, is now at war, might disregard the relations in which war between said Governments has place them; and whereas, I have received information that some of the citizens of Texas have already violated their duty in the remises, as good citizen —

Now therefore, I, Edward Clark, Governor of the State of Texas, do issue this my proclamation to the people of said State, notifying them that all communication of whatsoever character between them and the citizens of the States and Territories now at war with the Confederate States of America must be discontinued; that all contracts heretofore made between them are suspended, and all that may be made during the continuance of said war, and until treaties of reciprocity are establish, will be void. It will be regarded as treason against the Confederate States of America and against the State of Texas for any citizen of said State to donate, sell, or in any manner exchange any property or commodity whatsoever with any citizen or citizens of either said State or Territories now at war with said Confederate States, without special permission from proper authority.

It will also be treasonable for any citizen of Texas to pay any debts now owning by him to a citizen of either of said States or Territories, or to contract with them any new debts or obligations during the continuance of war.

The statute of limitation will cease to run, and interest will not accrue during continuance of war.

The Executive deems it proper especially to warn all persons from endeavoring to procure title, in any manner, to property situated in Texas, and claimed by persons who are citizens of either of said States or Territories now at war with said Confederate States, or who have until recently or many now be citizens of Texas, or of the Confederate States, or of any of the States or Territories not including among those making war upon said Confederate States, and who have joined her enemies, as the Legislature may hereafter deem it property to provide for the confiscation of such property.

If there be citizens of the State of Texas owning such debts, the Executive would suggest that they deposit the amounts of the same in the Treasury of the State, taking the Treasury's receipt therefore. The United States are largely indebted to the State of Texas, and it may be determined by the Legislature of said State at some future time, that such deposits shall be retained, until the United States has satisfied the claims now held by Texas against her. Citizens of either States or Territories, now at war with the Confederate States will not longer be permitted to visit Texas during the continuance of such war without passports issued by authority of the Executive of the Confederate States, or of this State. And if such persons are now within the limits of Texas, they are hereby warned to depart within twenty days of this date or they will be arrested as spies; and all citizens of this State of Texas are warned from holding any friendly communications whatsoever with such persons.

The Executive has issued this proclamation, impelled by belief that public safety required it, and he relies upon the people to sustain him; and to aid him in discovering and bringing to justice and lawful punishment anyone who many disregard his duty as herein set forth.

In testimony hereof, I have hereunto signed my name and caused the great seal of the State to be affixed, at the city of Austin, this the eighth day of June, A. D., 1861, and in the year of the Independence of Texas the twenty-six and of the Confederate States the first.

By the governor:

s/Edward Clark

Bird Holland, Secretary of State

Figure I

This law authorized the forced removal of those not complying. Anyone remaining within the Confederate States after the forty days who had not taken the oath of citizenship or declared his intention to become a citizen would be, "Liable to be treated as alien enemies".[44] To inform the citizens of the law on August 14, 1861, President Jefferson Davis issued a proclamation giving those forty days to either take the Confederate oath of allegiance or leave the country.[45] [Figure II, p 121].

Issued at the same time, as the proclamation was a set of regulations dealing with how to treat any aliens remaining in the Confederate States. These regulations authorized the arrest and either forced removal or trial of the person as spies of those not taking the oath of allegiance.[44] [Figure III, p 123]

The second law was passed on August 13, 1861 and amended on February 15, 1862. It provided details regarding how an alien property and other assets could be seized and sold. It was very detailed and spelled out in great length how this property could be seized.[45]

News of the new Confederate laws and Davis' proclamation spread quickly through Texas and the Texas Hill Country. Despite some recent claims, "news of its [the laws and proclamation] might not have filtered out to the more remote German-speaking communities until 1862", newspapers in San Antonio, Austin, and New Braunfels published the laws and proclamation by late August 1861. The San Antonio *Daily Ledger and Texan* published them on Thursday, August 29, 1861 and its sister paper the *Weekly Ledger and Texan* published them on Saturday, August 31, 1861. The *San Antonio Herald* published them on Friday August 30, 1861. The Austin *Semi-Weekly News* published them on Monday August 26, 1861 and the *State Gazette* published them on August 31, 1861. The laws and proclamations were even published in German by the *Neu Braunfelser Zeitung* on Friday September 6, 1861.[46]

PROCLAMATION BY
President Confederate States of America
Whereas the Congress of the Confederate States of America did by an act approved on the 8th day of August, 1861, entitled 'An Act Respecting Alien Enemies' made provision that proclamation should be issued by the President in relation to alien enemies, and in conformity with the provision of said act —

Now therefore I, Jefferson Davis, President of the Confederate States of America, do issued this my proclamation and I do hereby warn and require every male citizen of the United States of the age of fourteen years and upward now within the Confederate States and adhering to the Government of the United States and acknowledging the authority of same and not being a citizen of the Confederate States to depart from the Confederate States within forty days from the date of the proclamation. And I do warn all persons above described who shall remain within the Confederate States after the expiration of said period of forty days that they will be treated as alien enemies.

Provided however that this proclamation shall not be considered as applicable during the existing war to citizens of the United States residing within the Confederate States with intent to become citizens thereof, and who shall make a declaration of such intention in due form, acknowledging the authority of this Government; nor shall this proclamation be considered as extending to the States of Delaware, Maryland, Kentucky, Missouri, the District of Columbia, the Territories of Arizona and New Mexico and the Indian Territory south of Kansas, who shall not be chargeable with actual hostility or other crimes against the public safety, and who shall acknowledge the authority of the Government of the Confederate States.

And I do further proclaim and make known that I have established the rules and regulations hereto annexed in accordance with the provisions of said law.

Given under my hand and seal of the Confederate States of America at the city of Richmond on this 14th day of August, A. D. 1861.

s/Jefferson Davis

Figure II

Regulations Respecting Alien Enemies
Confederate States of America

The following regulations are hereby established respecting alien enemies, under the provisions of an act approved 8th August 1861 entitled 'An Act Respecting Alien Enemies:'

1.   Immediately after the expiration of the term of forty days from the date of the foregoing proclamation it shall be the several district attorneys, marshals and other officers of the Confederate State to make complaints against aliens or alien enemies coming within the purview of the act aforesaid, to the end that the several courts of the Confederate States and of each State having jurisdiction may order the removal of such aliens or alien enemies beyond the territory of the Confederate States or their restraint and confinement, according to the terms of said law.

2.   The marshals of the Confederate States are hereby directed to apprehend all aliens against whom complaints may be made under said law and to hold them in strict custody until the final order of the court, taking special care that such aliens obtain no information that could possible be useful to the enemy.

3.   Whenever the removal of any alien beyond the limits of the Confederate States is ordered by any competent authority under the provisions of said law the marshal shall proceed to execute the order in person or by deputy or other discreet person in such manner as to prevent the alien so removed from obtaining any information that could be used to the prejudice of the Confederate States.

4.   Any alien who shall return to these States during the war after having been removed therefore under the provisions of said law shall be regarded and treated as an alien enemy, and if made prisoner shall be at once delivered over to the nearest military authority to be dealt with as a spy or as a prisoner of war, as the case may require.

Figure III

The citizens of the Hill Country read all of these newspapers so they were well aware of laws and Davis' proclamation. Many took the required oath. Among those taking the oath were thirty-one Comfort men, many whom were later identified as insurgents. Two took the oath on May 20, 1861, another on May 24, followed by a fourth on May

27, 1861. Two others took the oath on July 1, one on July 8, and another on August 19, and August 26. On September 30, 1861, twelve Comfort area men took the oath. On December 30, 1861, another seven took the oath and on January 27, 1862 two other man took the oath. On February 17, 1862, two more took the oath.[47]

The two who took the oath on May 20, 1861 included Ferdinand and Wilhelm C. Schultze. Wilhelm F. Geissler took the oath on May 24, 1861 and Ferdinand Simon on May 27th. August Faltin and Gottfried Stieler took the oath on July 1, 1861 and Ludwig Lange on July 8, 1861. Louis Boerner took the oath on August 19, 1861, and Henry Schwethelm took the oath on August 26, 1861. The twelve Comfort men who took the oath on September 30th included Paul Hanisch, Gottfried Stieler, Anton Bohnert, Alexander Brinkmann, Anton Heinen, Louis Strohecker, Edward Lieck, Charles Bruckisch, Charles Hilke, Emil Serger, and Frederick Dietert, Jr. The seven who took the oath on December 30th included: William Heinen; Frederick Dietert, Sr., Charles Schmidt; Charles C. Maertz; John Stecher, John Karger, and Hubert H. Heinen. Theodore Bruckisch and Charles Brinkmann took the oath on January 27, 1862. Christian Jacob and Otto Ludwig took the oath on February 17, 1862. [Appendix AI, p 685][48]

But not everyone took the oath. Many of the militant Unionists refused. One Nueces Battle survivor explained why, "The time allowed for their departure was short; nobody cared to pay a fair price for land and other property that when forfeited to the Confederacy would be auctioned off and sold for a song. Without money in Mexico, the nearest place to which he must go, what could a refugee hope for?" The Unionist felt if he took his family with him, how could he support them? If he left them, what hope did he have they would escape the scalping knife? All things considered, "he resolved to stay, not to abandon his property, not to desert his family and leave it to the prey of the savage." It was his and other Unionists hope that they could just wait out the war, "Certainly if he stayed quietly at home attending to his business and doing nothing against the Confederacy, he would not be molested. Arguing thus, very few of those to whom the proclamation applied obeyed its commands."[50]

The problem was the Hill Country Unionists did not, "stay quietly at home attending to their business and doing nothing against the Confederacy." According to one historian, "Perhaps the shrill struggle over loyalty to slavery cause across Texas would not have been driven to fever pitch had the strongest-willed abolitionists in Sisterdale held their tongues." By June 1861, two months before Davis' Proclamation and the 'Alien Laws', a small group of Freethinkers and Forty-Eighters organized a 'political front' organization. This 'political front' organization now known as the Union Loyal League, but called "The Organisation" by its members. The early part of Phase I, an insurgency, was well underway in the Hill Country.[51]

Map C -- Confederate and State Posts in 1962

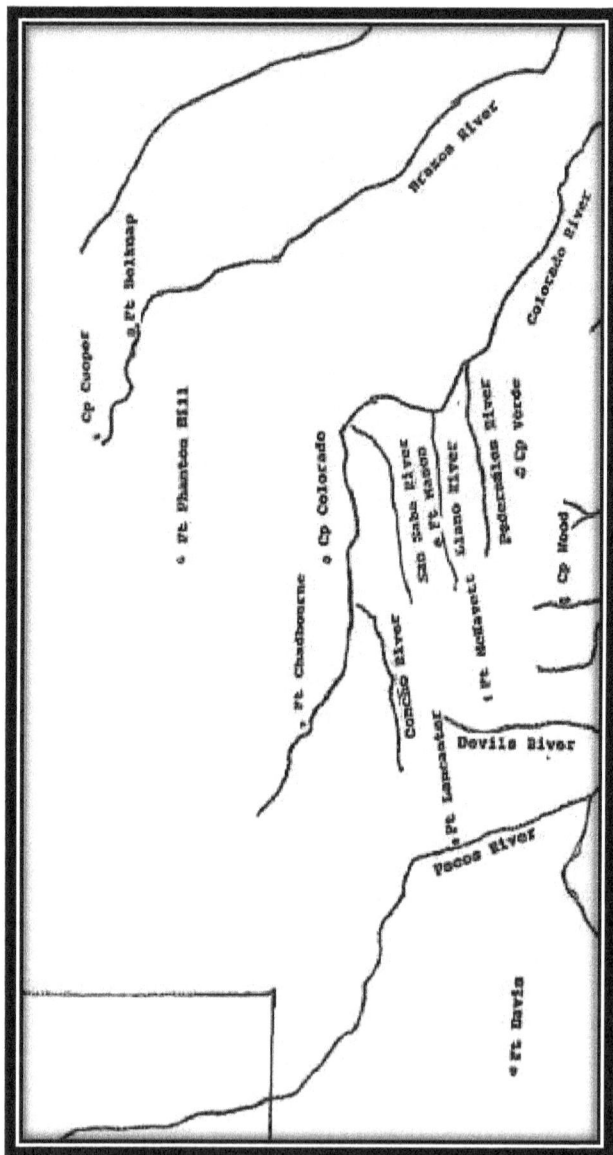

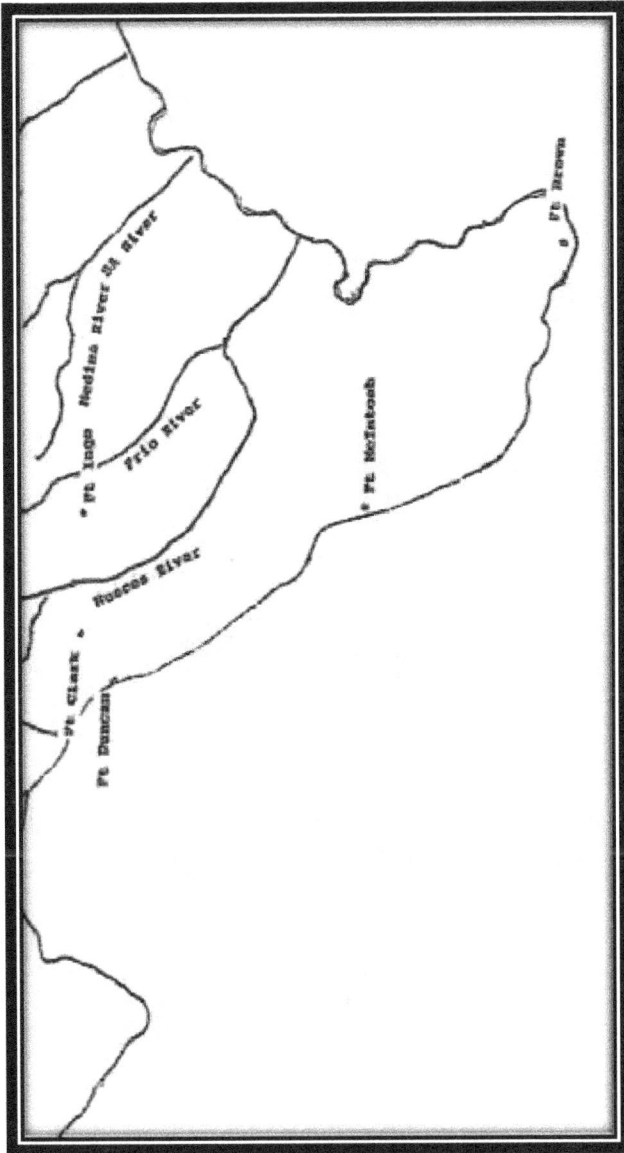

131

Map D – Union Invasion Routes and the Free State of Texas

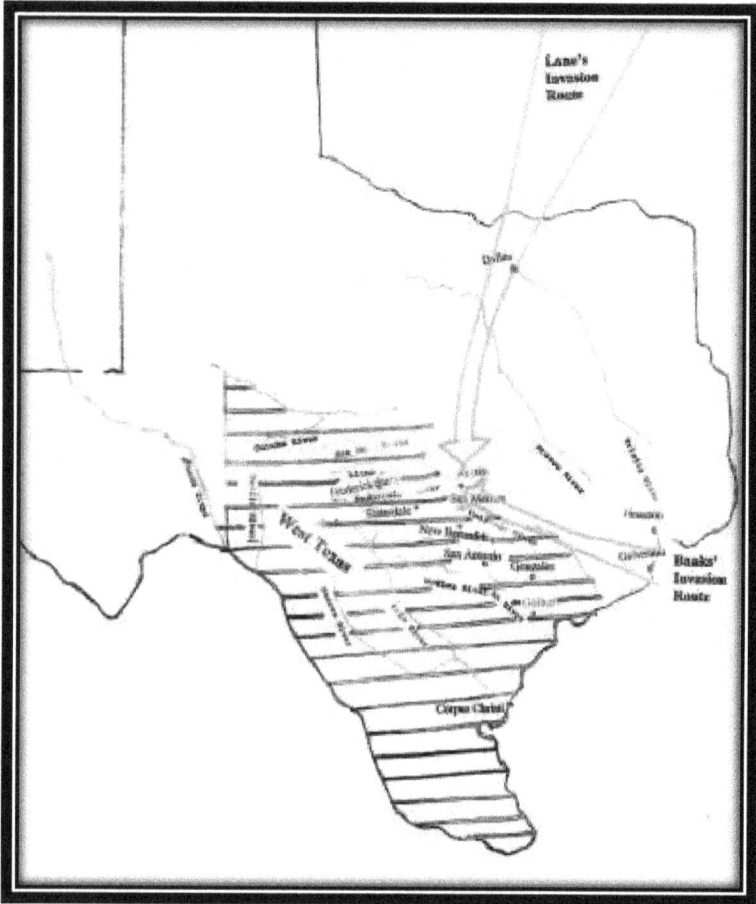

Endnotes – The Killing Starts
Unionists Prepare to Oppose the Confederacy

1. Eighth U. S. Census of 1860: Mortality Schedule 3 – Persons Who Died During the Year Ending 1st of June, 1860 For The State of Texas, p 41 and 69 establishes the date this incident took place in April, 1860. It shows both Frederick Louis and Gottfried Treibs were killed by gunshots. Unfortunately, Dennis Kingston's death is not shown on the 1860 Texas

Mortality Schedule. The book *Blanco County History* by John Stribling Moursund, (Nortex Press, Burnet, Texas), 1979, p 431, has a copy of the Blanco County Mortality Schedule 3, which provide more data on Frederick Louis.

2. Dennis Kingston was born about 1833 in Ireland. His enlistment papers show his occupation as a baker prior to enlistment. He and Martha Shallenberger were married about 1857/1858. Their son, William Lewis, was born on April 23, 1859 in Mason County. William L. [known as Bill] later owned over 30,000 acres in Jeff Davis/Reeves counties. Bill Kingston died on January 28, 1956. Martha Shallenberger was born in 1842 in Illinois. After Dennis Kingston's death, she married Oscar Splittgerber in 1860. Martha died in 1885. John M. Hunter was born on September 2, 1821 in Tennessee. He was one of three Hunter brothers or cousins who settled in Gillespie County. John M. Hunter married Sophia Ahrens on March 9, 1848 in Gillespie County. He was the first Gillespie County Clerk and a very prominent merchant. He died on September 5, 1870 in Gillespie County. Dennis Kingston's Military Service Records located in Record Group 94, Microcopy No. 233 "Registers Of Enlistments In The United States Army 1798 - 1914" Roll 24, Volume 50, July 1853-December 1854, 116, National Archives, Washington, D. C.; *The Menard County History: An Anthology*, compiled and edited by Menard County Historical Society, (Published by Anchor Publishing Company, San Angelo, Texas), 1982, pp 276, 601-602; Tenth U. S. Census, 1880 Menard County Texas Census, p 133A; Twelfth U. S. Census 1900, Jeff Davis County Texas, Census E.D. 37, Sheet No. 1; "Kingston Family" located in *Jeff Davis County, Texas,* by Lucy Miller Jacobson and Mildred Bloys Noreb, (Fort Davis Historical Society, Inc. Fort Davis, Texas) 1993, pp 453–454. This account says Dennis Kingston arrived in New York in 1855. Since his army records show he enlisted in the army in September, 1854, he had to have arrived in 1854, not 1855; *Julius Theodore Splittgerber (1819 – 1897) Volume Two: His German Ancestors and American Descendants*, Compiled by Mary Lewis Turner, (The Watercress Press, San Antonio, Texas), 1997, p 102; Gillespie County Historical Society *God's Hills, Volume* I, pp 61-62.

3. "Burning of Hunter's Store" by H. R. v. Biberstein located in Wisseman *Fredericksburg ... The First Fifty Years*, p 47.

4. This Biberstein account must be evaluated with close scrutiny as it is likely biased due to the fact the Biberstein and Schellenberger families were 'avowed enemies'. "The First Lynching in Fredericksburg" by H. R. v. Biberstein located in Penniger *Fest Ausgabe fuenfzigjaehrigen Jubilaeum der Deutfchen Kolonie Friedrichsburg*, pp 47-48.

5.   Frederick Louis was born about 1823 in Hessen. He arrived in Texas on the *St. Pauli* in 1847. He was a member of a Freethinker group known as the *Die Vierziger* and one of the original settlers of Boerne. Edward Maier was born on August 7, 1832 in Fulda, Hessia. His father, Franz Andreas, and a brother and a sister, arrived in Texas on the *Gerona* in 1845 from Fulda. Edward arrived later in 1854. On February 6, 1860, he became the Gillespie County sheriff. From 1862 until 1866, he was the district clerk. Edward Maier died on October 1, 1884 in Gillespie County. There is a great deal of doubt if this story is completely true. Part of the Kingston family believes Dennis Kingston was killed "in an attempt to break up a fight between two of Kingston's friends." The 1860 Texas Mortality Schedule seems to support this belief. It shows Frederick Louis was "shot while attempting to reconcile a quarrel between two others." The 1860 Texas Mortality Schedule shows a second man, Gottfried Treibs of Fredericksburg, was shot and killed at the same time. Gottfried Treibs was born about 1842 in Prussia, the son of Ann Mary and Gottfried Treibs. The family arrived in Texas on the *James Edward* in 1846 from Kirchberg. Seventh U. S. Census, 1850, Victoria County Texas Census, p 234; *A New Land Beckoned – German Immigration to Texas, 1844 – 1848*, by Chester W. Geue and Ethel H. Geue, (Genealogical Publishing Company, Inc., Baltimore, Maryland), 1982, p 118; Biesele *German Settlements*, p 173; *Blanco County History* by John Stribling Moursund (Nortex Press, Burnet Texas) 1979, p 431; Gillespie Historical Society *God's Hills*, Volume I, pp 117 – 118; Menard County Historical Society *The Menard County History* pp 353; 1860 Texas Mortality Schedule, pp 41 and 69; Wisseman *Fredericksburg ... The First Fifty Years*, p 20 and Seventh U. S. Census, 1850 Gillespie County Texas Census, p 311.

6.   Samuel Schellenberger was a Pennsylvania 'Dutchman' of Swiss origin. He was born in 1812 and married Elisabeth Rice about 1832. They had seven children; Louis, born in 1834; Ellen Evaline, born in 1837; Martha, born in 1842; Mary Elisabeth, born in 1845; Clara, born about 1853; Dorothy, date of birth not known; and Mary Jane, who died young. The family arrived in Texas about 1848 and arrived near Fort McKavett about 1852. After Samuel's death, Elisabeth married William McDougal. On August 6, 1866, he was killed in an Indian raid. Clara Shallenberger, the daughter of Elisabeth and Samuel, was wounded in the same raid. Elisabeth died in 1886. The Bieberstein family are Caroline Schuchard and Hermann von Bieberstein. Hermann was born in July, 1824. The family arrived in Texas on the *Elisa and Charlotte* in 1846 from Gorlitz, Silesia. They settled in the Gillespie and Mason County area. In February 1861, Hermann was the commander of a Mason County home guard company. On July 26, 1864, he was appointed/elected the commander of a militia company in the Second Frontier District. He later served as the county surveyor for both Mason and

Gillespie Counties. He died on December 15, 1907. Turner *Splittgerber Family History*, 61; Wisseman *Fredericksburg ... The First Fifty Years*, p 48; Menard Historical Society *Menard County History*, pp 601–602 and *Fort McKavett – A Texas Frontier Post* by Jerry M. Sullivan, (Published by Texas Parks and Wildlife Department), n.d., p 25; Geue *New Land Beckoned*, p 81; *Fort Mason – Training Ground For Generals*, by Jerry Ponder, (Ponder Books, Mason, Texas and Doniphan, Missouri), 1997, pp 103, 119–120, 132–134, and 153–154; and *Der Friedhof Cemetery* Compiled by Armand A. Lindig, (Published by The City Cemetery Association, Fredericksburg, Texas), 1990, p 88.

7. Wisseman *Fredericksburg ... The First Fifty Years*, p 48.

8. "Was a Survivor of the Nueces Battle" by Albert Schutze 'Frontier Times', Volume 2, No 1 (October 1924), p 25; Ransleben *100 Years*, p 24; and *Kerr County Texas 1856 – 1976*, by Clara Watkins, (Published by Hill Country Preservation Society Inc., Kerrville, Texas) 1975, p 68.

9. Ron Tyler, et al, eds., *Handbook of Texas*, Volume I, p 351 and Volume III, p 587. The two differ in the total number of votes for secession. Volume I, states 168 while Volume III, states 166; Biggers *German Pioneers*, p 57; Curtis *History of Gillespie County*, p 57. Curtis' Thesis is a good example of how this myth is accepted in the academic community. A member of her thesis committee was Dr. Rudolph L. Biesele, a leading authority on German Pioneers in Texas. He approved her thesis; Sansom *Battle of Nueces* p 2; Ransleben *104 Years*, p 80; *Texas: The Lone Star State*, by Rupert N. Richardson, Ernest Wallace, and Adrian Anderson, (Prentice-Hall, Inc., Englewood Cliffs, New Jersey), 1981, and *The Shattering of Texas Unionist*, by Dale Baum, (Louisiana State University Press, Baton Rouge, Louisiana), 1998, p 231.

10. Ron Tyler, et al, eds; Handbook *of Texas*, pp 759–760, Volume 3, pp 438–439; Richardson *Texas, The Lone Star*, p 226; "Unionist in Texas, 1856 – 1861" by Frank H. Smyrl, "Southwestern Historical Quarterly", Volume LXVIII, (October 1964), pp 191–192; Ransleben 100 Years, p 80; Biesele *German Settlement*, p 206; and Baum *Shattering of Texas Unionist*, pp 65 – 67.

11. Baum *Shattering of Texas Unionist, pp 65 – 67.*

12. James Webb Throckmorton was born on February 1, 1825 in Sparta, Tennessee. He came to Texas in 1841 and by 1851was elected to the Texas legislature. He was a member of the Secession Convention and one of eight to vote against secession. Despite his membership in the Texas Union Party after Texas seceded he was one of the first to take the Confederate oath of allegiance. He raised a company for Confederate service and was later a

state brigadier general in command of the First Frontier District. After the war, he briefly served as governor, but was removed by Federal General Sheridan as an 'impediment to Reconstruction'. He died on April 21, 1894. Buenger, *Secession and the Union in Texas, p 67* and Ron Tyler, et al, eds., *New Handbook of Texas*, Volume 6, pp 485 – 486.

13. Commanding Department of Texas, OR, Series I, Volume I, p 583; *Capt'n John—Story of a Texas Ranger*, by Frankie Davis Glenn, (Nortex Press, Austin, Texas), 1991, p 3; Siemering's *Germans During Civil War*, May 8, 1923; "Sam Houston and the Civil War" by Andrew Forest Muir, contained in 'Texana', Volume VI, No. 3 (Fall 1968), pp 282-287; Letter Sam Houston, Austin, March 29, 1861, to Colonel Waite, U. S. Army, San Antonio, Texas, OR, Series I, Volume 1, p 551; *Texas, The Dark Corner of the Confederacy: Contemporary Accounts of the Lone Star State in the Civil War*, Edited by B, P. Gallaway, (University of Nebraska Press, Lincoln, Nebraska) 1994, p 82.

14. R. H Williams *With the Border Ruffians*, (University of Nebraska Press, Lincoln, Nebraska), p 163. Before U. S. military commanders in Texas surrendered to Texas troops, Governor Houston advised he was prepared to assist them in protecting the military posts and property. Houston's Letter January 20, 1861; Muir *San Houston and the Civil War*, pp 282-287 and Buenger *Secession in Texas*, p 176.

15. Muir *Sam Houston and the Civil War*, pp 282-287.

16. Elisha Marshall Pease was born on January 3, 1812 at Enfield, Connecticut. He arrived in Texas in 1835 and was an early supporter of Texas independence from Mexico. He held various positions in the Republic of Texas government. He was elected Texas governor in 1853. In 1867, Federal General Sheridan appointed him Texas government. He resigned shortly afterwards. Pease died on August 26, 1883. George Washington Paschal was born in Greene County, Georgia on November 23, 1812. He came to Texas about 1847. Paschal was both an attorney and a journalist. After the Civil War, he moved to New York. Paschal died in Washington, D. C. on February 16, 1878. Eber Worthington Cave was born on July 14, 1831 in Philadelphia. He came to Texas in 1853 and worked as a journalist. Cave was appointed Secretary of State in 1860. After Texas seceded he assisted in raising several companies for Confederate service. He died on March 28, 1904. Benjamin Holland Epperson was born in Amite County, Mississippi in 1826. He arrived in Texas by 1847. He was active in Texas politics from 1851 until the war and was a member of the Texas Union Party. After Texas left the Union, Epperson actively supported the Confederate government. He died on September 5, 1878. James Pearson Newcomb was born in Amherst, Nova Scotia on August 31, 1837. His

family moved to Texas in 1839 and by 1846 lived in San Antonio. He established the *Alamo Express*, a pro-Union newspaper in August, 1860. After the Knights of the Golden Circle [a semi-secret pro-slavery organization] destroyed his paper he fled to Mexico. He served the Union Army briefly as a scout in the California area. He returned to Texas in 1867 and purchased an interest in the *San Antonio Express.* He died on October 16, 1907. Edmund Jackson Davis was born in St. Augustin, Florida on October 2, 1827. He arrived in Texas in 1838. In 1853, He was elected district attorney and in 1854, was appointed a state judge in Brownsville and the southwestern district. When the war started, he fled to Mexico. He was appointed a Union colonel and raised the First Regiment [Union]Texas Cavalry. He was captured by elements of Duff's Regiment in 1863, but released as he was on Mexican soil when taken prisoner. In 1865 he was appointed a Union brigadier general, the highest rank obtained by a Texan in the Union Army. After the war, in 1869, he was elected Reconstruction Governor of Texas by a little more than 800 votes. He was a strong believer in disenfranchisement of former Confederates and other 'radical' polices. He was defeated for Governor in 1873 by Richard Coke. Davis died on February 7, 1883. Andrew Jackson Hamilton was born on January 28, 1815 in Huntsville, Alabama. He arrived in Texas in 1846. In 1849, he was appointed Attorney General of Texas. Hamilton was elected to the U. S. Congress in 1859. He was a member of the Travis County Union Party. In mid-1862 he fled to New Orleans and later to Washington, D. C. President Lincoln appointed him military governor of Texas. After the war, he was appointed provisional governor of Texas. Hamilton died in Austin on April 11, 1875. John Leal Haynes was born on July 3, 1821 in Bedford County, Virginia. He arrived in Texas during the Mexican War. In 1863 he was appointed a Union colonel and given command of the Second Regiment [Union] Texas Cavalry. He returned to Texas after the war and died on April 2, 1888. Ron Tyler, et al, eds., *New Handbook of Texas*, Volume 5, pp 80, 112-113; Volume 6, pp 485-486; Volume 1, p 1046; Volume 2, pp 878-879; Volume 4, pp 526-527 and 989-990; Volume 3, pp 427-428 and 517; *Mexican Texans in the Union Army* by Jerry D. Thompson, (The University of Texas at El Paso), 1986, pp 13-14; *Vaqueros in Blue and Gray* by Jerry Don Thompson, (Presidia Press, Austin, Texas), 1976, pp 85-86; and "Texans in the Union Army 1861—1865" by Frank H. Smyrl 'Southwestern Historical Quarterly', Volume LXV (October 1961), p 235.

17. For more information on the removal of Federal troops from Texas see The *Exodus of Federal Forces from Texas 1861* by J. J. Bowen, (Eakins Press, Austin, Texas), 1986. There are few books that deal with Texas frontier defense during the War Between the States. One is *Frontier Defense in the Civil War: Texas's Rangers and Rebels* by David Paul Smith (Texas A and M University Press, College Station, Texas) 1992.

18. James Paul was born about 1822 in England. He arrived in Texas about 1847 and settled in Medina County. Paul married Regina Solms on April 5, 1852 in Medina. He was the commander of the Medina County Castle of the Knights of the Golden Circle. His company assisted in the surrender of Federal General Twiggs in February 1861. His company became a home guard company and served for about six months. His date and location of death is not known. William T. Harbour was born about 1833 in Mississippi. The date he arrived in Texas not known. He was the captain of the February 1861 home guard company from Kerr County. He was elected 2nd lieutenant in Davis' Company and led the detachment from the company in the pursuit. He distinguished himself during the Nueces Battle. He re-enlisted in Hunter's Company on February 29, 1862. It is not known what happen to him after the war. Smith *Frontier Defense in Civil War,* p 22; *San Antonio Alamo Express,* March 6, 1861; *San Antonio Weekly Ledger and Texan,* Saturday, April 20, 1861; *San Antonio Alamo Express*, March 24, 1861; *San Antonio Alamo Express,* April 10, 1861; *Austin State Gazette,* Saturday, April 13, 1861; *San Antonio Alamo Express*, April 1, 1861; *Austin State Gazette*, April 20, 1861; Letter, Charles Montague, Justice of the Peace, Bandera, Texas, July 19, 1961 to Governor Clark, Governor Clark's File, TSA, Austin, Texas; Muster Roll (MR), Captain William T. Harbour's Company, March 5, 1861, TSA, Austin, Texas; Report Captain W. T. Harbour to Governor of Texas, dated May 27, 1861, A.G.C., TSA, Austin, Texas; MRs, Lieutenant James Paul's Company, dated March 4, 1861 and May 16, 1861, A.G.C., TSA, Austin, Texas; and Williams *Border Ruffians,* pp 167–173.

19. Ibid.

20. William A. Blackwell was born on April 11, 1817 in Missouri. He arrived in Texas about 1850 and settled in Blanco County. He was elected captain of the Blanco Home Guard Company on May 4, 1861. Blackwell died on October 4, 1898, in Blanco County. Charles Henry Nimitz was born on November 8, 1826, in Bremen. He arrived in the United States in 1844 and in Texas in 1846. He was a Gray. Nimitz married Sophie Dorothea Mueller on April 8, 1848 in Gillespie County. He was the grandfather of Fleet Admiral Chester Nimitz of World War II. Charles Nimitz died on April 28, 1911 in Gillespie County.Philip Braubach was born on July 28, 1829 in Wiesbaden, Duchy of Nassau. He arrived in Texas and Gillespie County in 1850. Braubach was elected captain of a Gillespie Home Guard Company on February 7, 1861. It served for a year. He was Kuechler's first lieutenant in the December 1861 Unionist company. The Confederates arrested him in June 1862. He escaped and conducted guerrilla operations along the Texas Mexican border un May 4, 1864 when he enrolled in the First Regiment [Union] Texas Cavalry and elected captain of Company H. Braubach died on June 30, 1888 at San Antonio. Harbour's MR; Muster Roll, Captain

William A. Blackwell's' Company, May 4, 1861, TSA, Austin, Texas; Muster Roll, Captain George Freeman's Company, organized on August 24, 1861, dated March 1, 1862, TSA, Austin, Texas; and Muster Roll, Captain George Freeman's Company, November 14, 1861, TSA, Austin, Texas; Muster Roll, Captain Charles H. Nimitz's Company, July 31, 1861, TSA, Austin, Texas; "Minutes of Meetings of Gillespie Rifles on February 23, 1862, and March 29, 1862" located in Case Number 183, Gillespie County District Clerk's Office, Fredericksburg, Texas; Muster Rolls, Captain Philip Braubach's Company, dated February 7, 1861, May 25, 1862, August 25, 1861; November 25, 1861 and February 25, 1862, TSA, Austin, Texas.

21. For the disposition of the 1st and 2nd Regiments Texas Mounted Rifles see General Order No. 8, Headquarters, Troops in Texas, San Antonio, May 24, 1861, *Austin State Gazette*, June 8, 1861.

22. Smith *Frontier Defense in Civil War*, p 190.

23. Smith *Frontier Defense*, p 42; and General Order Number 1, Headquarters Frontier Regiment, Texas Rangers, Austin, February 1, 1862, A.G.C., TSA, Austin, Texas. For a discussion of how Texas viewed the Frontier Regiment would eventually become a Confederate regiment and its relationship to Texas, especially in regarding to the cost to keep it in the field, see *Six Decades in Texas or Memories of Francis Richard Lubbock, Governor of Texas in War- time, 1861-63*, Edited by C.W. Rains, (Ben C. Jones and Company Printers, Austin, Texas), 1900, pp 357-376.

24. Smith *Frontier Defense*, 42; and General Order Number 1, Headquarters Frontier Regiment, Texas Rangers, Austin, February 1, 1862, A.G.C., TSA, Austin, Texas.

25. Ibid.

26. Ibid.

27. *History of Medina County, Texas* by the Castro Colonies, Heritage Association, Inc., Published by the Castro Colonies Heritage Association, Inc., Castroville, Texas copyright not dated, pp 9 – 10 and Oral interviews by the author with David Paul Smith, April 6, 1997.

28. Records, 31st Brigade District, A.G.C., TSA, Austin, Texas; Quarterly Returns, 31st Brigade, July 1, 1862 and October 1, 1862, A.G.C., TSA, Austin, Texas; and Eighth U. S. Census, 1860, Comal County, p 175b.

29. Ibid.

Chapter 2 – The Killing Starts

30. Smyrl *Texans in the Union Army*, 65; Smyrl *Unionism in Texas*, pp 172-195 and "Union Sentiment in Texas 1861 – 1863" by Claude Elliot, 'Southwestern Historical Quarterly', Volume L, (April 1947), pp 448-477.

31. *Civil War Recollections of James Lemuel Clark and the Great Hanging at Gainesville Texas in October 1862* by James Lemuel Clark, edited by L. D. Clark, (Republic of Texas Press, Plano, Texas), 1997, p 21; Elliot *Union Sentiment in Texas*, pp 449-477; McCaslin *Tainted Breeze*, pp 64-64; James Lemuel Clark *Recollection,* pp 30-38; and New Handbook of Texas, Volume 1, Pages pp 308 and 600. Also see Austin's *State Gazette,* Saturday, February 2, 1861.

32. *Tainted Breeze – The Great Hanging at Gainesville, Texas 1862* by Richard B. McCaslin, (Louisiana State University Press, Baton Rouge, Louisiana), 1995, 63-64. Also see James Lemuel Clark *Recollection*, 30-38; and Elliot *Union Sentiment in Texas*, pp 449-477.

33. McCaslin *Tainted Breeze* pp 64-66; James Lemuel Clark *Recollection*, p 30 and Elliot *Union Sentiment* pp 449-477.

34. Richardson Texas, *The Lone Star*, pp 238-239; McCaslin Tainted *Breeze*, p 152 and James Lemuel Clark *Recollections*, pp 30–38. For an example of Texas citizens believing such an invasion was to take place see *A Texas Cavalry Officer's Civil War: The Diary and Letters of James C. Bates* Edited by Richard Lowe, (Louisiana State University Press, Baton Rouge, Louisiana), 1999, pp 74 and 201-202.

35. Ibid.

36. George Duncan Hancock was born on April 11, 1809 in Tennessee. He arrived in Texas in 1835 and fought at San Jacinto. He served in several military campaigns against Mexico. Hancock moved to Austin in 1845 and was a member of the Republic of Texas House. He was strongly against secession. In 1873, he helped organize the Texas Veterans Association. Hancock died on January 6, 1879, in Austin. Morgan Calvin Hamilton was a brother of A. J. Hamilton. He was born February 28, 1809 in Alabama. He arrived in Texas in 1837 and settled in Austin. He was another who was strongly against secession. In 1870, Hamilton was elected to the U. S. Senate. He died on November 21, 1893. The Officers and Men of the February 25, 1862 Capital Guards included: George Hancock, captain; George H. Gray, first lieutenant; William M. Smythe, and James W. Hancock, second lieutenants; Theodore V. Coupland, first sergeant; Ch. H. Fox, second sergeant; M. F. Bell, third sergeant; Georg R. Scott; fourth sergeant; James M. Reid, fifth sergeant; John Buttery, first corporal; William R. Robinson, second corporal; F. Steupy, third corporal; William M. Preece , fourth corporal; T. H. Arlett, Charles Springer, Fred Blum; F.

Raats; E. Pressler, Julius Schuetze, C. Domske, Thomas Sterzing, and W. D. Ezell. Sr., musicians; and R. A. Alexander, D. W.C. Baker, Daniel Bennett, C. Boatman, William Bozeth, A. G. Buddington, Gustav Boehn, John Bauers, B. C. Berrett, E. S. Berry, F. Berndt, T. M. Cole, Frank H. Coupland, R. D. Carr, A. C. Davis, John Dougherty, Andrew Davis, W. D. Ezell, Jr., C. V. Ezell, Thomas J. Hatch, John Hemminger, A Hensinger, J. A. Hancock, J. E. Hudson, A. J. Hamilton, Charles Johnson, Georg W. Jordan, R. M. Johnson, J. G. Lame, William Lee, John T. Lee, Alfred H. Longley, R. N. Lane, J. D. Morrison, A. Morrison, R. Manthe, P. A. Meyer, G. W. Mills, Thomas W. Nolen, Richard L. Preece, W. C. Phillips, C. Pecht, C. Payne, James Philips, W. W. Roberts, F. Reichen, W. F. Robinson, S. B. Reid, Ed Swisher, B. O. Stavely, J. Seigmond, August Sutor, C. Schaffer, J. M. Swisher, Swante Magnus Swenson, C. Spaulding, J. W. Thompson, William M. Vance, H. Voght, T. M. Warren, B. Wachter, William Gardner, John T. Graves, L. H. Luckett, and A. Nelson; privates. At least fourteen of these men joined the Union Army: Frank H. Coupland, Theodore V. Coupland, John Dougherty, William Gardner, James W. Hancock, Thomas Hatch, John T. Lee, Alfred H. Longley, Thomas W. Nolan, James Philips, Richard L. Preece, William M. Preece, James M. Reid, and William M. Smythe. Another two fled in July: A. J. Hamilton and George H. Gray. Martin *Texas Divided,* 22. Also see the *Austin State Gazette,* Saturday, February 2, 1861; Muster Roll of the Capital Guards, George Hancock, Captain, 26th Brigade February 25, 1862, A.G.C., TSA, Austin, Texas; Barr CMC, Volume LXII, (July 1969), 94-95; "The German Citizens Were Loyal to The Union" by John W. Sansom, 'Hunter's Magazine', Volume II, November 1911; *Colossal Hamilton of Texas: Andrew Jackson Hamilton Militant Unionist and Reconstruction Governor* by John L. Waller (Texas Western Press, The University of Texas at El Paso), 1968, p 36; *San Antonio Herald,* August 16, 1862; and *San Antonio Herald,* March 22, 1862.

37. McCaslin *Tainted Breeze,* pp *64 – 65.*

38. Martin Texas *Divided,* p 65; Ron Tyler, et al, eds., *New Handbook of Texas,* Volume 3, pp 430 and 438; 430; and Muster Roll of the Capital Guards, George Hancock Captain, February 25, 1862, TSA, Austin, Texas.

38. *Austin State Gazette, April* 13, 1861 and San *Antonio Tri-Weekly Alamo Express* February 20, 1861, February 23, 1861 and February 25, 1861.

39. *San Antonio Tri-Weekly Alamo Express* April 11, 1861; Webb, *Handbook of Texas,* Volume II, p 275 and Sibley *Lone Stars and State Gazettes,* pp 296-297.

40. Stephen Schwartz was a Federal/Union soldier who was captured in May 1861 by the Texas militia. He was one of many Federal soldiers who were at first placed on a type of parole and allowed to mix with the civilian

population. They were later confined at Camp Verde and other frontier
posts. The prisoners were exchanged later in the war. *Twenty-Two Months
a Prisoners of War* by Stephan Schwartz, (A. F. Nelson Publishing
Company, St. Louis, Missouri), 1892, pp 41-52 and Martin *Texas Divided,*
pp 113-114.

41. Texas Governor's Proclamation Book, TSA, Austin, Texas and *San Antonio
Herald,* June 8, 1861.

42. See The *Free State of Jones: Mississippi's Longest Civil War, by* Victoria
E. Bynum, (The University of North Carolina Press, Chapel Hill, North
Carolina) 2001; *Guerrillas, Unionists, and Violence on the Confederate
Home Front,* edited by Daniel E. Sutherland, (The University of Arkansas
Press, Fayetteville, Arkansas), 1999; and *Enemies of the Country: New
Perspectives on Unionists in the Civil War South,* edited by John C. Inscoe
and Robert C. Kenzer, (The University of Georgia Press, Athens, George),
2001. For other accounts of Unionists in Texas see, *Brush Men and
Vigilantes* by David Pickering and Judy Falls, (Texas A and M University
Press, College Station, Texas), 2000.

43. 'An Act Respecting Alien Enemies', approved August 8, 1861, OR, Series
II, Volume II, pp 1368-1369 and 'An Act to Alter and Amend an Act
Entitled "An Act for The Sequestration of the Estates, Property, and Effects
of Alien Enemies and for Indemnity of Citizens of the Confederate States,
and Persons Aiding the Same in the War with The United States", August
13, 1861, OR, Series II, Volume II, pp 932-944.

44. Act Respecting Alien Enemies.

45. Proclamation by Jefferson Davis, President Confederate States of America,
August 14, 1861, OR Series II, Volume II, pp 1368-1369.

46. Regulation respecting Alien Enemies, August 14, 1861, OR, Series II,
Volume II, p 1370.

47. Law Regarding Seizure of Alien Property, p 932-933.

48. Thorp *Historical Friction,* p 78; San *Antonio Daily Ledger and Texan*
August 29, 1861; San *Antonio Weekly Ledger and Texan,* August 31, 1861;
San Antonio Herald, August 30, 1861; *Austin State Gazette,* August 31,
1861; and *Neu Braunfelser Zeitung,* September 6, 1861.

49. There was not a uniform method of implementing the requirement.
Therefore, it is not possible to find a single source that shows who did or
did not take the required oath. Kerr County Probate Book A, pp 46–48, pp
50–54 and Kerr County Naturalization Records, pp 92-109. This record was

kindly provided to the author by Gregory Krauter of Comfort, who at a great expense located and obtained the ledger.

50. Ibid.

51. Sansom *German Citizens*.

52. Paraphrase of Sansom's quote in Sansom's *German Citizen* and Sansom *Battle of Nueces*, p 3 and Conzen *Clash of Utopias*, p 52.

CHAPTER 3 – Hill Country Unionists and Insurgents Organize

As the secession vote showed, the Texas Hill Country contained many whom were opposed to leaving the Union. However, the anti-secession vote did not mean that all were Unionists. Many of these votes showed a desire to preserve the Union. One major reason was the frontier posts provided a ready market [and hard cash] for the settler's crops. Another was Federal troops provided protection against Indian attacks. Many opposed secessions because they felt it would leave them without protection from Indian attacks.[1] The Unionists were divided into two major groups. One was the moderate who wanted to remain as neutral as possible so as not to cause Texas and Confederate counteractions. Most of the moderates gave some support to the Confederacy, even taking the oath of allegiance. In the German communities, these were also known as the Grays. The second group was the radicals or militants, who would become the insurgents. They wanted to take steps to actively oppose Texas and Confederate governments and force a reunion with the Federate Government. In the German communities, these were also known as the Greens. The strength of the Hill County insurgent movement was the Germans settlers who arrived as part of the *Adelsverein* led by the ultra-radicals of 1854 and militant Unionists.[2]

Medina County had a large and active Unionist movement centered along the Medina River and in the towns of Castroville, D'Hanis, New Fountain, and Quihi. In Kerr

County, there was strong Unionist sentiment in the west and northwest parts of the county. Most militant insurgents were located near Comfort, in that part of the county that became Kendall in 1862. Bandera County had a Unionist element and at first it appeared it would be one of the most radical. Blanco County Unionists were mainly located in the western part of the county, bordering Gillespie County, and in the northeast part of the county, bordering Travis County. The most militant insurgents were in the southwest part of the county, near Sisterdale, which became part of Kendall County. The militant insurgents at Sisterdale, Comfort, and along South Grape Creek in Gillespie County were mainly the Forty-Eighters and Freethinkers.

One of the major achievements of the Hill Country Unionists was the creation of Kendall County in January 1862 out of the northwest part of Comal County, the southwest part of Blanco County, and the eastern part of Kerr County. Mainly Germans, many of who were very politically oriented and strongly supported Texas remaining in the Union, settled this area. The settlements included: Sisterdale; Curry's Creek; Comfort; and Boerne.[3] [Map E, p 197].

Many of the moderates or Grays took a-wait-and-see attitude, not believing the Confederacy would last. A typical moderate's reaction was described when he said, "As for my self, although the reports of all that was going on which penetrated our mountain fastnesses, orally and through occasional newspapers, stirred by blood and tempted me to openly align myself on the Union side, the conviction that I was but one against many." However, "the attractions of a home blessed by a charming young wife and the work I had to do as a farmer, stockraiser and occasional surveyor, sufficiently curbed my patriotism and kept me neutral in action, if not in speech."[4] Even after secession most of the Hill Country Unionists did not believe the Confederate States would maintain itself for long. As one Anglo Hill Country Unionist put it, "believing the secession movement a mere rebellion that would be quickly suppressed, they deemed neutrality their wisest course ... None knew the depth of the rebellion."[5]

Early Confederate battlefield success led to this Unionist's reaction, "1861 was a year of changes and surprises. The Confederate armies were so uniformly victorious in the first battles as to seemingly justify the belief that the Union was doomed, and that the Southern Confederacy would become an accomplished fact." He went on to state, "influenced by this belief many pronounced Unionist ... were persuaded into a more or less honest recantation of former views and into seeming sympathy with the Confederates."[6]

An Anglo Unionist frontiersman wrote in his homespun dialect his feelings and experiences, "I had some redhot secesh neighbors who wound [sic] frequently drop in to talk on this subject in order to find out my politicle views and I was frank in giveing my opinion in regard to the matter." He explained, "I never went to town but what I was tackled by someone and I spent my opinion freely, thinking every man had a right to do so, always advocating the Union principles and was pounced on by someone as though I was the only Union man in the county. Thare was plenty others, but they kept their mouths shut as I should have done." He began to realize maybe he should not be quite so vocal, or as he said, "I saw that I had the sow by the wrong ear but then I should have sided in and been content had they let me alone and not bulyraged and teased me and called me a black republican. One man, a large stock owner, said to my face that a Union man should not live in the county in peace." I say, "How are you going to help yourself. What are you to do." Pressure continued mount on him to tone down his opinion, "All this time I was waited on, worned [sic] and threatened, but I told them that I never would take the oath until I was forsed to do so, and it went on this way for some months."[7]

A description of the time states, "The Civil War began and some friends became enemies. The hill country was torn asunder." It explained that most of the German settlers had come to Texas because they wanted freedom from the churches [The Freethinkers] and Forty-Eighters [who had taken part in the German Revolution of 1848] both had a firm belief in maintaining the Union. The other Hill Country insurgent support came from many of the Anglo frontiersmen. They had immigrated to the area from both northern and southern states. These pioneers were non-slave holders and worked the land by either farming or stock rising. Others were in some trade such as the mercantile business. Many arrived after Texas independence and statehood and did not feel the strong emotional ties to Texas. This account claims the country was divided three ways, the

Southerners, Germans who supported the Union, and bushwackers who used the war as an excuse to rob and murder. "The bushwackers were both German and American who left their homes in stealth to waylay and murder."[8]

There were some Anglo Texan Unionists who saw the future destruction. Among these was Noah Smithwick, a veteran of the Texas Revolution, who lived in nearby Burnet County. By late April 1861, he sold what property he could at a great loss. The remainder he turned over to his nephew, John Hubbard, who was a "Unionist [and] decided to stay and face the consequences; a decision that cost him his life." Smithwick and 34 others left for California.[9]

The moderate Unionists were in the majority in Bandera, Blanco, Kerr, and Medina counties, especially after Kendall County was formed. The insurgents were in the majority in Gillespie and Kendall counties. These two areas took drastically different methods of anti-Confederate activities and Unionist support. This was a major reason the counties of Gillespie, Kendall and part of Kerr later felt the brunt of Confederate steel and the counties just north, east, and south did not.[10]

At first the Unionists in all six counties of the Hill Country took some common action. The most important was joining and thereby attempting to control the home guard; the militia units of the 31st Brigade District; and the local companies of the Frontier Regiment. The moderates of Medina County, led by Forty-Eighter Henry Joseph Richarz from Prussia and thirty-nine-year-old Jacob Haby of Alsace, were especially effective as they fully controlled their authorized militia battalion of the 31st Brigade. The insurgents of Gillespie and Kendall Counties, led by Forty-Eighters Philip Braubach, Rudolph Radeleff, Fritz Tegener, Ernest Cramer, Freethinker and *Vierziger* Jacob Kuechler, and *Adelsverein* settlers Valentine Wilhelm Hohmann, and William Feller, were also were very effective in controlling their home guard and 31st Brigade units. However, they went a step further and attempted to control the full-time company of the Frontier Regiment authorized from their area. When that failed, they created an 'unauthorized' military battalion dedicated to their cause.

Unfortunately, while the moderates were willing to say home and do nothing against the Confederacy hoping they would not be molested, the insurgents were not. They began to organize an insurgency. Their activities plunged the Hill County into violent reactions from Texas and Confederate military forces which eventually resulted in the deaths of over fifty men. By the time that Texas seceded and joined the Confederate States of America, the Hill Country Unionists, both Germans and Anglos, were actively involved in local political events. They controlled the governments of Bandera, Medina, and Gillespie counties and greatly influenced Blanco and Kerr county governments. The home guard companies were organized in early 1861. The militant insurgents controlled one of the two companies in Gillespie County, that of Philip Braubach. In the south, it could also count on the Comfort *Turnverein*. In Blanco County, they greatly influenced one of its two companies, that of George Freeman's Pedernales Cavalry Company. In Kerr County, about a third of William Harbour's Company were insurgents. In Bandera and Medina, the citizens refused to join an 'authorized' home guard unit and formed 'unauthorized' units, which were Unionist in nature. By the spring of 1861, the insurgents controlled three military units, one *Turnverein* unit, and greatly influenced two other home guard units. Only the home guard companies of Nimitz in Gillespie County and Blackwell in Blanco County were not controlled or greatly influenced by insurgents. This gave them a substantive military element. By whatever action, by early 1862 the Unionists in all six counties had a military force and it appeared the entire six-county area was about to become a battlefield. Whether a moderate or militant they faced a common problem. Just what should they do and how far should they go in opposing the Confederate government?

In June 1861, just three months after Texas seceded and only two months after the war started, the German militants [the Greens] organized a formal political organization known today as the Union Loyal League [*Des Organisator*]. They used the methodology described by political science and military science as an insurgency.[11] It is of special interest to note while the militant Germans were organizing *Des Organisator*, Bandera citizens strongly considered some form of a 'Union Association' likely in concert with Travis Unionists. Joseph B. Miller, the Bandera Assessor and Tax Collector, reported to C. R. Johns, the state comptroller, that a "paper was shown him, with a request to sign it, purporting to be a Union association with a good many names on it to include [A. J.] Hamilton."[12]

There is no primary source regarding the organization of *Des Organisator*. John W. Sansom is the single source most historians use. Sansom's testimony is not a primary source about the League because he was never a member. Nor did he attend any of its meetings. His information regarding *Des Organisator* is secondhand, provided him by insurgents much later. Nevertheless, he is the best we have. It is not known who provided Sansom the information about *Des Organisator*, but most likely to have been Fritz Tegener and Eduard Degener. When he wrote his account in 1905 he made the statement that he expected the leader of the party [meaning Fritz Tegener who died in 1900] or the Chief of the Union Loyal League [meaning Eduard Degener who died in 1890] to make some type of report from their point of view. But since they had not and most were dead or feeble, he deemed it his duty and privilege to do so. There is one clue as to who may have provided Sansom some information. On the back of a photo of Fritz Tegener, owned by Anne Stewart of Comfort, is a hand-written note that says it was Tegener who provided the information to Sansom.[13]

The major points Sansom tells us about the organization of the Union Loyal League are; "shortly after the promulgation of the Ordinance of Secession, a "Union Loyal League" was organized in June 1861, by representatives from sections named." "Only eighteen persons were present at the first and initial meeting, but these were the chosen representatives from as many different sections." "Each of these [representatives] bound himself by a solemn vow not to bear arms against the Federal government." "Each [representative] was appointed by the body as a committee of one whose duty it was to persuade others to join the League and make the same pledge." "the eighteen dispersed and went to work so diligently and with such success, that on July the 4th, 1862, not less than five hundred male Unionists met on Bear Creek in Gillespie County and proceeded to perfect the organization. Among other measures taken up was the organization of three companies, to wit: The Gillespie County Company ... The Kendall County Company ... and the Kerr County Company."

One of the most controversial bits of information Sansom provides is; "Its objects and purpose, not to create or encourage strife between Unionist and Confederate sympathizers, but to take such action as might peaceably secure its members and their families from being disturbed and compelled to bear arms against the Union, and to protect their families against the hostile Indians."[14] A close examination of Sansom's information about *Des Organisator* reveals it is extremely biased and in many cases incorrect. The major examples are the purpose of the League, when the League organized its military arm, that *Des Organisator* disbanded and offered an assurance to the Confederates that they were not a threat; that after the Nueces battle any member of the Unionist group that was later captured was immediately executed; and data regarding the time when President Davis issued his proclamation giving forty days to leave the Confederacy.[15]

Despite problems, Sansom provides valuable information about the organization of *Des Organisator*. An analysis of Sansom's information established the insurgents used the methodology of an insurgency. In Phase I, a secret 'hard-core cell' is organized. We have no firm data on who were members of the League's 'hard-core cell', but some assumptions can be made and developed in to a hypothesis. Next, the insurgency organizes a larger 'political front'. The 'political front' was *Des Organisator*. Sansom's data shows the Union Loyal League was such a 'political front'. Next, the 'political front' creates some type of military force it used for 'self-defense'.

The individual who provided Sansom the data on the *Des Organisator* gives us all the elements to show the League meets the definition of an insurgency. First, Sansom states *Des Organisator* was organized in June 1861. Second, he says there were eighteen present at the "first and initial meeting." Third, he reports these eighteen were "chosen representatives from as many sections." Fourth, each member of the League took a "solemn vow". Fifth, each member was appointed "as a committee of one whose duty it was to persuade others to join the League and make the same pledge." Last, the so-called objects and purposes are all 'self-defense'. All these actions fit the methodology of Phase I of an insurgency.[16]

The date of June 1861 is very important. This was three months after Texas seceded and two months after the war started. Neither Texas nor the Confederate governments had taken action against the insurgents at this time. The only overt act was the governors' proclamation of June 1861. There was no outside danger to the insurgents. Based on this fact it is clear the League was a political anti-Confederate and pro-Union organization. It was a proactive, not a reactive organization; in other words was designed to take action against the Confederate and Texas governments, not because they had done anything against the Unionists. The date and time in which *Des Organisator* was organized fits the methodology of Phase I of an insurgency.

The number of original members of *Des Organisator*; eighteen is important. Any larger number would be difficult to control and keep secret. Eighteen men didn't just get together one day and say, "Let's organize a pro-Union, anti-Confederate group." There was someone or several people who had the idea and put it to work. This someone was the 'hard-core cell' discussed in the Insurgency Model.

The third important data Sansom provides is these eighteen were "chosen representatives from as many sections." No eighteen separate localities knew about a secret organization being organized, met and elected a representative. So how were they selected? The fact that eighteen men from different localities were selected shows advance planning. Someone had to have chosen them. That someone was the 'hard-core' cell. In the methodology of organizing a 'political front,' it will claim to represent several localities and uses the façade of being democratically elected by the areas it claims to represent. The claim that *Des Organisator* original members were democratically elected and represented the various areas fits the methodology of Phase I of an insurgency.

The "solemn vow" described by Sansom also demonstrates *Des Organisator* methodology of an insurgency. For Phase I to be successful it is critical the government be unaware of the existence of an organized subversive political foe. A secret and often blood oath, whereby an agreement is made to kill any member who exposed the cell, is the tool most often used. It is clear from various existing data this oath was a blood oath. A later member of the League talks about the oath. He says, "We all had taken the pledge." The existence of the oath was later common knowledge. A strong secessionist stated in 1893, "At the beginning of the war the Union Germans met and went into a secret oath-bound organization." A descendant of one of the Unionist in 1997 had this to say: "Each person took a vow to never betray the United States of America." Several members of *Des Organisator* who later wrote about their experiences, or had articles written 'for them,' state the pledge was close to these words. Sansom says the vow was, "not to bear arms against the Federal government." Another source, written in 1894 by a German-Texan in German states tht the Unionists, "had vowed to remain true to the Union." A third account, credited to a member of the League paraphrases Sansom: "The Union Loyal League, whose members were pledged not to take up arms against the Federal government." A German-speaking native of Gillespie County in a 1944 academic study, using German-language sources concluded, "The League had secret signs of recognition and all were sworn to secrecy as to what had taken place at the meetings. Any member who turned traitor would be shot on sight."[17]

August Siemering was certainly a member of the 'hard-core cell' that organized *Des Organisator* in 1861. He says *Des Organisator* established, "certain secret codes [and] committed themselves to strict secretiveness." Another German-speaking researcher who interviewed many League members in 1909 concluded, "In Gillespie County, in which Fredericksburg was situated, there was a secret organization in behalf of the Union, and it was stated that any member who became a traitor would have been shot at sight." This threat was carried out on July 5, 1862 when *Des Organisator* executed Basil Stewart because they believed he was a spy. The 'solemn vow' Sansom says *De Organisator* took was a blood oath and fits the methodology of Phase I of an insurgency.[18]

A major problem in researching the Nueces Battle and related events is the lack of primary sources. As discussed, most historians treat Sansom's account of the organization of *Des Organisator* as a primary source, which it is not! A second source most historians treat as a primary source is an account written in 1924 and entitled, "Was a Survivor of the Nucces Battle." A third is the *Dallas Morning News* article of May 1929. Interviewers of the two sources wrote the articles. They contain many errors that historians have treated as facts. Of all known primary source documents, there are few in number, written in German and translated into English. Some of the German originals are known to contain more data than the English translated copy. These German-language sources were designed for consumption by very limited German-speaking audiences. Most of these were written prior to 1900. Even today the descendants of the early Unionists are extremely reluctant to speak to *Auslandere* [Outsiders] about information they known.

The fifth item Sansom provides, which fits the methodology of Phase I of an insurgency, is that each member was appointed "as a committee of one whose duty it was to persuade others to join the League and make the same pledge." It is extremely important in the early stages of an insurgency to protect the identify of the membership, especially the key members. This is accomplished by compartmentalization, or the use of cells. Each cell is known only unto it. The members of a cell do not know the identify of another cell's members. This is done in case one member of a cell is a spy or captured and forced to talk. The only members he knows is that of his cell. Therefore, the entire 'political front' is safe. The most common method is for each cell to recruit independently of the others. Sansom says each original member was appointed a committee of one whose duty it was to persuade others to join the League. Thus, *Des Organisator* had eighteen separate cells whose members were unknown to each other. Only the head of each cell knew the identity of the heads of other cells. To further keep the entire membership secret and thus safe, once a cell reaches a certain strength it splits; each individual is appointed head of a new cell. These cells began recruiting new members.[19]

The most controversial and most often quoted item in Sansom's accounts are the purpose of *Des Organisator*. Almost every writer and historian who has written about the Nueces Battle quotes Sansom to show *Des Organisator* was a peaceful organization and no threat to Texas and the Confederacy. Another, ostensibly "primary source" is found in the book *A Hundred Years of Comfort in Texas,* written in 1954 and updated in 1974. In this account the author, supposedly quoting from a 1929 article in the *Dallas Morning News,* has August Hoffmann, a survivor of the battle saying; "During the first year of the *Burgerkrieg* (Civil War), things were pretty quiet in Gillespie County. Our three-county organization, which was effected to keep the peace as far as Indians and border ruffians was concerned as well as keeping locals toned down in case of agitation ... We just wanted to keep things peaceful in the Grenzland (border areas), although most of our residents sided with the Union."[20]

There are major problems in the various accounts in *A Hundred Years of Comfort in Texas.* The author attempts to give the appearance the accounts were taken from previously published sources or that he conducted detailed interviews with participants. He also attempts to portray the accounts as unbiased, which they are not. In this example, the author claims this is a quote from a newspaper article. A detail reading of the newspapers finds no such quote. The quote in the newspaper articles says, "Up here in Gillespie County, we didn't notice the war much for the first year, although seven-eights of us were on the Union side." In a 1925 letter, Hoffmann said, "The first year the county didn't suffer much

as there were very few that volunteered for military service."
This is but one of many examples where later historians have
misquoted previously published accounts in order to present
an account more favorably to their bias viewpoint.[21]

A characteristic of the 'political front' of an insurgency
is to disguise or conceal its real purpose. Instead, the purpose
is stated as something much less radical. There are several
facts to consider in determining the objects and purpose of
Des Organisator. Fact; *Des Organisator* was organized only
three months after the war began. Fact; at that time, there
was no outside threat to the Unionist. Fact; *Des Organisator*
was organized as a secret anti-Confederate organization.
Fact; *Des Organisator* was organized long before any
thought of a Confederate draft. Fact; according to both
Sansom and Hoffmann there was little conflict between
insurgents and Secessionists when *Des Organisator* was
organized. Fact; there were more than sufficient Texas and
Confederate troops in the area to protect against Indian
attacks. Given these facts, why it was necessary, in June
1861, when *Des Organisator* was organized, to create a
secret organization to "peaceably secure its members from
being disturbed" when there was no threat? Why was it
necessary at that time to create a secret organization to keep
its members from having "to bear arms against the Union"
when there was no action being taken to force anyone to bear
arms? The answer is that there was no necessity at all –
except to organize a secret political and military organization
which the insurgents would use against the Secessionists and
to protect them against any counteraction. In clear terms, *Des*

Organisator was organized to oppose the Confederacy in both political and military arenas. They were getting ready for armed conflict. If *Des Organisator* had been a 'peaceful' organization, then there was no need for secrecy.

The real purpose of *Des Organisator* has been clouded in half-truths and incorrect information for over 150 years. Recent uncovered documents establish *Des Organisator's* real purpose. As a current Gillespie historian says, "But, by 1861, the year that the Texas vote approved secession and withdrawal from the Union, such German-Texan leaders as Jacob Kuechler of Friedrichsburg and Eduard Degener of Sisterdale began backing a plan to secede, *by force if necessary* [emphasis added], from the Confederacy and to create the *State of West Texas* [emphasis added], which would be a part of the United States."[22] Eduard Degener even wrote down the boundaries for this *State of West Texas:* "Commences at a point in the Gulf of Mexico three miles from the shore opposite the middle of the main channel of Pass Caballo, thence up the middle of said channel and of Matagorda Bay to the mouth of Colorado River, thence up the main channel of said river, with its meanders to the point where said river is intersected by the thirty-second parallel of North latitude, thence along said parallel to a point ( ) miles west from said river, thence in a straight line to the junction of the Pecos River and Rio Grande, thence down the main channel of the Rio Grande, with its meanders, to the Gulf of Mexico, thence along parallel to the shore of the Gulf of Mexico, three miles from the land to the place of beginning."[23] [Map D, p 132]

The last part of Sansom's apologia, that *Des Organisator* was designed "to protect [its members and their families] against the hostile Indians,"[24] does not stand up under serious scrutiny. In June 1861, there were more than sufficient 'authorized' military units in the Hill Country to provide protection. First were the home guard state troops. Gillespie County had two companies: Braubach's Company and Nimitz's Company. To the east was Blackwell's Company. To the south, in Kerr County was Harbour's Company. Second, there were the Confederate companies. At Fort Mason there was a company from the 1st Regiment Texas Mounted Rifles. At Camp Verde in Kerr County there was a company from the 2nd Regiment Texas Mounted Rifles. This makes a total of four companies of state troops and two companies of Confederate troops in the area of *Des Organisator*. So how was it a small secret group of eighteen were going to further 'protect their families against hostile Indians'? The answer of course is they were not! The insurgents needed a military arm to protect them from the state and Confederate troop's counteraction and to assist the League in declaring a free state of West Texas.

The insurgents were familiar with some form of the Insurgency Model, even if they may have been short on theory. At least some of them knew the writing of Clausewitz, who visited in many of their parent's home and in the early 1830s, was writing about insurgencies. He acknowledges in *On War* (1832) his military principles did not cover insurgencies. Writers, such as Don Heinrich Tolzmann in *The German-American Forty-Eighters 1848-*

1998 (1997) tell about the Forty-Eighters using the term insurgency in many of their writings.[24] The Forty-Eighters and Freethinkers even had a manual explaining how to conduct an insurgency. It was the *Manual for the Patriotic Volunteer on Active Service in Regular and Irregular War; Being the Art and Science of obtaining and Maintaining Independence* by Hugh Forbes, a Forty-Eighter. It was first published in English in 1854.[25]

Siemering compares the organization of *Des Organisator* to that of the Ruetli League in *William Tell*, an 1805 German dramatic classic by Friedrich von Schiller. It is about the Swiss 1307 struggle [insurgency] for independence from Austrian domination. Siemering's reference to this play, especially the oath the Swiss 'insurgents' took on the Rutli, provides an insight into the thoughts and minds which German intellects took during their 1848 Revolution, including those who fled to Texas and formed *Des Orgaisator*. The play is based on actual events of the Swiss liberation movement by the 'common people' against tyranny and alien intrusion. The play's theme is repeated in the organization and operation of *Des Organisator*.

The major theme that so closely duplicates the activity of *Des Organisator* is the insurgent's meeting in the Ruetli and forming a secret league. The Ruetli League invoked a principle which in special circumstances, justified homicide. Part of the guiding principle of the Ruetli League was before they acted there must be no hint of revolution. The Ruetli League had to be kept secret in order to be able to spring a surprise attack on the Austrian governors. Each Ruetli League members took a secret oath. Part of the philosophy of the Ruetli League was that rebellion incitement had to be vigorously expressed and justified, but at the same time there could be no excess that included unnecessary bloodshed. The Ruetli League believed action was the only preventive against further ravages. All members must abstain from private actions before the date when group scores were settled. The Ruetli League understood that the enemy forces were well-armed and, like those of the Confederacy, would not be subdued easily. A Ruetli League method was a surprise attack by well-armed citizenry, or the 'common man', who had to be organized by an elite group, represented in the play by the character Werner Stauffacher. In the case of *Des Organisator*, the elite group consisted of the Freethinkers and Forty-Eighters.

The Ruetli League operated on a principal that they must have a scapegoat on whom to blame their actions or failures. If any one of the Rutli League decided to 'go it alone' all members of the League would have to pay the price, therefore no one man was allowed to do so. It accepted the notion that once an individual resorted to violence his nature changed and the individual lost something of himself, but this was not the case for the League. Violence was acceptable within the scope of the goals of the League. Violence, killing, or even murder was all right as long as it was for the good of the League. Any killing outside of the League was murder and, therefore, condemned. All Ruetli League characteristics were found in *Des Organisator*.[26]

The organization of the Ruetli League takes place in Act Two, Scene Two. It follows an insurgency Phase I. For example, cells of ten men from different locations or villages organized the Ruetli League. They met in a secret place that government forces could not easily reach. In the play, high rocks and woodlands in the Ruetli, or meadow, surrounded it. For the Union Loyal League, it was at the head of Bear Creek. In both cases, each cell came by different routes. They came armed. They both took a secret and sacred oath. The Ruetli League required all members to raise their right hand and thrust their swords into the ground in front of the League leader. Each cell was from a separate geographic location; exchanged greetings with the entire group, but remained with their cells.

They did not mix. *Des Organisator* used the same methodology. They organized by counties. The Ruetli League formed a new league based on a previous one. *Des Organisator* was also a new league, based on a previous organization *Der freie Verein* [the Free Union] of Sisterdale. The reason new leagues were based on older ones was so it could feed upon and gain strength from the older one. In both the Ruetli League and *Des Organisator*, each member pledged his life, his all for the group or league; a blood oath. The Ruetli League ceremony called for the members to raise three fingers when they raised their right hand, which became a sign of recognition. We do not know what was the ceremony of *Des Organisator*, but we know it had secret recognition signs. Both the Ruetli League and *Des*

Organisator required each member to enlist his friends into the new league.[27]

According to Sansom, *Des Organisator* originally had eighteen members. No surviving record names these eighteen. However, a hypothesis can be made based on actions of likely members. First, they were all Germans. Sansom says they were notably from the counties of Gillespie, Kendall, and Kerr, and in localities of Medina, Comal, and Bexar. There is little doubt Eduard Degener was one of the original members. Sansom identifies him as 'Chief of the Union Loyal League' and as 'Head of the League's Advisory Board'. Degener was both a Freethinker and a Forty-Eighter. He was a member of *Der freie Verein* at Sisterdale and an organizer of the 1854 San Antonio Convention and chosen as their delegate to the national convention of the *Bund freie Maenner* in St. Louis. He was certainly a member of *Des Organisator* 'hard-core cell.'

August Siemering was also an original member. He was both a Freethinker and a Forty-Eighter. He was the secretary of the Sisterdale *Der freie Verein*. He was an organizer of the San Antonio Convention and served as one of its secretaries. Siemering was part of the inner-circle of the 1854 ultra-radicals named the Committee of Correspondence. Siemering was a member of the 1854 *Bureau*. He was one of the men the Gillespie County Rifles sanctioned for Unionist activities in February 1862. There can be little doubt August Siemering was a member of the League's 'hard-core cell.'

Another original member of *Des Organisator* who had also been a member of The *Bureau* was Louis Schuetze, who now lived in Fredericksburg. Schuetze was a Forty-Eighter as well as, a Freethinker. He was a secretary of the San Antonio Convention.[28]

Philip Braubach was very likely an original member. He was the commander of one of Gillespie County's two home guard companies. It would have been critical to have him a member since he gave *Des Organisator* a military element. He was one of four Gillespie men sanctioned by Nimitz's Gillespie Rifles for their Unionist activities in February, 1862. He was also certainly a member of *Des Organisator* 'hard-core cell'.

Two other Fredericksburg Unionists were among those sanctioned by the Gillespie County Rifles and so likely to have been original members of the League. They were Ferdinand W. Doebbler and Frederick Lochte. Doebbler was a Freethinker and a Forty-Eighter; Lochte a Freethinker and an open Unionist.[29] Other Freethinkers and Forty-Eighters or sons of Forty-Eighters who were very likely members include Rudolf Radeleff, August Duecker, and Ferdinand Ohlenburger of Gillespie County; Frederick Tegener of Kerr County, Ernst Schwethelm and Gottlieb Bauer of Comfort; Julius Dresel and Oskar von Roggenbucke of Sisterdale; and J. A. Staehely of Comal County. Fritz Tegener was likely a member of *Des Organisator* 'hard-core cell'.[30]

Three members of *Die Vierziger* (the Forty), the Freethinker group that arrived in Texas in 1847, Dr. Ferdinand von Herff, Philip Zoeller, and Jacob Kuechler, were also likely original members. Kuechler was the Gillespie County Surveyor and very influential in the area. He was certainly a member of *Des Organisator* 'hard-core cell'. Zoeller lived near Sisterdale and like Kuechler, was very influential. Herff was a prominent San Antonio medical doctor.[31]

A critical element in an insurgency is that the political front has some type of military element. At first, *Des Organisator* did not have to create a military arm; one was already in place in the form of Braubach's Gillespie Home Guard Company and the Comfort *Turnverein.* The Comfort *Turnverein* was organized by 1860. Almost of the young men of the area were likely members. The Comfort *Turnverein* marched in local parades and other social events. They conducted long marches in the nearby countryside. The names of many of the hills and valleys around Comfort received their names from these *Turnverein* marches.[32]

Philip Braubach's Home Guard Company was composed of 40 members, most of who were Unionists. [See Appendix E.] At least nine were members of the August 1862 fleeing insurgent group. This nine-man group included Valentine Hohmann, August Duecker, Peter Gold [also member of Kuechler's February 1862 Company], Carl Graff, Peter Jacoby, William Klier [also member of Kuechler's Company], Mathias Pehl, Michael Tatsch, and Peter Tatsch [also member of Kuechler's Company]. At least two other were members of Kuechler's February 1862 Company. These two included Christoph Feuge and Wilhelm Juenke. Three other members of Braubach's Company were later killed as being pro-Union bushwhackers; Peter Burg, John Blank, and William Feller. Toward the end of its terms of service Braubach's Home Guard Company came under criticism. In mid-December 1861, a prominent Gillespie citizen wrote to the state comptroller complaining about Braubach's Company. His letter said, "This company is useless, has done nothing." The letter also reported that the chief justice refused to sign the company's last pay voucher.[33]

Beside Braubach's Company, the insurgents also greatly influenced William T. Harbour's Kerr Home Guard Company. It also had 40 members. At least eight are known to have been militant Unionist. These include Thomas Ingenhuett, Thurman T. Taylor, Henry Schwethelm, Theodore Bruckish, Ernst Cramer, A. B. Nelson, Amil Schreiner, and James T. Taylor. The Unionist in Bandera and Medina Counties, while not part of the insurgency, had unofficial home guard units that were mainly Unionists. The Anglo Unionists of nearby Blanco had also succeeded in infiltrating a new home guard company. This was the cavalry company of George Freeman, was organized on August 24, 1861, in the northern part of the county. It had about 70 members of whom five were militant Unionists and include Moses Snow, Edward King, Archibald B. Edon, William Lundy, and William F. Snow. At this point no Anglos were included in the German Unionist insurgency.

Thus far, Texas and Confederate authorities were not aware of the insurgency. The Unionists had done a good job obtaining Phase I. Daily they grew in strength. They awaited the Union invasion they believed was going to take place. From the time of its organization in June, 1861, until the early winter of 1861-62, *Des Organisator* was very successful in almost all of its undertakings. It gained a large following in the Hill Country. It controlled or influenced most of the areas military units. It had close ties with other Texas Unionist groups. *Des Organisator* either controlled the local governments or had great influence over them. The existence and most of the activities of *Des Organisator* were unknown to most local, state, and Confederate authorities.

Events began to change in the late fall and early winter of 1861-62. Despite early Confederate military victories, it now appeared the Union was on the verge of militarily defeating the Confederacy. General Grant had pushed his forces down the Mississippi and Tennessee Rivers and captured Fort Henry and Fort Donelson. Other Union forces captured Roanoke Island and parts of North Carolina. The war was not going well for the Confederates in the West. They had retreated from Missouri. The Union blockade of the Confederate seaports was causing major logistical problems for the South. The war in Texas was also looking bleak. After early successes in West Texas and New Mexico, Sibley's Brigade was defeated and retreated toward San Antonio. Only in frontier protection had Texas and Confederate forces been successful. During the first year of the war, Indian attacks had decreased.[34] The psychological effect of these Confederate defeats on both the Confederates and Texas Unionists was great. For the Confederates, it was the realization that the war was not going to be short, but long and costly. The Union was not going to let the South go their separate way. It was truly going to be a war for survival or total war. In addition, the Confederates concluded they had to commit their total efforts to winning the war. They could not tolerate a perceived Unionist threat in their rear and had to act to eliminate such a threat. For the Texas Unionists, the Confederate defeats signaled the Confederacy was on the verge of defeat and a Union Army would soon come to their aid. Unfortunately for the Unionists they were wrong.

The military threats resulted in the Confederates trying to protect its most vital area, which was northern Virginia. The need for manpower on the eastern battlefields became critical. The Confederate government wanted to move as many military units from Texas to the main theater of war as possible and to organize new units or re-enroll those whose terms of enlistment was about to expire. Among the Confederate units whose term of service was about to expire or likely to be deployed east were the 1st and 2nd Regiments Texas Mounted Rifles that were guarding the frontier post of Forts Inge, Clark, Davis, and Camp Hudson, and Camp Wood. Rumors of a possible Confederate draft began to circulate.[35]

State and Confederate military responded to the new situation. Texas created militia brigade districts. In Gillespie County, there were six companies of state militia, one in Bandera County, one in Blanco County, one in Kerr County, three state militia companies in Kendall County and in Medina County a four-company battalion. A new state unit, the Frontier Regiment, was deployed. Half a company was stationed in at Camp Davis in Gillespie County. The other half was just across the county line in Mason County. Half a company was at Camp Verde in Kerr County and the other half in western Bandera County. There remained Confederate units at Camp Mason and Camp Verde. Thus, by early 1862, there were eleven companies of state militia, two companies of full-time state troops, and at least two companies of Confederates for a total of fifteen companies to provide protection. But the problem for the insurgents was with the disbandment of the home guard companies their military arm was greatly reduced.

The state of Texas responded to the likely loss of 'regular' Confederate units and took action for an increased state force. This action came in the form of two laws. The first was to organize the entire state in case of invasion by dividing the state into militia brigade districts and requiring each male of military age to serve in these militia units. These brigade districts were, like the home guard companies, part-time soldiers and not designed for frontier protection. The second law created a full-time unit for frontier protection. In December, 1861, the Texas Legislature passed two laws creating the new military forces.

As stated before, the Hill Country was in the 31st Brigade District. Two regiments and one independent battalion made up the Hill Country state militia. These were the 2nd Regiment from Gillespie County and the 3rd Regiment made up of the counties of Bandera, Blanco,

Kendall, and Kerr. Medina County had an independent battalion of four companies.[36] [Appendix I].

The six companies of the 2nd Regiment from Gillespie County were organized by March 1862. The insurgent J. Rudolf Radeleff commanded Company A. Jacob Schmidt commanded Company B. The insurgent Valentine Hohmann commanded Company C. Jacob Dearing commanded Company D. The insurgent Jacob Kuechler commanded Company E. The insurgent William Feller commanded Company F. The regimental officers were: Christian Kothe, Colonel; Charles Weyrick, Lieutenant Colonel; and insurgent Jacob Luckenbach, major. Thus, of the six companies in Gillespie 2nd Regiment, insurgents commanded at least four.[37] [Appendix G]. The six companies of the 3rd Regiment were also organized by March 1862. A. J. Kecheville commanded Company A; from Blanco County Julius Schlickum, an 1854 Ultra-Radical and a major leader in the San Antonio Convention in 1854, commanded Company B from Kendall County. The insurgent Ottomar Labhardt commanded Company C from Kendall County. Insurgent Michael Lindner commanded Company D also from Kendall County. Thomas Saner commanded Company E from Kerr County. Braden Mitchell commanded Company F from Bandera County. The regimental officers of the 3rd Regiment being elected were all Unionist. They were the insurgent Fritz Tegener, Colonel; Unionist Julius Schlickum, Lieutenant Colonel; and insurgent Ernest Cramer, Major. Of the six companies in the

3rd Regiment, insurgents commanded at least three, those in Kendall County.[38] [Appendix H].

Within the area controlled by *Des Organisator* that being Gillespie and Kendall Counties, insurgents controlled at least seven of the twelve 'authorized' militia companies. This would indicate the insurgents were very successful in recreating a military arm after the home guard companies' term of service expired. While not in the League's area of control, the Unionists in nearby Medina County controlled a battalion of four militia companies. In the militia re-organization plan of December 1861, Medina County was authorized an independent battalion of four companies. Unionist Hubert Weynand commanded Company A. George Mayers commanded Company B. Unionist Blasins Kieffer commanded Company C. Unionist Jacob Haby commanded Company D. The militant Forty-Eighter Henry Joseph Richarz commanded the battalion. Its strength was at least 170 members. According to one Bandera/Medina secessionist, nine-tenths of this battalion were Unionists.[39] [Appendix I].

The second new state military force created was a full-time regiment designed for frontier defense. The Texas Legislature authorized this regiment, known as the Frontier Regiment, on December 21, 1861. The law authorized a regiment of ten companies of about one hundred men each. Nine of these companies were to be raised from the frontier counties. The tenth from the state at large. Two of the ten companies were to be made up of individuals from the Texas Hill County.[40]

The Hill Country insurgents saw these developments not only as an opportunity but also as necessary to either control or at least greatly influence these two frontier companies. It was also a way to avoid any likely Confederate draft. They took action to accomplish this task by having a Unionist as the enrollment officer for each of the two companies. The enrollment officer would be the one to approve membership and probably be elected commander. The League's leaders used one of the Insurgency Model techniques. The 'political front' had its members join local 'authorized' military units. This is done to receive training and equipment and bring these military units under the control of the insurgency, or at least neutralize its effectiveness if called upon by the government to fight the insurgency. The insurgents were already very effective in obtaining key positions in the area 31st Brigade District militia units. Given the fact the two full-time companies were going to operate in the insurgent areas of control, or influence, it became critical to at least neutralize them.

The first effort was not by the insurgencies. It was the Anglo Unionist, John W. Sansom, of Curry's Creek. In December, 1861, he was under consideration as the enrollment officer for the Frontier Regiment Company for Bandera, Blanco, Dawson, Edwards, Medina and Uvalde counties. This resulted in a major dispute among several citizens of the areas and within Sansom's family. One of his brothers-in-laws, James M. Patton, wrote the governor opposing Sansom's because he was a Unionist. In his letter, Patton quoted Sansom as saying Sansom would be, "God damned or will be God damned if he would ever fight against the Federal government (Meaning the Lincoln Government)." Patton went on to say Sansom asked him, "If I was willing to fight for Southern men's Negroes that if I was, I was a Damned old fool, that he would not do it: that he made his living thus far by the Sweat of his brow, and did not intend to fight for Southern men's property." For these reasons, Patton told the governor he was against Sansom's appointment.[41]

Patton may have been against Sansom's appointment, but others were in favor. Several pro-Confederate citizens of the Curry's Creek area wrote letters to the governor in defense of Sansom. These letters stated they believed Sansom to be a good Southern man and felt Patton's allegation was based on a personal matter between the two and recommended Sansom's appointment. Despite this support, Sansom was not appointed. Instead Charles de Montel, a known secessionist, was appointed enrollment officer and elected company commander. De Montel kept most of the area Unionist out of the company and made it a pro-Confederate unit. They had four officers, nine non-commissions officers and 106 troopers. Beside de Montel the officers were Thomas McCall, the sheriff of Bandera County as first lieutenant, Amos Valentine Gates, the Chief Justice of Blanco County, and Benjamin F. Patton, one of the men who had supported Sansom as enrollment officer. It turned out Patton was a Unionist.[42] [Appendix J]

In the League's major area of influence, the insurgents made their greatest effort to control the enrollment of the

Gillespie, Kerr, and Hays Company. The insurgent, Jacob Kuechler, who was already the commander of Company E, 2nd Regiment, 31st Brigade District, was appointed the enrollment officer for the company. He quickly began recruiting. The problem was he recruited only insurgents. By February 12, 1862, he enrolled seventy-four men and held elections. All four of the elected officers were insurgents. They included Kuechler as captain and Philip Braubach, the Gillespie County sheriff and the commander of the insurgent controlled home guard company as first lieutenant.

The two-second lieutenants were Moritz Weiss and Hugo Degener. Of the eight non-commission officers, two where in the fleeing insurgent group and another joined the Union Army. Of the seventy-four men on the February 12, 1862 muster roll, twenty were in the fleeing insurgent group. At least another eleven joined the Union Army. This means at least thirty-one of the seventy-four in Kuechler's Company took up arms against the Confederacy. An item of interest is that of the seventy-four names only three were Anglo. Two of these were non-commission officers. This is the first time that Anglos were included in a League organization.[43] [Appendix K].

If there was one act, which resulted in the insurgents losing their veil of secrecy and the authorities becoming aware there was a serious insurgent threat, the overt attempt at organizing this pure insurgent company was it. From this point forward counter-reaction to the insurgents can clearly be seen. This counter-reaction took place immediately. On February 13, 1862, two prominent Hill Country citizens sent letters to Governor Lubbock protesting the organization of Kuechler's Company. One was David H. Farr of Kerr County and the second was Frank Van der Stucken of Gillespie County. Their letters dealt with the fact area citizens, who were Confederate supporters, were not able to enroll in the company. These letters, and later petitions, claimed Kuechler secretly enrolled the company and only Unionists were enlisted.[44] Farr stated in his letter that he and others attempted to enroll in Kuechler's Company on the 1st and again on February 11th, but Kuechler would not meet with them. Farr said Kuechler and his 'German friends' left Fredericksburg to organize. They met at or near Comfort, 'another Dutch town' to enroll only Unionists. Farr also said that in order to get the required number of men, Kuechler was enrolling men from Blanco County who lived near Sisterdale. One was a man by the name of Degener. Farr described Kuechler as a "violent Union man [and] is thought to be a black republican."[45]

Van der Stucken stated in his letter that he knew of over fifty men who had attempted to enroll in Kuechler's Company, but could not find him. He also said Kuechler met with men from Comfort, but not a man from Hays County attended any of Kuechler's secret meetings because they had not been notified. Like Farr, Van der Stucken claimed that in order to fill the company with the required numbers Kuechler had illegally enrolled Blanco County men.[46]

Farr and Van der Stucken were not the only Hill Country citizens to complain to Governor Lubbock about Kuechler's Company. On February 14, 1862, thirty-eight

Kerr citizens petitioned claiming Kuechler enrolled only Unionists and that they, and some thirty Gillespie citizens, had been unable to enroll. They requested the company be disband and a new enrollment officer be appointed. Captain Philip Braubach, the insurgent Gillespie sheriff, threatened retaliation against those who signed a petition. Braubach said, "He would bring two hundred men to their doors which would make them talk different." Governor Lubbock disband Kuechler's Company and appointed Joseph Walker of Travis County as the new enrollment officer. Walker soon had more than enough volunteers for the new company. He directed them to assemble at Fredericksburg.[47]

Several excellent sources establish it was *Des Organisator* plan to control this full-time unit and make it an insurgent unit. Key League members are two of these sources: Ernest Cramer and August Siemering. Cramer wrote: "The young men of the counties joined and an intimate friend of by the name of Kohler [Kuechler] (of Darmstadt) was elected captain. He soon had a company together of all Germans. Now a party formed themselves in Freidrichsburg, composed mainly of Americans, and appealed to the Governor at Austin, declaring that our company was not legally organized and that they had not been given the opportunity to join. Our company had been formed of men gathered together with the understand that as soon as the Northern troops would come within reaching distance, we would join them. But it followed that our plans were overthrown in a most lamentable manner."[48]

August Siemering comments were much the same. He said: "The legislature had given the governor the authority to recruit a number of rangers for border duty and for the prevention of Indian attacks. A citizen of Gillespie County, Jacob Kuechler ... formerly a resident of Sisterdale ... received the authority to form such a company in Gillespie and surrounding counties. Immediately plans were made to organize this company from Union people in the Gillespie, Kerr, and Kendall Counties and to it use as a defensive measure against possible emergencies in the named counties. Other plans developed called for this company to be the base for other organizations, which then at the right moment can act on behalf of the Union. A few weeks thereafter the company was organized in Comfort and Jacob Kuechler was chosen as the captain. I received the position of first sergeant. With these actions we believed we were protected for a long time. But we had not considered the few Secessionists and spies, which were in the area. It was not long before an order came to dissolve the company and at the same time a man by the name of Walker from Austin appeared with the order from the governor to organize a different company."[49]

These two sources clearly establish the fact the insurgents organized this company as an insurgent unit for their own use. They believed the Union was going to invade shortly; having this unit would protect them until the invasion and afterwards it was to be used to assist the Union Army. Other sources supporting this view include several Gillespie citizens who heard about the purpose of Kuechler's Company. Charles Nimitz told what he knew about Kuechler and Braubach attempting to raise the company in secretary. Nimitz said it was customary when a company was to organize to give public notice. But when Kuechler's Company was organized no one, except those whom the insurgents wanted to join, could not find where to meet. The others were told falsehoods as to the time and place of the meeting. Instead of holding their meeting at Fredericksburg, Kuechler and Braubach held the meeting at Comfort, some twenty-two miles away.[50]

Frederick Fresenius was also one of the Gillespie citizens who heard reports about Kuechler's Company. He later told Confederate authorities when the company organized there were a great many who intended to join, but could not find out where or when to meet. Kuechler and the insurgents went about twenty miles from Fredericksburg and organized. One of the men who joined Kuechler Company told him their "object was to join and when the Yankees come they would lay down their arms." Fresenius said the men who joined Kuechler's Company were, for the greater part, ones who had shown their disloyalty since secession and loyalty to Lincoln and were now "positively against us."[51]

With the exposure of the insurgent's plan and success of the Gillespie insurgent's others began to speak out opposing their actions. On February 23, 1862, the Gillespie County Rifles passed a resolution denouncing the Unionist activity of three men: Philip Braubach; Ferdinand Doebbler and Rudolf Radeleff. A fourth man's name, that of August Siemering, was originally on the resolution but was marked out when it passed. All four of these men were likely original members of *Des Organisator.*

August Siemering told about how he learned of the meeting and how his name was removed from the resolution: "On a Sunday in a hotel in Fredericksburg a meeting was held, at which, by chance, I appeared without any knowledge of what was going one. To my surprise the question of whether or not certain persons, which were suspected to be Union leaders, were to be banished from the county or be warned, was discussed. The suspected leaders were Rudolf Rodlett [Radeleff] , Philipp Brauburg [Braubach], Friedrich Wilhelm Doebbler and myself. I protested vigorously against such actions, but accomplished nothing, other than the removal of my name from the list of the leaders and that warning was deemed appropriate for the other three."[52]

While the insurgents' plan to fully control the company of the Frontier Regiment failed and their actions resulted in strong criticism from other Gillespie County Germans, they still had hopes of having large numbers of men in and gaining leadership positions in the new company. Their attempt failed.

Complying with Joseph Walker's directions, the new company of volunteers assembled at Fort Martin Scott near Fredericksburg. The Hays County group left San Marcos at 9:30 in the morning of February 25th, headed by Henry T. Davis. They arrived in Fredericksburg on February 28, 1862, where they found the Kerr County group headed by William T. Harbour and the Gillespie County groups. There were two groups from Gillespie County. One was 'the German

Company,' presumably Kuechler's. The other was 'non-Germans' headed by James M. Hunter.[53]

Each of the three counties was to provide a third of the men. There were well over a third in each of the Hays and Gillespie groups. This required the men to draw lots to determine who would be allowed to enlist. Pieces of paper were placed in a container. Some had numbers; others were blank. Those who drew a number would be allowed to enlist, those who drew a blank would be sent home.[54]

One of the Hays County men was D. P. Hopkins who kept a diary for the first few months. Hopkins states that the first Gillespie man to draw was the "captain of the German Company." He drew a blank! This greatly upset the other men from "The German Company." The result was "the whole German outfit became indignant, and backed out." Another Gillespie citizen corroborates Hopkins comments. When testifying later against the insurgents, he said after Kuechler's Company was broken up the disloyal men did not join the new company and "left in a lump."[55]

The insurgent August Siemering tells about the failure of insurgents to enroll in the new company. He said Walker, "brought with him a number of Americans from Hays County, which he had already recruited for the company. With this the plan of the Germans was destroyed." Siemering claims the insurgent nevertheless still attempted to enlist in Walker's Company in mass, and through the selection of Kuechler to control the same, but this was not successful because Walker noticed the intentions and rejected a majority of those who applied. The whole thing came to a stop when even those, who had already been accepted, gave thanks for the honor and went their way.[56]

Siemering told of the immediate results of the insurgents' failure. "That's how instead of a German company there was an American one that instead of a Union company, a secessionist one formed and another threat was added to those already present ... The evil thing was that the small secessionist party was encouraged by these events and now became aggressive, when previously it had been quite passive."[57] From the insurgent's view, Siemering is absolutely correct. Up to this time, or about February and March, 1862, there had been little open or organized opposition to the Hill Country insurgents. This was especially true in Gillespie County. This allowed the insurgents to secretly organize and gave them a false sense of security. When they began to openly oppose the Secessionists, the result was concern and alarm, not only on the part of Secessionists, but moderate Unionists as well. They were concerned, as the Gillespie Rifles solution said, the militants were dangerous to the community and their overt actions would only result in strong Confederate counteractions.

This refusal by the insurgents to enroll in the new company was a major tactical error. They could have rallied behind another leader and elected him to a leadership position. Even if half of the Gillespie 37 men were insurgents, it still would have given them a significant influence in Davis' Company. This company was the major full-time military force in the area. By not joining the insurgents left a void, which secessionist filled. The new Frontier Company was mustered into service on March 4, 1862, at Fort Martin Scott without a large number of Hill County insurgents participating. Muster rolls show thirty-seven, Gillespie men, forty Hays men, and thirty-two Kerr men enrolled. Of those from Gillespie County, only nine had German surnames. Nine men from Kuechler's old company enrolled; Adam Blucher, Arthur de Cloudt as 4th Sergeant; Joseph Dengel, John Peter Mosel, Andrew Jackson Nixon II, Frederick Sauer, Arthur Striegler, Olfort Striegler, and Owe Striegler as bugler. The company had four officers, eight non-commissioned officers and 97 troopers. No insurgents were elected officers. Henry T. Davis of Hays Company was elected captain. A Gillespie County man, James M. Hunter was chosen as first lieutenant. William T. Harbour of Kerr County was elected senior second lieutenant and James L. Williamson of Hays County was elected junior second lieutenant.[58] [Appendix L]

The citizens of the area were now faced with many of the same problems as the rest of the state; a major war going on and people having to take sides. Being neutral, or hoping the war would be over quickly was no longer an option. What is of special interest is this growing conflict was not due to the introduction of any outside force. The only outside force was the 40 Hays County men who were part of the new frontier company. The conflict was between the Hill County citizens. Any threat or counter threat was among neighbors. If there had been radical Secessionists in the area, most had left early in the war and joined 'regular' Confederate units.

The conflict was between militants, the Greens, [now insurgents] and moderate Unionists, the Grays, supported by the few remaining Secessionists. As the secession vote had shown, there was no large base of secessionist support. Only 16 Gillespie citizens voted for secession while 398 voted against. In the Comfort area, only 15 voted for secession while 42 voted against. Now at least twice that many were signing petitions opposing the insurgents. In the remainder of Kerr County, 61 voted for secession and 15 voted against. In Blanco County 83 voted for and 170 against. Therefore, in the entire area, which was *Des Organisator* stronghold only 175 voted for and 625 voted against. Now despite this heavy vote against secession, which many claim as a pro-Union vote, the Unionist stronghold was speaking out against them and they felt they were the ones being threatened. Something must have happened to change public opinion since the secession vote. That something certainly was not that the war was going well for the Confederacy. It was not. The only thing that can be identified is *Des Organisator* began openly opposing the Hill Country, Texas, and Confederate governments. While the area citizens had no great loyalty or love for the Confederate cause, they were aware that any perceived overt threat or action against the Confederate cause would be met by strong counteractions.

That counteraction would be harmful to the Hill Country. Even Julius Schlickum said at his trail that friends of his were not "such fools, as to be guilty of joining such an organization [*Des Organisator*], which would only bring destruction on the people and bloodshed on the frontier."[59]

As Siemering points out the, Secessionists were gaining strength while the insurgents were losing theirs. The insurgents now came under criticism for their brazen actions. This was a major mistake which should never happen in a successful insurgency. Many began to come under criticism for their open support of the Union. Philip Braubach was one. Several Fredericksburg citizens noticed a change in his attitude and behavior. Among these was a man identified as 'Mr. Wolman', who was likely Baron von Woehrmann. He later testified he heard Braubach say the "South never could succeed." At first Braubach said, "He would keep neutral," but as time went on he "talked otherwise." Woehrmann said that Braubach began discussing politics with other like men and his associates were men who were opposed to the Confederate government. Another citizen, Hilmar von Gersdorff, also testified that Braubach always said something like "The South was too weak to win the war." Frederick Fresenius testified that he believed Braubach was disloyal to the Confederacy. Charles Nimitz testified he could not call Braubach loyal to the Confederacy and that Braubach was not regarded by any loyal men of Gillespie County as being loyal to the South. Nimitz said that Braubach was, "an intelligent man and I think his influence is great among the illiterate." Nimitz further said Braubach was "an active man that has something to say on all passing events. His associates for the last few months have been those whose loyalty is doubted." Engelbert Krauskopf also spoke up against Braubach, saying, "I have heard [Braubach] make remarks which led me to believe that he was opposed to us. He always went with the opposition or Union party." Krauskopf further said Braubach, "always meets with the party who halloes for the Union."[60]

A second militant member of the League, J. Rudolf Radeleff, also came under local criticism. Radeleff was a partner with the Hunter brothers in their Fredericksburg merchantile business which provided beef and other supplies to the military posts. Both Hunter brothers attempted to get Radeleff to tone down his anti-Confederate rhetoric. James M. Hunter testified he heard Radeleff make remarks, "often which I could draw no other conclusion than that he was opposed to the South." Hunter said Radeleff urged him to get rid of all Confederate money because, "as soon as the first battle was fought the money would go down." Hunter said he believed it was Radeleff's desire that the Confederates, "be whipped in the first great battle." On the question of slavery, Hunter remembered Radeleff always spoke out against it. It was Radeleff's outspoken opinion, according to James Hunter, that "Slavery was doomed ... thought free labour [sic] would pay better and was preferable." Hunter stated Radeleff, "never appeared to credit any favourable [sic] [newspaper] report to the South." Radeleff was, according to Hunter, "A man very much beliked where he lives, and in all other respects he is thought a great deal of influence over his neighbors." James Hunter attempted to get Radeleff to change. Hunter explained, "I often spoke to him, asked him to change his course, he said he could not, it was impossible." Hunter said these conversations took place up to about the last of December 1861 at their store in

Fredericksburg and with several others beside and always against remonstrance."[61]

The other Hunter brother, John M. Hunter, reported he and Radeleff had a great many arguments relating to the troubles. Radeleff always argued against him in favor of the Union. Radeleff always said that, "The North would certainly subjugate the South, there was no other chance." Hunter testified Radeleff was very much dissatisfied about the secession of the South. He often asked Radeleff to change his course, which Radeleff replied, "Can the Leopard changes his spots." Other Gillespie citizens spoke out against Radeleff. One was Dr. Wilhelm Keidel, one of the 1854 ultra-radicals. He said, "I don't think [Radeleff] is very loyal. We have often conversed on the state of Country." Keidel went on, "When we read news of our victories he [Radeleff] always doubted, until doubt was beyond reach. He never believed that the Confederacy would succeed."[62]

Another Gillespie citizen who spoke out was Charles Nimitz. According to Nimitz, Radeleff, "often argued against the course the [Confederate] government had taken, and against the basis of currency." Nimitz said Radeleff, "was opposed to the government itself. He did not believe the money worth what it called for." Nimitz further testified Radeleff made many arguments against the Confederate government and made bets against its ability to maintain itself. Radeleff's tendency was to bring the government into dispute. Nimitz further stated they had many disputes and Radeleff told him on one occasion, "that if he knew the sentiments of the people as well as he did, that I would not hoist the Confederate flag on my house."[63]

The insurgents now felt under siege. Regardless of where this threat came from, it was real to them and forced *Des Organisator* to consolidate their power, and for those who had not made a firm commitment to either side to now do so. Part of *Des Organisator* consolidation was a desire to make them fully self-sufficient. This is best explained by one insurgent who had not taken sides up to this point. Ernest Cramer, who later became a major *Des Organisator* military leader, stated in an October 1862 letter, "But as the agitation became stronger and stronger, we had to definitely take a stand for one side or the other. We ... naturally joined our friends in support to the North." "Threats continued to become more violent and finally reached the state where the men of the counties of Kerr – Kendall and Gillespie assembled to form a defensive organization."[64]

There are several insurgent accounts of forming this 'defensive organization,' including those of August Siemering, Fritz Tegener, Ernest Cramer, Julius Schlickum, John W. Sansom, August Hoffmann and others. All these sources provide some data on the organization of this military arm; when added together, provide a clear picture of the how and when the battalion was organized. The one most quoted, and the less reliable, is that of John W. Sansom, written in 1905. He says the insurgents met on July 4, 1862, on Bear Creek to "perfect the organization." However, individuals who were key *Des Organisator* members say it was much earlier. In 1875, Fritz Tegener, the commander of the insurgent battalion, said he was elected major, "at Bear Creek in the beginning of May, 1862 ... and there was never a second election." August Siemering, also writing in 1875, says, "The meeting of course had to be kept secret, and a place in the mountains on the watershed between Fredericksburg and Comfort was selected, where Bear Creek originates. ... The meeting took place in the spring of 1862." The most reliable is insurgent Captain Ernest Cramer, the leader of the Kendall Company, written just months after the event. He even gives an exact date. He says, "On the 24th of March a meeting was called at a well [or spring] named *Barenquelle* [Bear Springs] between Fredericksburg and Comfort." From these various sources, it is clear the League's Military Battalion organization took place in the spring and not on July 4, 1862. It took place on March 24, 1862. Other events support Cramer's description.[65]

The best overall source is August Siemering. He was at the meeting and a key leader of the League. "Unionists, due to the aggressive politics of the secessionist, now saw it as their duty to organize themselves and in case of an emergency [or act of aggression] to resist." Siemering explains how messengers went all over the county and surrounding areas to call supporters of the Union, both American and German, to a meeting. It "of course had to be kept secret, and a place in the mountains on the watershed between Fredericksburg and Comfort was selected, where Bear Creek originates." Here they were protected from discovery. He states the meeting took place in the spring of 1862 and representatives from the counties of Gillespie, Kerr, Kendall, Comal and Mason attended. Insurgent leaders decided to permanently organize the Union people by county under the command of a major. The individual counties were to be commanded by captains. Siemering reveals the name of the military leader, "Fritz Tegener from Kerr county, was selected as major, who was known as a decisive character ... In addition, certain secret codes were established, committed themselves to strict secretiveness and separated with the intention to complete the organization as soon as possible and to render appropriate reports."[66]

This comment makes it clear *Des Organisator* now included Anglos. The result was the organization of an Anglo, the Kerr Company. The statement by Siemering about Comal and Mason counties implies the League had plans to form companies in those counties, as well as in Kerr, Kendall, and Gillespie. Few sources were located dealing with Comal or Mason counties and the League. No companies in either of these counties were known to have been organized. It appears the insurgents of Comal County were members of the Kendall Company. The Mason County insurgents appear to have been members of the Gillespie Company. There are few references regarding Mason. One states, "While no mention has been found in any of the official correspondence to link anyone in Mason County with the Union Loyal League, it seems reasonable that some of the residents were at least in sympathy with their cause." When the Confederates declared the Hill Country in rebellion, one military task force was sent to Mason and surrounding counties. Its commander reported all the area residents took the required oath. It is of interest to note all the named counties are northwest of San Antonio. Medina County, which contained large numbers of Unionists is not named, nor is Bandera or Blanco County. It appears no plans were made to include these counties in *Des Organisator*. The Blanco German insurgents were likely included in either the Gillespie or Kendall Company and the Blanco Anglos were likely included in the Kerr or American Company. As will be discussed later, Unionists of Bandera and Medina counties as well as Bexar, would be included in a larger plan. Siemering's statement about plans to complete the organization as soon as possible supports Sansom's account that the battalion organization was 'perfect' at a later time and may explain the different dates reported by Siemering, Cramer and Tegener. However, according to this Siemering account Tegener was elected major at the first meeting. Cramer clearly states he was elected captain on March 24, 1862. It is also of special interest to note Siemering says appropriate reports were to be rendered. This supports the hypothesis that the League kept some type of written record. These accounts provide a great deal of vital information. When other data is added, a clear picture emerges. First, the organization meeting took place at Bear Creek in the spring of 1862, most likely on March 24th. Second, insurgents from the counties of Gillespie, Kerr, Kendall, Comal, and Mason met and formed a permanent military organization, a battalion, commanded by Fritz Tegener. Third, each county was to have a company commanded by a captain. Fourth, *Des Organisator* members established secret codes and oaths and the Basil Stewart's execution establishes it was a 'blood oath.'

The Cramer and Tegener accounts, added with Sansom's data allows us to deduce company officers' names. Cramer says, "Companies were organized and officers for the different districts were elected. My company, composed of 80 men, elected me captain for the Comfort district." Tegener says, "It was the assembly at Bear Creek in the beginning of May, 1862, where we chose at the same time Kuechler as captain for Gillespie County, Hartman as captain of the Americans, and Cramer ... as captain ... from Comfort and surrounding area of Kendall County." Sansom provides the names of the company officers. "Among other measures taken up was the organization of three companies, to wit: The Gillespie Company, Jacob Kuechler, Captain and Valentine Homan [Hohmann], Lieutenant; the Kendall County Company, E. Kramer [Ernest Cramer], captain, Hugo Degener, Lieutenant; and the Kerr County Company, Henry Hartman, captain, Phil G. Temple, Lieutenant."[67]

All references to this company in the German language refer to it as the American Company. Most non-German articles, which use Sansom's account as the source, refer to it as the Kerr Company. This company was composed of non-Germans from the Hill Country. The Germans of Kerr County belonged to the Kendall Company. Having elected officers, the three companies then elected Fritz Tegener as Major to command the battalion."[68] [Appendices M, N, O, and P].

Des Organisator now had its own military arm, a battalion of at least three companies, outside of the state militia force structure. The battalion strength is not known. Sansom says, "Not less than five hundred male Unionists met on Bear Creek in Gillespie County, met on Bear Creek and proceeded to perfect the organization." But that number appears to be too high. The actual strength was likely between 250 and 300.[69]

The nature and the consequences of *Des Organisator* military battalion have aroused considerable controversy. A modern Mason County historian says, "The anti-Confederates had ample time to leave the area in 1861, but instead, organized as a fighting force, an open invitation to the Confederates and an insane action, practically asking for abolishment." A secessionist wrote in 1893, "At the beginning of the war the Union German met and went into a secret oath-bound organization, to keep from taking action on either side of the conflict; and not to remain peaceful and law-abiding citizens, but to murder and pillage." "It was a snare set to catch Southern men; the pretended desire for peace meaning death and loss of property … In the early spring of 1861 [1862] the Union Germans formed a permanent organization."[70]

Even some of the radical Unionists were against the organization of *Des Organisactor*. For example, a Freethinker and Forty-Eighter ultra-radical leader of the 1854 San Antonio Convention wrote in 1862, that the belief the Union would soon raise its flag over Texas "prompted the Germans and Americans near Fredericksburg to form and organized a sort of secret brotherhood." He was told *Des Organisator* was mainly for protection against surprise attacks, burning of the settlements and hanging from the southern oriented parties. But, he understood the real purpose. He was approached to join and take command. "With fearful worry did I watch the unfolding of this association because I was too familiar with the circumstances not to see the danger and futility of this undertaking. I knew too well that a few backwoods men could not lead war against the State of Texas; a State who had about 15,000 men in service with weapons." This 1854 ultra-radical asked, "What did we have?"

He answered his own question, "Hunting guns, little powder, and if we were to battle, no line of retreat. In spite of all rumors, I could not believe in an early arrival of the Union Army. The Government of the U. S. could not send troops to Texas, as long as the situation in Virginia and Tennessee was so uncertain." He "emphatically" refused to be part of this undertaking and tried with all his power to reason and convince *De Organisator* leaders to reconsider this "foolish attempt." He repeatedly depicted to *Des Organisator* leaders the ruin of the settlement in the end. "In vain!"[71]

Even some of the Forty-Eighters were opposed to *Des Organisator*. One who arrived in Texas in October, 1849 stated his reasons for supporting the Confederacy, "I decided to vote for succession for these reasons: I had left Prussia being proscribed form my political opinions; I selected Texas for my future home with full knowledge of the institution of slavery existing here; I did not come here as a reformer, I came here to live with this people who received the <u>stranger unconditionally</u>; and felt, right or wrong, my place was with the people of Texas, to stand with them in upholding the cardinal principles of selfgovernment laid down in the Declaration of Independence July 4, 1776."[72]

Two of the major themes later writings of the insurgent's stress are the threats faced by *Des Organisator* and the resulting 'defensive organization' they formed. As discussed above, the only threat they were under at this time was from local citizens and not from any outside force. It is ironic that because of *Des Organisator* creation of a military arm to protect their interest, just the opposite took place. Instead of an insurgent military unit, a secessionist one was created. The otherwise small group of pro-Secessionists now had a powerful ally, a full-time military unit and support of many area moderate Unionists.

Des Organisator only had control of some militia units of the 31st Brigade District that were just being organized. *Des Organisator* reaction to the failure to control the new frontier company was to organize its military arm outside of any approved state structure. This new 'defensive organization' was the secret 'unauthorized' *Des Organisator* military battalion. This action is what sets *Des Organisator* apart from any other Texas Unionist organization and clearly establishes it as an insurgency. The League followed the model which political and military science terms an insurgency. It not only had a 'political arm' but also a 'military arm'. They were now well into Phase I and ready to enter Phase II of an insurgency.

Map E – Unionist Areas of Influence

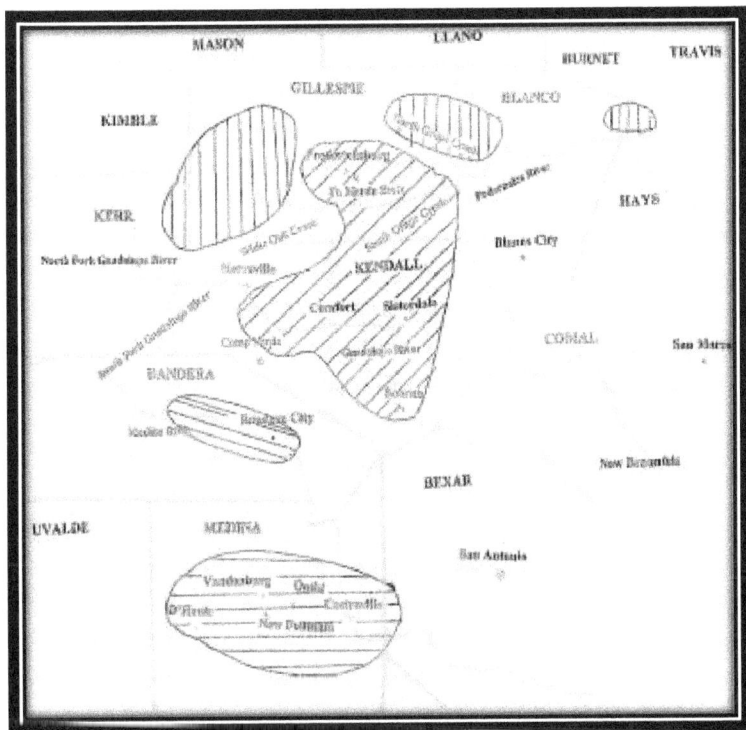

Endnotes -- Hill Country Unionists Organize

1. Biesele *German Settlements*, p 206 and Lich *The German Texas*, p 94.

2. Kaufmann *Germans in American Civil War*, 63-64; Wittke *Forty-Eighters*, pp 198-201 and Siemering's *Germans During Civil War*, May 8, 1923.

3. There was a major disagreement between the area citizens about the creation of Kendall County. This was especially so between the Unionist and the citizens in the Comfort area because Comfort recently became the county seat of Kerr County. Two separate petitions were submitted. One for the

creation; signed by 95 citizens mainly from the Boerne area. Eighty-Eight citizens, mainly from the Comfort area, singed the petition against the creation of Kendall County. List of 'Petitioners for The Formation Of Kendall County' and List of 'Petitioners Against The Formation Of Kendall County', located in "Keys To The Past" by Boerne Genealogical Society, Volume III, No. 2, April 1984, pp 3-4.

4.   Sansom *German Citizens*.

5.   Ibid.

6.   Ibid.

7.   *Now You Hear My Horn: The Journal of James Wilson Nichols 1820-1887*, edited by Catherine W. McDowell, (University of Texas Press, Austin, Texas), 1967, pp 143-145.

8.   "Louisa Naomi Kent and James Billings" by John Billings, "Stockman Family Newsletter, Silver City, New Mexico" n.d. Copy provided by Maxine Powell of Kerrville, Texas and *The Evolution of a State or Recollections of Old Texas Days* by Noah Smithwick, (University of Texas Press, Austin, Texas), 1983, p 250.

9.   Nicholas, Now *You Hear My Horn*, pp 143-145. For an account of Hubbard's death see *Burnet County History—A Pioneer History 1847-1979*, Volume I, by Darrell Debo, (Eakin Press, Burnet, Texas), 1979, 35-36 and "Deadman's Hole to be become county park landmark" in *Marble Falls Picayune, May 27, 1998.*

10.   Johnson *New Perspective*, p 187.

11.   Sansom *Battle of Nueces*, p 2.

12.   *Austin's State Gazette* Saturday, June 1, 1861.

13.   John William Sansom was born in Alabama in 1845. He arrived in Texas about 1839. In 1850, he moved to Curry's Creek in present-day Kendall County. Sansom acted as a guide for the fleeing Unionist group. After the battle, he made his way to Mexico and on to New Orleans where he joined the First Regiment [Union] Texas Cavalry. After the war, he returned to the Hill County and served as a Ranger captain. In 1905, Sansom wrote one of the few accounts of the Battle of the Nueces by a participate. John William Sansom died on June 19, 1920 in Bexar County. Webb and Branda Handbook *of Texas*, Volume II, p 567 and Volume III, p 854; "Memoirs Capt. John W. Sansom" by Alexander Brinkmann, Notes taken by Brinkman during interviews with John Sansom about 1918, copy provided by Gregory J. Krauter, Comfort, Texas and Sansom *Battle of Nueces*, p 1.

14. Sansom *Battle of Nueces*, pp 2-3.

15. Sansom Battle *of Nueces*, pp 2-3, and 11-12; Glenn Capt *John*, pp 8-9 and 14-15; and *John William Sansom's Battle of the Nueces* by Frankie Davis Glenn, (Frontier Series, Boerne, Texas), 1991, pp 5-6 and 13-15.

16. It is stressed that accounts of events which Sansom personally took part in are very accurate. However, accounts of events in which he relied upon others for information, are biased and some are incorrect. Thorpe *Historical Friction*, 78; Isaak *Scope and Methods of Political Science* and Sansom *Battle of Nueces*, pp 2-3.

17. Sansom *Battle of Nueces*, pp 2-3; Letter, August Hoffmann, dated September 1, 1925, Bankersmith, Texas. Copy provided by Gregory J. Krauter, Comfort, Texas; *Twenty-Seven Years on the Texas Frontier* by Captain William Banta and J. W. Caldwell, Jr., Published by L. G. Park, Council Hill, Oklahoma, 1893, Introduction; *The Fredericksburg, Texas Manuscripts* by Kenn Knopp, (Published by Kenn Knopp, Fredericksburg, Texas), 1997, Part I, p 7; Weber *Deutsche Pionier*, p 12; Raley's *Blackest Crime in Texas Warfare*; Siemering's Germans During Civil War, May 29, 1923; Heintzen *Fredericksburg, Texas During The Civil War and Reconstruction* and Benjamin *The Germans In Texas*, p 109.

18. Basil B. Stewart was born about 1835 in Scotland. The date he arrived in Texas is not known. The 1860 Gillespie Census shows him living in Henry Attrill's household. Stewart was a member of the Kerr [American] Company of the Union Loyal League's Military Battalion. In late June 1862, he informed a justice of the peace on some of the activities of the League. League members drew straws to determine who would execute him. Ernst Beseler was selected. He shot and killed Stewart on July 5, 1862, just north of Comfort. The account of Stewart's death is further evidence that the League used cells as described herein. This account says, "He knew by name only a few of the men who attended the meeting." This shows Stewart was most likely a member of a cell and thus not aware of the names of key members of the League. It is of special interest to note for a full year no League member broke his vow of silence. It is also of special interest to note even when the Confederates finally cracked down on the Unionist and arrested Eduard Degener, no one testified that he was the leader of the League. One reason is the compartmentalization of members. Siemering's *Germans During Civil War*, May 29, 1923 and Sansom *Battle of Nueces*, p 2.

19. Sansom Battle *of Nueces, p* 2.

20. *A Hundred Years of Comfort in Texas: A Centennial History* by Guido E. Ransleben (The Naylor Company, San Antonio, Texas) 1974, p 87; Riley *Blackest Crime in Texas Warfare*; Letter, August Hoffmann, Bankersmith, Texas, September 1, 1925 to his children, copy provided by Gregory J. Krauter, Comfort, Texas; Another translation of the letter can by found in "A German-American Pioneer Remembers: August Hoffmann's Memoir" Edited by David R. Hoffman 'Southwestern Historical Quarterly' Volume CII, No. 4 (April 1999).

21. August Hoffmann was born on December 9, 1842 in the Village of Lichtenau, Prussia. His mother died when he was nine years old. In 1854, August and his father, Traugott, arrived in New Orleans on the *Friedrich der Grosse*. They arrived in Gillespie County about 1855. August was a member of the League's Military Battalion and the last surviving member of the Unionist group. He died on March 4, 1935 in Gillespie County.

22. Ibid.

23. "A German State in the New World" The Adelsverein's Texas Colonization Scheme: 1842 - 1847" by Kenn Knopp 'The Journal', Volume XXVI, Number 2, Summer 2004, p 166.

24. *Constitution of the State of West Texas* written by Jacob Kuechler and Eduard Degener, American History Library of the University of Texas-Austin, Section 1: Boundaries of the State of West Texas as contained in Knopp *a German State in the New World*, p 166.

25. King Meusebach, p 10. For articles by or about Forty-Eighters and the use of the word Insurgency or Insurgents see: Siemering, *Germans During Civil War*, May 29, 1923 and Tolzmann *German-American 48ers 1848-1998*. There are some current researchers who are beginning to say that the League was conducting an insurgency? Oral Presentation Rich Hamby, University of Texas at the Permian Basin, "Texas Germans, and the Battle of the Nueces River" at the 1998 Texas State Historical Association Annual Meeting in Austin, Texas, March 6, 1998 and "Battle Or Massacre? The Incident on the Nueces, August 10, 1862" by Stanley S. McGowen 'Southwestern Historical Quarterly', Volume CIV, No. 1, (July 2000).

25. *Being the Art and Science of obtaining and Maintaining Independence* by Hugh Forbes, (De Witt and Davenport, New York, New York) 1855 [Reprinted by. AstroLogos New York, New York 2007].

26. *William Tell,* by Johann Christoph Friedrich von Schiller, Translated and Edited by William F. Mainland, (The University of Chicago Press, Chicago, Illinois), 1972, pp 34-35, 43-62, and 141153 and Rapoport *Clausewitz On War*, pp 35, 53, 256 and 389.

27. Schiller *William Tell*, xi-xxxiii, pp 34-35, 43-62, and 141-154.

28. For a source telling how Schuetze "was boldly outspoken, given to argument of ideas" see Gillespie Historical Society *God's Hills*, Volume II, pp 187-188. Three prominent members of *Der freie Verein* or the San Antonio Convention who may not have been members were Ernst Kapp, Dr. Wilhelm Keidel, and Julius Schlickum. Kapp owned a slave by 1859 and seems to have lost some of zeal for revolution. By 1862 Keidel also seems to have reduced his support for revolution as he was a member of Nimitz's Gillespie Rifles and signed the petition against Unionist actions. Schlickum wrote a long letter to his father-in-law in which he said he cautioned restraint. Sansom *Battle of Nueces*, pp 2-3; 1859 Blanco County Tax Roll, Nimitz's MR.; Letter, Julius Schlickum, Gulf of Mexico on board British Frigate 'HMS Hope', December 21, 1862 to 'Dear Father'; and Beseler *San Antonio Convention*, pp 247-261.

29. Friedrich Wilhelm Doebbler was born about 1826 near Berlin, Prussia. He fled Prussia about 1851 and settled in Gillespie County. Doebbler wrote several pro-Union newspaper articles for northern newspapers in the early 1860s. He operated a 'Beer Hall' in Fredericksburg that was a meeting place for Unionists. Duff arrested him in May 1862. Doebbler was tried and convicted by the Confederate Military Commission for disloyalty. On July 20, 1862, he along with two other men escaped and made their way to Mexico where they waited out the war.

Frederick Lochte, Sr. was born on October 17, 1809 in Hanover. He and his family arrived in Texas on the *Herkules* in November 1845 and settled in Gillespie County. Lochte was a merchant in Fredericksburg and refused to accept Confederate money as payment for goods. Duff arrested him in May 1862. Lochte was tried and convicted by the Military Commission. He was fined and released. Frederick Lochte, Sr. died on July 19, 1867 in Gillespie County. "Grapetown Cemetery Records" Compiled by Joyce Behr, 'The Gillespiean', Volume 3, No. 2; *Gone—But Not Forgotten: A Survey of Cemeteries in Boerne and Surrounding Areas, Volume I*, Compiled by Boerne Area Historical Preservation Society, (Published by Boerne Area Historical Preservation Society, Boerne, Texas), 1983, p 81; Geue *New Land Beckoned*, pp 88 and 117; Barr *CMC*, Volume LXXI, pp 258-260; Gillespie County Rifles Resolution; Eighth U. S. Census, 1860 Gillespie County Texas Census, pp 10 and 25; Lindig *Der Friedhof Cemetery*, pp 16 and, 53; and Duff's Report, p 786.

30. J. Rudolf Radeleff was born about 1829 in Holstein. The date he arrived in Texas is not known, but it was likely in mid-1850s. Radeleff was a business partner with the Hunter brothers in the mercantile business in Fredericksburg. Duff arrested him in May 1862. He was tried and convicted

by the Confederate Military Commission of Unionist activities. He was expelled from the Confederacy. Radeleff returned to Gillespie County after the war and elected Presiding Justice [county judge] in 1870 and 1872. He died about 1878 in Gillespie County.

August Duecker was born about 1829 in Prussia. He arrived in Texas with his parents on the *Texas* in 1853. Duecker was a member of Braubach's Home Guard Company, Kuechler's Company, and a group of insurgents known as the Luckenbach Bushwhackers. He was a member of the fleeing insurgent group and a participant in the Battle of the Nueces. After the battle, Duecker returned to Gillespie County. He was arrested in mid-1863 and indicted for Unionist activities. It is believed he escaped before he could be tried. Duecker died on April 19, 1894 in Gillespie County.

Ferdinand Ohlenburger was born about 1835 in Nassau. It is believed he was part of the Ohlenburger family that arrived in Texas on the *Adolphine* in 1859. He was a schoolteacher in Gillespie County. Ohlenburger joined Van der Stucken's Company in May 1862 but deserted, fled to Mexico and on to New Orleans where he enrolled in Company G, First Regiment [Union] Texas Cavalry. He was commissioned a 2nd lieutenant in Company H, 1st U. S. Regiment [Colored] Cavalry on September 22, 1865. He left the army in February 1866 and returned to Gillespie County where he was appointed sheriff in June 1866. He died on May 19, 1930 at San Antonio.

Frederick [Fritz] Tegener was born about 1833 in Prussia. He and his brothers, Gustav and Wilhelm arrived in Texas about 1852. They settled in the Comfort area in 1856. In 1858 he was elected the Kerr County Treasurer. Fritz's brother, Wilhelm, was killed about 1859 by 'outlaws'. Fritz Tegener was the military commander of the League's Military Battalion and wounded at the Battle of the Nueces. State Troops hanged his other brother, Gustav, on August 22, 1862. Fritz escaped and finally reached Mexico where he waited out the war. In 1866 and 1870, he was elected to the Texas legislature. Frederick Tegener died in 1901 at Austin in Travis County.

Ernst Schwethelm was born about 1815 in Hanover. He was the father of Henry Schwethelm who was in the battle of the Nueces. Ernst and his family arrived in Texas about 1851 and settled in the Comfort area in 1854. He was the captain of the Comfort Home Guard Company in February 1862. Gottlieb Bauer was born about 1808 in Duchy of Altenburg. He and his family arrived in Texas about 1853 and settled in the Comfort area in 1854. He was the father of Leopold Bauer who was killed at the battle of the Nueces and the father-in-law of Ernest Cramer and Ferdinand Simon, both of who were in the battle of the Nueces. He was on the list of Unionist to be killed in March 1864 by a group of pro-Confederate bushwhackers known as the *Haengerbande*.

Julius Dresel was born about 1833. He arrived in Texas on the *Louis* in 1848 and settled at Sisterdale. Dresel was very likely a member of *Der freie Verein*. Siemering claims he was arrested and tried by the Confederate Military Commission but his name is not included on the Commission's docket. After the war, Dresel moved to California and took over the operation of a Rhineland Vineyards in Sonoma County.

Oskar von Roggenbucke was born about 1811 in Prussia. He and his family arrived in Texas on the *Gessner* in September 1854 and settled in the Comfort area in 1854. He was likely a member of *Der freie Verein*. Roggenbucke was known among his fellow Forty-Eighters as "One of the most zealous pro-Union men in Texas." Two of his stepsons, Moritz and Franz Weiss were in the Battle of the Nueces and killed in October 1862 at the Battle of Rio Grande Crossing.

John Abraham Staehely was born about 1825 in Prussia. He was a Freethinker, but not a Forty-Eighter. Staehely arrived in Texas about 1846. He was a merchant in New Braunfels. Staehely was the first president of the *Demokratischer Verein* [political club] in Comal County. Eighth U. S. Census, 1860 Gillespie County Texas Census, pp 15, 15b and 32; Barr *CMC*, Volume LXXI, pp 260-276; Duff's Report, p 786; Wisseman *Fredericksburg ... First* Fifty Years, pp 53-56; Geue *New Homes*, pp 64 and 113, Gillespie Marriage Records, Book 1, p 19; Braubach's MRs; Letter, Jacob Kuechler, undated [c1887] to Honorable James Newcomb, San Antonio, Texas, copy in possession of author; Kuechler's MR; Oral Interviews with Peter Kleck, grandson of Sylvester Kleck, May 1, 1997 and October 4, 1997 and Letters Peter Kleck to Paul Camfield, Gillespie County Historical Society, May 8, 1997 and June 3, 1997 with copies of rosters [list] of 'Luckenbach Buschwhackers' and 'Luckenbach Rangers'; Lindig *Der Friedhof Cemetery*, p 16; Geue *New Land Beckoned*, p 89; Ferdinand Ohlenburger's Confederate Service Records, Record Group 109, National Archives, Washington, D. C.; Ferdinand Ohlenburger's Union Service Records, National Archive, Washington, D. C.; Eighth U. S. Census, 1860 Kerr County Texas Census, pp 70 and 73; Ransleben *100 Years*, p 24; Bennett *Kerr County*, p 75; Sansom *Battle of Nueces*, p 2; Additional BIO data on Fritz Tegener provided by Roland Hall of Fort Worth, Texas, a great-great grandson of Tegener's and Siemering's Germans During Civil War, May 29, 1923.

31. Ferdinand von Herff was born on November 29, 1820 in Darmstadt. He arrived in Texas ahead of most of the Forty, arriving on the *Galveston*. Dr. Herff was a very prominent surgeon performing many delicate operations on San Antonio citizens, frontiersmen and Indians. He assisted several Unionists jailed in San Antonio to escape. Dr. Herff died in San Antonio on May 18, 1912.

Philip Zoeller was born about 1819 in Hesse-Darmstadt. He arrived as part of the Forty on the *St Pauli* in 1847. After the breakup of the Forty settlement Zoeller settled near Sisterdale. He was likely a member of *Der freie Verein*. The Adolphus Zoeller who was a member of the fleeing insurgent group and later a captain in the Union Army, was likely a nephew. Philip Zoeller died in 1900 in Kendall County.

32. Comfort local folk stories.

33. Letter, A. O. Cooley, Fredericksburg, Texas, December 19, 1861 to C. R. Johns, Comptroller, Governor Lubbock's Files, TSA, Austin, Texas.

34. Smith *Frontier Defense*, p 42.

35. Another major problem was that most of the Confederate units in Texas had enlisted for only a year. That year was about to expire. There was a major effort to re-enlist these units, this time for the duration of the war. Many of the men who enrolled in Texas Confederate units did so in order to serve close to home or at least in the state.

36. *The Laws of Texas 1822 – 1897,* Compiled and Arranged by H. P. N. Gammel, (Gammel Book Company, Austin, Texas), 1898, Volume V, pp 11-21 and Report, Headquarters 31st Brigade, Texas State Troops, New Braunfels, Comal County, March 8, 1862, A.G.C., TSA, Austin, Texas.

37. There were two Jacob Schmidts in Gillespie County. It is believed this Jacob Schmidt was born on December 11, 1824 in Hessia. He and his family arrived in Texas about 1857. He was the 2nd sergeant in Nimitz's Gillespie Rifles. Jacob Schmidt died on September 22, 1905 in Gillespie County.

Valentine Wilhelm Hohmann was born March 10, 1805 in Saxony. He arrived in Texas on the *Neptune* in 1845. He was an early member of the Union Loyal League. Either his son or nephew, Valentine Hohmann, was the lieutenant of the Gillespie Company of the Union Loyal League. The young Hohmann was killed in October 1862 at the Battle of Rio Grande Crossing. Valentine Wilhelm Hohmann died June 17, 1891 in Gillespie County.

Jacob Dearing was born about 1818 in Bavaria. He arrived in Texas about 1852. The Union Loyal League likely targeted him as an enemy in the summer of 1862. Dearing commanded a company in the Third Frontier District in 1864. His date of death is not known.

William Feller was born February 5, 1826 in Nassau. He arrived in Texas with his parents on the *Arminius* in 1845. He was a farmer in Gillespie County. Feller was an early member of the Union Loyal League. The Fellers raised August Hoffmann a member of the fleeing insurgent group. In

January 1864, Feller was a member of the Unionist company of Louis Schuetze. Pro-Confederate bushwhackers hanged him on March 9, 1864.

Christian Kothe was born about 1822 in Brunswick. He arrived in Texas about 1851. Kothe didn't arrive in Gillespie County until about 1860. Data and location of death not known.

Charles Weyrick was born on August 3, 1819 in Saxony. He arrived in Texas on the *John Dethardt* in 1845. He was a member of Nimitz's Gillespie County Rifles in 1861. If he was a Unionist, he was a Gray. Weyrick was a member of Krauskopf's May 1864 Company of the Third Frontier District. He died on November 19, 1902 in Gillespie County.

Jacob Luckenbach was born on October 6, 1817 in Marienberg, Duchy of Nassau. He first arrived in Texas in mid-1836 but returned home. Jacob, his family and two brothers, returned to Texas on the *Johann Dethardt* in 1845. They were among the first settlers in Gillespie County. Jacob's brother, August, was a member of the fleeing insurgent group but escaped the battle. He was hanged just north of San Antonio in late August 1862. Jacob Luckenbach died in 1911 in Kendall County.

Lindig *Der Friedhof Cemetery,* pp 34, 77 and 93; *Kirchen-Buch; Church Records Book of the Verein-Kirche 1849 – 1870,* translated by Ella A. Gold, (Gillespie County Historical Society, Inc., Fredericksburg, Texas), 1986, pp 25, 33,38, 40, 51, 63, 68, 77, 101, 104, 129, 154, 163 and 169: Seventh U. S. Census, 1850 Gillespie County, Texas Census, p 311; Eighth U. S. Census, 1860 Gillespie County, Texas Census, pp 3, 4b, 16, 19, 19B, 29 and 31; Quarterly Returns, 31st Brigade, July 1, 1862, A.G.C., TSA, Austin, Texas; Sansom Battle *of Nueces*, p 13; Cramer's Letter; Ninth U. S. Census, 1870 Gillespie County, Texas Census, pp 346-347; Gillespie County Deed Records, Volume H, p 187; Muster Roll. Captain Jacob Dearing's Company, June 1, 1864, A.G.C., TSA,

Austin, Texas; Eighth U. S. Census, 1860 Gillespie County, Texas Census, p 30; Geue *New Land Beckoned*, pp 92,117 and 154; Braubach's MR; Muster Roll Captain Louis Schuetze's Company, January 1864, A.G.C., TSA, Austin, Texas; Nimitz's MR; Krauskopf's MR; Oral Interview with Dorothy Basse, Fredericksburg, Texas, May 1997 and Gillespie Historical Society God's *Hills* Volume II, pp 84-86.

38. A. Jacob Kecheville was born about 1821 in Virginia. He arrived in Blanco County in 1853 and settled on the Blanco River about 7 miles east of Blanco City. Kercheville's son was the first white child born in Blanco County. He was a member of Blackwell's Blanco County Minute Man Home Guard Company in 1861. He was Blanco's County third sheriff but was dismissed during Reconstruction. His date and location of death is not known.

Julius Schlickum was born about 1825 in Hessen Darmstadt. He was a Freethinker and a Forty-Eighter. He and his 16-year-old brother-in-law, Wilhelm Klier arrived in Texas on the *Franziska* in 1849. Schlickum was a major leader of the 1854 San Antonio Convention. Schlickum was opposed to organizing an insurgent armed force. Duff arrested him in June 1862. He was tried and convicted by the Confederate Military Commission for disloyalty. Schlickum and two other Unionist escaped on July 19, 1862 and made their way to Mexico. He died in late 1865 or early 1866 in Mexico.

Ottomar Labhardt was born about 1835 in Thurgau. The date he arrived in Texas not known. Labhardt enlisted in Kampmann's Company B, 3rd Regiment Texas Infantry on August 1, 1862. He deserted in late 1863 or early 1864 and fled to Union lines at Brownsville where he enlisted in Company I, First Regiment [Union] Texas Cavalry. He was mustered out of Union service on October 31, 1865 at San Antonio.

Michael Lindner was born about 1820 in Bavaria. It is possible Lindner was a Forty-Eighter. He arrived in Texas in the early 1850s and in 1853 settled in the Comfort area. Michael Lindner died in 1878 in Kendall County.

Thomas Saner was born on September 4, 1826 in Davidson County, North Carolina. He arrived in Texas in 1849 and Kerr County by 1854. Saner was a Kerr County Commissioner in 1856-59 and 1862-1866. He owned one slave in 1860. Saner was a member of D. H. Farr's Company in the Third Frontier District in 1864. Thomas Saner died on December 4, 1893 in Kerr County.

Braden Mitchell was born on October 25, 1834 in Prince William County, Virginia. Mitchell arrived in Texas and in Bandera County in 1856. He settled near Ten Mile Crossing, now named Mitchell's Crossing in his honor. Mitchell was captain of the Bandera Company in the Third Frontier District in 1864. He served Bandera County as deputy sheriff and in 1888 elected to that office. Braden Mitchell died on April 20, 1890 in Bandera County.

Ernest Cramer was born about 1836 in Bavaria. He was a Forty-Eighter. Cramer arrived in Texas on the *Iris* in 1860 and settled in the Comfort area in 1861. He was a member of Harbour's Home Guard Company in February 1861. Cramer was the commander of the Union Loyal League's Kendall Company. He was in the insurgent Unionist group and survived the Battle of the Nueces. He successfully escaped to Mexico where he waited out the war. After the war, he was customs collection at Eagle Pass. Cramer and his family moved to Idaho where he died about 1915.

Heritage of Blanco County, Texas by Blanco County News, (Curtis Media, Inc., Dallas, Texas), p 613; Blackwell's MR; Tenth U. S. Census, 1880 Blanco County Texas Census, p 322; Eighth U. S. Census, 1860 Kerr

County Texas Census, pp 68 - 68B, 72b; Gillespie Historical Society God's *Hills*, Volume I, p 170; New *Homes in a New Land: German Immigration to Texas 1847 – 1861* by Ethel Hander, (Genealogical Publishing Company, Inc., Baltimore, Maryland), 1982, pp 61, 91, 126 and 133; Ransleben *100 Years*, pp 17, 23-24, 26-27, 56, 59,102, 147 and 208; Schlickum's Letter; Ottomar Labhardt's Confederate Service Records, Record Group 109, National Archives, Washington, D. C.; Ottomar Labhardt's Union Service Records, National Archives, Washington, D. C.; *Gone – But Not Forgotten: A Survey of Cemeteries in Boerne and Surrounding Areas*, Volume II, by Boerne Area Historical Preservation Society (Boerne Area Historical Preservation Society, Boerne, Texas), 1985. p 30; *Confederate Veterans of Kerr County, Texas* by Shirley Smith, (Adam R. Johnson Chapter United Daughters of the Confederacy, Kerrville, Texas), 1991, p 42; Kerr *County Texas 1856 – 1956* by Bob Bennett (Naylor Company, San Antonio, Texas), 1956, pp 5, 7, 11, 25, 47, 58, 110, 115 and 297; *Bandera County Cemetery Records* Compiled by Mrs. Howard Denson, Mrs. Billy Burnes and Mrs. Howard Graves, (Bandera County Historical Survey Committee, Bandera, Texas), n.d., p 13; Hunter's *100 Years in Bandera*, p 8; *Veterans from Bandera County of All Wars* by Mrs. Howard Graves, (Bandera Printing Company, Bandera, Texas), 1978, p 28; Eighth U. S. Census, 1860 Bandera County Texas Census, p 434a; Quarterly Returns, 31st Brigade, July 1, 1862, A.G.C., TSA, Austin, Texas; Cramer's Letter; Harbour's MR; Ninth U. S. Census, 1870 Maverick County Texas Census, p 523 and Letter, Walter D. Kamphoefner, Department of History, Texas A and M University, College Station, Texas, February 26, 1995, to author.

39. Hubert Weynand was born on December 4, 1822 in Bigenbach, Prussia. He arrived in Texas on the *Bangor* in 1846 and settled in Medina County. Weynand joined the army during the Mexican War and returned to D'Hanis by 1850. He later served in Frank Haby, Jr's militia company and in George Robbins' Company in the Third Frontier District. Hubert Weynand died on July 21, 1906 in Medina County.

George Mayers was born about 1830 in Bavaria. He and his wife arrived in Texas in the late 1850s and settled in Medina County. His date and location of death not known.

Blasins Kieffer was born about 1835 in Alsace. He arrived in Texas about 1858. Surprisingly he is shown as a member of the Castroville Knights of the Golden Circle in 1861. Duff arrested him for Unionist activities. He was tried but charges were dismissed. He was elected a Medina County Commissioner in August 1862. He may have fled Medina County later in the war. His date and location of death not known.

Jacque [Jacob] Zacharie Haby was born in 1823 in Alsace. He arrived in Medina County in 1851 and helped establish the Haby Settlement, just north of Castroville. Jacob Haby died in 1899 in Medina County.

Henry Joseph Richarz was born on September 8, 1823 in Prussia. He was a Forty-Eighter. He and his wife arrived in New Orleans in 1849. They came on to Texas and settled in Medina. Duff arrested him in May 1862, but he was later released. He was elected Medina County Chief Justice in August 1862. Richarz was a Ranger captain after the war. He died on May 21, 1910 in Medina County.

Castro Colonies Heritage Association Medina *County History*, pp 13, 302, 463, 553-554; Muster Roll, Captain Frank Haby's Company, October 12, 1863, TSA, Austin, Texas; Muster Roll, Captain George Robbins' Company, March 31, 1864, TSA, Austin, Texas; Muster Roll, Captain George Robbins' Company, January 18, 1865, TSA, Austin, Texas; Eighth U. S. Census, 1860 Medina County Texas Census pp 302, 46 and 6; Muster Roll, Captain George Mayer's Company, August 31, 1862, TSA, Austin; By Laws Castroville Castle "K.G.C" [Knights of the Golden Circle], Approved August 16, 1861", Castroville Public Library; Report, Headquarters, 31ˢᵗ Brigade, Texas State Troops to Texas Adjutant General, June 28, 1862, A.G.C., TSA, Austin, Texas; Gammel Laws *of Texas*, Volume V, p 12; *Medina County, Texas Cemeteries, Volume II* by Carol Henley, (Published by Frances T. Ingmire, St. Louis, Missouri), 1985, p 3; Oldtimers *of Southwest Texas by* Florence Fenley, (The Hornsby Press, Uvalde, Texas), 1957, pp 290; *1850, 1860, and 1860 Federal Census's, Medina County, Texas* Compiled by Jim and Doris Menke (Medina County History Series, Hondo, Texas) 1998, p 25 and Letter, Captain Montel, Camp Ives, March 16, 1862 to Colonel Dashiell, Texas Adjutant General, Austin, Texas, A.G.C., TSA, Austin, Texas.

40. Gammel *Laws of Texas*, Volume V, pp 8-11.

41. James Madison Patton was born about 1811 in Alabama. He arrived in Texas on April 22, 1836, the day after San Jacinto. He first settled in Caldwell County. By 1860, he was living at Curry's Creek. On January 1, 1862, he enrolled in Company A, 15ᵗʰ Regiment Texas Cavalry and elected first lieutenant. He served until May 30, 1862 when he was discharged due to his age. His date and location of death is not known. Letter, James M. Patton, Austin, January 7, 1862 to Governor Lubbock, Governor Lubbock's Files, TSA, Austin, Texas; Webb *Handbook of Texas*, Volume II, p 7; Moursund *Blanco County Families*, pp 335-336; Eighth U. S. Census, Schedule 4, Production of Agriculture, 1860 Blanco County Texas and James Patton's Confederate Service Records, Record Group 109, National Archives, Washington, D. C.

42. These letters were from: Neill Robison; John Lawhon; J. C. Nowlin; J. S. Abbot; and another brother-in-law, Benjamin F. Patton.

Thomas McCall was born on June 9, 1832 in Ireland. Date arrived in Texas not known. Elected Medina County sheriff in August 1860. McCall re-enrolled in Lawhon's Company, the Frontier Regiment on December 29, 1862. It appears he was promoted to captain before the end of the war. Thomas McCall died on October 26, 1902 in Bexar County.

Amos Valentine Gates was born on August 29, 1825 at Washington on the Brazos, Texas. His father and grandfather were part of Austin's 'Old 300'. He arrived in the Blanco City area in 1856. He was elected Blanco Chief Justice in August 1860. In February 1862, he resigned and enrolled in Montel's Company, the Frontier Regiment. He later enrolled in a Confederate unit and rose to the rank of captain. Gates and his men captured the Union gunboat, *Harriett Lane* at Galveston. In 1866, he was again elected Blanco County Chief Justice but removed by the Reconstruction Government. Amos Valentine Gates died on November 5, 1912 in Blanco County.

Benjamin F. Patton was born about 1837 in Bastrop County, Texas. He served in several ranger companies with John W. Sansom before the war. Patton re-enlisted in Lawhon's Company, the Frontier Regiment on December 29, 1862 and elected second corporal. He deserted in March 1864 and John W. Sansom recruited him into the Union Army. Patton enlisted in Company I, First Regiment [Union] Texas Cavalry on April 1, 1864. He transferred to Company C on July 15, 1864. Patton deserted the Union Army on August 7, 1864. He was in the group that recovered the Unionists' bodies in August 1865. His date and location of death not known.

Letters, Neil Robison, Curries Creek, December 10, 1862 [1]; John Lawhon, Blanco County, January 7, 1862; J. C. Nowlin, Blanco County, n.d.; and B. F. Patton, Blanco County, n.d., Governor Lubbock's Files, TSA, Austin, Texas; Montel's MR; Cemetery *Records City Cemetery No. 4, Confederate Section, Bexar County, San Antonio, Texas* Recompiled by Mydena Brown and Helen Ione Kilborn (Published by Helen Ione Kilborn and Mydena Talley Burleson-Brown, San Antonio, Texas), p 18; Eighth U. S. Census, 1860 Medina County Texas Census, 6; Medina County Commissioner Court Minutes, Book 2, p 230; Moursund Blanco *County Heritage*, pp 484 - 485; Montel's MR; Ninth U. S. Census, 1870 Kendall County Texas Census, p 124 and Benjamin F. Patton's Union Service Records, National Archives, Washington, D. C.

43. Philip Braubach was born on July 28, 1829 in Duchy of Nassau. He was a Freethinker and a Forty-Eighter. He arrived in Texas on the *Neptune* from Wiesbaden in 1850 and settled in Gillespie County. He was elected Gillespie Sheriff in August 1860. He was one of those censored by the Gillespie County Rifles in February 1862. Duff arrested him in May 1862. The Confederate Military Commission tried and found him guilty of Unionist activity. He escaped on July 19, 1862, made his way to Mexico where he led a guerrilla unit until May, 1864, when he joined the First Regiment [Union] Texas Cavalry where he was elected captain in Company H. Philip Braubach died on June 30, 1888.

Moritz Weiss was born about 1836 in Prussia. His father was likely a Freethinker and a Forty-Eighter. The date the family arrived in Texas not known. His father died shortly after arriving. Moritz's mother married Oskar von Roggenbucke another radical/militant Unionist. Date and location of marriage not known. Moritz's name appears on the list of the Luckenbach Bushwhackers. He was a member of the fleeing Unionist group. Moritz and his brother, Franz, survived the battle but were killed on October 18, 1862, at the Battle of Rio Grande Crossing.

Hugo Degener was born about 1842 in Brunswick. He was a son of Eduard Degener, a Freethinker and Forty-Eighter. Eduard Degener was a major leader of the German 1848 Revolution and of the 1854 San Antonio Convention. He was also the head of the Union Loyal League. Hugo Degener was the lieutenant of the League's Kendall Company. He was in the fleeing insurgent group and killed at the battle site.

The two non-commission officers who were in the fleeing insurgent group included August Duecker and Louis Boerner. Gottlieb Sauer joined the Union Army. The 20 included: Jacob Kuechler; Moritz Weiss; Hugo Degener; Ernst Beseler; Hilmar Degener; Pablo Diaz; Joseph Elstner; August Graff; John George Kallenberg; Wilhelm Klier; Louis Schierholz; Christian Schaefer; Heinrich Steves; Wilhelm Tellgmann; Ludwig Usener; Gottlieb Vetterlein; Friedrich Weber; Franz Weiss; August Duecker and Louis Boerner. These 11 included: Philip Braubach; Gottlieb Sauer; Casper Fritz; John Klien; and his twin brother, John Adam Klien; Jacob Remeck; Carl Brinkmann; Rudolph Hartbeck; Anton Heinen; Joseph Leyendecker and Peter Tatsch. These 3 were: Adam Blucher who later was a member of Davis' Company; Josiah Reagan who was later a member of the Comfort Home Guard and A. H. Nixon. Reagan and Nixon were corporals. *The Saga of Captain Philip Braubach: As Told to Her by Sally _____*, by Anne Dornwell, An unpublished manuscript c 1986, copy provided by Gregory Krauter of Comfort, Texas; Eighth U. S. Census, 1860 Gillespie County Texas Census, p 17; Duff's Report, pp 785-787; Barr *CMC*, LXXI, (October 1967), pp 260-261; Schlickum's Letter; *A Texas Pioneer: Early Staging and*

A Perfect Reign of Terror

Overland Freighting Days on the Frontiers of Texas and Mexico by August Santleben, (Castro Colonies Heritage Association, Castroville, Texas), 1994, p 35; Philip Braubach's Union Service Records, National Archives, Washington, D. C.; Philip Braubach's Union Pension Records, National Archives, Washington, D. C.; Kuechler's MR; Ransleben *100 Years*, pp 24, 94-95, and 114; Eighth U. S. Census, 1860 Kerr County Texas Census, p 70; Cramer's Letter; Luckenbach Bushwhackers; Sansom *Battle of Nueces*, p 3; Webb *Handbook of Texas*, Volume I, p 282; Geue *New Homes*, p 62; Cramer's Letter and Luckenbach Bushwhackers.

44. David H. Farr was born about 1819 in Tennessee. The date he arrived in Texas is not known, but he is shown on a Texas census for the first time on the 1860 Burnet Census. Farr enrolled in Davis' Company, the Frontier Regiment, on March 4, 1862. He organized a Kerr County militia company in 1864 for the Third Frontier District. His date and location of death is not known. Eighth U. S. Census, 1860 Burnet County Texas Census, p 161; Davis' MR and Muster Roll Captain David H. Farr's Company February 13, 1864, A.G.C., TSA, Austin, Texas.

45. Letter, D. H. Farr, Kerrsville, Kerr County, February 13, 1862 to Governor Lubbock, Governor Lubbock's File, TSA, Austin, Texas.

46. Letter, Frank V. D. Stucken, Fredericksburg, February 13, 1862 to Governor Lubbock, Governor Lubbock's File, TSA, Austin, Texas.

47. These 38 included: C. C. Quinlan; Nathan Jackson; Isaac Lamb; Hiram Lamb; John Lowrance; James Coleman; A. P. Paul; Marcellon Dunham; F. P. Click; E. Whitley; P. E. Hampton; J. A. Hudson; A. J. Rogers; A. C. Farmann; J. H. Regges; R. F. Camford; M. A. Lowrance; Scott Kell; A. L. Williamson; J. W. Rogers; H. M. Burney; R. G. Farr; P. O. Lowrance; J. P. Stroops; D. B. Lowrance; J. D. Buchanan; William Long; E. W. Brown; J. D. Brown; Joseph Follord; Semon Blevins; E. A. Steel; R. W. Freemon; Freemon Myers; Daniel Arnold; P. M. Stanford; Aaron W. Jackson; and George Holliman. Gillespie County citizens also sent a petition to the governor. This petition has mysteriously disappeared from the Texas State Archives. Elliott in his article 'Union Sentiment in Texas 1861 – 1862' claims there were 74 signatures on the petition. Charles Nimitz says there were 28 signatures. The men who signed the Gillespie petition are likely the same 33 who signed the Gillespie Rifles Resolution and took the oath of allegiance to the Confederacy on February 23, 1862. These included: Oscar Basse; Henry Beckmann; Ewin Cameron; A.O. Cooley; N.M. Dennis; John E. Doss; A.G. Foster; Frederick Fresenius; Charles Human; James Hunter; John Hunter; Wilhelm Keidel; Wilhelm Kooh; August Kott; Engelbert Krauskopf; Adolph Lungkwitz; Albert Meinhardt; Charles Nimitz; Ottocar

Muller; Julius Ransleben; William Schneer; Thomas Smith; E. Staus; Emil and Frank Van der Stucken; Jacob Tatsch; William Wahrmund; Henry Wathersdorff; John Weber; Adolph Weiss; Louis Weiss; Frederick Wrede; and Levi Lamoni Wight. Petition, Citizens of Kerr County, To Governor Lubbock, dated February 14, 1862, Governor Lubbock's File, TSA, Austin; Barr *CMC*, Volume LXXI, (October 1967), pp 262-246; Kuechler's MR and Siemering's Germans During Civil War, May 29, 1923.

Joseph Walker was born about 1827 in Ohio. The date he arrived in Texas is not known. Kuechler's MR; Siemering's *Germans During Civil War*, May 29, 1923; and Eighth U. S. Census, 1860 Travis County Texas Census, p 291.

48. Cramer's Letter.

49. Kuechler's Muster Roll does not show Siemering as first sergeant. He is listed as a private. Siemering's *Germans During Civil War*, May 29, 1923 and Kuechler's MR.

50. Barr *CMC*, Volume LXXI, (October 1967), pp 262-264. It is of interest to note Kuechler's Muster Roll of February 12, 1862 was signed at Comfort, Texas.

51. Frederick Fresenius was born about 1826 in Prussia. Believe he arrived in Texas on the *Herschel* in 1849. He was the first lieutenant in Nimitz's Gillespie Rifles. He signed the Gillespie Rifles Resolution and testified against the Unionist at the Confederate Military Commission in San Antonio. His date and location of death is not known. Barr *CMC*, Volume LXXI, (October 1967), p 264; Eighth U. S. Census, 1860 Gillespie County Texas Census, p 17; Geue *New Land Beckoned*, p 70; Gillespie Historical Society *God's Hills*, Volume I, p 7 and Gold *Church Record Book of the Verein-Kirche*, pp 36 and 106.

52. At the time, there was only one hotel in Fredericksburg that being the Nimitz's Hotel owned by Charles H. Nimitz, the captain of the Gillespie County Rifles. Siemering's *Germans During Civil War*, May 29, 1923.

53. "The Diary of D. P. Hopkins", *San Antonio Express,* January 13, 1918; Eighth U. S. Census, 1860 Kerr County Texas Census, p 72b; Harbour's MR; Davis' MR; Williams *Border Ruffians,* p 247, and Sansom *Battle of Nueces*, p 9.

54. Hopkins' diary.

55. Desmond Pulaski Hopkins was born in 1838 in Tennessee. He arrived in Texas in 1838 with his family. He was elected the 1st sergeant in Davis'

Company. Hopkins went with Davis and most of the men from Hays County and re-enlisted in Duff's 14th Battalion where he served as an officer. After the war, Hopkins served a deputy U. S. marshal, district clerk of Hays County, Hays County Commissioner, and as the San Marcos Tax Assessor and Collector. His date and location of death is not known. Hopkins's Diary; Webb *Handbook of Texas*, Volume I, p 834 and Eighth U. S. Census, 1860 Hays County Texas Census, pp 9b and 16 and Barr *CMC*, Volume LXXI, (October 1967), p 266.

56.   Siemering's *Germans During Civil War*, May 29, 1923.

57.   Ibid.

58.   Henry T. Davis was born June 13, 1834 in North Carolina. He arrived in Texas in the mid-1850 and settled in Hays County. Davis married Susan T. Brownrigg on February 3, 1859 in Bexar County. He is shown as a merchant on the 1860 Hays Census. In early 1863, he and about half of his company enrolled in Duff's 14th Battalion Texas Cavalry. The company became Company F on February 6, 1863. Davis served the remained of the war in Duff's command and received a parole on July 31, 1865. The parole shows his rank as lieutenant colonel. After the war, Davis, moved to Orange County where he died on October 4, 1888.

James M. Hunter was born in 1829 in Tennessee. He was a brother of John M. Hunter and was one of three brothers who settled in Gillespie County. Hunter was elected captain of the company after Davis left. In early 1864 he was appointed a state major in command of the Third Frontier District. Hunter was later a Ranger captain as well as the county judge for Mason and Edwards counties. He died on August 31, 1907, in Mason County.

James L. Williamson was born in Alabama about 1824. He and his family arrived in Texas about 1851. His date and location of death is not known.

General Order Number 4, Headquarters Texas Frontier Regiment Camp San Saba, April 23, 1862, A.G.C., TSA, Austin, Texas and Muster Rolls, Captain Henry T. Davis' Company, dated March 4, 1862 and February 7, 1863, TSA, Austin, Texas; Eighth U. S. Census, 1860, Hays County Texas Census, p 10; Compiled Service Records, 33rd Regiment Texas Cavalry, Record Group 109, National Archives, Washington, D. C.; Henry T. Davis's Confederate Service Records, Record Group 109, National Archives, Washington, D. C.; Bexar County, Texas Marriage Records, Volume D2; *Orange County, Texas Cemetery Inscriptions* By Lorine Brinley, Published by Frances T. Ingmire, St. Louis, Missouri), 1985, p 78; Davis' MR; *Mason County's First Settlers 1840 – 1900* by Jerry Ponder (Published by Ponder Books, Mason Texas), 1997, p 5; *Mason County Historical Book* by Mason

County Historical Commission, (Dogwood Printing, Ozark, Missouri), 1986, p 111; *Mason County Historical Book Supplement II* by Mason County Historical Commission, (Dogwood Printing, Ozark, Missouri), 1994, p 2; Gold *Church Record Book of the Verein-Kirche*, pp 61, 74, 110 and 162; Muster Rolls Captain James M. Hunter's Company, December 24, 1862, February 28, 1863, April 1, 1863, October 31, 1863, A.G.C., TSA, Austin, Texas; Records Third Frontier District, A.G.C., TSA, Austin, Texas; and Seventh U. S. Census, 1850 Hays County Texas Census, p 4.

59. Barr *CMC*, Volume LXXI, (October 1967), p 254.

60. Engelbert Krauskopf was born on August 21, 1820 in Prussia. He arrived in Texas on the *Andacia* in 1846. Krauskopf was a well-known gunsmith. He was the 2nd lieutenant in Nimitz's Gillespie Rifles. In May 1864, he took command of a company in the Third Frontier District. Engelbert Krauskopf died on July 11, 1881 in Gillespie County. Barr *CMC*, Volume LXXI (October 1967), pp 261-269; Gillespie Historical Society *God's Hills*, Volume I, pp 107-108; Geue *New Land Beckoned*, p 112; Nimitz's MR; and Muster Roll Captain Engelbert Krauskopf's Company, May 1864, A.G.C., TSA.

61. Barr *CMC*, Volume LXXI (October 1967), pp 272-273.

62. Barr *CMC*, Volume LXXI (October 1967), pp 274-275.

63. Ibid.

64. Cramer's Letter.

65. Sansom *Battle of Nueces*, pp 2-3; Letter from Fritz Tegener to Herr August Duecker, Gillespie County, Texas, August 23, 1875. Copy provided by Gregory Krauter of Comfort, Texas; Siemering's Germans During Civil War, May 29, 1923 and Cramer's Letter.

66. Schiller *William Tell*, pp xi-xxxiii, 34-35, 43-62, and 141-153. For another account by August Siemering of a secret league in the Texas Hill County see *Ein Verstehl Tes Leben*, by August Siemering, (Published by August Siemering, San Antonio, Texas) 1876 and *The Hermit of the Cavern*, by May E. Francis, Naylor Printing Company, San Antonio, Texas), 1932, p 106; Siemering's Germans During Civil War, May 29, 1923; Ponder *Fort Mason, Texas,*126; Letter, Lieutenant Colonel Nat Benton, Commanding Expedition, Provost Marshal, *San Antonio Weekly Herald* August 30, 1862.

67. Valentine Hohmann was born about 1829 in Saxony. He was either a son or a nephew of Wilhelm Valentine Hohmann who commanded Company C,

2nd Regiment, 31st Brigade District. The date Hohmann arrived in Texas is not known. He was in the fleeing insurgent group and escaped the battle. Hohmann was killed on October 18, 1862 at the Battle of Rio Grande Crossing. His wife gave birth to a son, William Valentine Hohmann, two months after his death.

Henry Hartmann was born about 1821 in Ohio. He and his son, Amon, arrived in Texas in 1860. On October 26, 1861, he married the widow Mary Mathilda Lacy McDaniel in Gillespie County. On April 17, 1862, Hartmann deeded to his 13-year-old son his herd of sheep and goats [over 1,300 head], a mule and wagon, and all the improvements on his Rancho, situated on North Creek, a tributary of the Pedernales River. On September 30, 1862, the Gillespie Probate Court appointed Franz Petermann the guardian of Amon Hartmann because his "father was gone to parts unknown and is declared an enemy of the Southern Confederacy." Henry and Amon Hartmann fled Texas and reached New Orleans where on October 29, 1862 they joined Company A, First Regiment [Union] Texas Cavalry. Henry Hartmann was killed in February 1864 while on a "secret mission" back to Gillespie County. On April 21, 1864, his widow married Samuel Gibson, a pro-Confederate Bushwhacker. Amon Hartmann served in the Union Army until the end of the war. On October 21, 1866, he married Laura Dunn in Kerr County. They moved to California in 1873 and to Oregon in 1875. Amon Hartmann died on March 25, 1918 in Gilliam County, Oregon.

Philip G. Temple was born in 1835 in Ohio. He was a veteran of the Mexican War. The date he arrived in Texas is not known. Temple was a member of the Spring Creek Home Guard Company on April 22, 1861. He fled to Mexico and on to New Orleans were on October 27, 1862 he enrolled in Company A, First Regiment [Union] Texas Cavalry on November 6, 1862 he was appointed 1st lieutenant and on January 21, 1864 captain. He was wounded on June 25, 1864 at the Battle of Rancho Las Rucias. Temple returned to Gillespie County after the war. He became ill and returned to Ohio where he died on February 28, 1903

Cramer's Letter; Letter, Fritz Tegener, Austin, Texas, August 23, 1875 to Herr August Duecker, Gillespie County;Records, 31st Brigade District, A.G.C., TSA, Austin, Texas; Sansom *Battle of Nueces*, 3 and 13 Gold *Church Record Book of the Verein-Kirche*, p 68; Gillespie County Marriage Records, Book 1, p 358; Gillespie County Deed Records, Deed, dated April 17, 1862 from Henry Hartmann to Amon Hartman, Volume H, p 171; Gillespie County Probate Records, Volume C, p 202; Union Service Records of Henry Hartmann, National Archives, Washington D.C.; Union Service Records of Amon Hartmann, National Archives, Washington, D.C.; Gillespie County Marriage Records, Book 1, p 95; Union Pension Records of Amon Hartmann, National Archives, Washington, D. C.; Petition,

Kimbal [Kimble] County and Precinct VI, Gillespie County, April 22, 1862 to Governor F. R. Lubbock, Governor Lubbock Papers, TSA, Austin, Texas; Philip G. Temple's Union Service Records, National Archives, Washington, D. C. and Philip Temple's Union Pension Records, National Archives, Washington, D. C.

68. Ibid.

69. Sansom *Battle of Nueces*, p 3.

70. Ponder *Fort Mason, Texas* and Banta *Twenty-Seven Years*, p 185.

71. Schlickum's Letter.

72. *A Free Man*, The Journal, German-Texas Heritage Society, Austin, Texas, Volume XXXV, Number 3, Fall, 2013, pp 154 - 55.

CHAPTER 4 – The Frontier Regiment Deploys

As the insurgents were organizing their own military units, the state deployed the Frontier Regiment onto the frontier. Four companies deployed directly into the insurgent's area of influence or between them and Union officials in Mexico. The two companies taking up positions in the Hill County were Henry T. Davis' Company F and Charles de Montel's Company G. Two other companies deployed near the Mexican border and played a minor role against the insurgents; Thomas Rabb's Company A and John Dix's Company H.[1]

After being mustered-in during a blue norther on March 4, 1862, at old Fort Martin Scott just outside of Fredericksburg, Davis' Company took up field positions on March 14th. On that day, the company set up a camp near the mouth of Johnson Creek where it empties into the Guadalupe River. On March 16th, the company was divided into two detachments. Captain Davis commanded one detachment, which remained on Johnson Creek. Lieutenant Harbour, from Kerr County, commanded the second. Harbour's detachment was composed of Kerr County men and most of the Gillespie County men. It moved to its new location, Camp Llano, near the mouth of Rock Creek on the Llano River in Mason County. Camp Llano was officially established on March 29, 1862. The other half of Company F moved to Camp Davis, which was on White Oak Creek about four miles from its junction with the Pedernales River. Camp Davis was officially established on March 31, 1862.

It was April 6th before Davis' detachment moved to Camp Davis.[2]

Charles de Montel's Company G moved into its assigned area. On March 6, 1862, Captain Montel reported from a "Camp Near Guadalupe" in Blanco County that he had obtained some weapons and other supplies from the Chief Justice of Uvalde County and was getting more from Uvalde and Bandera counties. He informed his superiors that some of the citizens of Canyon de Uvalde had weapons and supplies belonging to the state and requested advice on how to appropriate them. Montel stated he was having difficulty in obtaining supplies, mainly because of the reluctance to sell to the state on credit. He reported several instances where he had to advance some of his private funds to avoid paying "exorbitant" prices.[3]

By March 14th, Montel assembled the detachments from Blanco County and from the 'upper Medina.' On March 15^{th,} his company "took up the line of march" and on March 16th reached Camp Ives, in southern Kerr County. Stopping here for a short time, he ordered Second Lieutenant Amos V. Gates with thirty-five troopers ahead to occupy Seco Arroyo near Seco Pass. Captain Montel informed headquarters that he planned to establish camps on the Frio and Nueces Rivers. Colonel Norris, the commander of the Frontier Regiment, established Montel's camps farther east. Montel established Camp Rio Frio about thirty miles north of Uvalde. The two main or official camps for Company G were Camp Montel, on the head of Seco Creek/Arroyo in Bandera and Camp Verde, [sometimes later referred to as

Camp McCord], about two miles below the old Federal post of Camp Verde, in Kerr County. These two camps were officially established on April 3, 1862.[4]

Captain Dix's Company H experienced difficulties getting to its assigned area because Dix had problems getting enlisted men who already had arms and horses. On April 3rd, the company assembled at San Saba Peak in southern Mills County. Captain Dix finally acquired the necessary arms and mounts. One source was the Chief Justice of Llano County who supplied him with three Yeager rifles and pistols. Half of the company was placed at Camp Dix, located where the San Antonio to Uvalde Road crossed the Frio River at Black's Water Hole. Camp Dix was established on April 4, 1862. The other half of the company was stationed at Camp Nueces located where the San Antonio to Eagle Pass Road crossed the Nueces River about five miles west of the town of Uvalde. Camp Nueces was established on April 14, 1862. But it was late April before the two camps were occupied.[5]

Captain Rabb's Company A reached its assigned areas more quickly. By April 16, 1862, they were in position. Half of the company was stationed at Camp Rabb, located where the San Antonio to Eagle Pass Road crossed Elm Creek in Maverick County. The other half of the company was stationed at Rio Grande Station, near old Fort Duncan and across the Rio Grande from Presidio del Rio Grande. Rio Grande Station was established on April 12, 1862, and Camp Rabb on April 18, 1862.[6]

Tensions quickly mounted between the insurgents and State troops. As soon as de Montel's Company arrived in their camps, he began receiving reports of insurgent military units operating in the area. On March 16, 1862, Montel sent the first of many reports regarding Unionists activity. He reported that three or four militia companies had been formed in Medina County. He described these units by saying "9/10 of [them] are not only Black Republicans by principal but [illegible] a number of the abolitionist who have elected the men of their own stamp to command them calculating to keep all together when they are ordered out and join the enemy the 1st opportunity they get." Captain Montel identified some of the Unionist leaders. He reported they were P. Kiefer and Jacob Haby, both from Castroville. Montel also identified a man named Richarz who lived on the Seco. He stressed these men were, "bitter enemies to the South." Montel stated he felt it was his duty to provide this information, but requested his name not be used as the source. The reason was his wife and children were, "down there among them unprotected and I feel they could harm my property some night."[7]

In another of Montel's early reports, dated April 13, 1862, he reported a Medina County Unionist plan to have the current Frontier Regiment become a Confederate unit and sent off to the war. The plan was than to petition the governor to authorize New Jersey-born George Henry Noonan, the Judge of the 18th Judicial District, and New Yorker Charles M. McDonald, a lawyer in Castroville, to raise a new regiment. The plan was for Noonan to be appointed as regimental colonel and McDonald as lieutenant colonel. This new regiment would have the mission of protecting the frontier. Montel said three-quarters of his men were ready to march east to join the war, but not for permitting "black republicans and enemies of the Country to arm themselves." Montel further informed his headquarters and the governor that Medina County Unionists were "getting very bold and the county was not a safe place for Secessionists ... several men from D'Hanis had sworn to fight against the state and Confederacy."[8]

Endnotes -- The Frontier Regiment Deploys

1. Henry T. Davis was born on June 13, 1834 in North Carolina. He arrived in Texas in the mid-1850s and settled in Hays County. Davis married Susan T. Brownrigg on February 3, 1859 in Bexar County. The 1860 Hays Census shows Davis as a merchant. In early 1863, he and about half of his company enrolled in Duff's 14th Battalion Texas Cavalry. The company became Company F on February 6, 1863. Davis served the remained of the war in Duff's command and received a parole on July 31, 1865. The parole shows his rank as lieutenant colonel. After the war, Davis moved to Orange County where he died on October 4, 1888.

    Charles S. [Scheidemontel] de Montel was born in Germany on October 24, 1812. He arrived in Texas by 1837 and settled in Medina County in 1844. He married Justine Pingenot on November 13, 1845 in Castroville. Montel and two others founded the town of Bandera in 1853. He was the first county clerk of Medina County and one of the first county commissioners in Bandera County. He represented Medina County in the Secession Convention. Montel was appointed the provost marshal of Bandera and Medina Counties when martial law was declared in May 1862. After his year as commander of Company G, Jefferson Davis appointed him commander of the streamer *Texas*. In 1881, he was again appointed a 'Ranger'. He died on August 3, 1882.

    General Order Number 4, Headquarters Texas Frontier Company, dated March 4, 1862 and February 7, 1863, TSA, Austin, Texas; Eighth U. S. Census 1860 Hays County, p 10; Compiled Service Records, 33rd Regiment Texas Cavalry, Record Group 109, National Archives, Washington, D. C.; Henry T. Davis' Confederate Service Records, Record Group 109, Nation Archives, Washington D. C.; Bexar County Texas Marriage Records, Volume D 2; *Orange County, Texas Cemetery Inscriptions* by Lorine

Brinley, Published by Frances T. Ingmire, St. Louis Missouri), 1985, p 78; Ron Tyler, et al, eds., *Handbook of Texas*, Volume I, pp 488-489; *A Brief History of Bandera County* by J. Marvin Hunter., (Bandera Printing Company, Bandera, Texas), 1949, pp 12-13; *100 Years in Bandera County* by J. Marvin Hunter, (Published by J. Marvin Hunter, Bandera, Texas), 1953, p 2 and *History of Bandera County Texas* by Bandera County History Book Committee, Curtis Medina Corporation, Dallas, Texas) n.d. ca 1987, pp 3–4.

Montel's Company letter designation was changed from Company G to Company D. For clarity sake, it will hereafter be referred to as Company D.

Thomas Rabb was born about 1830 in Texas. By 1860, he was living in Karnes County, Texas. He enrolled in Company A, the Frontier Regiment on January 13, 1862 and elected its captain. His company had at least one skirmish with League troops. When Duff's Battalion was organized in early 1863, his entire company joined and became Company D of Duff's battalion and later regiment. He remained with the unit until the end of the war. He returned to Karnes County after the war. He died at Deming in Luna County, New Mexico. on March 27, 1914.

John James Dix, Jr. was born on March 27, 1826 in Michigan. His parents arrived in Texas in the mid-1830s. Dix settled in the Corpus Christi area in 1845 and became a well-known surveyor. After the war, Dix was a state representative from the 38th District in the Twenty-Second Texas Legislature. John James Dix, Jr. died in 1910 in San Antonio.

Eighth U. S. Census, 1860 Karnes County, Texas, p 21; Thomas Rabb's Confederate Service Records, Record Group 109, National Archives, Washington, D. C.; Eighth U. S. Census, 1860 Nueces County, Texas, p 279; John J. Dix's Confederate Service Records, Record Group 109, National Archives, Washington, D. C.; Records of the Frontier Regiment, A.G.C. TSA, Austin, Texas, and Ron Tyler, et al eds., *New Handbook of Texas*, Volume 2, p 657.

2. It should be remembered that this was the company which the League attempted to control. Note that it was mustered into service before the League organized its military battalion. The town of Ingram is presently at this site. Hopkins' diary and Report Colonel James Norris, Headquarters Camp San Saba, April 23, 1862 to Colonel Dashiell, Texas Adjutant and Inspector General, Austin, Texas, A.G.C., TSA, Austin, Texas.

3. Canyon de Uvalde is the current Sabinal Canyon, which runs from northern Uvalde County into present Bandera and Real counties. Letter, Captain Montel, Camp Near Guadalupe, Blanco County, March 6 1862 to Colonel Dashiell, Texas Adjutant General, Austin, Texas, A.G.C., TSA, Austin.

4. Camp Ives was an abandoned Federal camp located about 4 miles north of Camp Verde on Turtle Creek in eastern Kerr County. Seco Pass is located in western Bandera County between Seco Creek and Rio Hondo Creek. Letter, Captain Montel, Camp Ives, March 16, 1862 to Colonel Dashiell, Texas Adjutant General, Austin, Texas, A.G.C., TSA, Austin, Texas and Norris' Report, April 23, 1862.

5. Colonel Norris, the regimental commander, never officially established Camp Rio Frio as one of the Ranger camps. Captain Montel continued to use this camp for at least a year. Letter, Captain Dix, Clinton, February 16, 1862 to Texas Adjutant General Dashiell, Austin, Texas, A.G.C., TSA, Austin; Letter, Captain Dix, In Camp near San Saba Peak, April 16, 1862 to Texas Adjutant General Dashiell, Austin, Texas, A.G.C., TSA, Austin, Texas; Norris' Report, April 23, 1862; Letter, Captain Montel, Camp Rio Frio, April 4, 1862 to Colonel Dashiell, Texas Adjutant General, Austin, Texas, A.G.C., TSA, Austin, Texas; Series of returns from Camp Rio Frio from Lieutenant Thomas P. McCall, the Frontier Regiment, March 1861 and 'Abstracts of Provisions' by 2nd Lieutenant J. C. Leach, Company I, Woods Regiment Texas Cavalry, March 31, 1863, File 401 – 835, Folder #8, A.G.C., TSA, Austin, Texas.

6. Letter, Captain Rabb, Fort Duncan, April 16, 1862, Eagle Pass to Adjutant General Dashiell, Austin, Texas, A.G.C., TSA, Austin, Texas and Norris' Report, April 23, 1862.

7. It is of special interest to note this was exactly what Kerr and Gillespie citizens in February said about Kuechler's Company. Montel's reports further confirm the Unionists belief Union troops were going to shortly arrive. Unable to identify P. Rufus. There was a Redus family of three brothers who lived at New Fountain, but they were strong Confederate supporters. Two of the brothers; George and John, were later members of Duff's Company of Partisan Rangers and likely in the group sent to Fredericksburg. There was also a Reus family in Medina County. This 'P. Rufus' may have been a member of the 'Reus' family. Jacob Haby commanded Company D of the Medina County Independent Battalion of the 31st Brigade District. This is Henry J. Richarz, the major and commander of the Medina County Independent Battalion. Montel's Report from Camp Ives.

8. George Henry Noonan of Medina County was born on August 20, 1828 in Newark, New Jersey. He arrived in Medina County in 1852 and settled near Castroville. Noonan was elected judge of the 18th Judicial District in 1862. His son, Nelson, served in Duff's Company of Partisan Rangers. In 1895, George Noonan was elected to Congress. George Henry Noonan died on August 11, 1907.

Mr. Donald was Charles M. McDonald, also of Medina County. He was born about 1838 in New York. McDonald enrolled in Company H, 32nd Regiment Texas Cavalry in May 1862 and elected 1st lieutenant. He was court martialed for Unionist activities in October, 1862. Captain Lytle, also of the 32nd Regiment, shot and killed him on June 15, 1863 at Castroville.

Letter, Captain Montel, Camp Near Camp Verde, April 13, 1862 to Colonel Dashiell, Texas Adjutant General, Austin, Texas, A.G.C., TSA, Austin, Texas; Webb *Handbook of Texas*, Volume II, pp 283-284; Bandera Handbook *of Texas*, Volume III, p 655; Compiled Service Records of Nelson Noonan, National Archives, Washington, D. C.; Charles M. McDonald's Confederate Service Records, Record Group 109, National Archives, Washington, D. C.; Probate Case Number 193, Medina County Probate Records and *San Antonio Semi-Weekly News,* June 18, 1863.

CHAPTER 5 – Conflict and Reversal

As stated before, *Des Organisator* now had its own military arm, a three-company battalion under the control of its 'Political Front.' The battalion was prepared to defend the League and its leadership against threats by the Gillespie Rifles or the two companies of the Frontier Regiment being deployed. More importantly, *Des Organisator* believed a Union invasion of Texas was imminent and was prepared to assist this invasion, which would result in them declaring the Free State of West Texas as was being done in West Virginia. The insurgents; Hill County Unionists, the San Antonio area, and Austin area Unionists were at the height of their power and influence. In the early spring of 1862, *Des Organisator* moved from the early states of Phase I of an insurgency to the later stages of Phase I and the early stages of Phase II, which was to take direct action against its enemies. This action was much more overt and would ultimately result in the Confederates applying overwhelming military force against the League. The insurgents were "not insane," as one modern historian puts it, and the Confederates were not just a group of "Bullies, bandits, and cut-throats". To each side, their actions were correct and seemed necessary for the goals and results they wanted to achieve.[1]

A major reason the insurgents felt they needed their own military force was because it appeared ripe for the long-expected Union invasion. This was due to the fact the war was not going well for the Confederates. The first major

Confederate military defeats took place in Tennessee in the winter of 1861-62. General Grant's push down the Mississippi and Tennessee Rivers resulted in the capture of Fort Henry on February 6, 1862 and Fort Donelson on February 16, 1862. In addition, the Union captured Roanoke Island and parts of North Carolina. By early March 1862, the Union appeared to be on the verge of several more major victories. A large Union army was poised in Northern Virginia, north of Richmond. A second was in northwest Virginia near Harper's Ferry. A third was in eastern Virginia on the Peninsula at Fort Monroe, also threatening Richmond and Norfolk. Confederates had retreated from Missouri. The Union blockade of the Confederate seaports was no longer just a nuisance; it was now a real danger.[2]

Besides these setbacks, Texas forces had to withdrawal from New Mexico. In October 1861, Sibley's Brigade marched from San Antonio to Fort Bliss in far west Texas. Its objective was the capture of Arizona and New Mexico Territories. The brigade had initial success, but by late March and early April 1862, the Texans were forced to withdraw back to San Antonio. These Union victories resulted in Texas Unionists showing open support of the North, especially in the Austin, Fredericksburg, and San Antonio areas.

The Unionists' belief in, and the Secessionists' fear of an invasion had been brewing for some time. Both Austin and San Antonio Unionists were openly cheering Union battlefield successes. In San Antonio, the Confederate military commander reported the Unionists were celebrating

Union victories by sending up small balloons and firing of guns. In his report of March 3rd, Colonel McCulloch, the commander of the Sub-District of the Rio Grande, headquartered in San Antonio, expressed his view that the Union would likely, "land in force on the coast, or invade us on the north," led by Unionist Jim Lane in the spring. An Austin newspaper reported that Lane had been given command of 25,000 men for such an invasion.[3] This was the invasion force that would allow the insurgents to declare for the Free State of West Texas. [Map D, p 132] In Austin, the Unionists were also openly celebrating Union victories. "It is reported and believed, that the Lincolnites of Austin did not even affect to conceal their pleasure upon hearing of our disaster at Fort Henry and Donaldson. That the traitors, whom the people have delighted to honor, actually hurrahed for the enemy!" According to one Texas historian, "In the minds of many Austenite it was no longer a question of whether Texas would be invaded but of when and where the attack would come." Some expected an assault at Galveston or elsewhere on the coast, but others looked for an invasion overland through northeast Texas.[4]

Colonel Henry McCulloch reported that San Antonio Unionists were trying to "break down" the currency of the Confederacy, by asking twice as much in Confederate currency as they would in gold or silver and demanding immediate payments for goods sold. For these and several other reasons, Colonel McCulloch stated martial law would soon be necessary.[5]

The Hill Country Unionists and insurgents opened a courier service with Union officials in Mexico. It operated from Fredericksburg, Comfort, Sisterdale, Boerne, Castroville, D'Hanis, and San Antonio to the Mexican border at Eagle Pass where it crossed into Piedras Negras. From Piedras Negras the messages were forwarded to Leonard Pierce, the U. S. Consul at Matamoros or to the U. S. Consul at Monterrey. There were no official United States officials in Piedras Negras, but Thomas P. McManns was a defacto consul there. Messages and northern newspapers were taken back to the Texas Unionists. Using this courier, insurgents and Unionists kept United States officials informed of their activities. On March 21, 1862, the U. S. Consul at Matamoros, Mexico, reported to the Secretary of State that, "The Union men in Texas are becoming bolder, and a battle is expected in the neighborhood of Austin and San Antonio."[6] To prepare for this possible 'battle' *Des Organisator* and Bandera, Bexar, Medina, and Travis counties made plans to join their armed units. The Unionists' military force was impressive. *Des Organisator* Battalion consisted of between 250 and 300 men. The Medina Unionists had the Independence Battalion of the 31st Brigade, which consisted of about 170 men. Bandera Unionists had about 35 men. In all, two battalions of eight companies consisted of at least 450 men. In addition to these 450, men the Austin Unionist Company numbered at least 89 men and were now part of the state military system. San Antonio and Bexar County also formed Unionist military units, based around the *Turnverein,* armed with shotguns, rifles, and pistols, with plenty of ammunition. One of these units consisted of "at least 75 men ... just waiting for the right time to act." In conservative numbers, the Unionists had at least 620 men under arms.[7]

This Unionists planning-and-coordination meeting took place on March 25, 1862, in San Antonio. Of this meeting insurgent Captain Ernest Cramer wrote, "The next day [March 25th] I was ordered to San Antonio for the purpose of joining our forces with the others from the different district. I found everything well prepared there."[8]

Important events followed the March 25 meeting. Just two days after the meeting, Colonel McCulloch learned of the existence of the San Antonio Unionists' military company and its connection with the Austin Unionist Party. Given the supposedly secret meeting this was surprisingly quick intelligence. On the very day of the meeting Colonel McCulloch warned his superiors that large numbers of Unionists were leaving the country for Mexico to join with other Unionists already there to "act in concert with men of like feelings about Austin, this place [San Antonio], Fredericksburg, and other points where they are still living among us."[9]

Four days after the meeting, the *San Antonio Herald* published an attack on Unionists activity in the form of an editorial on what it called 'Domestic Enemies'.

S A N A N T O N I O H E R A L D - March 29, 1862
Our late disasters have had one good effect to say the least.

They have so [illegible] Yankee sympathizers throughout the South as to induce them to express their sentiment in such a manner as to be fully understood, and to render future watchfulness on the part of loyal men a necessity. It is true there was nothing in our misfortunes calculated to shake the faith of [illegible] men in the success and permanency of the Confederacy. They know that the Southern people could not be conquered and that the Northern would soon conquer themselves by their enormous expense, even if they should not be stopped by our armies — But Southern Lincolnites are not overburdened with sense, and they verily believed their Northern friends were victorious — that the war was over; the South subdued and that they and theirs were safety landed in Abraham's bosom.

Entertaining these views, they were elated and communicating. They made statements that showed where their hearts were; and that will be remembered. A Union flag hosed over each Unionist's house in this city could not have indicated more who the enemies are that we are harboring in our community, then did the [illegible] expression that escaped them in the recklessness of their supposed triumph!

Let them be remembered. Let them be compelled to do their full share towards the expenses of this war, and in times of safety and prosperity be treated with the contempt which Tories deserve from patriots. But should danger [illegible] our state should contemplation of war enable Lane of Kansas or any other Yankee robber to invade Texas, and spread terror and devastation around them then it will be necessary to dispose of the lurking enemies in our midst. Fortunately, there will be not difficulty in determining who they are. Circumstances may render it necessary that the action of loyal citizens should be prompt and decisive.

As all events, it is certain that if a Northern army should invade Texas domestic enemies would not be permitted to run at large, and communicate intelligence to the invaders or assist them in any way. Men of principles have no excuse for being here; they had ample time to go to their northern friends and having chosen to remain, they must abide the con-sequence. Should there be no invasion of Texas, they may be safe from personal danger, although objects of contempt amongst loyal people, but should our State be invaded, self-defense, one of the best laws of nature, will demand that they be put out of the way of doing harm before the near approach of the enemy. Domestic foes are vastly more dangerous than foreigners. This fact is well understood by men of experience and reflection and it is preposteriors to suppose that each other merit or will receive the sympathy or protection of loyal citizens in times of alarm and danger. When our men go out to fight the invaders, will never by guilty of the folly of more dangerous foes behind them.

We do not intend these remarks as threats but as statements for the consideration of those misguided men, whose feet stand on slippery places — and who are not probably aware of the turpitude of their conduct, or the imminent peril of their position. They are known and unless they immediately renounce their errors, and in good faith expose the seeds of their [illegible], there is no telling how soon it may be necessary to treat them as Tories (and how) Tories and traitors are treated by all nations in times of war.

Figure IV

The editorial warned Unionists that the Texas and Confederate governments knew of the possibility of the Union invasion overland by James Lane and by General Nathaniel P. Banks on the coast. The editorial warned the Unionists, "It will be necessary to dispose of the lurking enemies in our midst", referring to the Unionist and they would be "treated as 'Tories' in the manner 'Tories' and traitors are treated by all nations in times of war."[10]

It was also on March 29, that German-language leaflets appeared in San Antonio informing everyone, "The Revolution has begun." On March 31, 1862, Colonel McCulloch reported German notices in prominent locations. A poor English translation of the notice stated:

Germans brothers are your eyes not opened yet? After rich took every picayune away from you, and the paper is worth only one-half what you so hard earned, now that you have nothing left, now they go about and sell you, or throw you out of employment for Dunhauer, who left his wife and children, wants to do the same with you to the poor you might leave. Now is the time to stay [slay?] the heads of Dunhauer,

Maverick, Mitchel, and Menger to the last bone. We are always ready. If the ignorant company of Newton fights you, do as you as please. You will always stay the God damn Dutchman. Do away with that nuisance, and inform everybody the revolution is broke out. It is a shame that Texas has such a brand. Hang them by their feet and burn them from below.[11]

Ernest Cramer used some of the same words in his October 1862 letter. Its says on March 24, 1862, "The Union Flag was hoisted and the outbreak of a revolution was momentarily expected." The March 1862 San Antonio German-language leaflet said, "Inform everybody the revolution [has broken] out."[12]

Colonel Henry McCulloch reported he was prepared to deal with the Unionists' actions. Beginning in early March, as soon as he was placed in command of the Sub-District, he urged martial law be declared to deal with the Unionists. He pressed his case both to the Confederate Department of Texas and to the Texas Governor. He stressed he had the necessary force to put down any Unionist action, but to do so he would have to interfere with civil laws. However, he was fully prepared to take any action necessary to put down "the malignant acts of these cowardly traitors." He identified the three major areas of Unionists activity as San Antonio, Austin, and Fredericksburg. McCulloch felt it was time to apply force and destroy the Unionists. He recommended they, "be crushed out, even if it has to be done without due course of law, or the country—the section in which I am stationed to protect and in which my family reside—will suffer." Colonel McCulloch did not realize what he was facing was a fully-organized insurgency, one which had developed into Phase II with a military force of more than two battalions of four hundred men, one that had just met with Unionists leaders from San Antonio, Medina County, Fredericksburg, Austin, and Kendall County. He was "fully aware of the responsibility of the steps I have taken, and how much it perils my reputation as an officer and how much it exposes my person and my domestic interest—my home, my wife and little ones—to the malignant act of these cowardly traitors, but I believe it my duty to my country, and in her case I will to peril my all." In preparing for military action, McCulloch ordered most of his 1st Regiment Texas Mounted Rifles to assemble at San Antonio. Open military action was about to begin between the Unionist and 'regular' Confederate military forces.[13]

Major bloodshed was prevented, at least for the time being. Henry McCulloch was promoted to brigadier general and given a new command. Newly appointed Brigadier General H. P. Bee replaced him on April 24, 1862. It fell to General Bee to deal with the insurgents and other Unionists. It would be General Bee who would act. It would be James M. Duff, not Henry McCulloch, whose reputation would be imperiled.

An incident that took place at D'Hanis on April 8, 1862, showed the high level of tension between Hill Country Unionists and Confederate troops. After Company F, 2nd Regiment Texas Mounted Rifles assembled in San Antonio it passed through D'Hanis. What happened next would be hotly disputed afterwards. Captain Samuel Richardson, the company commander, was ordered to arrest Squire Boone, a resident of Medina County for Unionist activities. Richardson directed First Lieutenant Thomas B. Ragsdale to arrest Boone. Ragsdale found Boone in D'Hanis, but Boone escaped. Ragsdale had his men search D'Hanis but could not find Boone. On April 15th, Boone gave himself up in San Antonio.[14]

According to D'Hanis Unionist Joseph Finger, the Confederate company arrived in town on April 8th and camped near Joseph Ney's grocery store. A number of Confederate soldiers were drunk and rode through town firing their pistols "at random in the streets." While walking from his home to the post officer Finger was, "Forcibly arrested, insulted, his hair torn, knocked over [his] head and shoulders with cocked pistols, and then [threaten to be] shot down and afterwards conveyed in the camp without giving any cause therefore and there guarded as prisoner all during the cold night."[15]

Finger identified the Confederate leader as Lieutenant T. B. Ragsdale. He asked Ragsdale for some blankets, but was refused. Finger stated Ragsdale was drunk. The next morning Finger showed Ragsdale his papers proving he was a lieutenant in the local militia company. He also produced papers showing he had taken the oath of allegiance to the Confederacy. Finger was released.[16]

Private William W. Heartsill, a member of Company F, saw things quite differently. Heartsill wrote in his diary on April 8th "At sunset we reach Dhanis and camp. Lieut

Ragsdale took into custody five Unionists. The boys will long remember Dhanis. 'Turn loose the prisoners and suppress the riot'." On the next day, April 9th, "A good day's march and reach Castroville, and camp on the Medina River. Lt R sets at liberty the prisoners as he has no positive proof against them; they go their way rejoicing." Joseph Finger forwarded a report of the incident to State Brigadier General Robert Bechen on April 12th. He demanded an investigation into the incident and, "Satisfaction rendered to me, otherwise I am bound to apply for my discharge as [an] officer of the Texas State Troops as I cannot serve anymore in corps were such an outrage can be committed on an Officer and a loyal citizen of the Confederate States." His commander, H. J. Richarz, who himself would later be arrested for Unionist's activities, indorsed Finger's report. Richarz said he was present at D'Hanis when the incident occurred but was unable to interfere as, "a number of the rangers were drunk very violent and fighting among themselves." There is no record of an investigation being conducted, but Joseph Finger remained a lieutenant in his company.[17]

The March 24th San Antonio Texas Unionist's meeting resulted in state and Confederate leaders learning more about the Unionist's actions. Upon return from the meeting, League leaders notice that there was a problem. Ernest Cramer summed it up: "The spirit and enthusiasm that we had in our original organization was not there [in San Antonio]. I was told to await further orders. In my heart I knew that without help from the Northern forces we were helpless. I cannot deny that in San Antonio I lost much confidence. However, I did not allow myself to become disheartened. I did begin to realize however that in what I had untaken I had placed much too much faith and trust in other people. The others from the various districts who had been called at the same time that I had, returned to their homes with the same feeling of hopelessness."[18]

The Union Loyal League did not launch any major military operations, but did develop several plans which had major impact on the pro-Confederates and Grays of the Hill Country. One of these plans was for overt action to free the Federal prisoners of war at Camp Verde. These were the Federal prisoners captured by Texas and Confederate forces in May, 1861, soldiers who had been guarding the Texas frontier. There were two contingencies. The first, if the Union invaded Texas, the prisoners would be freed, join the League and assist the invading Union force. Second; free prisoners who would then either help fight the Confederates or flee to Mexico. One of the Camp Verde prisoners described this plan; "One day there was rumor afloat that several hundred Union men were concealed in the vicinity of our camp." The prisoners were told that within a week, at the longest, everything would be ready for their escape. The insurgents would fully arm them after their escape, then together they would cross the Rio Grande and report to the U. S. Consuls at Matamoros or Monterrey, Mexico. "They still further cautioned us, that we should hold ourselves in readiness, watching day and night, and always hoping that the present be our last day in Prison." However, "the Rebels soon found it out, so of course, they had to disperse." The

Confederates now took better precautions: "Sentries were then posted around Prison town, citizens were forbidden to enter, or to have any intercourse with the prisoners." There were still a few visitors allowed but, "If the Rebels at Camp Verdee [sic] had any suspicion that any of the visitors were Union men or Union sympathizers, they were always escorted through the camp by an officer, accompanied by three or four soldiers."[19]

The threat of the insurgents organizing a breakout remained a fear for the Confederates. By early spring of 1862, they dispersed the prisoners. About 330 of them were sent to Fort Mason, Fort McKavett, Camp Colorado, Fort Chadbourne, and Camp Cooper. August Siemering confirms this Unionists plan and the Confederate response. He stated that one of several actions taken by the Confederates to "counteract and render the Unionists harmless" was the "hasty movement of Federal prisoners at Camp Verde to Fort Mason, as letters had been intercepted support the supporting the Unionist in the Hill Country." The stationing of some of the prisoners at Fort Mason resulted in another Unionist attempt to free them.[20]

The window of opportunity for *Des Organisator* and other Hill Country Unionists to start a revolution was slammed shut by late April, 1862, with the placement of Texas and Confederate troops in the Hill Country, along the Mexican border, and around San Antonio and the decline of Unionists support. This was especially true for the Travis County Unionists, as the League and other Unionists looked to them for leadership. This lost of support was due to the

Union armies failure to achieve victories in the east. The League again had overestimated their abilities and support, as they had done in 1854.

Three other events, all within a thirty-one-day period [March 26th to April 25th], caused the League to reassess their goals and revert their insurgency back to Phase I. The Travis County Unionist Party disbanded, the expected Confederate draft became a reality, and the long-awaited Union invasion finally took place, but not on the Texas coast. Instead, on April 25th, a Union army landed near New Orleans. General Benjamin F. Butler commanded the ground force, not Nathaniel P. Banks.

Captain Cramer wrote of this reassessment and *Des Organisatpr* new "wait and see" policy. It was not an easy thing to do because some militants wanted to take offensive action. "And so we all came to the decision to maintain peace and quietness. It was no small thing to do. Every evening we met but delayed action from one day to another." Cramer's comment is interesting as it shows there was still a *Des Organization* League element that wanted to take direct action, but this action was "delayed from one day to another."[21]

The breakup of the Travis County Unionist Party was reported in both Austin and San Antonio newspapers. The *State Gazette* says the Unionists of Austin have given up their opposition to the South and their admiration of Yankee despotism. It says every American in the place is now in favor of the South. We are rejoined to hear it and hope that future developments may justify the correctness of the Gazette's opinion.[22]

While the newspapers were wrong that everyone in Austin was now a supporter of the Confederacy, the Austin Unionist Party had indeed disbanded. This news had a major impact on the League, as Cramer explained, "At last on the 15th of April we were informed that the Union party in Austin had been forced to disband." *Des Organisator* had been built on the strength of the Travis County Party and looked to them for overall guidance. When that failed, Cramer "realized that we would be forced to leave our homes, and our only hope was to be able to reach Mexico." The San Antonio *Turnverein* "refused to give up their flag [but did] dissolved and most of its members fled." These were major blows to *Des Organisator*. One can almost hear the desperation in Cramer's voice when he wrote, "We had taken our stand but all our plans had to be abandoned, our hope so inconceivable lost – we could not await the outcome." Cramer's comment makes it clear that *Des Organisator* looked to the Austin Union Party for overall leadership. It also makes it clear *Des Organisator* were therefore part of some overall plan. An interesting point is the date Cramer gives, April 15th, as the date of Austin Union Party breakup. It shows *Des* Organisator had to abandoned whatever overall plan they made, and that members of *Des Organisator* would have to flee to Mexico. This was before any Confederate martial law or other 'crackdown.' Given that at this point a major *Des Organisator* leader says he knew all this as early

April, then why did *Des Organisator* wait until August before attempting to flee? [23]

More damaging news followed. The long-rumored Confederate draft became a reality. On April 16, 1862, President Davis signed the Confederate Conscription Law stating that all males, except those exempted, between eighteen and thirty-five years of age were subject to the draft. The draft did not even exempt members of home guard or militia units from being called into Confederate service as many had hoped. The new draft law did allow a thirty-day grace period for men to enlist in the unit of their choice and be treated as 'volunteers' which among other things gave them a $50 enlistment bounty and allowed them the right to elect their officers.[24] The Confederate Conscription law had an immediate effect. There was a rush to join already organized Confederate units and to organize new units. One of these new Confederate units which benefited was the newly-organized 32nd Regiment Texas Cavalry. It quickly filled all ten authorized companies. It began organizing in March, 1862. By May, it was deployed in and around San Antonio. The regiment headquarters and most of its companies were encamped just east of San Antonio on Salado Creek. Two companies were deployed against the Unionists. One was Company H, replacing many of the units of the 1st and 2nd Regiments Texas Mounted Rifles after they were withdrawn from the abandoned Federal post west of San Antonio. Company H had detachments stationed at Fort Clark, Fort Inge, and Camp Hudson. The other company was Company K. It was much more of a direct threat to *Des*

Organisator. Company K was deployed in mid-February to Camp Verde in Kerr County. Its mission was to guard the Federal prisoners located there, as well as to provide the area protection from Indians and other threats.[25]

The second 'regular' Confederate unit, Joseph Taylor's 8th Battalion Texas Cavalry, was composed mainly of discharged troops from the 1st Regiment Texas Mounted Rifles. The battalion was organized in late April at Fort Mason, just north of Fredericksburg.[26] [Map C, p 130] The Unionists viewed the four Frontier Regiment companies, especially Davis' and de Montel's Company H and Company K of the 32nd Regiment, and Taylor's Battalion as a direct threat to them. The deployment of elements of the Frontier Regiment and Confederate cavalry units in the Hill County increased tension between state and Confederate military forces and insurgent military forces. Many Hill Country insurgents believed the mission of these units was to, "hunt down Union men ... especially ... in the settlements North of San Antonio."[27]

One of the newly-organized units would have long-lasting psychological impact in the Hill Country. Almost daily, local newspapers were reporting new companies organized for Confederate service. On May 10th, the *San Antonio Herald* described several such companies. The newspaper said this about one, "Another very fine company of cavalry has just been formed and organized in this city— Mr. James Duff Captain, J. R. Sweet 1st Lieut; Edwin Lilly and Richard Taylor 2nd Lieuts. This makes about the 20th company that San Antonio has provided the service."[28]

Another instantaneous effect of the draft was the almost complete stripping of members from newly-organized militia units. For the League, this was another damaging blow. Those of draft age were now faced with the real possibility of being forced into Confederate service. Siemering told of one reaction: "Within Gillespie, Kerr, Kendall, Mason County etc. every attempt to raise a volunteer Confederate company had failed. However, the realization that one of these days they would be drafted became more evident. However, there was an opportunity to delay this threatening evil."[29]

Siemering saw this 'opportunity' as where men could enroll in a local unit that he claimed the governor authorized specifically for home defense and replacing Confederate units stationed on the frontier. Frank Van der Stucken, whom Siemering claimed was a 'unionman,' commanded this company. The result was that, "the unit was made up of men who almost all belong to the Union party." What appears to have taken place is many of the Gillespie draft-age men decided to enroll in a company of their choice as 'volunteers,' hoping it would remain on the frontier. This included many men who were members of the League's Gillespie Company. About half of the men who joined Van der Stucken's Company were moderates [Grays] and about ten percent were militants [Greens] including Siemering. He was elected one of the company's second lieutenants. Despite his later claims that he was 'drafted,' he enlisted on May 1, 1862, during the period men could enlisted and be treated as a 'volunteer.' This enrollment of Siemering is very confusing. He was certainly an original member of the League. At this time, the Confederate had taken no action to 'force' anyone to enlist. Why would one of the most radical League members turn his back on it? It may be that he and other Unionists enrolled in a delayed attempt to control a Confederate unit. This is one of the techniques an insurgency uses, but none of Siemering's later writings claim this as a reason. It may also have been in the insurgents' mind that the 'reign of terror' had begun, as expressed by Cramer when they learned the Travis Union Party had dissolved. The problem with Siemering's account is that while he correctly summarization the conditions for raising a state unit, these units were the companies organized for the Frontier Regiment. All the Frontier Regiment companies had been enrolled by March. The company authorized from Gillespie County was Davis'. There was no authority to enroll another full-time state unit in the period Siemering is talking about. It could be Van der Stucken was hoping to raise a replacement company for Kuechler's. There is some evidence that he began enrolling as early as March. But by the time it was enrolled, the company was clearly a Confederate company. This can be established from the 31st Brigade District records. Van der Stucken was a militia officer of the brigade. He resigned to recruit a 'volunteer' company. A 'volunteer' company was not a state unit, it was a company for Confederate service. It is difficult to understand how Siemering could believe Van der Stucken was a 'unionman' since as early as February, 1862, he was clearly anti-Unionist. He was one of the men who wrote Governor Lubbock complaining that Kuechler was enrolling only Unionists. He was a member of Nimitz's Gillespie Rifles and signed the resolution condemning Braubach, Doebbler, Radeleff, and Siemering. He was one of the Gillespie men who signed the petition requesting Kuechler's Company be disband. Siemering claims Van der Stucken "suddenly changed from a unionman to a bitter secessionist." This claim does not match the facts. There is no evidence, except that claimed by Siemering, Van der Stucken was ever a 'unionman'. A likely reason for this claim is that Siemering enrolled in Van der Stucken's Company during the 30-day grace period, and to later justify his actions, he claimed it was believed the company was for home guard and not to fight against the Union Army. It is a fact that six of the members of Van der Stucken's Company had belonged to Kuechler's Company; August Siemering, Frank Jung, George Ottmars, Jacob Schneider, Charles Schwarz, and John Walther. Six others deserted and joined the Union Army: Henry Cramm, Ferdinand Ohlenburger, William Stockman, William Nickel, August Pfannensteil, and William Pfannensteil.[30]

On May 7, 1862, the company was mustered into Confederate service as Company C, Taylor's 8th Battalion Texas Cavalry. It was stationed at Fort Mason with two missions; guarding the Federal prisoners sent there from Camp Verde and protecting the area from Indian raids. [Appendix Q]. State Brigadier General Robert Bechen spoke for many in a May 29, 1862, letter when he said Texas and Confederate authorities were pleased that at long last Gillespie County had men in Confederate service.[31]

The idea that Texas Unionists were feeling threatened is support by one of the effects of the conscription law; that many German draft-age men fled Texas for Mexico. One New Braunfels diary states, "In the later war years all young Germans were pressed into the war service and woe to those who would try to flee to Mexico to avoid military service, often motivated their parents: at the border they were seized and shot."

As early as May 5, 1862, the U. S. Consul General at Matamoros, Mexico was reporting large numbers of new refugees. "The crowds of refugees from Texas do not diminish in the least, although it is very difficult, owning to the strict watch kept upon their movement, for them to get out. Many are arrested; some are hung, others are taken and pressed into service." In another part of his communication, he reported on the conditions at San Antonio and Austin. "At Austin there is a strong Union party, headed by Ex-Senator [A. J.] Hamilton, [who] will resist all attempts that may be made by the rebels to subdue them. At San Antonio, nearly all the stores and closed and many of the merchants are new residing here, waiting patiently for the time to come when they can return."[32] Two weeks later the U. S. Consul General at Monterrey, Mexico, also reported on the condition and numbers of refugees reaching him, "I am confident no portion of the United States has been so badly oppressed as the Union men of Texas." He went on to say that he could learn of nothing being done to relieve the refugees or avenge the oppressors. "I have received reliable information that within one week 6 of these refugees have been hung on the frontier of Texas on trees and left hanging." He expressed a common plea from both Union officials and Texas Unionists that was for the North to send an armed force to help the Texas Unionists "Let me urge that a force be sent onto this frontier; it may not be very large, if they have plenty of arms, I am assured that there could be 3,000 enlisted from Texas as soon as it is known."[33]

Des Organisator reaction to the Confederate draft was to try to keep as much control of their influence as possible and keep what men they could out of Confederate service, but they clearly were on the defensive. Intimidation became an insurgent weapon. Charles Schwarz had already succumbed once when he joined Kuechler's December 1861 Company. Gillespie Sheriff Philip Braubach had an arrest warrant for Schwartz. He told Schwartz if he joined Kuechler's Company he would not be arrested. Schwartz agreed. After Kuechler's Company was disband, Schwarz planned to enlist in Van der Stucken's Company. Braubach told Schwarz he would "make the warrant good" and arrest him if he enrolled. Schwartz attempted to get Braubach to accept a security note from his father-in-law to keep from being arrested. Braubach refused. Schwartz enrolled and was immediately arrested and placed in jail. Captain Van der Stucken paid the fine out of Schwartz's enlistment bonus money due Schwarz and got him released from jail.[34]

Arresting Schwarz was not the only anti-Confederate act of intimidation committed by Braubach and other insurgents. They openly beat individuals who supported or befriended a secessionist. The main insurgent gathering place in Fredericksburg was Doebbler's Beer and Wine Shop. They openly discussed anti-secessionist plans while drinking and toasting Lincoln and the Union government. There were Northern newspapers available for reading and it was even reported that a Union flag flew from atop the building. Insurgents who met there included: F. W. Doebbler, Philip Braubach, and Frederick Lochte. On May 18, 1862, a group of insurgents including Braubach, were in the shop. They were drinking and openly making anti-secessionist comments such as claiming they would 'whip' any secessionist. Oscar Basse, who had loaned a horse to Lieutenant James M. Hunter, of Davis' Company entered. The insurgents severely beat him. Sheriff Braubach did not interfere. One of Basse's friends came to his aid and 'was insulted' and beaten by the same group. A third man was beaten and they pulled a fourth man "from his horse and broke his head." Another account says a "man was killed." It was also on or about the same day that Braubach got into some type of altercation with Joseph Poetsch and warned him not to align himself with the Confederates. The League even conducted a military raid. In mid-May 1862, they raided the farm of R. A. Gibson in the Spring Creek Settlement in western Gillespie County. While no one was hurt, the League burned Gibson's barn and fences because his sons had joined Texas and Confederate units. The 'reign of terror' was also being directed toward Confederate supporters.[35]

Keeping their draft-age men out of Confederate military service was as much a major concern for the

Bandera and Medina Unionists as it was for the League. They devised a unique tactic: any time a Confederate enrollment officer arrived, they sent their men into the mountains 'scouting' for Indians or cattle. Besides keeping its members out of the draft, it also placed armed Unionists on military operations. This caused concern for Texas and Confederate officials, but there was little they could do; these parties were authorized state militia units.

By late May, *Des Organisator* had recovered somewhat from the shock of the draft and the fact the Union invasion of Texas did not take place. They again hoped that the Union might launch an attack on the Texas coast from New Orleans, which would cause a "quick release [of the Unionists] from the arms of the Confederacy." A Union naval demonstration at Galveston rekindled their hopes. On May 17th and again on the 19th, the captain of the frigate U. S. S. *Santee* sent a message to the Confederate military commander at Galveston saying that within a few days naval and land forces of the Union would appear off the town of Galveston. To prevent unnecessary blood and destruction of property, he demanded an immediate surrender of the city. This proved to be a ruse but it did cause major concern on the part of Confederate authorities, as well as renewing Unionist hopes.[36]

Another cause for hope was the fear of a Union invasion expressed in local newspapers: An invasion of the State probable—at or near Corpus Christi. It is a spot from which the whole of Western Texas may be reached ... Let them come. They will be met with a warm reception from our rangers.[37]

The insurgent's morale further rose when they received news that New Orleans might be a jumping off place for a Texas costal invasion, perhaps as close as Corpus Christi. Militarily they were still in good shape. *Des Organisator* completed its battalion organization. Bandera and Medina Unionists still controlled their 31st Brigade militia units. They had great influenced on one local Confederate company; that of Van der Stucken. So far there had been no action to force them into Confederate service. They were ready to move again into Phase II of their insurgency, or ready to start military operations. Maybe a free West Texas was still possible.

First indications of the Unionists' new aggressiveness came from Captain Montel and his Company D, Frontier Regiment. In mid-May, Bandera and Medina Unionist undertook a letter-writing campaign to Governor Lubbock, complaining of the actions of the Frontier Regiment. Montel responded to this criticism by blasting the letter writers as "croakers" and challenged their motives. He stressed that since his company arrived there was not single case where the raiding Indians succeed in getting off with their "spoil." Montel called the complaints "fabrications and false hood." He asked if any more such complaints reached the Frontier Regiment's headquarters, "for the sake of justice send me the complaint and its author."[38]

Montel reported on other Unionist activities. He told his headquarters that maybe it was time to give the

"Committee of Vigilance and Safety of San Antonio a warning!" Montel is referring to the Vigilantes and Vigilance Committee of San Antonio headed by Asa Mitchell. It conducted numerous executions in San Antonio and surrounding area. A matter which greatly disturbed Montel was that within the previous five days, Unionist courier activities had increased between Fredericksburg, Castroville, D'Hanis, Comfort, and Boerne, telling all the Unionists to 'arm up' as they were planning to unite their forces. This report supports the fact the Unionists had a courier service to communicate among each other, as well as with Union officials in Mexico. This indicates *Des Organisator* had recovered from the despair expressed by Cramer and were again hoping the Union would come to their aid, an invasion which they would in turn support. Montel told of an incident where one Unionist courier spoke with a German in his company, "I hope you belong to the good cause—we are soon going to have a chance to join our friends and must make ready now—and be united." When the courier was asked to what side he belonged, the reply was, "The North!" Montel was not the only Texas or Confederate commander to report that Unionists were attempting to unite their force in some type of military organization. Captain Duff also did so in his June 1862 report. Montel further informed his headquarters the Unionists were again in communications with the Union prisoners of war. While the exact word Montel used cannot be determined, this clearly shows Montel received news that the Unionists again were communicating with the Federal prisoners held at various frontier camps. In July, *Des Organisator* again developed plans to free them. Unknown to Montel, while the Travis County Union Party had dissolved, it had not gone away. The key leaders, to include A. J. Hamilton, had fled to the mountains west of Austin and were debating their next move. One option was to form the various Unionist groups into a larger military organization.[39]

State and Confederate leaders now were ready to take decisive action against those who opposed the Confederacy. Besides open Unionists actions, many merchants and citizens were not accepting Confederate paper money at face value. There remained open support of the Union in Austin, Fredericksburg, and San Antonio. In late March, Colonel Henry McCulloch had called for a declaration of martial law to enforce the laws of the Confederacy and stop anti-Confederate activities. On April 24, 1862, Brigadier General H. P. Bee replaced McCulloch. Bee took immediate action against the Unionists. On April 28th, he declared martial law throughout Bexar County. To ensure the Unionists understood the martial law requirements. Bee had it published in both English and German. The noose was beginning to tighten around the Unionists; fear spread.[40]

Sergeant Thomas C. Smith, of the 32nd Regiment Texas Cavalry, whose company was one of several used to announce and enforce martial law, reported what it was like in San Antonio the day martial law was declared: "This morning we all were up and ready as soon as we got our breakfast. We wanted [sic] awhile expecting to receive ammunition but the rest of the companies all formed we being the last one, brought up the rear or 'tail end.' The Battalion then formed by two's and marched to town. We went sometimes on a gallop and then slow. So on until we arrived just in the suberbs [sic] of town and were commanded to halt. Here we staid [sic]for about half an hour and then commenced the march again. Got into town and went by fours on a gallop through the City. Men women and children rushed to the sidewalk and window looking on in wonder. Brigdt. Genl H. P. Bee then came in and declare [sic] Martial Law throughout the City and County."[41]

Trooper Smith's description about the first weeks of martial law detailed why it was necessity. He said the cost of goods was "extravagant" and basic articles were "extorted." Smith further said the merchants would not take Confederate notes except at a third of its face value and than only if the troops spend at least half of the note. Smith explained for the first two weeks of martial law it was necessary for each man to pull guard once every week to "keep the traitors from avoiding the authorities." Smith's final comment summed up his feelings about the citizens of San Antonio as, "Don't think there are many good honest Southern people in Town."[42]

Some major points of Bee's Bexar County martial law declaration included appointing James R. Sweet, the mayor, as provost marshal; ordering all white males, above the age of sixteen to report to the provost marshal office to register and sign an 'un-conditional' oath of faith to the Confederate States of America; requiring any person who arrived in San Antonio after May 21, 1862, to explain the purpose of his business; requiring anyone who left the boundaries of Bexar

County to have a passport; requiring all places that sold liquor to be closed from 8 p.m. until 6 a.m.; and treating any attempt to devalue the Confederacy currency as an act of aggression. The San Antonio newspapers gave full coverage to the martial law proclamation. They reported General Bee personally read the proclamation at the Military Plaza in the presence of "a fine body of Military and large number of our citizens." At the close of the reading he addressed, "a few most touching and eloquent words to the assembled soldiers, remarking that on them he depended for the good fruits of the measure just announced, and exhorting them to set an example to the citizens by their good conduct in living up to the requirements of the proclamation."[43]

Bee's martial law declaration had a minor, though adverse, effect on the Hill Country Unionists and insurgents. What happen next had a profoundly adverse effect. On May 31, 1862, Confederate General Paul O. Hebert, the Confederate military commander of the Department of Texas, declared martial law over the entire state. Both Texas Governor Lubbock and General Bee had urged General Hebert to make this statewide declaration. Hebert's declaration stated in part:

"Every white male person above the age of sixteen years, being temporarily or otherwise within the aforesaid limits, shall upon a summons issued by the provost marshal, promptly present himself before said provost marshal to have his name, residence and occupation registered, and to furnish such information as may be required ... Provosts-marshal shall order out and remove from their respective districts all disloyal persons and all persons who presence is injurious to the interest of the country.

All orders issued by the provosts-marshal in the execution of their duties shall be promptly obeyed. Any disobedience of summons emanating from them shall be dealt with summary punishment.

Any attempt to depreciate the currency of the Confederate States is an act of hostility; will be summary punishment."[44]

Ernest Cramer told of the martial law effect. "Now San Antonio and 8 days later our whole district was put under the laws of Confederacy. Everyone must appear before the 'Provat' within ten days and take the oath to support and be true to the Comfort division of the Confederacy on Penalty of losing all properties."[45]

To inform the citizens of the requirements of martial law Confederate authorities again published the notice in local newspapers. The proclamation was printed in English, Spanish, and German newspapers. In addition, General Bee issued orders regarding the implementation of Martial Law.[46]

Three days after the proclamation of martial law, orders were issue to enforce the Confederate draft law. On June 2, the Confederate Department of Texas issued General Order Number 46. It announced that the Governor of Texas had placed all state military and civil officers at the disposal of the Confederate government for the purpose of enforcing the Confederate draft. General Order Number 46 required each of the state brigadier generals in command of one of the brigade district to appoint an enrollment officer for each county within their district. These county enrollment officers were directed to comply with any regulations from the Confederate War Department.[47]

A provost marshal was appointed in each county, as well as a Confederate enrollment officer. Joseph Graham was appointed provost marshal and the enrollment officer in Kendall County. James M. Starkey was appointed provost marshal and enrollment officer in Kerr County. Charles de Montel was the provost marshal in Bandera and Medina counties. The name of the Medina County enrollment officer is not known. O. B. Miles was the Bandera County enrollment officer. The names of the Blanco County provost marshal and enrollment officers are not known. The name of the Gillespie County provost marshal is not known, but Charles Nimitz was the enrollment officer.[48]

The insurgents chose at just this time to hold a mass meeting. The *San Antonio Herald* of May 31, 1862, reported in a tongue-in-cheek tone, on this meeting: "It is reported that the young men of [Gillespie County] had a meeting recently, in which they declared that as they were better farmers than the old men, they would remain at home, and let the old men go to the war. The latter held a meeting and replied that as the young men refused to fight, they also would remain at home.[49]

Now that General Bee had troops and authority to deal with the Unionist, he took direct action. One of the first things he did was to dispatch military units to some locations to insure the requirements of martial law and Confederate conscription law were properly circulated and enforced. These included Gillespie, Blanco, and Kendall Counties. The Hill Country Unionists and insurgents were about to feel the first bite of Confederate steel. For the League and their supporters, a 'reign of terror' was getting nearer.

CONFEDERATE STATES OF AMERICA
Headquarters Department of Texas
Houston, May 30, 1862
GENERAL ORDER NO. 15

I. The following proclamation is published for the information of all concerted.
PROCLAMATION
II. I P. O. Hebert, Brigadier General Provisional Army Confederate States of America do proclaim that Martial Law is hereby extended over the Sate of Texas.

Every white male person above the age of sixteen years, being temporarily, or otherwise within the aforesaid limits, shall upon a summons issued by the Provost Marshal promptly present himself before said provost marshal to have his name, residence and occupation registered, and to furnish such information as may be required of him. And such as claim to be aliens shall be sworn to the effect that they will abide by and maintain the laws of this State and the Confederate States as long as they are permitted to reside therein and that they will not convey to our enemies any information whatever or do any act injurious to the Confederate States or??? to the United States.

Provost Marshals shall order out and remove from their respective districts all disloyal persons and all persons who presence is injurious to the interest of the county.

All orders issued by the Provost Marshal in the execution of their duties shall be promptly obeyed. Any disobedience of summons emanating from the shall be dealt with summary punishment.

Any attempt to depreciate the currency of the Confederate States is an act of hostility; will be treated as such and visited summary punishment.

No interference with the rights of loyal citizens, or with the usual routine of business, or with the usual administrative of the laws, will be permitted, except when necessary to enforce the provisions of this proclamation.

By order of Brigadier General P. O. Hebert, P. Army C. S.
Commanding Department of Texas
Samuel Boyer Davis, Capt. and A. A. G.
HEAD QUARTERS, SUB. MIL DIST OF THE RIO GRANDE

San Antonio, July 21st, 1862

By virtue of General Order No 7, from the Head Quarters of the "Trans. Miss. Dist. South of Red River" bearing date July 8, 1862, as follows:

Head Quarters, Trans. Miss. Dist
South of Red River
San Antonio, July 8th, 1862

GENEAL ORDER, No. 7.

1. Brig. Gen. H. P. Bee, is charged with the execution of Martial Law thro' out the State of Texas.

2. He will appoint such Provost Marshals as he may deem proper and establish all necessary regulations for their guidance.

3. Provost Marshals will report by letter to Brig Gen. Bee, at San Antonio for instruction, when in doubt as to their duties of jurisdiction.

By order of
Brig. Gen. P. O. Herbert.
C. M. Mason, Capt. and A. A. Gen'l

The Provost Marshals throughout the State of Texas, are hereby directed to observe the following regulations:

1. No arrest will be made unless based upon two respectable citizens; except in cases when the suspected party is liable to escape before such basis for action can be obtained.

2. When parties have been arrested, whether upon affidavits or otherwise, the Provost Marshall by whose order the arrest was made, shall cause the accused to be brought before him for examination with the least possible delay, together with such witnesses as can be procured to establish any charge alleged.

3. The examination shall be conducted as far as, in accordance with the statute law of the State for Justices Courts. No *unnecessary* delays however for a strict and technical compliance with the details of said statute must be permitted; and only such will be extended as the Provost Marshals shall deem conducive to the ends of justice.

4. If the facts adduced upon the examination are such as will establish the guilt of the accused, they will be sent

before the nearest Military Commission with the accused. The testimony of each witness must be subscribed on said transcript; and the names and residence of all parties connected with the transaction, with the time and place, as definitely as possible, must be set forth and the whole certified by Provost Marshal.

VI. Prisoners will be securely confined, but no harsh measures must be resorted to, unless rendered imperatively necessary for his safe custody.

VII. PASSPORTS

Provost Marshals may issue passports to loyal citizens, who have taken the oath of allegiance, to pass and repass within the limits of the State, at discretion. They are not authorized under any circumstances to grant passports to persons liable to military duty under the Conscript Law.

No charge will be made for Passports.

Offices of the Confederate army traveling under orders, and showing them, will not require other passports.
Form of Passport.

THE CONFEDERATE STATES OF AMERICA.
Jurisdiction of _____County, Texas. No _____

These are to request all persons in authority and all others whom it may concern,
To let a citizen of and a resident of going by
To, about affairs pass safety and freely on said route without give him any hindrance, but, on the contrary, affording him all manner of protection, so far as necessary.

Witness my hand, at my office in this day of 186.

Provost Marshal
 Signature of bearer:

_____

DESCRIPTION:

Age _____ Eyes _____ Face _____
Height _____ Nose _____ Complexion _____
Weight _____ Mouth _____ Peculiarities _____
Hair _____ Chin _____

1. A Monthly return of the disbursements, all accounts for expenses and all requisitions for funds to defray them, but be sent to these Head Quarters for approval by Brig Gen H. P. Bee. There is neither rank nor pay attached to the office of Provost Marshal but the actual expenses will be paid.

2. The Governor of the State having placed the State troops at the disposition of the Confederate authorities, Provost Marshals are directed to call on them to assist in the discharge of their duties, when no Confederate troops are in the immediate vicinity.

3. Provost Marshals are to exercise a wholesome and efficient service over all travelers—to see that they are provided with proper passports, and to hold all persons without such evidence of their loyalty, to a strict accountability.

They will regulate the sale of all intoxicating liquors whenever the latter becomes necessary. They will regulate and establish a tariff of prices on all articles of prime necessity, whenever the disloyal and avaricious render this course necessary in exorbitant charges.

4. When travelers present themselves before a Provost Marshal within this state, it is not necessary for a new

passport to be issued. The Provost Marshal will simply endorse the old passport.

Should it, however appear that the first passport was issued without the oath of allegiance having been administered, Provost Marshal will administer the oath, register the party, and issue a new passport.

5. A register of all persons who have taken the oath, and all to whom passports have been issued, must be kept by the Provost Marshal. Said register shall show place of birth, place of residence, age, occupation, general description, and State or Country, of which a citizen.

6. By General Order No. 45, Department of Texas, all officers commanding troops are required to comply with requisitions made upon them for aid or assistance by the Provost Marshal. This does not confer authority upon the Marshals to order any officer. The officer is himself responsible for the mode and manner of complying with and executing the requisitions.

7. Provost Marshals will exercise their authority with judgment and discretion, and with the least annoyance possible to good and loyal citizens. No arbitrary or tyrannical acts will be tolerated, and upon proofs being furnished of the same, the Provost Marshal so offending will the summarily dismissed and otherwise punished under the law.

By order of Brig. Gen. H. P. BEE
E. F. Gray,
Major and A. A. Gen'l

Figure V

Endnotes – Conflict and Reversal

1. Ponder *Fort Mason, Texas*, pp 124-125 and Raley *Blackest Crime in Texas Warfare.*

2. *The Civil War Day by Day* by E. B. Long, (Doubleday and Company, Garden City, New York, various pages.

3. Jim (James) Lane was a senator from Kansas whom it was believed would lead the Union forces attacking from Kansas. McCulloch's Report, March 3, 1862 and *Austin State Gazette*, February 1, 1862.

4. *San Antonio Herald*, March 29, 1862, and "A 'Very Muddy and Conflicting' View: The Civil War as Seen from Austin, Texas" by David C. Humphrey in 'Southwestern Historical Quarterly', XCIV (January 1991), p 380.

5. Henry Eustace McCulloch was born in Rutherford County, Tennessee on December 6, 1816. He and his brother, Ben, arrived in Texas in 1835. Henry was an early Texas Ranger and fought with Captain Jack Hays' Ranger Company during the Mexican War. When the War Between the States began, he was appointed a Confederate colonel and given command of the 1st Regiment Texas Mounted Rifles. In 1862 he was promoted to brigadier general. Henry McCulloch died on March 12, 1895.

   Report H. E. McCulloch, Colonel, 1st Regiment Texas Mounted Rifles, C.S.P.A., Commanding S.M.D. of Rio Grande, San Antonio, Texas, March 3, 1862, OR, Series I, Volume 4, pp 701-702 and Ron Tyler, et, al, eds., *New Handbook of Texas*, Volume 4, pp 385-386.

6. Williams' *Border Ruffians*, p 233; Thompson *Vaqueros in Blue and Gray*, p 104; Federals *on the Frontier: The Diary of Benjamin F. McIntyre 1862-1864* Edited by Nannie M. Tilley, (University of Texas Press, Austin, Texas), 1963, p 352; Dispatches from United States Consuls in Matamoros, Mexico 1826-1906, Volumes 7-91; Dispatches From United States Consuls in Monterrey, Mexico 1849-1906, Volume 1; Report, James Duff, Captain, Commanding Company of Texas Partisan Rangers, Headquarters Camp Bee, San Antonio, Texas, June 23, 1862, OR, Series I, Volume 9, pp 785-787; and Letter, Captain Montel, Camp Verde, May 31, 1862 to Colonel Dashiell, Texas Adjutant General, Austin, Texas, A.G.C., TSA, Austin, Texas. While it is clear Union officials in Mexico knew of the Texas Unionists activities there is no record they led the Unionist to believe an invasion was close. Letter Leonard Pierce, U. S. Consul, Matamoros, Mexico to the Hon. W. H. Seward, Secretary of State, Washington, D. C., March 21, 1862, Dispatches from United States Consuls in Matamoros 1826-1906, Volumes 7-9.

7. The fact that San Antonio Unionists organized a military unit is supported by one of the ultra-radicals Forty-Eighters of 1854. Julius Schlickum was quoted during his trial before the Confederate Military Commission as saying he knew of an organization in San Antonio, but believed they were mostly boys and could easily be put down.

San Antonio and Bexar County were in the 30th Brigade District commanded by Brigadier General James M. Duff. The Bexar County Unionists were members of these state militia units. It is not known how many they controlled. There were 24 companies of about a 1,000 men each in the 30th Brigade. These companies were assigned to one regiment and three independent battalions. The nine companies assigned to the 1st Regiment included: Alamo Grays, commanded by Captain M. G. Cotton; Dashiell Guards, commanded by Captain Z. Van Ward; Crockett Guards, commanded by Lieutenant Richter; Company No. 2 of Precinct No. 1, commanded by Captain W. R. Henry; Company No. 3 of Precinct No. 1, commanded by Lieutenant G. W. Caldwell; Company No. 2 of Precinct No. 2, commanded by Captain R. A. Henson; Company No. 2 of Precinct No. 3, commanded by Captain L. Iwonsky; Company No. 3 of Precinct No. 3, commanded by Captain George Horner; and Company No. 4 of Precinct No 3, commanded by Captain E. A. Florian. The three independent battalions were 3rd Battalion, 4th Battalion, and 5th Battalion. The seven companies of the 3rd Battalion included: Mission Guards, commanded by Captain L. Castro; Calaveras Guards, commanded by Captain M. Delgado; Graytown Pioneers, commanded by Captain J. W. Gray; Company from Precinct No. 11, commanded by Captain Sierra; Medina Guards, commanded by Captain S. L. Stanfield; Company from Precinct No. 13 and part of Precinct No. 9, commanded by Captain E. B. Pue; and Company from Precinct No. 16, commanded by Captain Charles Keiser. The three companies of the 4th Battalion included: Company from Precinct No. 22, commanded by Captain Edward Gallagher; Company from Precinct No. 27, commanded by Captain A. Guggar; and Company from Precincts No. 10, 14, and 19, commanded by Captain G.H. Judson. The five companies of the 5th Battalion included: Company from Precincts No. 1 and 2 of Wilson County, commanded by Captain Edwin Lilly; Company from Precincts No. 6 and 7 of Wilson County, commanded by Captain Creed Taylor; Company from Precinct No. 5 of Wilson County, commanded by Captain Herrera; Company from Precincts No. 3 and 4 of Wilson County, commanded by Captain Ximenes; and Company from Precincts No. 12 and 14 of Bexar County, commanded by Captain Ewell. If the San Antonio Unionists followed the example of the League and organized pure Unionist companies, they used the San Antonio *Turnverein* as they did in1854 when they armed themselves and provided protection of the *San Antonio Zeitung* newspaper. McCulloch's Report, March 27, 1862; Barr *CMC*, Volume LXXI (October 1967), p 253; Orders

No. 6, Head Quarters 30th Brigade, Texas State Troops, San Antonio, April 2nd, 1862, TSA, Austin, Texas; Siemering's Germans During Civil War, May 8, 1923 and Ron Tyler, et al, eds., *New Handbook of Texas*, Volume 2, p 683 and Muster Roll of the Capital Guards, George Hancock, Captain, February 25, 1862, TSA, Austin, Texas.

8. Cramer's Letter.

9. McCulloch's Report, March 25, 1862 and McCulloch's Report, March 31, 1862.

10. *San Antonio Herald,* March 29, 1862.

11. McCulloch's Report, March 31, 1862.

12. Cramer's Letter.

13. McCulloch's Report, March 25, 1862 and McCulloch's Report, March 31, 1862.

14. Samuel J. Richardson was born about 1828 in Kentucky. The date he arrived in Texas is not known. By 1860, he was a prominent Van Zandt County citizen. He enrolled in the company on April 19, 1861 and elected its captain. Richardson remained as company commander until the end of the war. His date and location of death are not known.

Squire Boone was born about 1815 in Tennessee. Date he arrived in Texas is not known. By 1860, he was living in Medina County. Boone was not a likely Unionists. When Texas seceded, he was a member of a pro-Confederate Vigilante group known as the Medina Committee of Safety. Boone's date and location of death is not known.

Thomas B. Ragsdale was born about 1828, location of birth not known. He enrolled in Company F, 2nd Regiment Texas Mounted Rifles on April 19, 1861. He died on April 3, 1861, at St. Louis, Missouri while a prisoner of war.

San Antonio Herald, April 19, 1862; Eighth U. S. Census, 1860 Van Zandt County Texas Census, p 40; Samuel J. Richardson's Confederate Service Records, Record Group 109, National Archives, Washington, D. C.; Eighth U. S. Census, 1860 Medina County Texas Census, 29 and 'Constitution of the "Committee of Safety" of Medina County' provided by Jim Menke of San Antonio, Texas and Thomas B. Ragsdale's Confederate Service Records, Record Group 109, National Archives, Washington, D. C.

15. Joseph Finger was born on May 25, 1816 in Oberentzen, Alsace. He was among the first settlers of D'Hanis in 1846. Finger was the 1st lieutenant of Company A, Medina County Independent Battalion, 31st Brigade District. Joseph Finger died on August 27, 1886.

Joseph Ney was born about 1833 in Prussia. He arrived in Texas in August 1846. Ney was a prominent merchant in Medina County. He was also a member of the pro-Confederate Vigilante 'Medina Committee of Safety' although considered a Unionist. Among one his many businesses were furnishing hay to Fort Clark. He died in 1882.

Report, Joseph Finger, First Lieutenant, Beat Company #3, Medina County, D'Hanis, Medina County, April 12, 1862 to Brigadier General Robert Bechem, A.G.C., TSA, Austin, Texas; Castro Heritage Association *Medina County History*, pp 258-259 and 437-438; Quarterly Report, 31st Brigade, July 1, 1862, A.G.C., TSA, Austin, Texas; Eighth U. S. Census 1860 Medina Census, pp 18 and 437-438 and Medina Committee of Safety.

16. Ibid.

17. William Williston Heartsill was born on October 17, 1829 near Knoxville, Tennessee. He arrived in Texas in 1859. Heartsill enlisted in Company F, 2nd Regiment Texas Mounted Rifles on April 19, 1851 and remained with the unit until the end of the war. His book *Fourteen Hundred and 91 Days in the Confederate Army* is one of the accounts of a soldier's life in the Confederate Army.

Fourteen Hundred and 91 Days in the Confederate Army: A Journal Kept by W. W. Heartsill, Or Camp Life: Day-By-Day, of the W. P. Lane Ranger, by W. W. Heartsill, edited by Bell Irvin Wiley, (Roadfoot Publishing Company, Wilmington, North Carolina), 1992, p 6; Joseph Finger's Report and Captain H. J. Richarz's Endorsement of Finger's Report.

18. Cramer's Letter.

19. Schwartz *22 Months*, pp 89-90.

20. Schwartz *22 Months*, pp 89-90 and Siemering's Germans During Civil War, June 1, 1923.

21. Martin *Texas Divided*, p 115.

22. Cramer's Letter.

23. *San Antonio Herald*, March 29, 1862.

24. Cramer's Letter.

25. Fehrenbach *Lone Star*, p 362 and Webb *Handbook of Texas*, Volume I, pp 394395.

26. *The Dead Men Wore Boots: An Account of The 32ⁿᵈ Texas Volunteer Cavalry, CSA, 1862-1865*, by Carol L. Duaine, (San Felipe Press, Austin, Texas), 1996, pp 21-31; *Clear Springs and Limestone Ledges: A History of San Marcos and Hays County* Compiled by Hays County Historical Commission, (Nortex Press, Austin, Texas), 1986, pp 114-117 and Compiled Service Records, 32ⁿᵈ [also known as 36ᵗʰ] Regiment Texas Cavalry, Record Group 109, National Archives, Washington, D. C.

27. Joseph Taylor was born about 1815 in South Carolina. It is not clear when he and his family arrived in Texas. The 1860 Harrison Census shows him as a medical doctor. On July 20, 1861, he was appointed the Surgeon for McCulloch's 1ˢᵗ Regiment Texas Mounted Rifles, a position he held until April 1862 when the regiment's term of service ended. Four of the companies elected to remain on active duty. Joseph Taylor was elected major and commander of the new battalion on June 1, 1862. In May 1863, Taylor's 8ᵗʰ Battalion and Yager's 3ʳᵈ Battalion were consolidated and renamed the 1ˢᵗ Regiment Texas Cavalry. Taylor was not appointed/elected one of the new officers. It does not appear he served in any other Confederate unit during the war. His date and location of death is not known.

Compendium of The Confederate Armies:Texas, by Stewart Sifakis, (Published by Facts On File Inc., New York, New York), 1995, p 58; *Texas in the War 1861 – 1865* Compilation by Brigadier General Marcus J. Wright, CSA and Annotated, Compiled and Edited by Colonel Harold B. Simpson, U. S.A.F., (Ret.) (Hill Junior College Press, Hillsboro, Texas), 1965, pp 32, 35 and 125 Compiled Service Records, 8ᵗʰ Battalion Texas Cavalry, National Archives, Washington, D. C.; Eighth U. S. Census, 1860 Harrison County, Texas, p 462; Compiled Service Records, McCulloch's 1ˢᵗ Regiment Texas Mounted Rifles; Joseph Taylor's Confederate Service Records for both 1ˢᵗ Regiment Texas Mounted Rifles and 8ᵗʰ Battalion Texas Cavalry; McGowen *Horse Sweat and Powder Smoke,* p 62.

28. *San Antonio Herald*, May 10, 1862.

29. Siemering's Germans During Civil War, June 1, 1923.

30. Compiled Service Records of Taylor's 8ᵗʰ Battalion Texas Cavalry, Record Group 109, National Archives, Washington, D. C.; Kuechler's MR Records, 31ˢᵗ Brigade, A.G.C., TSA, Austin, Texas; and Compiled Service Records

of the First and Second Regiments Texas [Union] Cavalry, National Archives, Washington, D. C.

31. Records, 31st Brigade, A.G.C., TSA, Austin, Texas.

32. Unpublished diary of Elise Wuppermann copy in possession of the author. This is one of the earliest references to hanging of Unionists. Apparently, the consul had not learned the Austin Unionist Party had disbanded. Report, Consulate of the United States of America, Matamoros, Mexico, May 5, 1862 to Hon. W. H. Seward, Secretary of State, OR, Series I, Volume 9, p 686.

33. Report, Consulate of the United States of America, Matamoros, Mexico, May 23, 1862 to Hon. W. H. Seward, Secretary of State, OR, Series I, Volume 9, p 686.

34. Charles Schwartz was born about 1839 in Hessia. His date of arrival in Texas is not known. Schwartz was a member of Kuechler's Company in December 1861. He enrolled in Van der Stucken's Company on May 3, 1862. His date and location of death are not known.

    Barr *CMC*, Volume LXXI (October 1967), pp 265-266; Eighth U. S. Census, 1860 Gillespie County Texas Census, p 19; Kuechler's MR and Carl Schwartz's Confederate Service Records.

35. Duff's Report, pp 785-786; and Letter, R. A. Gibson, Camp Davis, March 31, 1864, A.G.C. TSA; CMC, Volume LXXI (October 1967), 264; and CMC, Volume LXIII (July 1969), p 84.

36. Siemering's Germans During Civil War, June 1, 1923 and Message, Henry Eagle, Captain, Commanding Naval Forces off Galveston, Texas, May 17, 1862 to The Military Commandant, Commanding Confederate Forces, Galveston, Texas; Message, Henry Eagle, Commanding Naval Forces off Galveston, Texas, May 19, 1862 to The Foreign Consuls residents of the Town of Galveston, Texas OR, Series I, Volume 9, pp 710-711.

37. *San Antonio Herald,* May 31, 1862.

38. Montel's Report of May 31, 1862.

39. Ibid.

40. General Order Number 525, Headquarters, Sub-Military District of Rio Grande, April 28, 1862, War Department Collection of Confederate Records, Record Group 109, Orders and Circulars, Department of Texas, Entry 105, National Archives, Washington, D. C. Copy provided by Professor Richard Selcer, Ft. Worth, Texas.

41. Thomas Crutcher Smith was born on December 12, 1843 in DeWitt County, Texas. He enrolled in Company G, 32nd Regiment Texas Cavalry on March 30, 1862 and served with the company until the end of the War. He was later the DeWitt County Clerk and received his law licenses in January 1871. Smith died on May 17, 1913 in Waco, Texas.

    Here's Yer Mule-The Diary of Thomas C. Smith, 3rd Sergeant, Company G, Woods, Regiment, 32nd Texas Cavalry, CSA, by Thomas C. Smith, (The Little Texan Press, Waco, Texas), 1958, pp 11-12.

42. Smith *Here's Yer Mule*, pp 11-12.

43. James R. Sweet was born in 1818 in Nova Scotia. He arrived in Texas about 1849. In the 1850s, he was a San Antonio alderman. The 1860 Bexar Census shows him as a merchant. At the time of Bee's martial law declaration, he was the San Antonio mayor. In early May 1862, he resigned as mayor and on May 4th joined Duff's Company of Partisan Rangers where he was elected 1st lieutenant. Sweet remained Duff's second in command until the end of the war. After the war, he fled to Mexico where he remained until the end of Reconstruction when he returned to San Antonio. In 1873, Sweet purchased the *San Antonio Herald*. James R. Sweet died on December 12, 1880 in San Antonio. Sub-Military District of Rio Grande General Order Number 5; *San Antonio Herald-Extra*, April 28, 1862; *San Antonio Herald*, April 28, 1862; Branda Handbook *of Texas*, Volume III, p 943 and James R. Sweet's Confederate Service Records, Record Group 109, National Archives, Washington, D. C.

44. Paul Octave Hebert was born on December 12, 1818 in Iberville Parish, Louisiana. He graduated first in his 1840 U.S. Military Academy class. In 1852, he was elected governor of his home state. He was commissioned a Confederate brigadier general at the start of the war. On May 28, 1862, he was appointed commander of the Trans-Mississippi District south of the Red River. In November 1862, he was relieved of command, largely due to his declaration of statewide martial law. Paul O. Hebert died on August 20, 1880 in Iberville Parish, Louisiana. One of the incorrect criticism levied against these provost's marshal was they were Herbert's "own band of henchmen" or they were "ruthless and despotic ... the most hated men in Texas." A leading Texas historian says they were likely "draft dodgers" and a "petty tyrant." The provost marshals of Kendall and Kerr Counties were the chief justices. Both were will liked and respected by both Secessionists and Unionists. The Gillespie County provost marshal was likely either the chief justice, William Wahrmund, or Charles Nimitz, the Confederate enrollment officer, both prominent and respected men.

    General Order Number 45, Headquarters, Department of Texas, Houston, Texas, May 30, 1862, OR, Series I, Volume 9, pp 715-716; Webb *Handbook of Texas*, Volume I, 792; "The Wrong Side of the River" by David F. Crosby in 'Civil War Times Illustrated', Volume 36, No. 7, February 1998, pp 48-49 and Oral Presentation by T. R. Fehrenbach as the "Nueces Encounter Symposium" on March 22, 1997 in Fredericksburg, Texas.

45. Cramer's Letter.

46. *San Antonio Herald,* June 7, 1862; Neu *Braunfels Zeitung, June* 20, 1862 and *San Antonio Semi-Weekly News,* June 26, 1862.

47. General Order Number 46, Headquarters Department of Texas, Houston, June 2, 1862, OR, Series I, Volume 9, p 717.

48. James Monroe Starkey was born on February 16, 1820 in Tennessee. He arrived in present-day Kerr County prior to the creation of the county. Starkey served as chief justice [county judge] in 1862-1863. James M. Starkey died on April 30, 1891 in Kerr County.

    Orlando B. Miles was born about 1824 in New York. The date he arrived in Texas is not known. He was the Bandera Chief Justice. He was the 1st lieutenant in Mitchell's 1864 Company of the Third Frontier District. His date and location of death is not known.

    "A Few Facts of Boerne History" by Edith Gray, January 1983, located in the Vertical File under title "Local History", Pioneer Memorial Library, Fredericksburg, Texas; Ron Tyler, et al, eds. *The New Handbook of Texas*, Volume 4, p 603; Letter, Charles de Montel, Camp McCord, July 21, 1862 to Colonel J. Y Dashiell, Adjutant and Inspector General, Texas State Troops, Austin, Texas, A.G.C., TSA, Austin, Texas; Bennett *Kerr County*, p 111: *The Ranger Companies of Bandera County* by Earl S. Hardin, Jr. (Printed by Earl S. Hardin, Jr., Bandera, Texas), 1995, p 162; Gillespie Historical Society *God's Hills*, Volume 1, pp 148-149; Watkins *Kerr County*, pp 109-113; Smith *Confederate Veterans of Kerr County,* p 45; Hunter *100 Years in Bandera*, pp 23 and 63-64 and Grave *Bandera Veterans*, p 38.

49. It is possible it was at this late May 1862 meeting when the League 'perfect the organization' as described by Sansom. This means Fritz Tegener was elected major in May as he states in his 1875 letter. This would also match Siemering's account when he says it was in the spring when the meeting was held. Even if it was late May before the League's Military Battalion was formed, it still was well before July 4th and not a reaction to martial law as claimed by Sansom. *San Antonio Herald,* May 31, 1862.

CHAPTER 6 -- Confederate Troops Arrive - *Des Organisator's* Reaction

The unit General Bee sent to the Texas Hill Country to proclaim martial law was the recently organized Partisan Ranger Company commanded by thirty-five-year-old Scotsman Captain James Duff. Duff had arrived in the United States on November 1, 1848 and enlisted in the U. S. Army on January 6, 1849. He served his 5-year term and was discharged on January 6, 1854, at Fort McIntosh, Texas as a sergeant. He was the post sutler at Fort Belknap from 1856 until 1859. On December 4, 1856, he married Harriett Paul, daughter of post commander and future Union brigadier general, Major Gabriel R. Paul. By 1860, Duff moved his family to San Antonio where he received a large contract to furnish the U. S. Army with supplies. When the war began, he organized a state militia company and later commanded a state infantry battalion. In January 1862, Governor Lubbock appointed Duff a state brigadier general in command of the 30th Brigade District. In May 1862, he resigned and was appointed a Confederate captain.[1]

The company left Camp Bee, in San Antonio, on May 28, 1862. [Appendix R]. The *San Antonio Herald* reported the departure of Duff's Company on May 31: Capt. Duff's command left town last Wednesday morning, armed and equipped as the law directs; and it is whispered that the noncombatants in and around Fredericksburg will receive the honor of its attention.[2]

Duff described the march and arrival in Gillespie County in a report to General Bee: "Agreeable to Special Orders, No. 242, I broke up my camp at this place [Camp Bee] on the morning of 28th of May ultimo and took up the line of march for Gillespie County. I reached Fredericksburg, the county site, on the morning of the 30th." One of Duff's troopers in a letter to his wife reported the company arrived about 5'o clock in the morning. The sudden appearance of 'regular' Confederate troops must have been a surprise to the local citizens and especially the area insurgents.[3]

Duff established his camp east of Fredericksburg at the abandoned Fort Martin Scott which he named Camp Gillespie. Upon arriving, Duff reported the first thing he did was, "immediately proclaimed martial law as existing within the limits of the county, in Precinct No. 5 of Kerr County, giving six days to enable the citizens to report to the provost marshal and take the oath of allegiance." Duff followed the example used by General Bee in San Antonio: he rode his entire command of 100 men into town in a show of force and very dramatically read the proclamation of martial law. This was followed by a citizen of Gillespie County, Fritz Messenger, reading the proclamation in German. Captain Duff immediately ordered thirteen men to be placed on the different roads, streets, and lanes, and not to let any person pass without having a pass from the provost marshal. One of the myths is that Duff was the provost marshal. He was not appointed provost marshal until his second trip at the end of July. The name of the Gillespie County provost marshal is a mystery. General Bee appointed the chief justices of Kendall and Kerr County as respective provost marshals of those counties. It may be this was the case in Gillespie County. The Gillespie chief justice was William Wahrmund. The other likely individual was Charles Nimitz, who was later appointed the Confederate enrollment officer. In some counties, these two duties were combined. Thirty-one-year-old Private Robert H. Williams, a native of England, a member of Duff's Company, and who was a major critic of Duff, stated the purpose was, "To compel these people to take the oath of allegiance to the Confederate government." However, another trooper, Manuel Yturri stated, "I am indeed very well pleased with company, and all the men seem to be very kind and accommodating, all well behaved.[4] This requirement to take the Confederate oath of allegiance was not something new. The Confederate Congress had ordered it and President Davis had issued a proclamation in August 1861 ten months before. Each of those who had not taken the oath could have been treated as 'spies' or forced removal from the Confederacy.[5]

Williams'described Fredericksburg and its inhabitants as, "a town of about 800 inhabitants, almost all of them Germans, and Unionists to man." Duff, like Williams, also told of his initial opinion of the inhabitants. He was not as critical as was Williams. "I found the people shy and timid. I visited with a part of my company several of the settlements and explained to the people the object of our visit to their county." However," according to Duff, "in a few days they displayed much more confidence in us, and in a corresponding ratio more desire to serve the government."[6]

Duff encountered more serious problems than just shy and timid inhabitants. The first was being unable to buy forage for his mounts. He learned the citizens who were favorable to the Confederacy had sold what they had. The others would not sell for Confederate paper money. Duff took direct action. He ordered one of his officers, twenty-one-old Lieutenant Edwin Lilly (later infamous as the officer who ordered the wounded insurgents shot), to visit Frederick Lochte, who unknown to Duff was an original member of the Union Loyal League. Duff merely described Lochte as a "wealthy merchant who had purchased most of the forage, but would not sell for [Confederate] paper currency, and inform him that I required fifty bushels of corn." The visit by Lilly convinced Lochte to sell his corn for Confederate paper money. This ended anyone refusing to take Confederate currency, or as Duff said, "After some little hesitation [Lochte] agreed to furnish [the corn]. After this I had not difficulty in getting forage and all other necessary supplies."[7]

Most of the Gillespie County citizens took the required oath of allegiance within the six days given by Duff, but according to Private Williams, "some took to the mountains rather than perjure themselves." Ernest Cramer, the commander of the Kendall League Company told about the citizens taking the oath. He wrote, "Excepting a very few, all took the oath, and also betrayed their officers. All officers had to immediately flee for their lives. And now, with a number of my friends, I departed into the mountains. There were Kuchler – two brothers by the name of Degener …

Franz and Moritz Weiss, Ernest Beseler, Wilhelm Telgman, Amil Schreiner." On June 3rd and while awaiting the end of the six-day grace period, Duff received orders from General Bee to send a detachment to Medina County to arrest "certain citizens" and break up the Unionist's courier service. Based on the way this requirement is written in Duff's Report, it appears General Bee named Medina County individuals he wanted Duff to arrest. Private Williams was part of the detachment. Duff left Fredericksburg with about forty men. Most of the men, to include Duff, only went as far as Camp Verde in Kerr County. There they tried to, "overawe, or convert into Southerners, more Germans of Northern proclivities," according to Williams. Beside the military post at Camp Verde, there was a small settlement of Zanensburg [now Center Point] located about 7 miles' northeast of Camp Verde. The 1860 Kerr Census shows 13 families receiving mail at Zanensburg. About half of those had German surnames. What is interesting about Williams' account of stopping in the Camp Verde area is that there was a Confederate company guarding Union prisoners at Camp Verde. This was Company K, 32nd Regiment Texas Cavalry, commanded by Captain Stokely Holmes. The company had been at Camp Verde since February 1862. Captain Holmes was known as a strong Secessionists and actively searched for Unionist and Union sympathizers. Williams and two other men were sent to Medina County where they arrested about ten "supposed disloyalties." One of those arrested was H. J. Richarz, the commander of the Medina County Independent Battalion of the 31st Brigade District. On July

25th, The Confederate Military Commission released him for lack of evidence. Others arrested included: Blasins Kieffer, Joseph Wipff, Valentine Haass, two men by the names of Capeans [likely Francois Joseph Carle] and Bergs [likely Jean Jacques Biry]. The Confederate Military Commission released all by July 30th. At the most six individuals were arrested, not ten, as claimed by Williams. Those arrested were sent to San Antonio. They arrived on Saturday, June 7th. A San Antonio newspaper reported their arrival: "We learn that a portion of Capt. Duff's Company arrived in our city on Saturday last, from Fredericksburg, via Castroville, bringing in several prisoners, with them. These prisoners, we presume, are persons who have not yet been convinced that the South is in earnest."[8]

Duff returned to Fredericksburg. The six-day grace period was over. He now set about finding and arresting known insurgents. At first, he found few local citizens willing to openly provide him with intelligence on insurgents. Again, Duff took direct action. "I found beyond doubt that the few citizens of the place who were friendly to this Government did not posses' moral courage enough to give information to the provost marshal of the sayings and doings of those who are unfriendly." Duff consulted with Lieutenant James Sweet his second in command and the former San Antonio mayor. They decided to summon some of the prominence citizen to meet with them. One of the men Duff and Sweet met with was Charles Nimitz. The relationship between Duff and Nimitz grew; he is the man R. H. Williams later describes as one who provided information to Duff. "They obeyed the order and made affidavits in regard to certain citizens of the county, viz: Sheriff [Braubach], Captain [Kuechler] (State troops) F. W. [Doebbler], a grocery keeper, and Mr. F. Lochte, merchant." When Duff used the term Captain Kuechler (State Troops) he was referring to the fact Kuechler was the captain of Company E, 2nd Regiment, 31st Brigade District. Duff was unaware how correct he was in the use of the term 'captain.' Besides being captain of Company E, Kuechler was also captain of the Gillespie Company of the League's Military Battalion. Braubach and Doebbler were two of the men denounced by the Gillespie Rifles the previous February and were original members of the League. The affidavits made were sufficient basis on which Duff issued warrants for their arrest. Duff detached parties for this purpose and succeeded in arresting Doebbler. The others all had left town.[9]

A housewife living in Fredericksburg recorded the arrest of Doebbler. She wrote to her husband, a Confederate soldier, on June 15, 1862, "The soldiers have left and taken with them Mr. [Doebbler] in chains. They would not trust him to this Jale [jail]. They sayed they was going to take him to Sanantonio." This same woman told about Duff's search for other Unionists. "One day they came in and surrounded that house by the Mill and jumped in the windows six-shooters and knives in hand but found no body there. Some thinks they got one of them but they do not know. The other made his escape." These searches by Duff's troopers found some of those who had "left town." Lieutenant Lilly captured Frederick Lochte, and Corporal Henry Newton followed

Philip Braubach to Austin where he was arrested. Duff was not able to find and arrest Jacob Kuechler.[10] *Des Organisator* made a deadly mistake by allowing their key members to be identified and mostly arrested. This should never happen if an insurgency is to be successful. The Confederates had taken the exact action to destroy or at least greatly hurt the insurgency.

The San Antonio Semi-Weekly News of June 16, 1862 described the arrival of Philip Braubach and the other prisoners in the city: Capt. Duff's Company continues to send to our city more prisoners from Fredericksburg and vicinity. Last week, among the prisoners from there, arrived here the Sheriff of Gillespie County, carrying along with him a ball and chain. We learn he had underway a body of 'home guards,' intended for frontier defense no doubt, in case Abe's forces land in Texas.[11]

Captain Duff described the four he acquired statements against as, "These men are all inimical to our country and possess a vast amount of influence among the laboring and agricultural classes. ... their absence from Gillespie County will tend more to make the people of that county united in favor of our Government than anything else." Duff expressed regrets he was unable to find and arrest Jacob Kuechler. He stated Kuechler was the only one of the four he was looking for who had not taken the oath of allegiance to the Confederacy. In his report to General Bee, Captain Duff said, "I may be allowed to suggest that steps should be taken to arrest Captain Kuechler. He is a man of great influence; a German enthusiast in politics and a dangerous man in the community."[12]

Duff turned to his attention to Precinct Number 5 of Kerr County, which was the western part of Kerr County where the majority of the Anglo insurgents lived. This area also included the western part of Gillespie County. Here he found "a few men who are bitterly opposed to [the Confederate] Government." He identified the inhabitants as frontiersmen. He further identified the Anglo insurgents as "renegades from justice from other States, and men who will not fail to injure a political or personal enemy whenever an opportunity offers." To make his case, Duff reported these insurgents had "burned the entire fences of an old man because his sons had gone to the war and because he was a good Southerner." Other sources identify this "old man" as Robert A. Gibson. Duff reported the Kerr County Unionists of Precinct Number 5 were "headed by an old man by the name of [Hiram] Nelson." The captain was careful about making sure Nelson and his followers received the requirement to take the oath of allegiance in sufficient time to report to the provost marshal and take the required oath. Nelson and his followers, "failed to do and Nelson sent [Duff] a defiant message." After the six-day grace period, Duff sent a detachment of state troops from Davis' Company to arrest Nelson. But Nelson "had taken to the cedar-brakes and escaped."[13]

James Duff now experienced the first of what was to become an almost continuous effort to 'demonize' him as a Confederate tyrant. This was the spreading of a rumor that he and his company had murdered a woman in Fredericksburg. Duff paid little attention to the rumor, except to report to General Bee that an effort had been made to "create a feeling in the community against my company by manufacturing and circulating the basest falsehoods in regard to it." Others came to Duff's defense. The *San Antonio Herald* in its June 11th issue said Duff's Company had been falsely accused of mistreating a woman in Fredericksburg. The paper went on to report no opposition had been shown to martial law and all was well in Fredericksburg.[14]

Insurgents and bad press were not the only problems Duff had in Gillespie County. He lost a horse, possibly to thieves. The June 21, 1862 *State Gazette* carried an ad from Duff offering a $50.00 reward for the return of his horse. The horse was described as a "blood bay 15 ¼ hands high, a star on the forehead, with a white snip on the nose."[15]

Unbeknown to Captain Duff at the time, he almost had trouble with Anglo insurgents from east Gillespie County and northern Blanco County. He later reported, "Information reached me which led to the conviction that these men, or a majority of them, with some of the rabble had gone to Fredericksburg armed and equipped to endeavor to raise a party to fight my command."[16]

Captain Duff identified some of the leaders and members of the group, which had tried to organize to fight him. "The names of the leaders of this party are Prescot, King, Howell and two brothers, or father and son, by the name of Snow." Duff identified another member of the insurgent group as named Eaton [Edom]. Edom who was the man who started the rumor about Duff's Company mistreating the woman in Fredericksburg. At the time, at least five of the insurgents; Edmund A. King, John T. Howell, Moses Moran Snow, William F. Snow, and Arch B. Edom were members of Freeman's Blanco Home Guard Company, the Pedernales Cavalry Company. All lived near North Grape Creek in Blanco County. Prescott was likely thirty-three- year- old Aaron Prescott, a native of Maryland. Moses M. Snow and William F. Snow were killed in January, 1864, during 'the Bushwhacker War'.[17]

On June 11, 1862, Duff and his company of rangers left Gillespie County and moved to Blanco City, county seat of Blanco County, where he declared martial law. Duff received a much warmer reception in Blanco County than he had in Gillespie County.

In referring to Blanco County he said, "Here I found the great majority of the people friendly, enthusiastically so, to the Confederate States Government."[18] It was in Blanco County that Duff received detailed information on Unionist plans. Beside the news about plans to fight his command Duff, like Montel, learned the A. J. Hamilton Unionists were trying to create a larger military organization of the various Unionist military units. The Blanco Unionists had sent Edmund King to meet with A. J. Hamilton in Austin to discuss the plans for organizing the military force. Hamilton fled Austin and was hiding at his brother's ranch, located on the Travis and Blanco county line. Here Hamilton was meeting with the "disaffected of Kerr, Gillespie, Llano,

Travis, Blanco and neighboring counties." John T. Cleveland, who owned a ranch near Cypress Mill in northern Blanco County, and some of "the best citizens of Blanco" informed Duff that a large number of "strangers" were gathering at the Hamilton Ranch. As many as twelve had stopped at Cleveland's house within the past two or three days looking for the Hamilton Ranch. Cleveland paid for passing this information to Duff as Unionists burned his home a few months later.[19]

The Confederates had already experienced problems with this area of Travis County. Shortly after the war began, Confederate recruiters attempted to enrolled men from the area. When they approached the home of 62-year-old Kentucky native William Martin Preece (a relative of Abraham Lincoln) near Bull Creek, gunfire erupted from the house where Martin and five of his sons were waiting. The gunfire cut the bridle of one rider and grazed the head of a second. The recruiters quickly withdrew and no further effort was made to recruit the Preece boys. As tensions mounted between Unionists and Secessionists, the threat of open warfare became more likely. About 20 to 40 Travis Unionists calling themselves the Mountain Eagles and centered on the Preece brothers, forted up on the top of a large mountain peak near Martin's Well. According to Preece family stories and local folklore, a fight took place between a Confederate partisian ranger company and the Mountain Eagles in June 1862 in which 30 'Secesh' and three Mountain Eagle Unionists died. No official record of this incident has been located, but nevertheless the peak is now called Dead Man's Peak.[20]

Edward Slessinger, a justice of the peace and notary from Bexar, arrived at Blanco the same time as Duff's Company. Slessinger was sent to administer the oath of allegiance. He sent notices for every male citizen to come to his camp and take the oath of allegiance to the Confederate government. "Nearly all came in promptly and took the oath – a few however, refused and the cavalry scoured the county for them. A few evaded the cavalry and took to the brush. Mr. Eden, [sic] one of the men captured, was shamefully treated."[21]

One of the Blanco County citizens who still refused to take the oath was James Wilson Nichols. He later told about Duff's arrival and administering the oath, "Thare [sic] was a man appointed in every county to administer the oath of allegiance to the Confederate government and Co'l Duff was apointed [sic] for this purpose in Blanco County." Duff and Slessinger set several dates for the citizens to come in and take the oath, but Nichols continued to work in his blacksmith and wood shop saying he would never take the oath until "forsed [forced] to do so." After four days, Nichols claimed everyone except himself and one other man had taken the oath. In the meanwhile, local citizens held several mass meetings in support of the Confederacy. After one of these meetings, Nichols was informed of the date Duff and Slessinger would administer the oath to him. Again, Nichols refused. The next day as he was working in his shop Duff and two other men arrived. Nichols thought to himself,

"Gone up now," meaning he feared he might be hanged. The three told Nichols the reason for their visit. His reply was, "Good gentlemen, I have said I would not take the oath without I was forsed to and I concidder [sic] a forse-put so out with your oath." Nichols took the oath and Duff and the two men with him left.[22]

Having gotten Nichols and the other citizens of Blanco to take the oath, Duff and his company left Blanco County for Boerne, the county seat of Kendall County, on June 18th. They arrived on the evening of June 19th and made camp. The next day Duff proclaimed martial law and advised the citizens of the requirements.[23]

One of the citizens of Boerne who watched the arrival of Duff's Company was an 1854 ultra-radical, Julius Schlickum, commander of Company B, 3rd Regiment, 31st Brigade District, Texas State Troops and in the process of being elected lieutenant colonel of the 3rd Regiment. He expressed concern that he might be arrested to an acquaintance, thirty-seven-year-old Georgia native Erastus Reed, who represented the area in the Ninth Legislature. He first told Reed that he would flee to the mountains to prevent arrest, but later said as he had done nothing wrong he would stay in Boerne. Schlickum also spoke to Joseph Graham, chief justice and later provost marshal, about his fear. Schlickum told Graham he heard he was to be arrested and asked for his advice. Schlickum went on "That he might have sung some Yankee songs, but that he was drunk, and did not think it treason." Graham, who was about to be appointed provost marshal, replied that Schlickum was accused of being an Abolitionist. At which point Schlickum replied, "I was brought up in Europe and my views and yours differ." Duff arrested Captain Schlickum on June 20th because he, "has been bitterly opposed to us and who I have reason to believe took an active part in forwarding expresses and information to Federal prisoners at Camp Verde and the disaffected citizens of his own and adjoining counties." Schlickum was always in possession of news from the eastern battlefields at least two days in advance of the mail, which in Duff's mind meant he got his news from the Unionist couriers. Schlickum was also very active in local political affairs. As Duff explained, "He controls from 100 to 150 votes in Kendall County."[24]

Schlickum's own account provides more details of his arrest, "I too had warnings from dependable sources about my imminent arrest. However, I felt free of any guilt and believed I had nothing to fear. I was not a supporter of the South, but I had broken no law and was not involved in anything that would speak against me." His wife, Therese, was ill with "a terrible infection in the abdomen and suffered great pain." Schlickum's was fearful and worried awaiting the arrival of a doctor, who lived 12 miles away. To add to his problems, their maid had left several days earlier and they were without domestic help. He stayed awake the night of June 19th. Early in the morning he went outside. Their living quarters were about 55 feet from his store, situated on a hill. He looked out and saw two horsemen stopped at the gate, carbines in hand: farther out he saw others, about 30 Confederates encircling his house. He knew why. Quietly he went back into the house, to prepare his "deathly ill wife for what was about to come." He did it as tenderly as possible. He returned to the gate, where at this time the commander had arrived. Schlickum knew him; Joseph Graham, newly-appointed appointed provost marshal. Schlickum extended his hand. Graham said to Schlickum, "Mr. Schlickum, I'm sorry, but I have orders for your arrest." Schlickum replied, "I was waiting for you already yesterday." When Graham became aware that his prisoner's wife was sick, he allowed him back into the house, until Captain Duff arrived to dispense further orders. Graham's men occupied the house. Around noon Duff arrived. Schlickum offered $25,000 in bail, if Duff would let him remain in the house until the doctor arrived and examined Therese. Duff denied the request and arrested Schlickum, who said bitterly, "Spare me the words about my feelings when I had to leave my sick wife and crying children behind."[25]

On the morning of June 20th, Captain Duff received Special Orders Number 299 from General Bee's headquarters, directing him to return to San Antonio from Boerne. This sudden change in orders resulted in many subsequent writers saying Duff was recalled because of the atrocities he committed. This was not the case. There is no evidence that General Bee was displeased with Duff's actions. The fact is, Bee began to realize the depth of the Unionists' control and wanted Duff's evaluation. On June 23, Bee responded to Duff's intelligence and dispatched other Confederates units to find the Unionists meeting at Hamilton's Ranch. Duff and his company left Boerne on the afternoon of the 20th and arrived back at Camp Bee at noon on Saturday, June 21, 1862, "looking hale and hearty, but tolerably well dusted, and somewhat sunburnt [sic]," said one newspaper.[26] [Map F, p 301]

Captain James Duff and his company of Partisan Rangers were in the Hill Country from May 30th to June 20th, a total of 22 days. During this time, he, or members of his company, visited Blanco, Gillespie, Kendall, Kerr, and Medina Counties. They arrested ten men: H. J. Richarz, Blasins Kieffer, Joseph Wipff, Valentine Haass, Francois J. Carle, Jean J. Biry, all of Medina County; Philip Braubach, F. W. Doebbler, Frederick Lochte, Sr., all of Gillespie County; and Julius Schlickum of Kendall County. All those arrested in Gillespie and Kendall County were convicted by the Confederate Military Commission of Unionist activities. The same commission released all those arrested in Medina County. The facts show that during the twenty-two day stay in the Hill Country Captain Duff was very careful not to abuse his authority.

Many later claimed that Duff's trip was the start of a "reign of terror, arresting innocent citizens, and invading the dignity of the homes." One local professor in modern times has stated that, "in the rugged territory north of Kerrville, many neither heard of Duff nor his instructions and in any event were disinclined to make an arduous trip to comply; accordingly, when approached by Duff's men, they received rough treatment." The 'Duff Bashers' are relentless in their accusations, claiming that, "Duff bullied leading citizens, forcing them to give affidavits against suspected Unionists.

The latter he then arrested and sent for trial to San Antonio." They dismiss the results of his mission as, "Finding no real threat to the Confederacy, his command then returned to San Antonio, having accomplished little more than to antagonize a few harmless frontiersmen." Another states, "Roving bands of partisan rangers charged about the countryside and, in the name of the Confederacy, hanged innocent men before the eyes of their wives and children. Indiscriminately the rangers 'burned the cabins of settlers whose loyalty to the South was thought suspect, and as they hustled the poor souls out into the night, they shouted their own Rebel Yell: 'Give 'em a quick look up a tree!'" The official caretakers for the 'Treue der Union' Monument say in one of their handouts, "Immediately upon arrival [Duff] began to arrest prominent citizens. This was followed by the burning of crops and homes and other various atrocities in the name of martial law."[27]

The facts do not support any of these allegations. No houses were burned, no one was killed, and no one was unfairly arrested. All those arrested were given 'due process' under martial law. The three Gillespie County men arrested were insurgents and key leaders of the League. Braubach was a militant and would subsequently take up arms against the Confederacy. Likewise, Doebbler was a militant. Insurgents were openly meeting in his Beer and Wine Shop – where they were beating secessionist supporters. Braubach and Doebbler were certainly original members of the Union Loyal League. Fellow citizens of Gillespie County had condemned Braubach and Doebbler for Unionist activities and warned them to stop long before Duff arrived. Lochte refused to accept Confederate paper money as ordered by the statewide proclamation of martial law. He not only refused to accept Confederate money, he did so just after a Confederate military force arrived and announced any such refusal was an act of treason. The 'Reign of Terror' such as it was, began with the insurgents targeting their opponents: the raid on Gibson's farm, the intimidation of Charles Schwarz and Oscar Basse, the beating of pro-secessionist men in Doebbler's Beer Hall. These actions provoked Duff's arrest of key League members.

Before Captain Duff arrested anyone, he obtained affidavits from other Gillespie citizens regarding their Unionist activities. The citizens from whom Duff obtained the affidavits were the ones who had earlier warned the Unionists to stop their activities. No one was 'bullied.' Those who gave affidavits included leading citizens such as Charles Nimitz and the other members of the 'Gillespie Rifles.' Captain Duff may have assured those who gave affidavits that they would be protected. These same citizens testified against the insurgents during their trial before the Confederate Military Commission. The four were provided a trial with the opportunity to present a defense, which they did. Lochte was found guilty of refusing only to accept Confederate paper money for which he was sentenced to 30 days in jail and a small fine. His punishment under the circumstances could have been much worse. The six Medina County men Duff arrested were not so obviously Unionists, but after the war, Richarz, Kieffer, and Wipff all claimed to have been Union men. Biry and Haass were also almost certainly Union men, but died before they could publicly state so. In fairness to the Confederates, all the Medina men were released after a 'legal procedure' failed to find sufficient evidence of any guilt. Most, if not all, of those from Medina County were Unionists. However, unlike the Gillespie insurgents, they had not openly violated Confederate law.

If Captain Duff's Report of June 23, 1862 is read without bias, it presents a clear picture of what he found on his trip. It also shows how scrupulous he was in following 'lawful procedures.' A crucial, overlooked portion of his report notes that a group of Unionists had tried to organize an armed resistance to Duff. Both Duff and Montel reported the Unionists were attempting to consolidate their armed forces into a joint military organization. Events were moving into the latter stages of Phase II of the insurgency. The declaration of martial law and the arrival of Captain Duff and his company forced many of the militant Hill Country insurgents to 'take to the mountains.' The cover story used by many of those fleeing was that they were just going on 'hunting parties' or 'cattle hunting' trips. Insurgent Captain Ernest Cramer described the effect martial law had on the League and its military battalion: "We formed a hunting party, hunted in the mountains, and were hunted and chased by soldiers. But we knew the country and every secret path and hiding nook too well to allow ourselves to be caught."[28]

The creation of the new Confederate units gave the Confederate military the necessary armed force to back up the decree of martial law. General Bee quickly applied this force. The Unionists felt the sting of martial law. Many of them understood their lives were in danger. One of these was A. J. Hamilton. Hamilton and his followers fled to the hills west of Austin, where he continued his Unionist activities, which included administering Federal loyalty oaths to other beleaguered Unionists." As a result of intelligence gathered by Duff, General Bee knew that many Texas Unionist leaders were meeting in western Travis and eastern Blanco County. He immediately ordered Confederate forces into that area to assist Robert J. Townes, a member of the Eighth Legislature and local judge who was the Travis County provost marshal to find and arrest them. On June 23rd, the same day as Duff prepared his report, Bee ordered the companies of Captain Joel K. Stevens and Captain Lewis A. Maverick from Wood's 32nd Regiment Texas Cavalry to move immediately to Colonel Flex Kyle's ranch on the Blanco River. The two-company task force was under the command of Captain J. K. Stevens. Once he received the provost marshal's orders, he was to, "proceed to execute them promptly." Bee told Stevens, "These orders will be carried out with the force under your command if sufficient for that purpose; if found insufficient you will immediately dispatch a courier to these Head Quarters with the information and stating the strength of the reinforcements required." Bee's orders directed, "All persons arrested by virtue of the orders of the provost marshal of Travis County will be conveyed to Austin and turned over to that office."[29]

Captain Stevens and his two-company task force did not find and arrest Hamilton or his fifteen-man group. They had fled Texas for Matamoros, Mexico despite a large reward offered and a botched kidnapping attempt. Other Unionists hid in the mountains for over a year before fleeing to Mexico and later joining the Union Army. Among these were the Preece brothers. Thirty-year-old Richard Lincoln Preece enrolled in Company D, the First Regiment [Union] Texas Cavalry on July 23, 1863. Wayne Pulaski Preece waited until his 18th birthday on January 8, 1844 and on February 12, 1864 enrolled in Company E, First Regiment [Union] Texas Cavalry. Twenty-five-year-old William Martin Preece, Jr. fled in 1863 and enrolled in Company A, First Regiment [Union] Texas. On February 20, 1865, he was promoted to second lieutenant. In Matamoros, Hamilton and his followers boarded Union navy ships and offered their services to the Union.[30]

The arrival of Hamilton and his party in Matamoros was reported in the July 31, 1862 *Fort Brown Flag:* Andrew Jackson Hamilton and seventeen misguided noodles with revolvers and bowie knives worth more than the renegade tribe, made their appearance in Matamoros on Sunday, and were very promptly disarmed and put under surveillance by the authorities.[31]

The activities of Hamilton and his group, while in Matamoros, were reported to Confederate authorities. One such reported stated, "Having been on business in this town [Matamoros] I was introduced to two men who said they were going to Washington to receive men and money, to wit,

Mr. Jack Hamilton and a Mr. Haynes who are to raise and equip men for the Lincolnists." Hamilton had a letter from U. S. Secretary of State William H. Seward, authorizing anyone who had the letter be treated as Seward's agent. The report went on, "The U. S. Consul Mr. Pierce, Dr. Lee, and a Mr. Pertrn [sic] are doing all they can to induce the Confederate soldiers to desert. Pierce gives them money and Pertrn gives them employment." Pierce was doing all he could to foment quarrels between Mexican authorities and Confederate authorities in Brownsville. One of the ways Pierce was doing this was continuing telling the Mexican 'Commandante' that the Confederate soldiers were coming at night to kidnap him and other American citizens. Pierce had also paid one of the Confederate soldiers to bring a U. S. Flag to him in Matamoros. "Mr. Pierce has around him eighty men."[32]

Leonard Pierce reported the arrival of Hamilton and his fifteen-man group, "The Hon. A. J. Hamilton from Texas arrived here on the 19 July with fifteen men he having escaped from Austin on horseback." Hamilton's group crossed the Rio Grande about two hundred miles above Matamoros. Shortly after the crossing, a party of men from Captain Santos Benavides Confederate Company disguised as a Mexican Guard crossed and succeeded in taking one of his men into Texas with them, they supposing at the time that they had taken Andres J. Hamilton from whom a large reward is reported to have been offered. After some difficulty, Pierce was able to get Hamilton and about six of his party on the Federal schooner *N. Berry* on the evening of

July 21st. The schooner got underway early the morning of the 22nd but stopped offshore waiting the arrival of its captain. By evening, there was a rumor that the Confederates were fitting out a vessel to capture the *N. Berry.* Fearing this was true, the first mate left without the captain.[33]

Pierce was able to get other Texas Unionists safely out of Matamoros and on board the U. S. steamer *Montgomery,* commanded by Charles Hunter, which was on blockade duty at the mouth of the Rio Grande. One of Hunter's reports, dated June 16, 1862, told about the many refugees reaching Matamoros and his ship, "The poor refugees still come to use, one, two, or three each day. They flee for their lives, leaving everything." One of the men Hunter told about was a Confederate solider who got a six-day pass to go to Matamoros to collect a debt. When he got to Matamoros, he deserted. Another man, also a Confederate soldier, plunged in and swam across the Rio Grande under a hail of gunfire. Of the forty men Hunter had on board, one was a lawyer, another a judge, and the third an influential politician; all going to see the U. S. President to suggest the immediate occupation. "There is a large number of Union men in the State who only want arms and protection to organize themselves and drive the secessionist out." reported Hunter. Leonard Pierce added his name to the going list of individuals pleading for the Union to send troops to Texas. Pierce said, "A force of one thousand men could conquer the entire Line of the Rio Grande as most of the German and the old U. S. Soldiers would immediately turn at sight of the Union flag."[34]

While A. J. Hamilton, E. J. Davis and many other Texas Unionists fled to Mexico, the majority of the Hill Country Unionists – especially the leadership of the League – did not. But they did leave their homes and go into hiding.

Duff's arrival had a devastating effort on *Des Organisator* military battalion. Many of the battalion members not only came in and took the oath; they also gave the authorities the names of their insurgent officers. This almost totally destroyed *Des Organisator* capability to wage warfare. Only days before, the League was thinking in terms of undertaking offensive operations, now they were faced with a life and death struggle for survival. Instead of moving into the latter part of Phase II of their insurgency, they suddenly were forced back to Phase I, or reorganizing a military arm and keeping their political element from being discovered. [Map G, p 302]

Captain Cramer tells about taking his wife to live with his brother-in-law, Ferdinand Simon, who had taken the oath on May 27, 1861. He had prudently sold his property to his father-in-law, Gottlieb Bauer, on March 4, 1862. He describes what he did next. "And now, with a number of my friends, I departed into the mountains." Those who went with him included Jacob Kuechler, a member of *Die Vierziger* in 1847. He had also been the one who attempted to organize an insurgent company for the Frontier Regiment and who Duff had attempted to arrest. Jacob Kuechler was the commander of the Gillespie Company of the League's Military Battalion. A second member with Cramer was Ernst Beseler, a twenty-year son of Carl Beseler, a Forty-Eighter and Freethinker, who arrived on the *Franziska* in 1848. Two of Eduard Degener's sons, Hilmar and Hugo, fled with Cramer. Hugo was the lieutenant of Cramer's Kendall Company of the Leagues' military battalion. Franz and Moritz Weiss, another pair of brothers, were in the group. Moritz was the oldest, born in 1836. Franz was not yet twenty-years old, having been born in 1843. Their father was a Freethinker and Forty-Eighter who died enroute to Texas. Their mother married Oskar von Roggenbucke, "one of the most radical" Freethinkers and Forty-Eighters in Texas. Twenty-nine-year-old Wilhelm Tellgmann watched his brother Charles march off to war as a Confederate soldier. But, Wilhelm followed his father's Forty-Eighter beliefs and was an early member of *Des Organisator*. All would fight in the Nueces Battle and most would be killed. They were, "all educated young men of fine families."[35]

This group of about ten insurgents had a base camp near the head of the South Fork of the Guadalupe River. Another group of about fifteen to twenty had a base camp about six miles southwest of the current community of Albert in Gillespie County. The site of their base camp is now known as "Bushwhack Springs". They gave themselves the name "Luckenbach Buschwacher." The known members of the 'Luckenbach Buschwacher' included Sylvester Kleck, August Duecker, Adolph and Louis Ruebsamen, Jacob Kusenberger, August Schoenewolf, August Hoffmann; Joseph and Christian Poetsch, Adolph Vater, August Luckenbach, Philip Beck, Conrad Bock, Christian Schaefer, and Heinrich Weyershausen. The two Weiss brother, Moritz and Franz, are also listed with the group as well as Cramers' group on the Guadalupe. There were other groups which hid at various locations, but their names and exact locations cannot be determined. One group was made up of Anglo members of *Des Organisator* military battalion. They hid in several areas of western Gillespie and Kerr Counties. One witness testified at the Confederate Military Commission that some 70 or 80 Hill Country men had fled to avoid conscription. As part of his testimony he said, "They are in the mountains, on the Leona or Pedallis [Pedernales River]." Charles Nimitz testified nearly all of the former members of Kuechler's disbanded company were "now in the woods." Another Gillespie citizen, Frederick Fresenius, testified that a few of the former members of Kuechler's disbanded company had joined other companies, but, "the balance cannot be found." Confederate Captain Frank Van der Stucken, also of Gillespie County, testified, "The bulk of the men who composed Kuechler's Company are now in the woods." Ottocar Miller, still another Gillespie citizen, testified that the majority of Kuechler's former company were, "now out in the woods." Erastus Reed, a citizen of Kendall County, testified that Julius Schlickum told him that, "he knew there were over one hundred men out in the woods." Some of these Unionist groups were members of the Medina Independence Battalion of the 31st Brigade District. They were also on 'hunting trips' and thus avoiding Confederate enrollment officers.[36]

Both Captain Davis and Captain de Montel of the Frontier Regiment began expressing concerns about the strength of these Unionist groups operating in the mountains. One of the first of these reports was by Captain Montel. On June 16th, news reached him that a group of what was first believed to be Indians was operating on the upper Medina River. A ten-man detail under the command of Second Sergeant G. W. Hill was sent to scout the area. The next day, the 17th, Montel received another report of a second group of what was again believed to be about twenty Indians crossed the West Frio River and were headed toward the Medina River. Montel was concerned; if they were Indians it was likely the two groups were headed for "their former ground of depredation, Blanco and Kendall Counties." Montel sent a detail of sixteen men under the command of Second Lieutenant Gates after the second group. The two groups turned out to be armed white men. Lieutenant Gates reported finding a nine-man group, who were, "well mounted and armed to the teeth." They informed Gates they were part of a larger party who had gone higher up the country and they belonged to the German settlement on the Medina River above Castroville. They were indeed the parties seen, but they were just out "cattle hunting." Gates could do little else but return to his camp and let the armed party on their way. A few days later Second Lieutenant B. F. Patton also of Montel's Company G, stationed at Camp Montel on the Seco Creek, heard of a group of Indians roaming the mountains. He went after them with a twelve-man detachment. No Indians were found but "plenty of Union men in the mountains" were.[37]

Captain de Montel did not believe the armed groups were just out "hunting cattle." He reported to his headquarters that these parties were part of Captain Haby's Company D of the Medina County Independent Battalion "who were ordered out by their officers to keep conscription officers from finding them." Montel felt this group would join "Lincoln's forces if an opportunity should offer." Montel again warned his headquarters that, "a number of other [Unionist] parties of Medina County still remain in the mountains."[38]

General Bee and Governor Lubbock would have read with alarm a June 7, 1862 issue of a New York newspaper which stated Texas Unionists were about to strike and restore Texas to the Union. The article was reprinted in a later issue of the *San Antonio Herald*, which added its comments to the New York article;

MORE UNION GAS.

NEW YORK, June 7—The New York Tribune editorial says; "We learn through a private channel in which we bonafide, that the Unionists of Texas will soon be heard from. We understand that their arrangements for restoring their State to the Union have been quietly [illegible] and that have [illegible] thrown the old flag to the breeze under Gen. Houston. We cherish strong hopes that the rebels of Texas will soon turn up missing, and the Uncle Sam will have possession of the State. We await for the tidings with lively interest."[39]

The *San Antonio Herald* added its comments; "The arrangements of the Unionists of Texas to restore the State to the Union were made no doubt with more quietness than discretion. The 'arrangement' did not work well, and the old flag, instead of being thrown to the breeze, has been thrown to the dogs, whilst the said Unionist are generally in the Confederate armies, having long since become utterly disgusted with the vandals and their cause. The rebels of Texas, instead of turning up missing, are turning up in all the battle fields where the invading vandals may be found and it is the latter who are 'missing' whenever they encounter the deadly assaults of the said rebels."[40]

By the end of June, General Bee wanted units of the Frontier Regiment, as well as, Confederate units to be more aggressive when patrolling and finding Unionists. On June 30th, he requested Captain Montel send more armed scouts out. Montel complied and sent Lieutenant Gates and a twelve-man detachment back into the mountains. This time they found no Unionists because the Unionists had learned to avoid the company's patrols.[41]

Map F – Route of Duff's Company, May & June 1862

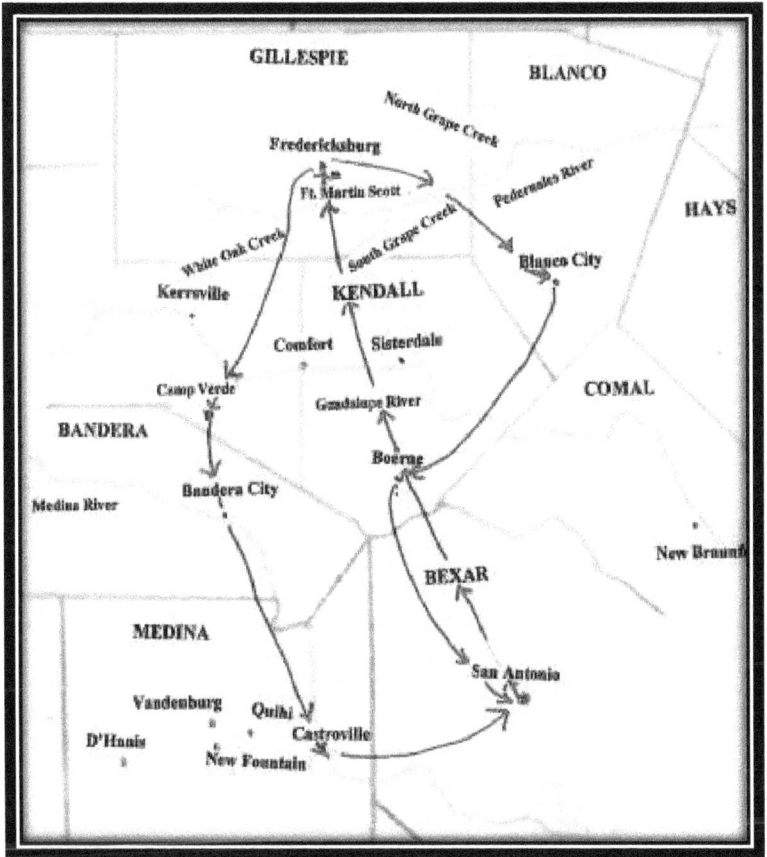

Map G – Insurgent Base Camps & Confederate/State Posts

Endnotes – Confederate Troops Arrive - Des Organisator's Reaction

1. James M. Duff was born in May 1827 in Logierait Parish, Perthshire, Scotland. He arrived in the United States on November 1, 1848 at the Port of New York. Duff joined the U.S. Army on January 6, 1849. He served his 5-year term of enlistment and was discharged on January 6, 1854 at Fort McIntosch, Texas as a sergeant. He was the post sutler at Fort Belknap, Texas from 1856 until 1859. He married Harriett Paul, the daughter of the

post commander, Major Gabriel R. Paul, a future Union brigadier general, on December 4, 1856. Duff bid on and received several contracts to furnish the U.S. Army with corn and other supplies. By 1860 Duff, moved his family to San Antonio, Texas where he received a larger contract with the U. S. Army. When the Civil War started, he raised a state militia company. He was elected a state lieutenant colonel in command of an infantry battalion that helped capture elements of the 8th U. S. Infantry in May 1861. In January 1862, the Texas governor appointed Duff a state brigadier general was and given the mission of organizing the 30th Brigade District. After accomplishing this mission, Duff resigned and in May 1862 was commissioned a Confederate captain. He raised a company of Partisan Rangers. His command was increased to battalion size in November 1862 and in May 1863 it became the 33rd Regiment Texas Cavalry. After the war, Duff, along with several other former Confederate officers, fled to Mexico. He returned to the United States in 1877, this time to Denver, Colorado where he operated a major investment company. He returned to England in the mid-1880s where he died on April 16, 1900 in London.

Letter, Robert and Penelope Fenwich, Perth, Scotland, dated February 5, 1998; *A History of Young County, Texas* by Carrie J. Crouch, (Texas State Historical Association), 1956, p 26;.Naturalization Papers, Bexar County Texas District Court, Fall Term 1858, dated September 5, 1858; *Three Months In The Southern States* by Lieutenant Colonel Arthur J.L. Fremantle, (University of Nebraska Press, Lincoln, Nebraska), 1991, p 53; Eighth U. S. Census, 1860 Medina County, Texas, p 5; Enlistment papers, James Duff, Records of the Adjutant General's Office, Record Group 94, Enlistment Papers, United States Army 1798 – 1914, National Archives, Washington, D. C.; James M. Duff's U. S. Army's Service Records, Record Group 94, Microfilm Roll M232-22 "Register of Enlistment in U. S. Army 1798 – 1914, July 1826 – October 1850", 65, National Archives, Washington, D. C.; Young County, Texas Marriage Records, Book/Volume 1, 2, December 3, 1856; *Fort Belknap Frontier Sage: Indians, Negroes and Anglo-Americans on the Texas Border* by Barbara A. Neal Ledbetter, (Published by Barbara Ledbetter, Graham, Texas), 1982, pp 43, 47, 49, 58, 61, 68 and 78; *Creoles of St. Louis* by Paul Beckwith, (Nixon-Jones Printing Company, St. Louis, Missouri), 1893, pp 25-32; *The U. S. Army and the Texas Frontier Economy 1845 – 1900* by Thomas T. Smith (Texas A and M University Press, College Station, Texas) 1999, pp 205, 207 and 212; Eighth U. S. Census, 1860 Bexar County, Texas, p 410; Report of Captain James Duff, Texas Troops of the capture of a company of Eighth U. S. Infantry near San Antonio, Texas, San Antonio, Texas April 23, 1861; Report of Colonel Van Dorn, C. S. Army, of the surrender of the U. S. troops in Texas, and of his subsequent operations, Headquarters Troops in Texas, San Antonio, Texas, May 10, 1861; OR, Series I, Volume 1, pp 572-574; General Orders No. 5, Headquarters Troops in Texas, San Antonio, May 13,

1861, OR Series I, Volume 1, pp 634635; Muster Roll of Captain James Duff's Company of Riflemen for the month of December 1861, January 1862, File 204, A.G.C., TSA, Austin, Texas; Letter James Duff, San Antonio, Texas February 3, 1862 to Colonel J. Y. Dashiell, Adjutant and Inspector General, Texas State Troops, File 401-828-7, A.G.C., TSA, Austin, Texas; *Terrell's Texas Cavalry: Wild Horsemen of the Plains in the Civil War* by John Spencer (Eakin Press, Austin, Texas), 1982, pp 66-67, 95 and 159; "What Really Happened on the Nueces River" by Richard Selcer and William Paul Burrier in 'North and South', Issue #2, January 1998, pp 56-67; James M. Duff's Confederate Service Records, Record Group 109, National Archives, Washington, D. C.; *Denver_Republican*, May 10, 1899; *History Of Denver,* Edited by Jerome C. Smiley, (The Denver Time – The Times-Sun Publishing Company, Denver, Colorado), 1901, p 968; Series of Letters from July 1996 to August 1996 containing research on James M. Duff, from Patricia A. Kemper, Golden Colorado; Series of Letters from December 2001 to March 2002 containing research on James M. Duff from Joanne M. Gonsalves, Littleton, Colorado; and Compiled Service Records of 33rd Regiment Texas Cavalry, Record Group 109, National Archives, Washington, D. C.

2. *San Antonio Herald*, May 31, 1862.

3. Duff's Report, pp 785-787.

4. Robert Hamilton Williams was born in England in 1831. He came to the U. S. in the early 1850s and to Texas in 1860. He was very critical of Duff and many other Texas Confederate leaders. He was a member of the pursuit force and took part in the Battle of the Nueces. He left Duff's command in mid-1863 and led a Frio County militia unit the remainder of the war. He returned to England after the war.

    Early Fredericksburg and Fort Martin Scott by Roy J. Betzer, an unpublished manuscript, n. d., p 4, copy in possession of author; Duff's Report, p 785 and Williams *Border Ruffians*, p 232.

5. Act Respecting Alien Enemies; Proclamation by Jefferson Davis, President Confederate States of America, August 14, 1861, OR Series II, Volume II, pp 1368-1369 and Regulation Respecting Alien Enemics, August 14, 1861, OR, Series II, Volume II, p 1370.

6. Duff's Report, p 785.

7. Edwin Lillywas born about 1840 in Georgia. The 1860 Bexar Census shows him as a law student. By early 1861, he was practicing law in San Antonio. He was the captain of the militia company from Precincts 1 and 2 of Wilson County. Lilly joined Duff's Partisan Ranger Company on May 7, 1862 and

was appointed/elected junior second lieutenant. He was the commander of the detachment from Duff's Company in the pursuit force and identified as the individual that executed the wounded prisoners after the Battle of the Nueces. Lilly was appointed regimental quartermaster with the rank of captain when the command became the 33rd Regiment Texas Cavalry. He fled to Mexico after the war. His date and location of death is not known.

Eighth U. S. Census, 1860 Bexar County, Texas, p 357; William *Border Ruffians*, pp 249-250; Edwin Lilly's Confederate Service Records, Record Group 109, National Archives, Washington, D. C.; San *Antonio the Weekly Ledger and Texan*, April 20, 1861; *Escape from Reconstruction,* by W. C. Nunn, (Texas Christian University Press, Ft. Worth, Texas) 1956, p 133 and Duff's Report, p 785.

8. Williams *Border Ruffians*, p 232; Duff's Report, pp 785-786; Eighth U. S. Census, Kerr County, Texas, pp 74-75; *San Antonio Semi-Weekly News,* June 9, 1862; and Barr *CMC*, Volume LXXI, (October 1967), pp 256 and 277 and Volume LXXIII, (July 1969), pp 83 and 90.

9. According to Duff's Report of June 23, 1862, he wrote a separate report regarding his activities in Medina County. That report has not been located. Lieutenant Sweet was his second in command. Two of these 3 men [Braubach and Doebbler] were the same as had been denounced the previous February by the Gillespie Rifles as dangerous to the community. It is interesting Duff did not obtain the name of the third man, Rudolf Radeleff. It may be Radeleff had already fled Gillespie County as he was arrested in Cibolo a few days later by Confederate troops. Williams *Border Ruffians*, p 237 and Duff's Report, p 786.

10. The 1860 Kerr Census shows Henry M. Newton as 25 years old, a store keeper at Camp Verde, born in the Cherokee Nation. He enrolled in Duff's Company on May 7, 1862. Newton was promoted to captain in August 1864 and remained with Duff's Regiment until the end of the war. He received a parole on August 15, 1865 in San Antonio. Newton died on January 31, 1918, in Bexar County.
    Reminiscences and Civil War Letters of Levi Lamoni Wight, Davis Bitton, Editor (University of Utah Press, Salt Lake City, Utah), 1970, pp 113-114; Duff's Report, p 786; Eighth U.S. Census, 1860 Kerr County, Texas Census, p 74; Cramer's Letter; and Confederate Service Records of Henry M. Newton, Record Group 109, National Archives, Washington, D. C.

11. *San Antonio Semi-Weekly News,* June 16, 1862.

12. Duff's Report, p 786.

13. It is interesting to note General Bee did not send Duff and his company into Kerrsville [now known as Kerrville] and the remaineer of Kerr County. Duff's trip to the Camp Verde area in southern Kerr County may also have included the eastern part of Kerr County and Comfort in the very western part of Kendall County. Nelson was Hiram Nelson who was born about 1800, likely in North Carolina. He and his family arrived in Kerr/Gillespie County from Illinois about 1859. The 1860 Gillespie Census shows Nelson living in Gillespie County. Duff's Report, 786; Gibson's Statement; and Eighth U. S. Census, 1860 Gillespie County Texas Census, p 22. The information of this census is not correct. For the correct data see Seventh U. S. Census, 1850 Jefferson County, Illinois Census, p 363.

14. The newspaper account of this event is very hard to read. It is not clear if Duff was accused of just mistreating a woman or actually murdering one. Duff's Report, 786 and *San Antonio Herald,* June 11, 1862.

15. *Austin State Gazette,* June 21, 1862.

16. Writers and historians have conveniently overlooked this part of Duff's report. It shows there was a Unionists element prepared to engage in armed conflict with the Confederates as early as June 1862. Duff's Report, p 786.

17. Duff's Report, pp 786-787; Eighth U. S. Census, 1860 Lampasas County, Texas, p 180; Freeman's MR; Moursund *Blanco County Families*, pp 114-115, 218-219, 255-256, and 412-414; 1860 Blanco Agriculture Census; Hopkins' diary; *A History of Blanco County* by John W. Speer (The Pemberton Press, Austin, Texas), 1965, pp 31-33. This reference incorrectly says these deaths took place in 1863. In fact, they took place in early 1864. See Smith *Frontier Defense*, pp 156-157 and Report, Captain C. Dorbandt, Burnet County, Texas, January 23, 1864 to Texas Adjutant General, A.G.C., TSA, Austin, Texas; Freeman's MR and Moursund *Blanco County Families*, pp 412-414.

18. Duff's Report, pp 786- 787.

19. This is the second report that the Unionists were trying to create a military organization. The first was Montel on May 31, 1862. See page 264. While Duff was correct about such an organization beginning created, he was wrong about which group of Unionists were organizing the military organization. Duff reported it was Hamilton and other Anglos, but the Hill Country Unionists had already created such an organization; the Union Loyal League.

The location where Hamilton and the other Unionists met was near Hamilton Pool, now a Travis County Preserve.

Captain John Treadwell Cleveland was born on October 1, 1798 in Salem, Massachusetts. The exact date he arrived in Texas is not known. Cleveland was living in Blanco County by the time it was organized in 1856. He owned a ranch near Cypress Mill, which was on the Blanco-Travis County line. He died on January 14, 1875.

Guy Hamilton is likely Morgan Calvin Hamilton, a brother of A. J. Hamilton. John W. Sansom confirms Unionists to include A. J. Hamilton, were meeting at Guy Hamilton's Ranch at this time.

Slessinger testified against Unionists at the Confederate Military Commission. Duff's Report, pp 786-787; Moursund *Blanco County Families*, p 63; Sansom *German Citizens*; Barr *CMC*, Volume LXX, (July 1966), p 634: *Austin State Gazette*, December 28, 1863; and Waller *Colossal Hamilton*, pp 35-36. Morgan C. Hamilton's Ranch is located about 25 miles west of Austin on Highway 71, and named "Hamilton's Pool".

20. William M. Preece was born on May 1800 in Kentucky. He married Mary Elizabeth Giddens on March 22, 1825 in Pile County Kentucky. They were the parents of at least 12 children. He arrived in Texas by 1845 and settled in western Travis County. William M. Preece died on September 14, 1870 in Travis County. Texas Tales by Mike Cox *Fredericksburg Standard-Radio Post*, December 7, 2005, and Preece Family History http://genforum.genealogy.com'preece'messages'56.html.

21. Mr. Eden is the same man Duff named as "Eaton". The two accounts of capturing Edon/Eaton are very different. This account says he was "shamefully treated" while Duff's account says Eaton "was so worthless that I took no steps in the matter." Blanco County News *Heritage of Blanco County*, Texas, p 10 and Duff's Report, pp 786-787.

22. James Wilson Nichols was born about 1820 in Tennessee. He arrived in Texas with his parents on December 16, 1836 and in Blanco County in 1859. Nichols died in1891 in Kerr County. Nichols *Now You Hear My Horn*, pp 146-147 and Ninth U. S. Census, 1870 Blanco County Texas Census, p 350

23. Duff's Report, p 787.

24. Erastus Reed was born about 1825 in Georgia. He arrived in that area of Blanco County that was ceded to Kendall about 1859. Reed represented Bandera, Blanco, Comal, Concho, Llano, Gillespie, McCulloch, Mason, Medina, Menard, Medina, San Saba, and Uvalde counties as a state senator during the Ninth Legislature in 1861-62. He was a member of W. E. Jones' 1864 company in the Third Frontier District. He moved to San Antonio after the war and operated a furniture store. Erastus Reed died in June 1870 in Bexar County.

Barr *CMC*, Volume LXX, (April 1967), p 637; Duff's Report, p 787; *Senate Journal of the Ninth Legislature of the State of Texas: November 4, 1861 – January 14, 1862,* Compiled and Edited from the Manuscripts in the Texas State Archives by James M. Day, (Texas Library and Historical Commission), 1964, p 2; Speer *History of Blanco County*, p 435 and 1860 Blanco Agriculture Census.

25. Schlickum's Letter.

26. Letter, Headquarters Sub-Military District of the Rio Grande, San Antonio, June 23, 1862 to Captain J. K. Stevens, Record Group 109, War Department Collection of Confederate Records, Entry 105; General and Special Orders, Headquarters, Sub-Military District of Rio Grande May – October 1862, Box Number 77, National Archives, Washington, D. C.; *San Antonio Herald*, June 28, 1862 and Duff's Report, p 787.

27. Edwards Story *of Fredericksburg*, 35; Dallas *Morning News*, November 23, 1997; "The Dogs of War Unleashed The Devil Concealed in Men Unchained" by Joe Baulch, 'West Texas Historical Association Year Book' Volume LXXIII, 1997, p 129; "Mainstreams—The 'Last Gentlemen's War' Also Had Its My Lai" by Edwin A. Roberts, Jr., of Comfort, Texas, 'The National Observer', May 31, 1973, copy provided by Margie Morgan of Kerrville, Texas; and Comfort Handout.

28. Cramer's Letter.

29. Robert J. Townes was born on February 20, 1810 in Amelia County, Virginia and arrived in Texas in 1836. He became Texas Secretary of State on September 8, 1862 and held that office until May 2, 1865. Robert J. Townes died on October 3, 1865 near Mt. Bonnell.

    Joel K. Stevens was born about 1829, likely in Texas. Prior to the war he was a medical doctor. He organized a company from Bexar and surrounding counties on March 22, 1862 and it became Company C, 32nd Regiment Texas Cavalry. Stevens died on June 8, 1864 as results of wounds received at the Battle of Yellow Bayou in May 1864.

    Lewis Antonio Maverick was born on March 23, 1839 in San Antonio the son of Mary Ann Adams and Samuel A. Maverick. He enrolled in Company E, Wood's 32nd Regiment Texas Cavalry on March 29, 1862 and elected captain. Lewis Maverick was promoted to major and assigned to General DeBray's staff in the summer of 1864. He was seriously wounded during the Battle of Blair's Landing. Lewis A. Maverick died of his wounds on June 6, 1866 in San Antonio.

Felix Kyle was born on April 18, 1829 in Hawkins County Tennessee. He arrived in Texas prior to 1850. Kyle owned at least one slave. Kyle died on August 18, 1902 in Caldwell County.

Martins *Texas Divided*, pp 66-67; Webb *Handbook of Texas*, Volume I, p 792; Joel K. Stevens' Confederate Service Records, Record Group 109, National Archives, Washington, D. C.; Lewis Maverick's Confederate Service Records, Record Group 109, National Archives, Washington, D. C.; Duaine *Dead Men Worn Boots*, p 103; *Miscellaneous Texas Newspaper Abstracts – Deaths*, Volume 2, Compiled by Michael Kelsey at al (Heritage Books, Inc., Bowie, Maryland) 1997, p 214; *They Came to San Antonio* S. W. Pease, (Published by S. W. Pease, San Antonio, Texas) c1975, p 714: Moursund *Blanco County Families*, p 270; Blanco County News *Blanco County Heritage* pp 9-13; Bee's Orders to Stevens and Waller *Colossal Hamilton*, p 36.

30. Richard L. Preece was born on December 23, 1833 in Pile County Kentucky. He married Mary Catherine Shannon on December 23, 1866 in Travis County. They were the parents of 12 children. Richard Preece died on November 18, 1906.

    Wayne Pulaski Preece was born on January 8, 1845 in Texas. He married Margaret Elvira Shannon about 1870 in Travis County. Wayne Preece died on August 16, 1905. William Martin Preece, Jr. was born on March 4, 1838 in Kentucky. He married Anna Melissa Rutledge on September 19, 1869. His date of death is not known.

    Cox 'Texas Tales'; Richard L. Preece's Union Service Records, Record Group? National Archives, Washington, D. C.; Wayne P. Preece's Union Service Records, Record Group 6, National Archives, Washington, D. C.; and William M. Preece, Jr.'s Union Service Records, Record Group 6, National Archives, Washington, D. C.

31. This group contained: A. J. Hamilton; E. J. Davis; John C. McKean, who was captured in Mexico, and tried before the Confederate Military Commission; and George Gray, who was the chief justice of Travis and Hamilton's brother-in-law. Testimony during the Military Commission establishes William Montgomery, Buck Rogers, and John Sweringer were also in the group. John Sansom identifies Theodore V. Copeland, Alfred Langle, Frances A. Vaught, Jeff G. Lilly, James Sneed, and Jesse Stance as being in the group. It is likely John Haynes and George Hamilton were also members. Martin's *Texas Divided*; *San Antonio Herald*, August 16, 1862; Barr *CMC*, Volume LXII, (July 1969), pp 94-95; Sansom *German Citizens* and Waller *Colossal Hamilton*, p 36.

32. Unable to identify Dr. Lee and Mr. Pertrn. Leonard Pierce, the Union Consulate in Matamoros, mentions a Dr. W. J. Moore a refugee from Corpus Christi as "having spend most of his time" with Pierce. Pierce also mentions a Mr. H. A. Peeler as one who could take charge of his office in his absence. Pierce also says there is a Mr. Hardin with him who was refugee from the interior of Texas. Pierce reported the number of Union men from Texas was growing also daily due to martial law and the draft. On March 24th, he reported there were 50 men. By April 8th, this number grew to over a 100 and that the city "presents the appearance of an American town."

    Letters, Leonard Pierce, U. S. Consul, Matamoros, Mexico to Hon. W. H. Seward, Secretary of State, Washington, D. C., dated March 24, 1862; April 8, 1862; April 14, 1862; and May 17, 1862, Dispatches from United States Consuls in Matamoros, 1826-1906, Volumes 7-9, January 1, 1858 – December 28, 1869 and Letter William Devine, Matamoros, April 20, 1862 to Colonel John S. Ford, Governor Lubbock's Papers, TSA, Austin, Texas.

33. Letter, Leonard Pierce, U. S. Consul, Matamoros, Mexico to Hon. W. H. Seward, Secretary of State, Washington, D. C., dated August 26, 1862,

    Dispatches from United States Consuls in Matamoros 1826-1906, Volumes 7-9, January 1, 1858 – December 28, 1869; Waller *Colossal Hamilton*, 36 and *Sam Antonio Herald*, July 27, 1862.

34. Charles Hunter became a Midshipman on April 25, 1827. He was appointed a naval commander on June 9, 1862. Hunter retired from the U. S. Navy on June 21, 1865 as a captain. He died on November 22, 1873.

    Report, Charles Hunter, United States Steamer Montgomery, *June* 16, 1862, OR, Series I, Volume 15, p 522; Charles Hunter's U. S. Naval Service Records, National Archives, Washington, D. C.; and Letter, Leonard Pierce, U. S. Consul, Matamoros, Mexico to Hon. W. H. Seward, Secretary of State, Washington, D. C., dated March 24, 1862, Dispatches from United States Consuls in Matamoros 1826-1906, Volume 7-9, January 1, 1858 – December 28, 1866.

35. Both Franz and Moritz Weiss had been members of Kuechler's February 1862 Company. Both would be killed in October 1862. The Degener brothers would both be seriously wounded at the Battle of the Nueces and executed by Lieutenant Lilly.

    Wilhelm Tellgmann would also be seriously wounded in the battle, but continued fighting while the last survivors withdrew. Lieutenant Lilly would also execute him.

Ernst Beseler was the man that drew the 'short straw' and executed the alleged spy Basil Stewart. Beseler was the second Unionist killed during the battle.

Amil Schreiner would also die a heroic death. After Fritz Tegener was wounded, Amil Schreiner led a counter-charge in which he was killed.

Cramer's Letter; Simon's Oath of Allegiance to Confederate States of America, May 29, 1861, Book A, p 40-41, Kerr County Probate Records; Deed Ernest Cramer to Gottlieb Bauer, March 4, 1862, Book 1, p 50-51, Kendall County Deed Records; Eighth U. S. Census, 1860 Kerr County Texas Census, pp 70 and 74; Ransleben *100 Years*, pp 24,66, 91, 94-95, 109 114 and 124; Kuechler's MR; *The Carl Beselers of Welfare, Texas* by Lucille V. Whiteturkey, Oklahoma City, Oklahoma, an unpublished manuscript, ca 1995, copy provided by Gregory Krauter of Comfort, Texas; Geue *New Homes in New Land*, pp 54 and 138; Kuechler's MR; Sansom *Battle of Nueces*, p 7; Kendall County Probate Records, Case #1, Volume D-1, pp 0, 1, 3, 6 and 7: Siemering's Germans During Civil War, May 29, 1923; Eighth U. S. Census, 1860 Kerr County Texas Census, p 75b and Raley *Blackest Crime in Texas Warfare*.

36. Oral Interviews with Peter Kleck, grandson of Sylvester Kleck, May 1, 1997 and October 4, 1997; and list Luckenbach Bushwhackers and Luckenbach Rangers; Fredericksburg's *Radio Post*, October 2, 1952 and Barr *CMC*, Volume LXXI, (July 1967), pp 253-255 and 260-266.

37. Letter, Captain Montel, Camp McCord, July 7, 1862 to Colonel James Norris, Commander, the Frontier Regiment, A.G.C., TSA, Austin, Texas and Montel's MR.

38. Montel's Report of July 7, 1862.

39. *San Antonio Herald*, July 5, 1862.

40. Ibid. Also see *Clarksville Standard*, July 7, 1862.

41. Montel's Report of July 7, 1862.

CHAPTER 7 – *Des Organisator* Strikes Back

One of the reasons General Bee ordered increased efforts to find the insurgents who had fled into the mountains was that he learned they were again making plans to free the Federal prisoners of war at Fort Mason. These prisoners had been moved from Camp Verde because of an earlier insurgent plan to free them and were now guarded by Captain Frank Van der Stucken's Company C, of Taylor's 8th Battalion Texas Cavalry. The plan to free the prisoners developed sometime in late May or early June, 1862, as *Des Organisator* moved into Phase II of its insurgency. This was shortly after Union forces defeated Confederate Brigadier General Sibley's forces in New Mexico and were moving east into western Texas. It was not known how far east these Union troops would move. This again gave hope to the insurgents that Union troops might come to their aid. It was not to be. Union forces sent scouts as far east as Fort Davis, but the main force stopped at Fort Bliss. The plan to free the Union prisoners called for *Des Organisator* Battalion, along with insurgents in Van der Stucken's Company, to seize the fort and release the prisoners. The prisoners would be armed with weapons taken from the fort and with others provided by the League. The two small cannons at Fort Mason would provide them with additional firepower.[1]

August Siemering, now an officer in Van der Stucken's Company, described their intentions after that. "A plan was developed in which the organizations which had been established in the different counties would take over the fort

[Fort Mason], arm the Federal prisoners and together either go to Mexico or join the Federal troops in New Mexico." Enough weapons, ammo, provisions and transportation were readily available, making the march possible. "Two small mountain cannons at Fort Mason were available." The plan was feasible and could have been executed except that Captain Van der Stucken learned of it. He immediately reported the plan to his superiors in San Antonio and took steps to prevent any escape attempt. Part of the translated quote says, "This plan would have probably been realized if the captain of the company had not suddenly changed from a unionman to a bitter secessionist." It is very doubtful this was the case. Van der Stucken's prior actions clearly demonstrated he was a secessionist and not a Unionist.

Therefore, the likely reason the plan was cancelled was that Van der Stucken learned of the plan, not him changing sides.[2] It is of interest to note this discussion by Siemering is an example where actual members of the League talk about their '*Organisator*'. In all these cases, they refer to it as '*Des Organisator.*' At no time, did they use the term 'Union Loyal League.'

No record of Van der Stucken's report to General Bee exists today. One of his troopers, Levi Wight, told of learning that the prisoners might have weapons. His account in broken English says: "At ft Mason my first duty as a soldier was with the company guard prisners [sic] of war of which them was about 700." The Confederate quarters were about a half mile from the prisoner's camp. The sentries were on duty for twenty-four hours at time, taking turns walking their post. They received information some of the prisoners might have arms. This was a major concern because of the large number of prisoners and the relatively small number of guards. No additional troops could be provided so Van der Stucken's Company were on their own to disarm the prisoners. Wight continued, "If worst come to worst we would have to fight it out against odds of about 6 to 1 without experience." "This looke wolley to us, but we were sworn to obey our superours and it would be a slam to be licked by prisners [sic] of war." The guards were marched by fours to the camp, and formed a line of battle with arms at the ready while their officers searched for arms. None were found. "If one of those prisoners [sic] had had an old goverment pistol and ran towards [us] and bursted [sic] a cap and hollowed come on boys, we would have thrown down our guns and made good time to our quarters and the prisners would have ben [sic] armed sure enough and we would have ben disarmed and at the mercy of our prisoners," said Wight. [3]

After this episode, Captain Van der Stucken was ordered to apprehend some of the more "dangerous American Unionmen." Complying with his orders Van der Stucken attempted to arrest these "dangerous Unionmen," but was not successful. He selected a few of his more trusted men and proceeded to the home of one of these Anglo Unionists. They arrived at the Unionist house, left their horses outside and went inside to search for him. The man had been warned and was hiding nearby. When he saw Van der Stucken and his men go inside, he quickly grabbed Van der Stucken's horse and rode away firing the captain's pistol, which had been left hanging on the saddle horn, shouting, "Hurrah for the United States!" In a letter written by Van der Stucken on February 29, 1864, he states he was part of the Confederate element attempting to arrest Unionists near Fort Mason from May 7th to July 4th, 1862. His letter says, "My Company was at Fort Mason Texas from May 7th to July 4th, duty in command of the post and property, apprehending disloyal persons, [and] conscripting ... I was an active party with my Company in the breaking up and arrest of Buschwachers, in that portion of the country." [4]

Captain Stokley M. Holmes, the commander of Company K, 32nd Regiment Texas Cavalry, who was guarding Federal prisoners at Camp Verde, also arrested some Unionists. It is not clear how many he apprehended. According to John W. Sansom, Joseph Graham, the Kendall County provost marshal and his deputy, John G. O'Grady succeeded in restraining the soldiers under Holmes from acts of violence in Kendall County.[5]

Because of the fear *Des Organisator* might again attempt to free POWs, General Bee directed they be moved to San Antonio. Van der Stucken's Company escorted the prisoners from Fort Mason to San Antonio. One member of the company described the decision to move the prisoners. "It was considered unsafe and expenciv [sic] to keep the prisners [sic] at ft Mason and we was orderd to San Antonio." Bee also directed the two Fort Mason cannons moved to San Antonio to keep them out of insurgents' hands. Lieutenant Siemering was placed in charge of the cannons. Rumors soon spread he was part of an insurgents' plan to capture the guns. Siemering and a few men were to take the cannons directly to San Antonio using teams of oxen. When they reached Fredericksburg, Siemering decided to stay for a short visit. He directed the cannon detail to continue and wait for him at a certain point. When he arrived at the rendezvous point neither the men nor the cannon were there. Siemering went on to San Antonio without the guns. When he arrived, rumors spread he allowed them to be captured. This proved untrue as a short time later the cannons arrived. But the rumor caused great concern among Texas and Confederate authorities.[6]

On July 4, 1862, the prisoners from Fort Mason, escorted by most of Van der Stucken's Company, moved to Camp Verde in southern Kerr County. They remained at Camp Verde a few days. Then along with Holmes' Company K, 32nd Regiment Texas Cavalry, escorted the prisoners on to San Antonio. The date of July 4, 1862, is important in establishing that John W. Sansom was wrong when he said it was on that date at least 500 Unionists met on Bear Creek and organized the military battalion. The movement of Federal prisoners took place at that time. Captain Van der Stucken's Company, along with several other Confederate units were moving the prisoners from northern posts to Camp Verde. They arrived at Camp Verde on July 4, 1862. Thus, the date of **July 4, 1862 was the only day when some 500 insurgents could not have met because of the large numbers of Confederate troops moving in that exact area.**[7] From its apex of March, 1862, until early July 1862, the League had been battered, rebounded and battered again.

The expected Union invasion took place in Louisiana, not Texas. The Confederate draft followed within six weeks by the martial law proclamation and resulted in many of its members deserting the 'expected revolution,' betraying the 'hard-core' insurgents, and fleeing into the mountains. Here they regrouped, again hoping Federal troops would come to their aid. Again, they made plans, such as to free the Federal prisoners at Fort Mason. But their plans were compromised again and Confederate attempts intensified to find and arrest their leaders. They were frustrated. Insurgent Captain Cramer reflected this feeling when he said, "Early in July I had to go home ...It was then that I felt my circumstances bitterly. I could not sit quietly for one instant. I had to conceal myself as though I were a thief." The League had used intimidation to discourage its opponents and even overt acts such as beatings and burning fences against the Secessionists. Now the League was threatened as it had never been before. This fact is support by a Comal County German diary. It states, "Many patriotic songs were produced which are still in my possession, above all the Bonny Blue Flag, with the refrain: Hurrah! Hurrah! for the bonny blue flag, that bears a single star!" The insurgents' support base was being destroyed. They felt they had to act in self-defense.[8]

One of their first targets was Captain Van der Stucken. He and a few of his men that were not part of the escort party that moved the prisoners of war had stopped in Fredericksburg. Van der Stucken paid a visit to his home where he learned of an insurgent plan to ambush and kill him on the way to San Antonio. He sent word to some men whom he trusted to meet him. They left Fredericksburg under the cover of night and traveled a different route to San Antonio. They arrived at Boerne early the next morning and ate breakfast at a local hotel, after which they traveled on until they reached San Antonio later in the day.[9]

The story of how Van der Stucken escaped the plot to kill him was later told by one of his men, Private Levi Wight: "Capt. Stuckens [sic] and I with others of the Co was in Fredericksburg. Times was warm. Union men and Southern men of note and princeple was read [sic]hot. The union men had conspired to kill Capt Stuckens and their scheme was betried [sic] to him. He sent word for me to come to his house. I amediately [sic] complied. He disclosed the situation to me and said we must leave that nigh [sic] late and that every one must be prepared for an attacket [sic]. He had chosen certin [sic] ones to accompany him and the rest had permission [sic] to remain a few days in Fredericksburg. However the main rout [sic] to San Antonio was way laid, we learned afterwords, but we sliped [sic] out through another rout and pasd [sic] unmolested, arriveng [sic] at Boerne early morning. Halted and took breckfast [sic] at the Hotel and went in to San Antonio sometime in the day and joined our command."[10]

The next man *Des Organisator* selected for death was not as fortunate as Van der Stucken. This was thirty-seven-year-old Basil B. Stewart, a native of Scotland, who worked for Henry Attrill, a slave owner. Stewart had joined the American Company of *Des Organisator* military battalion.

He was executed in a well-planned League operation. There are several accounts of Stewart's execution, but all agree on the main points. Stewart was present at Bear Creek when the military battalion was organized, which means he was a member of the 'American Company' of the battalion. After martial law was declared, Stewart was one of the men Captain Cramer spoke about when he said, "Excepting a few, all took the oath, and also betrayed their officers." After taking the oath, Stewart signed some sort of affidavit telling what he knew about the insurgent plans. He apparently did not know the names of all the major insurgent leaders and their plans but he did know enough to cause major alarm among *Des Organisator* leadership.[11]

The insurgents drew straws to see who would execute Stewart. Ernst Beseler drew the short straw. On the morning of July 5, 1862, he lay in wait for Stewart in a narrow canyon along the road from Comfort to Fredericksburg about seven miles north of Comfort. Stewart and a twenty-two-year Negro slave belonging to Attrill were driving cattle north toward the Henry Attrill farm. As the trail climbed up a canyon between two mountains, Beseler took careful aim with his rifle and fired, killing Stewart instantaneously. He then triumphantly swung the rifle above his head. The Negro immediately fled to Attrill's farm, telling of the ambush.

The two primary accounts of Stewart's execution come from August Siemering and Jacob Kuechler. Both were written in the German language. Kuechler's is told in the German-language book, *Die Deutsche Pioniers Zur Geschichte des Deutschthums in Texas.* An English translation of the event states, "In the meantime, a young Scot named Stuart, who had been into the secret society, announced his intention of withdrawing from the society and from the flight to Mexico. He was judged to be a traitor to the cause and his death was decided upon. It fell to the lot of Ernst Beseler to carry out the execution. The well aimed shot of Beseler's and the death of the Scot were the signal to make a hasty start for Mexico." Siemering's account is also told in the German-language newspaper *San Antonio Zeitung.* It has much more information, "The traitor was among us at the meeting [on Bear Creek]. His name was Stewart and was an Englishman by birth. He lived on a farm of a rich compatriot, who was raising sheep close to Fredericksburg. This man had nothing better to do than to go to the justice of the peace and relay everything he had seen.

Fortunately, he knew only a few of the names, [of *Des Organisator* leaders] but he had revealed enough to cause general concern. Immediately, the leaders of the Union party got together and decided to execute the traitor. They drew straws as to do the execution. A young man from Kendall County drew the straw. The opportunity presented itself right away." This implies that the execution took place within days of the decision. Stewart was executed on July 5th. This means the decision was made a few days earlier, perhaps as late as July 4th. This would mean the League leadership had some type of meeting on July 4, 1862, as claimed by Sansom. Cramer's letter supports this hypothesis. He says he returned home in early July. This meeting in early July was a small one; not the one where 500 members met and organized the battalion. It was also likely the decision to kill was made at the same meeting. Another account says "Stewart, together with a Negro, [*The 1860 Gillespie County Slave Census shows Attrill owned a twenty-year-old male slave. This is likely the 'Negro' described in these account*] was busy taking cattle from Comfort to his friends' Attrill farm. The trail, as it climbed up the mountain, led through a long, narrow gorge. As Stewart was passing through this gorge, the deciding shot was fired, ending his life immediately. The Negro immediately fled the area and the avenger triumphantly swung his rifle above his head and left." The Siemering and Kuechler accounts are other examples where members of *Des Organisator* did not use that term when talking about the League. In Kuechler's case, he used the term 'society'. In Siemering's case, he used the term 'Union party' or *Des Organisator*. This is an interesting statement; that one of the reasons for Stewart's execution was he did not want to make the trip to Mexico. Stewart was executed on July 5, 1862. According to Sansom, the decision to flee was made in late July. Sansom's is the only account that says the decision was made in late July. Kuechler's account suggests that the decision to flee was made much earlier, in late June or very early in July. This is supported by the fact the League planned to free the Federal prisoners at Fort Mason and to either make their way to New Mexico or to Mexico. Both these accounts state the League leaders met and decided to execute Stewart. This helps prove the League's oath was a 'blood oath'. Because Stewart knew only a "few of the names" helps prove the League used a cell type organization thus limiting the knowledge of all of the League leadership.[12]

Henry Attrill attempted to find out who was responsible for the death of Stewart. He placed several advertisements in the San Antonio newspapers offering a reward for information about the person who killed Stewart. One of those ads in the *San Antonio Herald* of July 26, 1862 said:

500 DOLLAR REWARD

Whereas, Basil B. Stewart was murdered by some person or persons unknown on the morning of the 5th July, on the road between Fredericksburg and the town of Comfort, and about seven miles north of the latter place. Now I, the undersigned offer a reward of Five Hundred Dollars for such private information as will lead to the arrest and conviction of the perpetrators of this foul crime.

HENRY ATTRILL Wolf Creek, Gillespie Co.,
Post Office Fredericksburg.
July 10, 1862.[13]

The attempt to free the Federal prisoners, the attempted execution of Captain Van der Stucken, and the execution of Basil Stewart sent shockwaves through the Hill Country. The insurgents had drawn first blood. They had moved from mere intimidation to murder. 'The reign of terror' had begun in earnest and it was directed against Secessionists' supporters. Fear was now the dominant mood. Many of the area citizens who opposed *Des Organisator* appealed to Confederate authorities for protection. They agreed to testify against the insurgents at the Confederate Military Commission being conducted at San Antonio. General Bee convened this commission to try those arrested by Captain Duff and others. The Commission began hearing cases on July 2, 1862. At least twenty-five men testified during the first part of the trial, fifteen against the insurgents, and three testified for the insurgents. Another seven clearly did not desire to say anything harmful against them [14]

Des Organisator members or supporters tried in these July sessions included Julius Schlickum, Frederick Lochte, Philip Braubach, Rudolph J. Radeleff, and F. W. Doebbler. Every man was found guilty of Unionist activities. The Hill Country men whose testimony was clearly against the insurgents included Erastus Reed, Joseph Graham, James P. Waldrip, Hilmar von Gersdorff, von Woehrmann, Charles Nimitz, Frederick Fresenius, Engelbert Krauskopf, Frank Van der Stucken, Dr. Wilhelm Keidel, Ottocar Miller, Frederick Wrede, Charles Schwartz, William Wahrmund, James M. Hunter, and John H. Hunter. Those who testified on behalf of the insurgents included Christoph Rhodius, Emil Serger, and Charles Beseler. Those whose testimony showed they were intimated or didn't want to say anything against the insurgents included George Wilkins Kendall, Pastor Henry Basse, Frederick Dambach, Gustav Schleicher, Ernst Altgelt, Edward Kapp, and August Siemering.

August Siemering tells about the feelings toward Secessionists supporters and reactions: "Bitterness rose immeasurably in the Union counties after these incidents. One held the few Secessionists responsible for everything that happened and frightening threats were made against them. Fearing for their lives, they hurried to San Antonio and made sworn testimonies against the Unionists at the same time asking for protection of the military. This was promised them."[15]

The fear by insurgents, Grays, and Secessionists, became so great that in early to mid-July, nine area citizens prepared wills or deeds giving their property to friends or family members. These fatalistic nine included Friedrich Wilke, Casper Fritz, Henry Heimann, Heinrich Markwordt, Jacob Dearing, Ernst (Edward) Felsing, Ernst Cramer, Ferdinand Simon, and John W. Sansom. Wilke gave his property to his wife, Amelia; Fritz gave his to Elisabetha Staudt, his future wife; Heimann sold his to his father Gottfried for $230; Markwordt gave his to his wife M. Magdalena; Dearing gave his to his wife Elisabetha; Felsing gave his to his father Jacob; Cramer and Simon gave theirs to Gottlieb Bauer their father-in-law; and Sansom gave his to his wife. They felt things were coming to a climax.[16]

Captain de Montel continued his pressure on the Unionists in the mountains south and west of Fredericksburg and Kerrville. On July 6th, Lieutenant Patton returned from one of his scouts to the head of the Frio River finding no signs of Indian or Unionist. On the same day, de Montel received requests from the Kerr and Medina county provost marshals for additional scouts. In response Montel sent Sergeant Lawhon with a twenty-man scout that, "ranged the tributaries of Guadalupe and Pedernales [Rivers]." The scout found no signs of Unionists. In response to the Medina provost marshal, Sergeant Lawhon was sent with his twenty-man detachment to enforce martial law in Medina County. Montel reported a "number of conscripts and other prisoners taken by them and escorted to San Antonio." On orders from General Bee, Montel occupied the old Federal Camp Verde after Confederate troops left with the Federal prisoners to keep the post from being vandalized. The captain warned his headquarters that, "a number of other [Unionist] parties of Medina County still remain in the mountains." Montel added that at the current pace of scouting his mounts would not last another thirty days. He also reported his supply of ammunition was nearly exhausted; therefore, he was discontinuing such heavy patrolling.[17]

The heavy patrolling was also having an effect on the hiding insurgents. Eduard Degener, the leader of *Des Organisator*, wrote, "Young men having remained true to the Union, and having gone for months into the Mountains and ravines loafing, and followed up and chased by Confederate troops, like Indians. It is astonishing that their efforts in taking these men prisoners have not been crowned with success." Captain Ernest Cramer explains one of the reasons the hiding insurgents had not been captured: "We formed a hunting party, hunted in the mountains, and were hunted and chased by soldiers. But we knew the country and every secret path and hiding nook too well to be caught." Both Degener and Cramer expressed surprise and concern that the war had lasted so long. Degener said, "If the south is victorious, it may become necessary for the Germans to emigrate again. In what direction then?" Cramer expressed much more concern, "I never thought that this disastrous war could last so long and bring us all such grievous circumstances."[18]

Many Unionists now made the fateful decision to flee Texas by way of Mexico. The *San Antonio Semi-Weekly News* reported on one of these groups on July 14, 1862: WHO ARE THEY? —A person who arrived here last Saturday, from Laredo, gives the information that when near old Fort Ewell, on the Nueces River, he met some thirty men on horseback, well-armed and equipped. Coming into hailing distances he accosted them, but could get no reply. He supposed they were bound westward. He met them at night, and thinks they were all Americans. Who are they?[19]

'They' were part of a major Unionist exodus from Texas. Hundreds of draft-age men fled to U. S. Consuls at Matamoros and Monterrey. Both consuls reported "the crowds of refugees from Texas do not diminish" and in Texas "there exist a perfect reign of terror" and that "within one week 6 of those refugees have been hung on the frontier of Texas on trees and left hanging" because, "four of them were deserters and the other 2 alleged to be suspicious characters." No record has been found identifying these six. The consuls appealed to the Union government in strong language: "Must this thing be, that our best truest citizens must be turned into the streets, from the door of an American Consul upon a population poor themselves, not able to speak their language, also threatened by the soldiers from Brownsville for harboring them?"[20]

The hanging of Unionists was hardly a secret. On July 19th, the *San Antonio Weekly Herald* reported the shocking development, "Their bones are bleaching on the soil of every county from Red River to Rio Grande and in the counties of Wise [and] Denton their bodies are suspended by scores from the 'Black Jacks'." Again, no record has been located identifying this individual. Fear between both Unionists and Secessionists reached almost hysterical level. *The Herald* ran an editorial that was a clear warning to the Unionists:

"Some of the citizens of the frontier counties seem not yet to have learned that the State has seceded from the Union. They claim to be good and loyal subjects of Abe Lincoln's government. Living remote from correct sources of information and receiving their political instructions principally from such men as Jack Hamilton and his followers, they have imbibed very absurd notions of their own powers, as well as of their duties as citizens. They refuse to be enrolled under the Conscript law and to take the oath of allegiance to the government. They are generally men of neither intelligence, property, or character and evade the penalties due to their conduct by a migratory way of living, which renders their capture difficult and uncertain. Since the escape of some of their leaders to Mexico, they have assembled in small parties threatening and evading our military authorities thus exhibiting the characteristics of their class—arrogance and cowardice. Their numbers are insignificant compared with the loyal citizens, and their disgraceful career is evidently nearly at an end. Gen. Bee has issued a Proclamation exactly adapted to their case; under the provisions of which they will become peaceable citizens of the Confederate Government, or be held as enemies, their property confiscated and their persons arrested and dealt with to the utmost rigor of Martial Law."[21]

The most unnerving item to reach General Bee was a "very extraordinary" letter of July 25 from Captain Davis of the Frontier Regiment, stationed just west of Fredericksburg. Davis wrote: "Our country is getting in a deplorable condition, more especially this portion of the line. I am satisfied unless something is done soon our companies will have to consolidate, our camps are threaten by the Union men in the mountains, they say they must have provisions, and by Mr. Edward A. King an attack on the weak camps along this portion of this line they can supply themselves. I am confident beyond a doubt that they are consolidating in large bodies with what view nobody is able to tell. There are no Indians in this portion of the country, the patrol scouts have become worthless, they generally meet the Union men in larger numbers than themselves consequently they can do nothing. I am really afraid sometime they will get the Scouts horses, the patrols I send out current compete with from twenty-five to forty well-armed men, unless there are Confederate troops sent to this frontier to aid the Frontier Regiment, companies will have to be consolidated. The opposite of our government are getting so numerous on this line, that I think we will have to abandon Indian hunting and turn our attention to Yankee hunting, for it is apparent that no Indians are coming into a country where so many signs of white men is found. I hope our Col will look into the matter immediately, and place us in a condition to not only defend ourselves, but arrest those men and bring to [illegible]. I mainly make these suggestions to show the Col, the condition of affairs at this end of the line for which purpose I suppose our Major visit Head Quarters in person."[22]

Colonel Norris, the Frontier Regiment commander, supported Captain Davis' request. He immediately requested Confederate forces be sent to this part of the frontier to, "effectively route and disperse the forces of Lincoln and rid the country of them." Eduard Degener also felt the situation was spiraling out of control and armed conflict between Texas and Confederate forces and the "volunteers, for the purpose of carrying on guerrilla warfare" was close. In his words, "Civil War is on the eve of breaking out." He went on to point out he did not want to use his regiment for this purpose as it was just becoming effective against the Indians. He said he was, "willing to engage Mr. Lincoln's forces if I can do so without losing too much time from my regular work."[23]

Eduard Degener also felt the situation was spiraling out of control and armed conflict between Texas and Confederate forces and those he called, "volunteers, for the purpose of carrying on guerrilla warfare" was close. As previously stated, Degener felt, "Civil war is on the eve of breaking out."[24]

Endnotes -- *Des Organisator* Strikes Back

1.   Siemering's Germans During Civil War, June 1, 1923.

2. Ibid.

3. Levi Wight *Civil War Reminiscences/Letters*, pp 23-24 and Siemering's Germans During Civil War, June 1, 1923.

4. This account is an English translation of Siemering's story. It appears the insurgents who planned the captured of Fort Mason were the Anglo element of the League. Duff identified these as Prescot, King, the two Snows, Eaton, and Howell. Most of these were killed during the Bushwhacker War. The two leaders of the 'American' Company were Henry Hartmann and Philip G. Temple, both of who would have fled and join the Union Army. Hartmann was killed during a secret trip to the Hill County in January 1864. Mary Scott reported a similar incident. See pp 561-563.

    Duff's Report, pp 785-787; Sansom *Battle of Nueces*, p 3; Henry Hartmann's Union Service Records and Letter Frank V. D. Stucken, Camp Sidney Johnston, February 29, 1864 to General J. S. Slaughter, Chief of Staff, District of Texas, New Mexico, and Arizona, Houston, Texas, Record Group 109, National Archives, Washington, D. C.

5. Stokley M. Holmes was born about 1828 in Mississippi. The date he arrived in Texas is not known. He organized a Confederate Cavalry Company on February 7, 1862 and stationed at Camp Verde. It was mustered into the 32nd Regiment Texas Cavalry on June 1, 1862, as Company K. He was promoted to major on January 12, 1865 and to lieutenant colonel by the end of the war. He died in 1905 in Hays County.

    John G. O'Grady was born in 1828 in Westport, Ireland. He arrived in Boston in 1848 and joined the U. S. Army. He was stationed at Fort Duncan and Fort McKavett. After his release from the army, he opened a mercantile business at Fort McKavett. When it closed, he moved to present-day Kendall County. He served as Kendall County Assistant Provost Marshal. He was the 1st lieutenant in W. E. Jones' Company of the Third Frontier District in 1864. O'Grady died in 1879 in Kendall County.

    Sansom *German Citizens*; Stokley M. Holmes' Confederate Service Records, Record Group 109, Washington, D. C.; Sansom *Battle of Nueces*, p 13; Duaine *Dead Men Wore Boots*, pp 25 and 101; Webb *Handbook of Texas*, Volume 2, pp 668-669 and Muster Roll W. E. Jones' Company Third Frontier District March 1, 1864, A.G.C., TSA, Austin, Texas.

6. Levi Wight Civil *War Reminiscences/Letters*, pp 23-24 and Foreigners *in the Confederacy, by* Ella Lonn, (The University of North Carolina Press, Chapel Hill), 1949, p 427.

7. What is surprising is on the next day, July 5, 1862, a Unionist ambushed killed Basil Stewart on the same route Confederate troops used the day before. Captain Holmes actually left Camp Verde on July 16, 1862. Van der Stucken's Letter of February 29, 1864; Schwartz *22 Months*, pp 105 and 109 and Request for Forage by Captain S. M. Holmes for period 1 – 16 July 1862, Stokley Holmes' Confederate Service Records.

8. Diary of Elise Wuppermann and Cramer's letter.

9. Levi Wight *Civil War Reminiscences/Letters*, p 25.

10. Ibid.

11. Cramer's letter.

12. Heintzen's Thesis; Siemering's Germans During Civil War, June 1, 1823; Weber Die *DeutschePioniers*, pp 12-13; Newspaper article, "True to Union, Texas Germans Fell on Nueces" by Elmer Kelton in *San Antonio Light*, August 12, 1962. Copy provided author by Gregory J. Krauter, Comfort, Texas; Stewart *Death On The Nueces, The Minna Stieler Stories*, p 21; "The Fleeing Sixty—A True Story" by Elmer Kelton 'Ranch Romances Magazine', January 18, 1952, pp 88-90; Eighth U. S. Census, 1860 Gillespie County Texas Census, p 26 and Eighth U. S. Census, 1860 Gillespie County Texas Census, Schedule 2, Slave Inhabitants in Gillespie County Texas, p 405.

13. *San Antonio Herald,* July 26, 1862.

14. Barr *CMC*, Volume LXXI, (October 1967), pp 247-277 and Volume LXXIII, (October 1969), pp 241-274.

15. Siemering's Germans During Civil War, June 1, 1923.

16. Deed Records of Gillespie County, Texas, Volume H, pp 183-187, Deed Records of Kendall County, Texas, Volume 1, pp 46-53; and Kerr County Probate Records, Volume B, pp 58-59.

17. John W. Lawhon was born on October 2, 1822 in North Carolina. He arrived in Texas about 1856. Lawhon was one of the largest slave owners in Kendall County having 12 slaves in 1862. In December 1862, he was elected captain after Montel left the company. John Lawhon died March 27, 1883, in Kendall County. This request was received the day after Stewart was killed. The men Lawhon were looking for were likely the executors of Stewart.

Letter, Captain Charles de Montel, Camp Verde, August 3, 1862 to Colonel James Norris, A.G.C., TSA, Austin, Texas; Montel's Report of July 5, 1862;

Boerne Area Historical Society *Gone But Not Forgotten,* Volume I, p 73; Ninth U. S. Census, 1870 Kendall County Texas Census, p 123 and Lawhon's MR.

18. Letter, Eduard Degener, August 1, 1862 to Colonel von Bemewitz and Ernst Braunic [Bramigk-Coether], Barr *CMC,* Volume LXXIII, (October 1969), pp 249-259.

19. *San Antonio Semi-Weekly News,* July 14, 1862.

20. Letters, Leonard Pierce, U. S. Consul, Matamoros, Mexico to Hon. W. H. Seward, Secretary of State, Washington, D.C., dated May 5, 1862 and C. B. H. Blood, U. S. Consul, Monterrey, Mexico to Hon W. H. Seward, Secretary of State, Washington, D. C., dated May 28, 1862, OR, Series I, Volume 9, pp 684-686 and Dispatches from United States Consuls in Matamoros and Monterrey, Mexico, 1826-1906, Volumes pp 7-9, January 1, 1858 – December 28, 1869.

21. *San Antonio Weekly Herald,* July 19, 1862 and *San Antonio Herald,* July 25, 1862. This reference to the escape of some of their leaders to Mexico could mean the escape of A. J. Hamilton who reached Mexico on July 19th or more likely the escape from a San Antonio jail Radeleff, Schlickum, and Doebbler on July 18th

22. Letter, Captain Davis, July 25, 1862, to Colonel James Norris, A.G.C., TSA, Austin, Texas.

23. James M. Norris was born in Greenville District, South Carolina on November 13, 1819. He and his family arrived in Texas about 1841. On January 29, 1862, Governor Lubbock appointed him colonel and commander of the Frontier Regiment. Norris resigned in 1863. James M. Norris died of a stroke on April 21, 1874.

   Webb *Handbook of Texas,* Volume II, and Letter, Colonel James M. Norris, Commander of Texas Frontier Regiment, August 1, 1862, to Colonel Dashiell, Texas Adjutant General, Austin, Texas, A.G.C., TSA, Austin, Texas.

24. Degener's comment about the insurgents being guerrillas is clear proof the League understood what an insurgency was and the use of guerrillas to support it. Eduard Degener was the leader of the Union Loyal League and for such a comment by him is clear evidence the purpose of those in the mountains.

   Degener's Letter of August 1, 1862.

CHAPTER 8 – Confederate Crack Down

The insurgents had another made a major error, resulting in a rapid loss of their veil of secrecy. About the same time as news of Stewart's execution reached General Bee, he was given copies of several intercepted insurgent letters stating the insurgents were prepared to take military action. Several writers and historians point out that those letters showed there was a strong insurgent element and it was prepared to take military actions. The single source of this data seems to be August Siemering when he says, "In Kerr County the Federal prisoners were moved from Camp Verde to Fort Mason when intercepted letters made it known that there was an understanding between them and the Unionists in the area." Bee considered the evidence. Two of his captains, Montel and Duff, had reported that the Unionists were planning to form a military organization. Van der Stucken reported news of the plan to release the Federal prisoners at Fort Mason. Insurgents had conducted raids against farms in Gillespie County; they had threatened and in some cases, beaten Confederate supporters. The insurgents had killed Stewart and attempted to kill Captain Van der Stucken. Numerous Hill County citizens had asked for 'regular' Confederates troops to protect them. These same citizens were putting their lives in harm's way by testifying against the insurgents at the Military Commission. Captain Davis reported the insurgents were planning attacks on Frontier Regiment camps. The 'reign of terror', directed toward secessionist sympathizers, was at its height.[1]

The majority of the insurgents were foreigners, mainly Germans. Bee had only to look at the events of early 1861 in Missouri where the Germans had risen up and taken control of the state for the Union. As evidenced by newspaper articles, there was real fear of the same, or something similar, occuring in Texas. Now the Texas Germans, like the Missouri Germans, were organizing military units. The fear was of something like the San Patricio Battalion happening. Would the Texas Germans rally behind a leader such as Forty-Eighter Franz Sigel, the leader of the Germans in Missouri? Texas newspapers had reported on the influence of Sigel in Missouri. The head of *Des Organisator*, Eduard Degener, commented on the Texas fear of a 'Dutch' [sic] leader such as Sigel, "They are all [the Germans] considered Black Republicans, that means opposed to Slavery; and the fact Sigel with his Missouri Germans has had so many successes is very maddening to them." By this time the American-German press had lionized Franz Sigel as a military genius. Many Germans wanted to be able to claim they "*fought mitt Sigel*". Texan Ben McCulloch had clashed with Sigel in Missouri and had seen how Sigel was idolized by not only by his troops, but Germans in Texas.[2]

Bee applied what in military terms is called both 'indirect action' and 'direct action' against the Unionists and their sympathizers. The indirect action was in the form of increased patrols by both Texas and Confederate units stationed west and northwest of San Antonio to cut off escape routes. The direct action was in the form of two battalion-size task forces sent directly into the heart of the insurgent sanctuaries to destroy their armed force. One of the major myths is *Des Organisator* was not a threat to Texas or the Confederacy; "Its purpose was to maintain neutrality and peace, and to take no part in the war." This myth has perpetuated the theme that the insurgents were neutral and did not want to fight for either the Union or Confederacy. They only organized because of the Confederate harsh actions against them. To support this theme, historians quote Sansom when he says that up to the time *Des Organisator* organized its military units, "there had been but little friction between the Unionist and Confederates," as *Des Organisator's* purpose was, "to take such actions as might peaceably secure its members ... from being disturbed and compelled to bear arms against the Union" and the phase every historian uses to show why the military unit was needed, "to protect their families against the hostile Indians." As far as Indian protection was concerned Texas and Confederate authorities recognized and responded to the need.

There were over ten Texas and Confederate units in the area to provide such protection. The facts are *Des Organisator* organized shortly after the war began and its military arm was organized long before any Confederate actions against them. *Des Organisator* and its military battalion were designed to militarily oppose Confederate rule and as Eduard Degener said, were for the purpose of, "carrying on guerilla warfare."

Two other original members of *Des Organisator*, Ernest Cramer and August Siemering, give the purpose of the military battalion. Cramer said, "as soon as the Northern troops would come within reaching distance, we would join them." Siemering said, "At the right moment … act on behalf of the Union." Individuals organized into a secret society with a political arm and a military arm, opposed to the government, operating in an area, intimating individuals such as burning fences and beating up government supporters and executing 'spies' meets any known definition of an insurgency. What is surprising is not that the Confederates sent military force into the area to destroy the insurgents, is that it took so long to do so. The Unionist armed force consisted of about eight companies with a strength of about 450 men. The Bandera Company had 30 to 50 men. The Medina Battalion had about 170. The three *Des Organisator* companies had about 250 men. Other historians may be correct in saying that the insurgent force was much larger. For example, an article in the March 4, 1934 *San Antonio Express* states, "some 500 subscribed to the Union Loyal League alone." The number of insurgents that 'had taken to the mountains' was estimated between 75 to 200. [3] General Bee explained to the Headquarters First District of Texas his reasons for taking strong action against the Hill Country insurgents:

"Shortly after assuming command of this district information was received from various sources to the effect that the citizens of the northwestern counties of this State, or very many of them (being chiefly foreigners by birth), were greatly disaffected and were organizing and arming to resist the execution of the law known as the conscript act. In July information was received establishing the fact that Jack Hamilton and other traitors were un-questionably in arms against the Government and had assembled in the counties designated, their force being variously estimated at from 100 to 500."[4]

One of these Confederate units in the indirect action was Company H of the 32nd Regiment Texas Cavalry, commanded by Captain John M. Carolan, the former Bexar County District Clerk. It had detachments at Camp Hudson, Fort Clark, and Fort Inge. Carolan put his scouts on a higher state of alert with orders to arrest anyone they suspected of being pro-Union. Carolan's scouts pursued at least one group of Unionists to the Rio Grande, where they escaped.[5]

At the same time, July 29th, Bee requested de Montel's Company at Camps Verde, Montel, and Rio Frio to increase their patrols. Despite his concern about the condition of his horses, Montel ordered Lieutenant Gates and a twelve-man scout to again scout the heads of the rivers and streams of the area. Captain John J. Dix of Company H, the Frontier Regiment, increased his scouts in response to General Bee.

Dix's men from Camp Dix and Camp Nueces scouted the Canyon de Uvalde, Frio Canyon, and Nueces Canyon. These areas included the Patterson Settlement on the lower Sabinal, the Ware Settlement on the upper Sabinal, and the Wall Settlement near the junction of the West and East Prongs of the Frio River, and the abandoned Camp Wood in what is now Real Canyon.[6]

Captain Thomas Rabb of Company A, Frontier Regiment, located between Uvalde and Eagle Pass, responded by increasing his patrols. One patrol led by First Lieutenant John D. Crain patrolled from Camp Rabb to Dix's Camp Nueces, then north to Adams Ranch, about twelve miles north of Uvalde, and back to Camp Rabb, a distance of about ninety-seven miles. A second six-man scout, led by Fourth Sergeant Robinson, patrolled north from Rio Grande Station, along the Rio Grande then made a wide circle east and returned to camp, about fifty miles. The third scout led by First Sergeant W. J. H. Dees patrolled south of Camp Rabb to the Rio Grande, turned east to Los Moras Creek then back to Camp Rabb, about forty-one miles. A fourth six-man scout led by A. J. Ayhurst patrolled eighteen miles north from Camp Rabb, turned southeast to the Los Moras Creek, near Fort Clark, before returning to Camp Rabb, about forty-five miles. A fifth seven-man scout led by J. A. Hunter covered the area north from Rio Grande Station making a wide circle of about forty-five miles returning to Rio Grande Station. J. B. Heard led a sixth six-man scout south from Camp Rabb for about ten miles then turned east to Los Moras Creek and back to Camp Rabb, a distance of about forty-four miles. Few signs of Unionists were found. While these scouts found few signs of their foe, they did have an effect. It forced the insurgents to find a route not completely covered by Texas and Confederate patrols. The result was the difficult route through the steep hills and river bottoms of the Guadalupe, Frio, and Nueces Rivers.[7]

Captain Henry Davis of Company F, despite his fear of being attacked, kept his men out on scout parties as well. In late July, Lieutenant Harbour received word that a party of 'bushwhackers' was camped on Beaver Lake near the headwaters of Devil's River. He immediately ordered a scout to the area. After "several days of hard marching" the scout reached the lake. They found six men who claimed to be stock hunters. The state troops arrested them believing them to be part of a larger group that had fled south pursued by Captain Carolan's Company H. The Unionist group reached Mexico ahead of Carolan's men. Harbour and his men took their prisoners to Camp Davis. [Map H, p 357] One other effect of this scouting was the arrest of militant insurgent J. Rudolf Radeleff at Cibolo Settlement on the road between New Braunfels and Fredericksburg in Comal County by elements of Wood's 32nd Regiment Texas Cavalry.[8]

General Bee's direct action was in the form of two battalion-size task forces sent into the heart of the League's area of influence. Their mission was "to issue a proclamation declaring martial law, and requiring all good and loyal citizens to return quietly to their homes, and take the oath of allegiance to the Confederate and State governments, or be treated summarily as traitors in arms." Bee further stated he ordered them to, "Send out scouting parties into the mountain districts with orders to find and break up any encampments and depots as had been reported to exist there, and to send the families and provisions back to the settlements." [Map H, p 357] The insurgents gave a different mission to the two task forces. They claimed it was to, "apply such vigorous measures as deemed necessary to put down the rebellion." But, even the *Neu Braunfelser Zeitung* reported the Confederate mission was to require the citizens, "to keep up their military duties and to breakup [the insurgent] connections to [those in] the mountains." Lindheimer's newspaper went to state the Confederates "sent peaceful invitations several times to the *Buschklappears* to persuade them to return to their civil duties.[9]

The commanders of the two forces were Lieutenant Colonel Nathaniel Benton and Captain John Donelson. Benton's task force was from the 32nd Regiment Texas Cavalry. It was sent into the counties of Kimble, Llano, Mason, and San Saba. Donelson's task force was from a variety of units. It was sent into the counties of Blanco, Gillespie, Kendall, and Kerr. Both task forces had men with additional authority in the form of the positions of provost marshal for their assigned areas.[10] Benton's task force consisted of detachments from three companies of the 32nd Regiment Texas Cavalry. They were Captain Joel K. Stevens' Company C and Lewis Maverick's Company E, the ones Bee sent to assist the Travis County provost marshal in late June. 1862, and about forty men from Captain Eugene Millet's Company B. It totaled about two hundred and twenty men. [Appendix S]. As soon as Benton arrived he directed the citizens to take the oath of allegiance. He later described his approach; "I made known my mission, produced my authority, and stated I held in one hand the olive branch, and in the other a sword." Colonel Benton stated he found, "everything quiet, being warmly and hospitably received by the citizens who, after an explanation of my mission assured me that there would be no opposition of the requirements of the government ... and no one to my knowledge failed to do so or absented himself to any demands upon him."[11]

Colonel Benton made it known he would not become a party to any private quarrel or neighborhood difficulty. The local leaders and citizens told him, "He would have no difficulty in the discharge of his duty." The citizens admitted, "they had an opportunity in common with other sections to do their duty" which, according to Benton, "they all did promptly." Benton stated, all took the required oath and he could not identify any person or act that warranted him making an arrest. In the few cases where he was informed someone was opposed to Confederate authority, "it would only be a rumor." By August 20, 1862, Benton accomplished his mission and returned to San Antonio.[12] [Map I, p 358]

It would seem that most Mason County citizens supported Colonel Benton's mission. One Mason County citizen wrote a letter to the *San Antonio Weekly Herald* explaining Benton's actions and the compliance by the area citizens. This letter blamed any rumors to the contrary on "bogus inhabitants of Gillespie County." "Allow me through your paper to contradict certain false and slanderous reports now in circulation that this Country had men in the bush with the Lincolnites, as I can assure you as one who knows that such is not the case not one man has left Said Country for such purpose nor none will. In accordance with instruction from Col. Nat. Benton, Prov M., the Deputy Provost Marshal of the Country notified the people to come up to the work more readily, there was a perfect rush, everyone came the same day and at the same time. The conscript enrolling officer enrolled and swore [illegible] 20 conscript for the war and without even a murmur. I write this in order that the public mind may be set right as to the people of this County. We think the unfounded rumor came from the 'Bogus' inhabitants of Gillespie County, and we dislike for them to slander us until they set their own people right.

Mason, Aug 21st 1862

Respectfully yours,

G. W. Y."[13]

Gilbert Jordan, a descendant of German settlers, says in his book, *Yesterday in the Texas Hill County*, that the German settlers in Mason were fairly neutral. They were not slaveholders as they had come to a new fatherland for the promised freedom and opportunity, and they did not want to risk losing these in a war. Many of them joined home guard units and the companies of the Frontier Regiment. Some Secessionists resented the fact that these men did not serve in the Confederate Army. One secessionist came to Jordan's grandfather's house one day, pulled out his six-shooter and, pointing at the notches on the handle, said, "If you weren't a friend of mine, Ernst, you would be the next."[14]

Donelson's Task Force consisted of his Company K, 2nd Regiment Texas Cavalry, Duff's Company of Partisan Rangers and three detachments from Major Joseph Taylor's 8th Battalion Texas Cavalry. The 8th Battalion was made up of former members of the 1st Regiment Texas Mounted Rifles whose enlistment expired in April, 1862. The battalion was later consolidated with Yager's 3rd Battalion and formed the 1st Regiment Texas Cavalry. The three detachments were from Company A, James M. Homsley in command; Company B, James S. Bigham in command; and a composite detachment from Companies C, D, and E, John S. Williams in command. [Appendices T, U, V, W and X]. Donelson's task force reenforced Davis' Company, the Frontier Regiment at Camp Davis. It totaled about 340 men, if Davis' men are included, or over a 100 more than Benton's task force. The command element of Donelson's task force was different from Benton's in that Benton was both the provost marshal and troop commander. This was not the case in Donelson's task force. John Donelson was the troop commander and James M. Duff was the provost marshal. The reason for Donelson and Duff command arrangement is not clear, it may be because Duff was only a captain and the positions of both provost marshal and troop commander were beyond the scope of duties of that rank. But the logical reason is Donelson's captain date of rank was September 1861 and Duff's date of rank was May 1862. Donelson outranked Duff by eight months. Duff had no authority to issue orders to Donelson. It takes careful reading of the documents to understand this command arrangement. First are Bee's instructions. Bee says, "I appointed Captain Duff provost marshal for the counties composing the disaffected district and placed under his **control** [emphases added] four mounted companies, **commanded** [emphases added] by Capt. John Donelson." It is important that Bee said the companies where under Duff's **control** but **commanded** by

Donelson. The second documents are the Camp Pedernales Returns for the months of August and September 1862. Both show the military commander as Captain Donelson. [15] [Map K, p 301]

The task force arrived in Fredericksburg on July 31, 1862. Duff established his provost marshal office at Fort Martin Scott where he remained. The next day, August 1, 1862, Captain Donelson established a camp on the Pedernales River about ten miles southwest of Fredericksburg. They named this Camp Pedernales. R. H. Williams, a member of Duff's Company, says they marched by, "easy stages to Fredericksburg" which likely means at an easy pace. Williams also reports "that the morning after our arrival we marched out fifteen miles to the west of the town and pitched camp on a stream called the Pedernalio [Pedernales]." The date the task force left San Antonio and arrived on the Pedernales River has been incorrectly stated in all previous articles and books. R. H. Williams and John W. Sansom are the sources of the two incorrect dates. Williams says Duff's Company left San Antonio on July 19th and arrived about the 21st. Sansom's says he was approached by Eduard Degener on July 25th and asked to guide the fleeing Unionists. Historians and writers have incorrectly assumed Sansom means this was after Duff's force arrived in Gillespie County. That is not what Sansom said. It can be established that Duff's Company did not leave San Antonio on July 19th as stated by Williams. Williams' tells about a review held for General Hebert. He says it was after this review the company left for Gillespie County. Special

Orders 369 and 371, Headquarters Sub-Military District of the Rio Grande establishes the review took place at 6:00 p. m. on July 19, 1862. The orders name the five companies that took place in the review. The were Captains Newton's, Toole's, Penaloza's, Bose's, and Duff's. The *San Antonio Herald* issue of July 26, 1862, contains an article on the review. It identified the companies as Newton's, Bose's Toole's, Penaloza's, Duff's and Bell's and confirms that the review took place "last Saturday" [which was July 19th] on Alamo Plaza and says General Hebert was present and reviewed the troops. Duff's Company did not leave immediately after the review because of darkness. One would therefore conclude it left the next day, July 20th and given it took two days' ride from San Antonio to Fredericksburg arrived about July the 22nd, one day later than reported by Williams. The problem is the Camp Pedernales' Post Returns clearly say the post was not occupied prior to August 1, 1862. This would mean the earliest Donelson's Task Force could have arrived was about July 31st and it took the better parts of three days to find a suitable site for the camp. This would farther mean the task force did not leave San Antonio until about July 29. The reason the date of Donelson's Task Force arrival in Gillespie County is important is most historians and writers claim it was only after their arrived was the *Des Organisator*'s decision to flee made. This is incorrect. *Des Organisator*'s decision was made by July 25, and likely earlier.[16]

Donelson's task force planned to remain for about six weeks. R. H. Williams says they were told by their commanders that "1,500 'bushwhackers' mostly Germans, had taken to the mountains, and were plundering and burning the ranches of the Southern loyalists." Furthermore, the bushwhackers were said to be well-armed, and intended fighting their way northwards to join the Federal forces. Williams claimed several members of Duff's Company who knew the county and its people "didn't believe one-tenth of this yarn, but our leaders swallowed it whole, or professed to, and made great preparations to put down this formidable insurrection." This number of 1,500 quoted by Williams is an example of his consistent exaggeration of numbers. Historians have used this Williams statement to show how out of touch Confederate leaders were with the actual number of insurgents. An example is a November. 1997, newspaper article about a historian who is supposedly leading the current research on the event. It says, "Duff was advised that up to 1,500 'bushwhackers' roamed the hills, terrorizing Southern loyalists," At no time did any Texas or Confederate leader say the insurgents numbered anywhere near 1,500 men. General Bee reported the insurgent's number between 100 and 500. Later Williams reduced his estimate of insurgents to about 200. At least one Southern loyalists' ranch was attacked and his fences burncd, that of Robert A. Gibson. Others were beaten, others threatened with death and at least one, Basil Stewart, was killed. One of the insurgent's plans was to join with Union forces which they believed were going to invade Texas. Instead of intending to fight their way North as claimed by Williams, the insurgents were going south. [17]

Williams described Fredericksburg as a place where most of the inhabitants remained quietly in their homes, "though a certain number of misguided men had taken to mountains, enroute to join one of the Federal armies." Their numbers were variously estimated, "but as far as I could make out, they did not exceed a couple of hundred." One of the major criticisms of the Confederate decision to send troops into the Hill Country is there was not a major armed threat as the citizens had only armed and formed militia units for self-protection. Williams was one of the critics. However, here he admits the Unionist force "did not exceed a couple of hundred." Even if this means only 200, it was still a major armed threat and enough, claimed F. W. Doebbler, an original member of *Des Organisator*, to retake the area. It is of interest that of the "couple of hundred" Williams claims was in mountains he also claims the insurgent force in the battle was about 160 of which 60 were killed and 20 wounded. Sixty-nine was the actual number of insurgents in the battle of which 19, included the wounded, was killed. Williams provided one correct statement when he stated Captain Duff, "was appointed Provost Marshall with full powers to deal with the rebels." Another source quotes Duff as saying "the God Damn Dutchmen Are Unionist to a man [and] I will hang all I suspect of being anti-Confederate." During the 'Nueces Encounter' Symposium on March 22, 1997, Mr. Fehrenbach was asked his source for the quote about Duff threatening to hang all he suspected of being Unionists. In a letter to the author dated March 26, 1997, he was unable to provide an exact source. Based on several likely sources Mr. Fehrenbach provided it likely came from a German-language newspaper, either the *Neu Braunfelser Zeitung* or the San Antonio *Freie Presse fuer Texas*. As of this printing the author has not located the source. It is the author's opinion this quote, like many others, is incorrect, and if exists was printed in a biased article without a source. The only known primary source that could have reported such a quote is R. H. Williams. No such quote is in his book. The nearest quote is when Duff's Company arrived in Fredericksburg in May, 1862, and Williams said, "Fredericksburg was a town of about 800 inhabitants, almost all of them Germans, and Unionists to man." [18]

On July 31st, from his office at Fort Martin Scott, Duff, as instructed by General Bee, issued a proclamation. This was the second one he had issued in two months at Fredericksburg, and the third time the citizens were told of the requirement to take the oath of allegiance. The first proclamation was President Davis' in August, 1861. The other two were issued by Duff. This latest one gave the inhabitants three days to come in and take the oath to the Confederacy, "threatening to treat all those who failed to so as traitors, who would be dealt with summarily." Like the June proclamation, this one was spread by word of mouth and by detachments of Donelson's soldiers traveling to the various settlements. [19]

Captain Donelson established Camp Pedernales on the afternoon of August 1, 1862. It was located ten miles southwest of Fredericksburg, near the present Morris Ranch headquarters. The campsite was: "an ideal one; a gentle slope, dotted with majestic liveoak trees, and at the foot a clear running stream of coolest water, abounding in fish. Under a great rock, half-way up the slope, gushed forth a spring of delicious water, which went singing on its downward course to the river. From the summit of the rising ground the eye could range, in that clear atmosphere, over miles and miles of rolling prairie, green with lush grass after the rains, and dotted with clumps of timer, like some vast park of Nature's own making."

The actual encampment, Camp Pedernales, was little more than a circle of wagons guarded by extraneous sentries.[20] The Confederate units of Task Force Donelson consisted of a total of ten officers and 237 enlisted men. Donelson's Company had four officers and seventy-six enlisted men; Duff's Company had two officers and sixty-five enlisted men; and the detachments from Taylor's Battalion were three officers and ninety-six men. The breakdown of Taylor's Battalion was: Company A, 36 men; Company B, 17 men; and Companies C, D, and E, 46 men. Dr. George H. Doran of Donelson's Company was assigned as post surgeon.[21]

According to R. H. Williams, the area population did not report en masse to take the oath. "Very few of the outlying settlers came in to take the oath before the expiration of three days ... Probably because they were more occupied with procuring a living, and protecting their families from Indian raids, than with politics." He also expressed the opinion that many of the settlers had never heard of the proclamation until "they were arrested." Eduard

Degener, *Des* Organisator leader, who lived over thirty miles away in Sisterdale, contradicts the claim the population did not hear about the oath requirement. He wrote on August 1st, "Confederate troops coming from all sides ascend our mountains." Degener told about the requirements: "Martial Law to the utmost rigor, for all those who do not adjure allegiance to the United States, swear allegiance to the Confederate States." The reason few came in and took the oath was most had already taken it. This Williams' statement is often used to demonize Duff and show how unrealistic the requirement was to come in and take the oath. This argument just does not stand up under close examination. First, this was at least the third time the population was told it had to take the oath. Second, there was no way a force of over 240 men arrived and word not quickly spread as to why they were there. Third, the neighboring counties were able to find out about the requirement and meet the deadline. Fourth, the leaders of the county were very political; they and the population knew what was happening. [22]

At the end of the three-day grace period, Captain Donelson led 100 troops out looking for anyone who was suspected of being an insurgent. They scouted the upper portion of Gillespie County in search of "certain bands of traitors [and] arrested a great many." Williams later claimed, "presently, however, sinister rumors as to [Duff] intentions began to spread, and it was said, amongst other things, that he had given certain of his followers to understand that the wanted no prisoners brought into camp." Williams further states, "presently many parties were detailed to scour the country that rarely, if ever, brought in any prisoners." Yet he next says, "Amongst these, two parties of twenty-five each were sent out with wagons to bring in from the scattered ranches the wives and children of those who had taken to the mountains, and, I fear to harry [pillage] their homes. In four days they returned with the wagons full of prisoners—four or five men, and eight women with their little ones. The latter were sent on to Fredericksburg, and the former confined to the guard tent." An example of historians twisting or adding words to a source to further demonize Duff is Ella Lonn. Using this Williams account, she says, "Two parties of twenty-five each were dispatched with wagons to bring in from the farms the wives and children of those who had taken to the mountains. In four days **one party returned with ten Germans who were promptly lynched**, [emphasis added] the other with four or five men and eight women with their little children. These women and children were sent to Fredericksburg, and the men to the guard tent." Notice how Lonn just added the phase one **party returned with ten Germans who were promptly** lynched.[23] The Confederates knew the names of many of the insurgent leaders, probably through the complaints of the Fredericksburg Secessionists.

Williams adds that the insurgents were, "informed against by a Dutch tavern-keeper in Friedricksburg [sic] who was often out in camp drinking … and … who had private spites against most of them, which he took good care to pay off." At the time, there were only two German tavernkeepers in Fredericksburg. One was the insurgent F. W. Doebbler who had been arrested. The other was Charles Nimitz.[24]

Siemering claims the Confederates were "especially determined to catch Fritz Tegener, the known leader of the insurgent group. This however, [they] did not accomplish this mission. Instead [they] had the opportunity to capture Tegener's brother," thirty-two-year-old Gustav Tegener. One account of the near capture of Fritz Tegener says: The soldiers again visited Johnston's Creek, but found most of the settlers had fled to the mountains. Frederick Tegener alone they surprised, sleeping under the porch of his house, but awakened by the cries of his wife and the discharge of the muskets of his enemies, who fired fourteen shots after him. He fortunately made his escape. The house was ransacked, and all movable property taken. Other farms in the neighborhood were also searched, the families taken prisoners, and the houses burned down. Many Germans and Americans were arrested and imprisoned in Fredericksburg.[25]

Arrested at the same time as Gus Tegener were Sebird Henderson, Hiram Nelson, Frank Scott and their families. Hiram Nelson was one of the Kerr Unionists Duff attempted to arrest in May, 1862. He was the 'Nelson' who Duff attempted to arrest in May 1862 and who sent Duff 'a defiant message.' An account attributed to Scott's wife says the reason he was captured was he was at home recuperating from a wound where his big toe had been shot off. Their wives and children were kept in a small one-room log hut in Fredericksburg. Among the known wives and children were: Mary Scott [wife of Frank Scott] and her three children, Chloe, Benjamin, and Frank Jr.; Mary Tegener [wife of Gus

Tegener]; Polly Nelson [wife of Hiram Nelson]; and the wife of Sebird Henderson. Also likely held were Susan Tegener (wife of Fritz Tegener) and her two children, Texana and an unknown child. These were the "four or five men, and eight women with their little ones," described by Williams.

There are accounts, mainly oral ones in the Henderson family, that state beside Sebird Henderson, Hiram Nelson, Gus Tegener, and Frank Scott with two others were also captured; Howard Henderson and Allen Nelson. One version claims all six escaped. Another says all six were on their way to the guardhouse at Fort Mason and camped overnight. Howard Henderson awoke and was able to free his hands. He helped Allen Nelson get loose and they made their getaway. The escape made the guards, from Davis' Company, mad and in reprisal executed the other four. Facts do not support this oral account.

Fritz Tegener's wife was Susan Ann Eveline Benson. She was born about 1844 in Illinois. She and Fritz married on December 21, 1858, in Kerr County when she was 14 years old. Susan's two older brothers, William Thomas Benson and John Washington Benson were in Davis' Company. After Fritz fled, likely she believed he was dead as on February 1, 1863, Susan signed an agreement for someone to take care of her cattle for a third interest in any cattle increase. On April 20, 1863, she sold several of their lots in the town of Kerrsville [sic]. On April 23, 1863, Susan gave birth to a daughter, Emilie. On July 27, 1863, she married Fritz Schladoer in Kerr County. On September 10, 1863, she sold other lots in Kerrsville [sic] that belonged to her and Fritz Tegener. On April 29, 1865, Susan married Green Berry Wofford, a Confederate soldier, in Bandera County. Fritz Tegener returned to Kerr County after the war and on May 15, 1866, filed divorce proceedings and charged Susan with adultery. Tegener refused to acknowledge he was the father of Emilie. Tegener was granted his divorce. On December 6, 1869, Susan married Charles Bernard Curren in Kerr County. On February 24, 1870, she sold all of her livestock to her brother, James Benson. One account says they went to South Africa where Susan died about 1870. Susan's mother raised her daughter, Emilie Mexico. Based on the actions of both Gustav's and Fritz's wives, it is likely they both turned on the insurgents and were the sources of some of the Confederate information. [26]

Among others arrested was thirty-five-year-old Christian Dietert of Comfort, a native of Magdeburg Prussia. The account of his arrest states, "Toward evening of a day in the first week of August, 1862 a patrol of Duff's men came upon the Dietert cabin at Comfort … The soldiers arrested Dietert, placed him in their wagon, and headed toward Camp Lipan [Camp Pedernales]." That night Dietert's wife, twenty-nine-year-old Rosalie Hess Dietert, a native of Duchy of Saxe-Weimar, was unable to sleep due to the crying of her four children and her own worry. By the time the sun was up, she knew what she was going to do. She would go to the Confederate camp and personally plead with the commander to spare her husband's life and release him.[27]

She took her three youngest children to her husband's parents. Her oldest, a son named Gustav, six years old, she took with her. They mounted horses and "galloped off to track the soldiers." Rosalie Dietert knew the country well. From Comfort, she and Gustav rode up Cypress Creek, crossed over the Guadalupe-Pedernales divide and into Bear Creek Valley. Here they turned west and followed Bear Creek to the Pedernales. There was plenty of water to drink. For food, they ate ripe plums and cherries found along the trail, as well as some cornbread she carried. When they reached the Pedernales River, they crossed over to its north side and continued. After a ride of about twenty-four miles, she and Gustav reached the Confederate camp. She saw her husband and with "a cry of relief" she rode up to him. Mrs. Dietert and Gustav were immediately taken from their horses. In fluent English, she asked to be taken to the commander. This and other accounts use Duff's name as the Confederate commander.

He was not. Duff was the provost marshal and in Fredericksburg. The Confederate commander at Camp Pedernales was John Donelson. In the meeting the, "pleas of the brave pioneer and her little son softened the heart of that soldier of fortune." Being resourceful, she remembered the European wars and the adage that an army marches on its stomach. She spoke of the main advantage Donelson would gain by releasing her husband; being a miller, could furnish rations to the Confederate soldiers. At the same time, Mrs. Dietert explained his small children desperately needed him at home. "Fate was kind to the Dieterts and they were allowed to go home" said one historian.[28]

Another insurgent the Confederates captured and released was Robert Schaefer, a thirty-year-old son of a Forty-Eighter and native of Wurttemberg who arrived in the Comfort area in 1856. Schaefer was a member of the Comfort *Turnverein* and had been a member of the Harbour's Home Guard Company in February 1861. A group of Comfort insurgents were preparing to flee when they were surprised while standing in front of a mercantile store. The Confederates only seemed interested in Schaefer and did not bother the other men. He was arrested and about to be hanged when a Confederate lieutenant arrived and vouched for Schaefer's loyalty. He was released and the Confederates rode away, paying little attention to the other men who were insurgents and packing to flee.[28]

Another Comfort insurgent who was captured and released was thirty-five-year-old Forty-Eighter and Freethinker Adolph Rosenthal. He had a degree in mathematics from the University of Berlin, and was currently a Kendall County Constable and Justice of the Peace. According to his story, he was a member of the Kendall Company of *Des Organisator*'s military battalion when captured. A Kerr County historian described Rosenthal's actions; "When Texas seceded from the Union [in 1861] he enlisted in the local company of Unionists. He served as company clerk; however, when they left Comfort prior to the ill-fated massacre of many at the Battle of the [Nueces] he elected to remain at home." The Confederate forces sought to extract the names of his comrades, but he steadfastly remained silent, even after a noose was placed

around his neck and the rope thrown over a limb. He was
released unharmed in the end. On November 1, 1862,
Rosenthal was drafted in Duff's 14th Battalion and served
until the end of the war.[29]

The search for insurgents was abruptly reduced on
August 4th when Captain Donelson discovered that the
greater portion of them had left the country. Returning to
camp he dispatched Lieutenant Colin McRae with 94 fresh
men and horses in pursuit of them."[30]

Map H – Routes of Confederate & State Scouts, July 1862

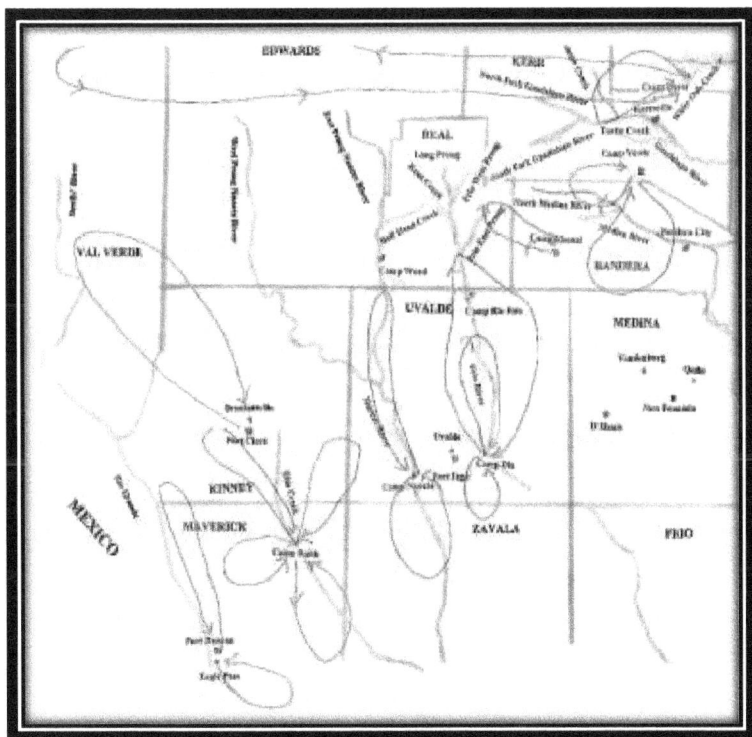

Map I – Task Force Benton's Area of Operation

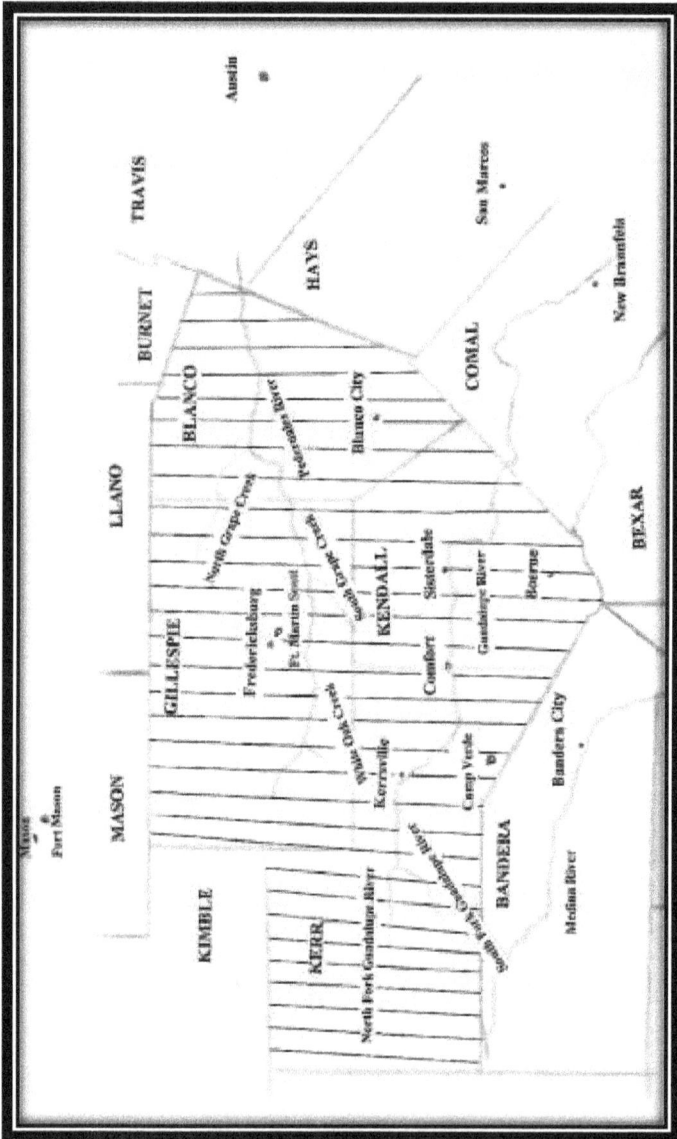

Map J – Task Force Donelson's Area of Operation

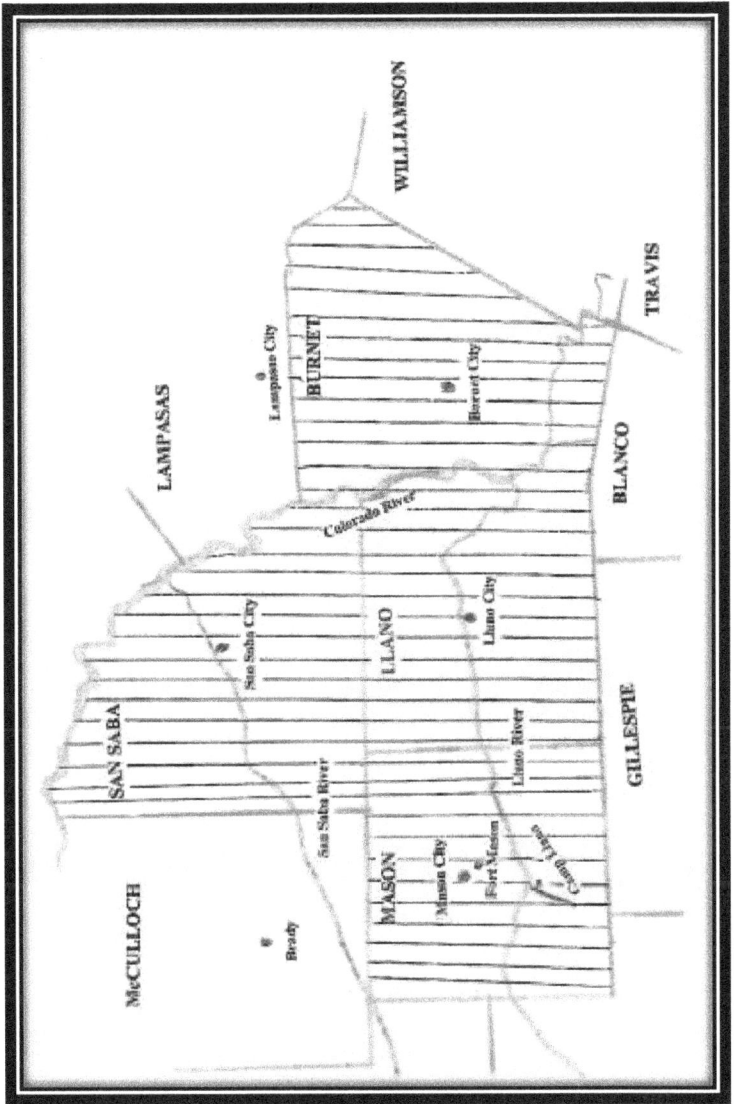

Endnotes -- Confederate Crack Down

1. Lick *Goethe on the Guadalupe*, p 69 and Siemering's Germans During Civil War, June 1, 1923.

2. Engle *Life of Franz Sigel*, pp 54-79; Biggers *German Pioneers*, p 57; Fehrenbach, *Lone Star*, p 363; Sansom *Battle of Nueces*, p 3 and *Austin State Gazette*, Saturday, June 1, 1861.

3. For a recent retelling of this false theme that the League was not a threat see *Defiant Unionists: Militant Germans in Confederate Texas* by Anne J. Bailey *Enemies of the Country: New Perspectives on Unionists in the Civil War South'*, Edited by John C. Inscoe and Robert C. Kenzer, (The University of Georgia Press, Athens, Georgia), 2001, p 214; Degener's Letter of August 1, 1862; Siemering's Germans During Civil War, May 29, 1923; Hover *San Patricio Battalion*, p 89; Kaufmann *Germans in Civil War*, pp 63, 72, 112, 114, 260 and 277-278; Moidenhauer *Battle of Cole Camp*, pp 15-16 and Degener's Letter of August 1, 1862.

4. Report Brigadier General H. P. Bee, Headquarters Sub-Military District of the Rio Grande, San Antonio, Texas, October 21, 1862, to Headquarters First District of Texas, San Antonio, Texas, OR, Series I, Volume LIII, pp 454-455 and *San Antonio Express*, March 4, 1934.

5. John M. Carolan was born about 1833 in Ireland. He arrived in Texas about 1851. He was the Bexar County Clerk when he resigned and recruited Company H, 32nd Regiment Texas Cavalry on May 15, 1862. John M. Carolan died on May 15, 1863 at Ringgold Barracks, Texas. Eighth U. S. Census, 1860 Bexar County Texas Census, p 390; John M. Carolan's Confederate Service Records, Record Group 109, National Archives, Washington, D. C.; Duaine *Dead Men Worn Boots*, p 30 and Banta 27 *Years*, pp 110-111.

6. Montel's Report of August 3, 1862 and Report, Captain John Dix, Camp Dix, August 1, 1862 to Colonel James Norris, Commander Frontier Regiment, A.G.C., TSA, Austin, Texas.

7. Rabb's Company had camps at Camp Rabb where the San Antonio and Eagle Pass road crossed Elm Creek in current Maverick County and at Rio Grande Station/Fort Duncan at Eagle Pass. John D. Crain was born about 1812, location not known. He enrolled in Rabb's Company from Karnes County in February 1862. His date and location of death is not known.

    Alfred Robinson was born on May 28, 1839 in Colorado County Texas. He enrolled in Rabb's Company from Live Oak County in February 1862. Robinson died on November 23, 1920 in Live Oak County, Texas.

W. J. H. Dees was born about 1814 in Mississippi. He enrolled in Rabb's Company from Karnes County in February 1862. His date and location of death is not known. A. J. Ayhurst was born about 1833, location not known. He enrolled in Rabb's Company in February 1862 from Karnes County. Rabb's Muster Roll shows his rank as private. Ayhurst's date and location of death not known. Joseph Anthony Hunter was born about March 1840, location not known. He enrolled in Rabb's Company in February 1862 from Karnes County. Rabb's Muster Roll shows his rank as private. Hunter died on March 21, 1922 in Nueces County.

Joel B. Heard was born in December 1841 in Alabama. He enrolled in Rabb's Company in March 1862 from Karnes County. Rabb's Muster Roll shows his rank as private. Heard died about 1925 in DeWitt County.

General Order Number 1, Headquarters Frontier Regiment, Texas Rangers, Austin, February 1, 1862, A.G.C., TSA Austin, Texas; Muster Roll, Captain Thomas Rabb's Company A, Frontier Regiment, Camp Rio Grande, June 30, 1862, A.G.C., TSA, Austin, Texas; John D. Crain's Confederate Service Records, Record Group 109, National Archives, Washington, D. C.; Alfred Robinson's Confederate Service Records, Record Group 109, National Archives, Washington, D. C.; Eighth U. S. Census, 1860 Live Oak County Texas Census, p 366; Rabb's MR and W. J. H. Dees' Confederate Service Records, Record Group 109, National Archives, Washington, D. C.; A. J. Ayhurst's Confederate Service Records, Record Group 109, National Archives, Washington, D. C.; J. A. Hunter's Confederate Service Records, Record Group 109, National Archives, Washington, D. C.; J. B. Heard's Confederate Service Records, Record Group 109, National Archives, Washington, D. C. and Report, Captain Thomas Rabb, Camp Rabb, August 1, 1862 to Colonel James C. Norris, Commander, Frontier Regiment, A.G.C., TSA, Austin, Texas.

8. Banta 27 *Years*, pp 110-111; Siemering's Germans During Civil War, June 1, 1923, and Webb *Handbook of Texas*, Volume I, p 347.

9. Bee's Instructions; *Biggers German Pioneers*, p 58; *New Braunfelser Zeitung*, August 25, 1862 and Sansom *Battle of Nueces*, p 3.

10. Nathaniel Benton was born in 1816 in Tennessee. He arrived in Texas about 1855 and settled in Guadalupe County. Benton organized a company from Guadalupe County in March 27, 1862 and elected its captain. When the company became a part of the 32nd Regiment Benson was elected the regimental lieutenant colonel. He lost his right arm at Blair's Landing and was released from service on January 12, 1865. He was elected Chief Justice of Guadalupe County in 1866 but was removed by the Texas Reconstruction Government in 1867. Nathaniel Benton died in 1889.

Eighth U. S. Census, 1860 Guadalupe County Texas Census, p 317; Nathaniel Benton's Confederate Service Records, Record Group 109, National Archives, Washington, D. C.; Duaine *Dead Men Wore Boots*, p 96; Lieutenant Colonel Nat Benton's Report *San Antonio Weekly Herald*, August 30, 1862, and Bee's Instructions.

11. Eugene B. Millet was born on April 25, 1830, in Washington County, Texas. He enrolled in Benton's Company on March 22, 1862, and when Benton was elected lieutenant colonel Millet was promoted to captain. He remained with the company until the end of the war. Millet died on October 18, 1915 at Los Angeles California.

    Benton's Report; Bee's Orders to Stevens; Eighth U. S. Census, 1860 Guadalupe County Texas Census, p 310; Eugene B. Millet's Confederate Service Records, Record Group 109, National Archives, Washington, D. C. and Duaine *Dead Men Wore Boots*, p 26 and 95.

12. Benton's Report.

13. *San Antonio Weekly Herald* August 30, 1862.

14. Yesterday in the Texas Hill Country *by Gilbert J. Jordan, (Texas A and M University Press, College Station, Texas),* 1979, p 15.

15. John Donelson was born about 1829 in Tennessee. He arrived in Texas about 1855 and settled in Live Oak County. He enrolled a volunteer cavalry company on September 23, 1861 and was elected captain. The company became Company K, 2nd Regiment Texas Mounted Rifles. He was promoted to major on October 8, 1862, but was forced to resign due to ill health on January 4, 1864. John Donelson died in July 1864 at San Antonio.

    Bee's Instructions; Eighth U. S. Census, 1860 Live Oak County Texas Census, p 365; John Donelson's Confederate Service Records, Record Group 109, National Archives, Washington, D. C.; San Antonio Genealogical and Historical Society Index *Wills and Inventories of Bexar County*, p 26; Compiled Service Records of 8th Battalion Texas Cavalry, Record Group 109, National Archives, Washington, D.C.; Sifakis Compendium of Confederate Armies for Texas, p 38, 46, and 58 and Post Returns, Camp Pedernales, for the months of August and September 1862, Documents # 292 and 293, Record Group 109, National Archives, Washington, D. C.

16. Williams Border *Ruffians*, p 235; Sansom *Battle of Nueces*, p 3-4; Special Orders Number 369, July 16, 1862 and Special Orders Number 371, July 17, 1862, Headquarters Sub-Military District of the Rio Grande, San Antonio, Texas, Entry 105, General and Special Orders, Headquarters Sub-Military District of the Rio Grande, May – October 1862, Record Group

109, National Archives, Washington, D. C.; *San Antonio Herald,* July 26, 1862; Post Returns Camp Pedernales for the months of August and September 1862; Compiled Service Records of Taylor's 8th Battalion Texas Cavalry; and Betzer *Early Fredericksburg*, p 4.

17. "Remembering Hill Country Bad Old Days" by Kent Biffle, *Dallas Morning News*, November 23, 1997, pp 44a-44b; Bee's Instructions and Williams *Border Ruffians*, p 235.

18. Williams Border *Ruffians*, p 232, 235, 239 and 248; Fehrenbach Lone *Star*, p 363 and Letter, T. R. Fehrenbach, San Antonio, Texas March 26, 1997 to author.

19. Williams *Border Ruffians*, pp 235-236, 239 and 248.

20. "An Encounter with The Partisan Rangers" by Victor Nixon, Jr., 'The Junior Historian of the Texas State Historical Association', Volume XXIV, No. 1, (September 1963), pp 8-9 and Williams *Border Ruffians*, p 236.

21. Camp Pedernales Post Return for the Month of August 1864 and Analysis of Compiled Service Records of Taylor's 8th Battalion Texas Cavalry.

22. Williams *Border Ruffians,* p 237 and Barr *CMC*, Volume LXXIII, (October 1969), p 250.

23. Camp Pedernales Post Returns for the Month of August 1864; Williams *Border Ruffians*, pp 236-237 and *Foreigners in the Confederacy* by Ella Lonn, (University of North Carolina Press, Chapel Hill, North Carolina), 1940, p 428.

24. Siemering's Germans During Civil War, June 8, 1923 and Williams *Border Ruffians*, p 237.

25. Siemering's Germans During Civil War, June 8, 1923. Gus Tegener who was born about 1830 in Prussia. The three Tegener brothers arrived in the Comfort area in 1854 and established a sawmill west of Kerrville near the current town of Hunt on a stream that bears their name. Gus Tegener married Mary Jane Payton on May 6, 1862. Gus Tegener was executed on August 22, 1862, on Spring Creek. His widow married John W. Helm on October 5, 1862, six weeks after his death. Eighth U. S. Census, 1860 Kerr County Texas Census, p 73; Kerr County Marriage Records, Volume A, p 51 and Volume B, p 32; Bennett's Kerr County, p 117 and Hopkins' diary.

Eighth U. S. Census, 1860 Kerr County Texas Census, p 73; Kerr County Marriage Records, Volume A, pp 16, 54, and 99; Bandera County Marriage Records, Volume 1, 2; Deed, A. J. Paul and Maria J. Paul to Susan E.

Tegener, April 20, 1863, Kerr County Deed Records, Volume B, pp 326-327; Articles of Agreement, Susan E. Tegener and Nathan Jackson, February 4, 1863, Kerr County Deed Records, Volume B, pp 319-320; Deed F. Schladoer and S. E. Schladoer to John E. Ochs, September 10, 1863, Kerr County Deed Records, Volume B, pp 337a-338b; Case Number 61, Kerr County District Court, May 31, 1866 and "Massacre Of The Germans in Texas", in *St Louis Republic*, January 2, 1863. Translation of the German Newspaper *Galveston Union*, contained in 'Rebellion Record a Diary of American Events', Volume 6, Edited by Frank Moore, (Arno Press, New York, New York), 1977, p 49.

26. Sebird [S. B.] Henderson was born 1805 in South Carolina. He, wife, and two nephews; Howard and John Henderson, arrived in Kerr County about 1856. Sebird Henderson was executed August 22, 1862, on Spring Creek.

    Hiram Nelson was born about 1800 in North Carolina. He and his family, to include a son Allen, later a Union soldier, arrived in the Spring Creek area about 1854. Hiram Nelson was executed on August 22, 1862, on Spring Creek. Charles Frank Scott was born about 1827, likely in Illinois. He arrived in the Spring Creek area by 1855. Frank Scott was executed on August 22, 1862, on Spring Creek.

    Howard Henderson linked up with the fleeing Unionist on August 8th.

    Kerr County Album by Kerr County Historical Commission, (Taylor Publishing Company, Dallas, Texas), 1986, pp 270-271; Henderson Cemetery Historical Marker Dedication Program, June 10, 1990, copy provided by Gregory J. Krauter, Comfort, Texas; *Story of My Mother's Life*, an unpublished manuscript by Rose Ann Harper, copy provided by John F. Koepke, Fredericksburg, Texas; *The Turknett Family: Descendants of Jacob c 1857 – 1816* by Frank D. Jenkins (Privately Published by Christine Knox Wood, Lubbock, Texas), 1981, p 32; Williams *Border Ruffians*, p 237; *Here's Harper 1863 - 1963* Harper Centennial Committee, Harper, Texas, (The Radio Post, Inc., Fredericksburg, Texas) 1962, p 12; Eighth U. S. Census, 1860 Kerr County Texas Census, p 82; Bennett *Kerr County*, p 117; Hopkins' diary; Seventh U. S. Census, 1850 Jefferson County Illinois Census, p 24; Duff's Report, p 786; Telephone Interview with Mrs. Temple Henderson of Mt. Home, Texas on February 17, 1997; Baulch *Dogs of War*, pp 131-132 and Biffle *Remembering Bad Old Days*.

27. Christian Dietert was born on August 24, 1827 in Magdeburg, Prussia. He and his father were likely Forty-Eighters. The family arrived in Texas on the *Franziska* in 1854. Dietert operated several sawmills along the Guadalupe River from Comfort to Kerrville. Christian Dietert died on November 27, 1902 in Kerr County. It is believed Christian Dietert was arrested after the Confederate pursued force left. The reason for this belief

is on August 2nd and August 4th he was issued a travel permit by the Kerr Provost Marshal to travel from Kerrville to Comfort with "a wagon and team." The likely reason for his arrest was Frederick Weber, a member of the fleeing Unionist group, worked for Dietert. Dietert may have been one of the individuals who took supplies to the Unionists prior to their departure.

Rosalie Hess was born about 1833 in Duchy of Saxe-Weimar. She was an orphan and emigrated with friends to Texas in 1852. She and Christian Dietert married on August 25, 1855 in Gillespie County. Rosalie Hess Dietert died on April 5, 1926 in Kerr County.

Nixon *Encounter with Partisan Rangers*, pp 8-9; Eighth U. S. Census, 1860 Kerr County Texas Census, p 72; Bennett *Kerr County*, pp 115 and 291; Geue *New Homes*, p 63 and *Cemeteries of Kerr County 1859 – 1976* Compiled by Arlene and Fred Tolman (Kerrville Genealogical Society, Kerrville, Texas), 1980.

28. Nixon Encounter *with Partisan Rangers*, pp 8-9; Eighth U. S. Census, 1860 Kerr County Texas Census, p 72; Gillespie Historical Society *God's Hills*, Volume II, p 0; Gillespie County Marriage Records, Book 1, p 160 and Tolman *Kerr Cemeteries.*

29. Robert Schaefer was born about 1832 in Wuerttemberg. He was likely a Forty-Eighter. Schaefer arrived in Texas about 1854. Schaefer and his family arrived in the Comfort area in 1856. He was a member of Harbour's Home Guard Company in February 1861. Schaefer was a member of W. E. Jones' Company of the Third Frontier District in 1864. He was a member of the party that recovered the Unionists from the battle site in 1865. His date and location of death is not known.

The identify of this Confederate lieutenant is not known. It could have been Lieutenant Harbour of Davis' Company. Schaefer had been a member of Harbour's Home Guard Company a year earlier, but it was not likely Harbour because he had just returned from a long scout to Beaver Lake. On August 4th Harbour and his detachment were dispatch with the Confederate pursuit force. Another possibility is the lieutenant could have been either Joseph Graham or John O'Grady, the Kendall County Provost Marshal and the Assistant Kendall County Provost Marshal. John W. Sansom gave both credit for saving the lives of several Unionists. Oral Account provided by Anne Stewart, a Comfort historian; Eighth U. S. Census, 1860 Kerr County Texas Census, pp 69-69b; Ransleben *100 Years*, pp 24 and 94; Bennett *Kerr County*, pp 50, 68 and 112; Harbour's MR; W. E. Jones' MR and Sansom *Germans Citizens.*

Adolph Rosenthal was born in Berlin on May 13, 1827. He received a degree in mathematics from the University of Berlin. He was a Forty-Eighter and a Freethinker. He arrived in Texas on the *Galliott Fora* in 1849.

He was a Kendall County justice of the peace and Precinct Number 4 Constable in 1862. Rosenthal was a member of the Comfort Home Guard in February 1862. On November 1, 1862, he enrolled or was conscripted in Duff's Company E, 14th Battalion Confederate Cavalry and served until the end of the war. Adolph Rosenthal was killed on January 17, 1891 in a fall off the San Antonio and Aransas Pass Railroad trestle east of Comfort. Kerr County Historical Commission *Kerr County Album*, p 393; Geue *New Homes in New Land*, p 122; Bennett *Kerr County*, Various Pages; Ransleben *100 Years*, Various Pages; Smith *Confederate Veterans Kerr County*, p 66; Harbour's MR and Adolph Rosenthal's Confederate Service Records.

30. Williams *Border Ruffians*, p 237 and Camp Pedernales Post Returns for the month of August 1862.

CHAPTER 9 – To Fight or Flee

Learning that 'regular' Confederate troops were being deployed into the Hill Country with the mission of destroying them, *Des Organisator*'s military arm, which Eduard Degener described as "for the purpose of carrying on guerrilla warfare," had a tough decision to make. Major Tegener learned Confederate troops were being deployed as early as July 20th while he was at work in his sawmill on Tegener Creek. He immediately called a meeting of *Des Organisator*'s political arm, the Advisory Board, to plan their response. This decision is mentioned in several insurgent accounts. Much discussion took place as to what course of action they should take. One idea was to carry on with guerrilla or insurgency warfare. This course of action had some advantages. Mainly, the mountainous terrain was favorable. The insurgents knew the land and had support from most of the inhabitants. However, there was concern about what would happen to their dependents. By July 25th, after much discussion and soul-searching, the decision was made to, "forsake the land, cross the Rio Grande into Mexico and offer themselves to the United States Army." All accounts by leaders of the fleeing group, or accounts attributed to them, agree on this point. August Hoffmann said, "Complying with this order sixty-five or sixty-eight men set out for the Rio Grande, intending to cross into Mexico whence they would sail for New Orleans and join the Union Army." Captain Ernest Cramer said, "We would go to Mexico where there might be a chance for us to join the Northern forces." Captain Jacob Kuechler said, "A Party of about 62 Union men met in camp on Turtle Creek, in Kerr County, to leave Texas and join the Federal Army by way of Mexico." The first written account of the event made public was in 1866. It stated the insurgents planned, "to proceed to Mexico and, if possible, to join the Union Army." Even today, some insurgent descendants proudly state their ancestors "were going to fight for the north." The idea of a free state of West Texas, was over, at least for a while. Major Tegener issued orders for the insurgents desiring to flee to assemble at Turtle Creek in southern Kerr County.[1]

The fact that *Des Organisator* learned of the decision to send Confederate troops into the area as early as July 20th shows that they had informants inside General Bee's headquarters. It is not known on what date Confederate deployment orders were issued, but the troops were not told until July 23rd and the deployment did not begin until July 29th. *Des Organisator*'s decision was made by July 25th, or four days prior to the Confederates leaving San Antonio and six days prior to their arrival in Fredericksburg. Most accounts claim the insurgents left their assembly area at Turtle Creek on August 1, 1862. The source credited for this belief is Sansom's 1905 Pamphlet. In it he says, "In the afternoon of the following day, August 1st, 1862, sixty-one of these, including myself and Major Tegener, set out for the Rio Grande." Two years later, in a letter to James T. DeShields, a noted Texas historian, Sansom corrects the date of departure and says they left on August 3rd. Eduard Degener's letter of August 1st confirms the date of August

3rd as the date of departure where he says his sons were not going to leave for the assembly area until after dinner. Another confirmation the group left later than August 1st is insurgent Captain Kuechler's letter where he gives the departure date as August 5th. Likewise, insurgent Captain Cramer's letter says it was later than August 1st when they left. Tegener learned that Confederate troops were on the way as early as July 20th, or eleven days before they reached Fredericksburg.

Des Organisator insurgents' decision to flee required them to prepare for the trip, which was going to take a long time as "the way led for about 150 miles through uninhabited wasteland and so demanded a goodly supply of provisions." The work of gathering the supplies had to be done in secrecy to prevent Texas and Confederate patrols from finding tracks to the insurgents' camp. How the supplies were obtained can be determined by examining the records of the Kerr County Provost Marshal. Those records show that nine men were issued travel permits on August 2, 1862; 45-year-old Ernst Schwethelm, the father of Henry Schwethelm, one of the fleeing insurgents; 36-year-old Gustav Steves, a brother of Heinrich Steves, Jr., another fleeing insurgent and the brother-in-law of Wilhelm Boerner, a fleeing insurgent; 28-year-old Fritz Metzger; 38-year-old Casper Real, whose brother assisted the insurgents once they reached Mexico; 24-year-old H. C. Sanger; 55-year-old L. Sanger; 22-year-old James S. Hope; and Charles Bergmann [Burgmann].[2]

Not all the insurgents met at Turtle Creek. It was an assembly area only and not one of their hiding areas, or base camps. The German companies had at least two such base camps. One was near the current community of Albert in southeast Gillespie County. The other was at or near the head of the South Fork of the Guadalupe River in western Kerr County. What likely happened was that *Des Organisator* members from Gillespie County and a few of the leaders assembled at Turtle Creek. There are several insurgent accounts which confirm that this Turtle Creek group went to the insurgent's camp on the Guadalupe River. Eduard Degener says the Turtle Creek group was going to be joined by 47 others. Hoffmann likewise implies such when he says there were, "33 from Gillespie County and 34 young men from Kendall County." The third confirmation comes from insurgent Captain Cramer who says, "At the well of Guadeloupe we met 40 more men."[3] [Maps G & K, pp 202,416].

On July 25th, as the supplies were being gathered, Eduard Degener went to see John W. Sansom at his home in Curry's Creek. Degener told about the decision to flee and that two of his sons were going to be in the group. He asked Sansom to go as "their guest" [meaning he was not a member of the League] and guide the group. Degener told Sansom "that if [he] wanted to go [he] could make [his] arrangements in time" as the group was assembling on Turtle Creek southwest of Kerrville on July 31st. Sansom, who at the time was the Kendall County sheriff, was already planning on going to Mexico alone. He changed his plans and agreed to accompany the members.[4]

While Eduard Degener was visiting Sansom, his sons were preparing to embark. They had a horse shod at the blacksmith shop in Comfort. They shared a packhorse with Wilhelm Tellgmann for the trip. Eduard Degener tells of his feelings seeing the departure of his sons, "One more dinner, then goodbye, who knows whether we will meet again. Not a cent of money in the Pocket, thrown amongst Indians and Mexican robbers, where Confederate money is of no more avail … Mama is baking the last bread for some time to come for her dear sons."[5]

There are several accounts that state the Governor or some other Texas or Confederate authority issued some type of proclamation giving all those who did not want to live under Confederate rule 30 days to leave the state. One account says at the time of the battle the insurgents still had 12 days remaining. A search of Texas and Confederate records located no such proclamation. The first time a reference was located saying anything about a proclamation was Jacob Kuechler's letter read at the twenty-fifth anniversary of the battle. In it he says, "Under the proclamation we had sufficient time to leave Texas," John Sansom in his 1905 Pamphlet makes reference to such a proclamation when he says, "Having read a proclamation from the Confederate Government announcing that all persons not friendly to it might leave the country," The two 1920s newspapers interviews with August Hoffmann and Henry Schwethelm also refer to such a proclamation. The 1924 Schwethelm interview says, "consequently when the Governor of the State issued a proclamation that all persons who would not take the oath of allegiance to the Confederate States would have to leave the State within thirty days." The 1929 Hoffmann interview says, "who were unwilling to take the Southern oath of allegiance were given thirty days to leave the county," and "although the time set in the Governor's proclamation had expired ten days before." So, at first glance it appears there are four primary sources: Kuechler, Sansom, Schwethelm, and Hoffmann, confirming that such a proclamation was issued and the insurgents were traveling under its authority. Yet, a detailed search of Texas and Confederate records found no such proclamation; not in the Governor's Proclamation Book, not in the Texas Adjutant General's Correspondence, not in the Governor's correspondence, not in any San Antonio, Dallas, Austin, New Braunfels, and Clarksville newspaper of the time period, not in the records of the State Military Board, not in any records of the Department of Texas on the Sub-Military District of the Rio Grande – nowhere. Since the earliest record was Kuechler's letter read twenty-five years after the event, other reliable primary documents were checked. In the one letter written before the battle, Degener makes no reference of any such proclamation. The proclamation is not mentioned in any of the trials held before the Confederate Military Commission from July to October 1862. In the two letters written shortly after the battle, Cramer's in October 1862 and Schlickum's of December 1862 – March 1863, no reference to any such proclamation was found. The next printed document was the January 1866 issue of *Harper's Weekly* that told about the battle and burial. Again, no proclamation was found. The same is true of Tegener's 1875 letter and Siemering's 1875 notes on which the 1923 articles were written. There was one final document to check. This was R. H. Williams' 1907 book. Surely it would have contained such a reference since Williams was critical of Duff, but again, no reference to any proclamation.

The only conclusion that could be made is **NO SUCH PROCLAMATION WAS ISSUED** [emphasis added]. A reexamination of the four primary documents that mentioned the proclamation was made. First, is Kuechler's. In other articles attributed to him a reference to a proclamation was located. It read, "It is not to be forgotten that the Confederation at the outbreak of the war granted each citizen who did not want to stay under the Confederate flag, 90 days to leave." This is referring to President Davis' Proclamation of August, 1861, in which it gave the requirement to take the oath of allegiance and become a citizen of the Confederacy or leave within forty days. Likewise, a check of Sansom's other writings or articles discovered a reference to a proclamation; again, it was to Davis' August 1861 Proclamation. This document reads "that there were many other self-confessed Unionists whom it might be well to watch and punish for their failure to leave the State, as by proclamation already issued all persons unwilling to take the oath of allegiance to Confederacy had been commanded to do. The time allowed for their departure was short: nobody cared to pay a fair price for land ... Arguing thus, very few of to whom the proclamation applied obeyed its commands. But we were no doubt kept under strict surveillance and spies were instantly amongst us, no vigorous campaign was undertaken again June 1862." Sansom is likewise referring to the Davis' Proclamation of August 1861, well before June 1862. A re-examination of the Schwethelm and Hoffmann articles revealed the person who conducted the interview added the comments about the proclamation; not what Schwethelm and Hoffmann actually said. This means that no primary source said that a proclamation had been issued in the June – July 1862 period during which insurgents were traveling. It was a trick of words. No writer or historian had taken the time to check the proclamation story for accuracy. When this fact was pointed out to insurgent's descendants a typical response was, "that even if the proclamation *was* issued in 1861, news of its contents might not have filtered out to the more remote German-speaking communities until 1862." A check of Kerr County Records shows citizens knew about the requirement, and began taking the oath in September, 1861. In conclusion, there was no proclamation issued in August, 1862, giving the insurgents time to leave the Confederacy![6] John W. Sansom, the two Degener brothers, and Wilhelm Tellgmann left for the assembly area on the evening of August 1, 1862. Major Fritz Tegener joined the insurgents at their assembly area on July 29, 1862 leaving his wife and two children "in comfortable circumstances, provided with money, notes, accounts, the use of a mill and other resources." Twenty-eight-year-old Heinrich Steves, a son of a Forth-Eighter and Freethinker, as well as a Comfort *Turnverein* also decided to flee with the group. He had been a member of Kuechler's December 1861

Company. Steves asked sixteen-year-old Fritz Schellhase to drive his mule team while he was away.

Schellhase asked him, "What are you going to do?" Steves replied he, "was going to escape and go to Mexico with the bushwackers [sic]." Schellhase agreed and Steves left for the assembly area. It was the last time Schellhase ever saw him.[7]

Not all the insurgents who were hiding in the mountains chose to flee. There were from 75 up to 200 League members hiding in the mountains. One of the ultra-radicals of 1854 reported about 200 insurgents fled to the mountains, "since the invasion of the Northern Troops didn't happen." Of these, only 64 fled for Mexico. The rest, "returned home and swore allegiance to the Confederate States and were promptly drafted into [Confederate] service." Sansom confirms that not all the insurgents that were at the assembly area departed for Mexico. He says, "Assembled there and recognizing Major Tegener as their leader were about eighty men. In the afternoon ... sixty-one ... set out for Mexico." This means about twenty insurgents did not make the trip and returned to their homes. It is likely these twenty-some-odd were those that later accounts claim left the insurgents camp on the evening of August 9th.[8] Those that returned to their homes and conscripted from Gillespie and Mason Counties included: twenty-one-year-old Carl August Arlt, a native of Prussia; Frederick Fresenius; twenty-two-year-old William Kuehn, a native of Hanover; twenty-one-year-old Johann Adam Meurer, a native of Nassau; twenty-year-old-old Sebastian Novian, also from Nassau; twenty-seven-year-old

Christian Probst, a native of Hanover; nineteen-year-old Adolph Quintel, a native of Hanover; twenty-five-year-old Frederick Rahe, a native of Prussia; twenty-nine-year-old Ernest Ludwig Theumann, a native of Hessia; twenty-five one-year-old Wilhelm Herman Toepperwein, and twenty-one-year-old Johannes Paul Toepperwein, both natives of Prussia.[9]

Those who returned to their homes and subsequently conscripted from Kendall County included twenty-five-year-old Alexander Brinkmann and his twin brother Charles Brinkmann, natives of Prussia, thirty-three-year-old William F. Geissler, a native of Austria and thirty-four-year-old Adolph Rosenthal. [10] A later account provides additional information on the location of the assembly area on Turtle Creek: "On August 1, 1862 Major Tegener and some 68 men, ... met on the banks of a tributary of Turtle Creek. ... Nearby settlers noticed the unusual numbers of horsemen coming into the area. They were so afraid of bushwhackers! ... and the creek, on the banks of which they gathered has since been known as Bushwhack Creek. Now the road which follows the creek has been name the Bushwhack Road."[11]

Sansom's account in particular has caused great controversy. He says when Degener visited him on July 25th he told him about the Union Loyal League (*Des Organisator*) being organized at an earlier meeting and only recently, "it was decided to disband the three Companies organized, as an assurance to the Confederate military authorities that no armed conflict was to be expected." Sansom's statement about *Des Organisator* disbanding has resulted in this statement being in all subsequent accounts of the Battle of the Nueces and further demonizes Duff and the Confederates. The problem is at the time of *Des Organisator*'s meeting and Degener's meeting with Sansom there was no Confederate military authorities in the area to inform they had disband and no armed conflict was expected. Two other items of some controversy need to be addressed. One is very minor; who was in command of the insurgent group? Siemering says it was, "The often-mentioned Jacob Kuechler, who knew all the hills, canyons, valleys and streams in this area, undertook the task of being their leader." Kuechler says, "I guided the party as I was acquainted with Western Texas." Julius Schlickum also says Kuechler was the leader. The statement by Siemering that Kuechler was the leader was one of the reasons Tegener wrote his 1875 letter. He took issued with Siemering and said the group was, "not under the leadership of Mr. Jacob Kuechler but that of Mr. Fritz Tegener." Which brings up the second controversial item, that is when was Tegener elected major? Many accounts state Tegener was elected major at Turtle Creek. An incorrect reading of Sansom's Pamphlet has caused this where he says, "Assembled there [Turtle Creek] and recognizing Major Tegener as their leader." Tegener clearly states the fact, "you made me the leader (major) at Bear Creek Spring … in the beginning of May 1862." Confederate troops did not arrive at Fredericksburg until July 31st, the first day League insurgent were assembling at Turtle Creek. Up to that time there were no Confederates in the area to assure "that no armed conflict was to be expected." At this time the Confederates were not aware *Des Organisator* had a military battalion of three companies.

The fact is *Des Organisator*'s Battalion, under the command of Fritz Tegener, was a paramilitary armed force on its way to join the Union Army, making it a legitimate military target. This battalion consisted of parts of two League military companies, the Gillespie Company under the command of Captain Jacob Kuechler and Lieutenant Valentine Hohmann and the Kendall Company under the command of Captain Ernest Cramer and Lieutenant Hugo Degener.[12]

The actual number of *Des Organisator* insurgents who departed for Mexico has always been something of a mystery. Various accounts say it was as low as sixty-one to a high of seventy-two. Sansom says "sixty-one, to include myself and Major Tegener, left for Mexico." Captain Kuechler gives the number as sixty-two. Hoffmann says in one account there were thirty-three from Gillespie County and about thirty-four from Kendall or a total of sixty-seven. In another account, he says there were "sixty-five or sixty-eight men set out for the Rio Grande." Henry Schwethelm, a twenty-two-year-old member of the group, native of Prussia, and Comfort *Turnverein,* gives the number as sixty-eight. Julius Schlickum and Captain Cramer both say sixty-eight. Siemering says there were seventy-two insurgents. The correct number for the Unionist group is provided by two primary sources. First, is insurgent Major Tegener; an 1875 newspaper article by Siemering resulted in Tegener writing a letter giving the correct number. Replying to Siemering's article, he said Siemering's number was not correct and the correct number was sixty-five, "because I remember the list, for I had all the names of our people in a pocket notebook in order to assign the guards for each evening." There are two other references to a list containing the insurgent names. W. J. Edwards, a member of Duff's Company, testified at the trial of Eduard Degener that they captured a list, which appeared to be a muster roll. He stated the last time he saw the list it was in the possession of Lieutenants McRae and Lilly. The second is a newspaper account, it says, "Their muster roll was found containing 69 names." The second primary source is Lieutenant McRae's Report of the Battle. He obtained his data from two sources. First, a captured list or muster roll and second, from Ferdinand Simon, a captured insurgent. McRae says the insurgent force "was composed of 63 Germans, 1 Mexican, and five Americans", four of the 'Americans' joining on the trail, a total of sixty-nine. When a name-by-name roster is compiled with information from families and other sources the number adds up to seventy-seven, however this is much too high. Sixty-five left, with another four joining them on the trail, make the total insurgent group at sixty-nine.[13] [Appendix Y].

Some Southern sympathetic writers later attempted to down play the significance of such a large anti-Confederate military force in the Texas Hill Country. Their theme is *Des Organisator* insurgents were just like, "many hundreds of Anglos [who] fled to Mexico to avoid the war, one tiny band of thirty Germans, all brand-new arrivals, in the State and all from the far west town of Comfort, tried to do the same thing." The 'band' was much larger then thirty, it was sixty-four and none were 'brand-new arrivals' in Texas.[14]

Among the insurgent Anglos, who fled as individuals or in small groups, were members of the 'American' or Kerr Company of the League's Military Battalion. Among the known members of the Kerr [American] Unionist Company the following enlisted in the First Regiment Texas Cavalry; Lieutenant Philip G. Temple enlisted October 27, 1862, John Strong enlisted October 27, 1861, Captain Henry Hartmann enlisted October 27, 1862, Asa P. Lacey enlisted May 1, 1864, Joshua C. Lacy enlisted October 27, 1862, Amon Hartmann enlisted October 27, 1862, Allen B. Nelson enlisted October 27, 1862, James Johnston enlisted October 27, 1862, and Jesse Starr enlisted on March 30, 1863.[15]

At least one *Des Organisator* member, eighteen-year-old Freethinker and Comfort *Turnverein* Casper H. Sueltenfuss, was unable to reach Turtle Creek in time to join the fleeing group. "He had intended to accompany the Comfort and Sisterdale men to Mexico, but was three days later in starting … The Confederates were on his trail and had tried several times to capture him by burning hay stacks and brush fences" where they believed he was hiding. Sueltenfuss eventually made it to Mexico where he worked as a checking clerk for an English merchant. He later met Captain Adolph Zoeller and joined the First Regiment [Union] Texas Cavalry.[16]

Two incidents took place either just prior to *Des Organisator*'s departure or the first night of their trip that resulted in the Confederates learning about their plans. The first incident explains how the Confederate learned of the departure of the insurgent group and involves the actions of an alleged Confederate 'spy' who was in the insurgent camp at Turtle Creek. Siemering says, "As fate wanted, just before the departure a German by name of [Burgmann], who worked in a mill in Sisterdale and who was thought of in general as a Union man, by chance found the camp." Siemering continues "because he was acquainted with most who were assembled [there] people spoke openly and he declared that he wanted to join the expedition, but that at first he must return home and would return in a few days." Captain Jacob Kuechler gives a different version. He says shortly after departing, the insurgents met Charles Burgmann and for some reason "Perhaps as a joke or out of hunger" took supplies from him. This resulted in Burgmann becoming angry and he "put the pursuers on the trail of his countrymen." The exact biographical data on Charles Bergmann or Burgmann is difficult to locate. One reason there were two other Charles Bergmanns in Comal/Kendall/Kerr Counties. The first time this Bergmann/Burgmann can positively be identified is the 1861 Kerr County Tax Roll. His name is spelled Bergman. He owned no land and paid a poll tax of 75 cents. The next record on which he is located is the Kerr Provost Marshall June 1862 List of Males in Kerr County not in the military. This record shows him as Charles Bergmann, age 36, [which means he was born about 1826] a shingle maker. On August 2, 1862, Charles Bergmann received a travel permit for the military jurisdiction. McRae's causality list shows Charles

Burgmann as a volunteer guide and wounded in three places, both thighs and an arm. After he recovered from his wounds on October 1, 1862, he enrolled in Duff's Company B. After the war, Bergman fled to Mexico.[17]

John Sansom also provides information about [Burgmann], "Later, but alas too late to be of service to us it was learned that a man by the name of Burgeman [sic], who had been at many of the meetings and had been confided in as a true Unionist, but betrayed to the Confederates the place of meeting and the route to be taken." Sansom claims Burgmann was not a resident of any of the counties from which members of *Des Organisator* came. "Subsequent developments gave us reason to believe that while associating with Unionists he was acting as a spy in the employ of the Confederates."[18]

The second incident that some accounts claim was the reason the Confederates found about the insurgents' departure was one of their horses got away from his insurgent owner. "On the first night of the meeting the horse of one to the troop members had become un-hobbled (sic) and made his way back to Tegener's Creek, where Tegener had a sawmill, and being followed by the owner until caught, was the cause of the trouble that occurred later." The owner went after the horse. Just as he reached Tegener's Mill, he ran into one of the Confederate scouts. The scout "had received some inkling of the departure of the Germans." They questioned him as to where he came from and why he was in that area. The insurgent "was somewhat 'green', having recently arrived from Germany." They finally exacted the truth from him about the insurgents. He was compelled to show them the camp where the insurgents had assembled, and the direction they had then taken. The scout took the prisoner back to Camp Pedernales.[19]

Des Organisator's reduced military battalion was ready to leave. They intended to cross the Rio Grande below the mouth of Devils' River. Their general route was "to strike southwest, crossing the Medina, Frio and Nueces River and passing the border not far from Del Rio." The number of horses and pack mules they had is very important in later determining when and how insurgents left the battle site. "The train started moving in the direction of the Rio Grande." It had "a small caravan of mules packed with provisions." One pack mule carried provisions for three men for three weeks. "The Germans had some 80 horses and traveled in regular formation." Given the fact that 65 members left for Mexico and there was a pack animal for every three members means there were about 21 pack animals, or a total of about 86 animals. The four Americans who joined on the trail would have brought the total to about 90 animals This is important because Lieutenant McRae's captured 83 head of horses. In addition to the 83 McRae captured, the five Americans left the battle site mounted on five horses, for a known total of at least 88 horses or mules.[20]

Just how well-armed were the insurgents is another point of controversy. Conventional wisdom has always said they were poorly armed. John Sansom is the source for this belief. "About forty of these [the insurgents] were fairly well-armed with muzzle–loading guns and six—shooters;

the others were poorly armed—one man having neither gun nor pistol." However, contrary to Sansom's recollection, one of the fleeing insurgents said they were, "fully equipped with rifles and six-shooters (the rifles mostly of German make)." The Confederates captured a total of 46 weapons. Insurgents who escaped from the battle took at least 46 weapons, bringing the total insurgents weapons to at least 92. This was a well-armed group.[21]

The insurgents left Turtle Creek on the afternoon of August 3, 1862, and traveled west to the South Fork of the Guadalupe River, entering the river near the present Crider Store. They turned south to the headwaters of the South Fork of the Guadalupe where the insurgent's base camp was located. They traveled about eighteen miles the first day. At the base camp, they joined up with the remainder of the insurgents. There are several references to linking up with the remainder at this base camp. First, is Siemering where he says the base camp was, "relocated to the upper Guadalupe." Second, is Cramer where he says after leaving Turtle Creek, "At the Well of Guadeloupe we met 40 more men." Everyone spent the night at the base camp. The next morning, the reduced two-company battalion left for Mexico.[22]

They traveled exceedingly slow for a group of purposeful men with no women or children to hold them up. They made only about ten to fifteen miles a day. The reason most often given for such a slow pace is that they did not believe the Confederates knew of their departure and therefore they were in no danger. With the large number of

Confederate patrols, it is hard to understand this naïve belief. This has given rise to the speculation the slow pace was deliberate and caused by a 'spy' in order to allow a pursuit force to catch up. This just isn't the case. The Schwethelm interviewer writes that Schwethelm and others were opposed to this, "unpardonable delay and openly expressed themselves about the matter. Tegener laughed at their fears, being overconfident that they were not being followed nor their whereabouts known." A careful reading of the Henry Schwethelm accounts shows that he stated his opposition to such a delay – if at all – only after the insurgents saw horsemen overlooking the camp on the afternoon of August 9th. Nowhere in his account does he say he expressed his feelings to Tegener earlier. The Sansom account also says he expressed concern only after seeing the horsemen, and not before. Several insurgents explain the reason for the slow pace. Jacob Kuechler says, "Our progress was slow, as the country was rough, and our pack animals were carrying heavy loads." Elsewhere he says, "On the way to Mexico their youthful spirits and carefree fun took precedence over caution and the real purpose of the trips. They took advantage of the time in joyous hunting and other amusements." John Sansom also says they did not think they were being followed. "Believing, as did the Major and most of the party that they had eluded the Confederate troops, known to be hunting for them, we pursued our way slowly and very much at our leisure … suspecting neither betrayal nor pursuit, Major Tegener while moving on steadily made not haste." Hoffmann states, "We were not in too big a hurry as we had plenty of time and we also spared our unshod horses over the rocky course." Schwethelm states they were in no hurry and they, "jogged along at the rate of only five to ten miles a day, spending a great deal of the time in hunting for game, which was plentiful enough. Some of them, being fond of wild honey, searched for hives of wild bees and took things easy in a general way." Schlickum describes the impact Jacob Kuechler had on the trip. He explained Kuechler was a courageous man who knew the woods well and "a very good hunter and also academically well learned." Schlickum says Kuechler "is the personified *Phlegma* 'don't hurry – don't push' [*Nur Nicht Draengeln*] He is a true *Urwald-Bummler* [stroller through primate forest and wilderness] and is unhurried [*Gemutlicheit*] he and his men roamed through the hills. There was no hurry!" Kuechler and his men hunted deer, harvested wild honey, cooked and roasted their meals at leisure, and in the evening sang songs.[23]

An examination of the insurgent's route offers another reason for the slow pace; to bypass the state and Confederate forts and camps. They traveled west below Camp Davis and above the Frontier Regiment camps at Verde, Montel, and Rio Frio. The route took them north and west of Camp Wood, north of Fort Clark and Fort Duncan/Rio Grande Station at Eagle Pass and below Camp Hudson. Plotting the locations of each nights' stop reveals that their pace was determined by two factors; the large number of mounts and waterholes. In every case, they stopped where they did because of water. Had they continued they would have been forced to camp without water, which they had still to do on the last night before reaching the West Prong of the Nueces River.

The insurgents left their base camp on the morning of August 4th. They traveled generally southwest to the head of the North Prong of the Medina River, a distance of about six miles. On the afternoon of August 5th, they reached the next waterhole, the West Prong of the Frio River, a distance of about eleven miles. They were now well into the mountains and in very rough terrain. On August 6th, they traveled about fifteen miles, reaching the head of Bullhead Creek. Water now became more plentiful. On August 7th, they moved down Bullhead Creek, which was flowing rich with fish, about sixteen miles reaching the East Prong of the Nueces River. They camped at the junction of Bullhead Creek and the Nueces East Prong. The community of Vance is now at this location.[24]

The next morning, August 8th, as they were preparing to leave, a small four-man party of Anglos joined them. They had left the settlements and traveled west until they reached the head of the East Prong of the Nueces River and turned south following the river. These four men included two brothers, twenty-six-year-old Thomas J. Scott, a native of Illinois, and his brother nineteen-year-old Warren B. Scott, a native of Iowa, both living in Kerr County. A third member was twenty-two-year-old William Hester, from Blanco County. The twenty-year-old nephew of Sebird Henderson, Howard Henderson, was the fourth man. All four were members of the American Company of *Des Organisator*'s military battalion. The four-man group agreed to travel with the insurgents at least as far as the West Prong of the Nueces and would later decide if they would go with them to Mexico. The first part of the trip on the 8th was pleasant, similar to the day before. They traveled along the banks of the Nueces East Prong for about twelve miles, reaching the location of the current town of Barksdale. They filled their canteens and watered their mounts then turned southwest and crossed a small stream now called Dutch Creek and into the roughest part of their journey.

They traveled about another seven miles when they stopped for the night. Ernest Cramer described the location as, "late in the evening, we came to a narrow 'draw' heavily covered with cedar growth. We stayed there, though without food for our animals."[25] [Map K, p 416].

Early the next morning they continued. After riding only about two miles they reached the West Prong of the Nueces and a large waterhole. It was August 9th. Cramer states, "We came to an open space – a meadow with fine grass and plenty of water. Not a stream but springs that seem to sink again. ... It was an outer arm of the Nueces." There were small groups of cedars scatted around the meadow. They estimated that Fort Clark was about 50 miles to the southwest, and about 45 miles from where the San Felipe joined the Rio Grande. "Our horses being hungry and tired we decided to rest there for the day and the next morning to continue on our way to the Rio Grande." Cramer's Company made their camp under the first group of cedars nearest the water. Kuechler's Company made its camp generally south of Cramer's.[26]

August Hoffmann reinforces Cramer's statement about the horses, "Then at the west fork of the Nueces, we halted at the waterhole for a day. Our horses being mostly unshod, had begun to go lame on the flinty mountain trails. So, it was decided to rest them, although one more day's travel would have put us safely across the Rio Grande."[27]

Sansom says, "No special precautions were taken against surprise." That afternoon, "Two men were detailed as guards, it was more for the purpose of keeping our horses together and on good pasturage than for protection of our party against the sudden and unexpected approach of enemies." On the 135th anniversary of the battle, the author, along with Uwe Herkenrath [aka Tim Darby], a Center Point archaeologist/historian visited and spent a night at the battle site. Mr. Darby has inspected the site several times. Since then he and the author have visited the site several more times. Based on these visits and artifacts found, the exact camp location was determined. [See Map L & M, pp 462-464]. The camp was located in an "an open prairie surrounded by mountains. Two small separate cedar groves, each surrounded by a circle of bushes, formed the only brush in the area. There groves were approximately 200 feet apart." The open prairie was on a curve of the Nueces. The river was to the north and made almost a **L**-turn so that its deepest part was to the east and southeast of the camp. To the south was a small creek filled with a timber strip. Major Tegener set his camp about 150 yards south and about 350 east of the **L**-turn and generally between Kuechler's in the south and Cramer's in the north. The entire camp was in an area about two acres. After establishing camp, the insurgents sent hunting parties out as, "deer, turkeys and other game were abundant in the country where we were, and hunting parties were going and coming all day."[28]

Meanwhile Lieutenant William Harbour and his scout had returned to Camp Davis from Beaver Lake about August 1st with the six prisoners. They were turned over to the Confederates at Camp Pedernales. No evidence could be found to establish they were Unionists, so they were released. Two of the men, Robert N. Robinson and J. K. Spears, enlisted in Duff's Company on August 1st and served until the end of the war. Lieutenant Harbour and his men remained at Camp Davis to rest for a few days. By August 3rd, they had joined the remainder of the company scouting for insurgents. About midmorning of August 4th, a ten-man scout, likely from Harbour's detachment and who "had received some inkling of the departure of the Germans", rode into Camp Pedernales with a prisoner. This prisoner was very likely the one captured at Tegener's Creek. He informed Captain Donelson the insurgent group were fleeing toward the Rio Grande and said he was willing to guide the pursuit force.[29] How the Confederates "had received some inkling of the departure of the Germans" is interesting. Two members of Harbour's detachment where Benson brothers, William Thomas, who lived in the Tegener household before enrolling in Davis' Company, and John Washington. Fritz Tegener was married to their sister Susan. Tegener's

Sawmill and home was on Tegener Creek. The Benson brothers visited their sister while in the area. There they learned that their brother-inlaw, Fritz and his group, had fled to Mexico.[30]

The exact identify of the prisoner who arrived in Donelson's camp, is not known. R. H. Williams' description comes the closest to identifying him. "One of the prisoners, an old soldier, and a friend of [Duff's] had been released, and he was to act as our guide and betray his friends, if possible, into our hands." The most likely identity is Charles Burgmann, the alleged 'spy'. It is known he was in the pursuit as a 'volunteer guide' and had worked for Fritz Tegener.[31]

However, there is another possibility. After Dennis Kingston's death, his widow married a German immigrant who, according to family stories, used the alias of *Bargeman* to join *Des Organisator*. These stories say the Confederates captured him; in order to save his life and revenge Dennis Kingston's death, he informed about the insurgent's departure. What gives this story creditably is Alex Brinkman, who prior to his death, was the leading expert on the Nueces events. He says, "Duff's spy and murderer, *F. Bauman,* that aided him in running down Union men and murdering them was a stranger in the country and aided in assassinating unarmed men at the Nueces Battle." Brinkman clearly identifies the 'spy's' given name not as Charles, which was *Burgmann's*, but as a *F*.[32]

Brinkman's *Bauman* is much closer to the *Bargeman* in the German immigrant family stories than to *Burgmann* or to *Bergmann* in the Siemering and Kuechler's accounts, and to the *Bergmann* who received a travel permit in August and was in the insurgent's assembly area. *Burgmann* or *Bergmann* is known to have been in the Nueces Battle, he was severely wounded and appears on McRae's casualty list. Due to the severity of his wounds, he was unable to take part in the execution of the prisoners, so he does not match the description of those helping kill the wounded guerrillas. It is also of interest that Sansom did not use the name Burgmann or Bergmann in his account, instead he used the name *Burgeman* and Sansom, like Brinkmann, says was "not a resident of any of the counties from which the members of our party came." Charles *Burgmann,* or *Bergmann,* arrived in Kerr County in 1860. He was a resident of the area and worked for Fritz Tegener. Therefore, he does not meet Brinkman or Sansom's description. It is the author's hypothesis that two men informed the Confederates about the insurgent's departure: *Burgmann* or *Bergmann* and *Bargeman/Bauman/Burgeman.*

Captain Donelson immediately ordered his second in command, twenty-seven-year-old native Texan, First Lieutenant Colin D. McRae with ninety-four, "fresh men and horses in pursuit" of the insurgents. McRae's scout was composed of detachments from each of the units at Camp Pedernales plus a detachment from Davis' Company. The detachment from his unit, the 2nd Regiment Texas Mounted Rifles consisted of twenty men. The detachment from Taylor's 8th Battalion consisted of twenty-five men and commanded by First Lieutenants James M. Homsley and

James S. Bigham. Second Lieutenant Edwin Lilly commanded the twenty-man detachment from Duff's Company. Second Lieutenant William T. Harbour commanded the twenty-nine-man detachment from Davis' Company. *Charles Bergmann* was one of the scouts and guides, the second one, likely *F. Bargeman,* making the pursuit force a total of ninety-six men.[33] [Appendix AA].

As in many of the events regarding the battle, the exact date the Confederates left Camp Davis and Camp Pedernales is in dispute. McRae's says it was the morning of the 3rd; this is not supported by other information. Sergeant Hopkins has no reference to any force leaving on August 3rd. He says on the 4th, "Some of the scouts having returned, a new detail was made." R. H. Williams' says it was on August 4th. Williams' account does not give the events by date. He uses the terms first day and so on. By working backward from the date of the battle, it can be determined he says the Confederates left on August 4th. "The day after the return of the wagons, one hundred of us, of whom twenty belonged to my company, were warned to prepare seven days' rations and to go on a scout into the mountains to find and attack the bushwhackers' camp." Williams goes on, "We all set off in high spirits, for we had soon tired of inaction, and here was a chance of a fight against men who really were in arms against our country, and were well-armed too."[34]

Accompanying McRae's Scout was at least one wagon carrying supplies. The first day of the trip from Camp Davis was along a road that was known at one time as Henderson Branch Road, but today as White Oak Road, to just past

Texas Ranch Road 783 where it turns into Rogers Trail. At this point, McRae's scouts turned south. They were now in the heart of one of the Anglo insurgent's strongholds. They continued and crossed, "a rather rough prairie country, in which we passed several small homesteads, ruined and deserted. At sundown we reached what had been a well-watered valley." Williams continues, "The owner, a Northern man named Henderson, had gone to the mountains, but his wife, also from the North, had been brought into camp with her numerous children." Williams described what they found at the Henderson farm as, "fenced and cultivated about twenty acres of good and on the side of the valley cleverly irrigated by the stream running through it. Now the crops were trampled and destroyed, and not a living thing was to be seen on the place; even the beehives in front of the comfortable log house were overturned and empty. The … furniture in the living-room, and the loom in the kitchen, had been smashed." The Henderson farm was that of Sebird Henderson, whom Confederates arrested the day before. His farm was located on a tributary of Johnson Creek, now named Henderson Branch, in northwest Kerr County, just inside the Kerr and Gillespie County line. The Confederates spent the first night at the Henderson farm.[35]

The next day, the 5th, the pursuit force traveled down Henderson Branch across Johnson Creek to the Guadalupe River about a mile west of the current town of Ingram and where Davis' Company had camped in early March. They turned west following the riverbed. They continued up the river past the 'region of settlements." Williams describes the

Guadalupe River here as, "beautiful, running clear and strong over a rocky bottom and between high cliffs crowned with giant cypress-trees." They stopped for lunch by a, "cool, shady pool, just deep enough for a delightful swim." The scout continued, the ground, "rising more and more, and growing more difficult for the wagon to follow." They stopped for the night along the river. The Confederate camp was east of the present town of Hunt. Due to the wagon, the pursuit force was traveling at a slow pace; about twenty-two miles in two days, still faster then the insurgents.[36]

The morning of August 6th, the scout reached the fork of the Guadalupe River. They turned south and went up the South Fork. That morning they crossed the insurgent's trail when they had left Turtle Creek and reached the Guadalupe River. At midday, after traveling about eight miles, McRae says they found the, "deserted camp of the men we were after. It was admirably situated in the midst of cedar-brakes, and had been left perhaps four or five days before, after being occupied for quite a month." McRae's also says they reached the trail, "on the morning of the 6th instant struck the trail of a party of horsemen." A member of Duff's Company, twenty-two-year-old Mississippi native, William J. Edwards, stated they, "Stuck [sic] the trail of the Bushwackers not far from the South fork of the Guadeloupe." McRae's scout found signs showing the insurgents, "Apparently meant business, for they had cut rude human figures on the trunks of some of the big trees, and had used them as targets for their rifle practice." According to Williams, the Confederates were able to determine a great deal from the deserted camp. First, they deduced that the enemy force totaled over a hundred men. Williams actually said the insurgents numbered about 160 men, another of his exaggeration of numbers. The insurgent force totaled 65 men, but due to the fact not all the insurgents joined the fleeing group the number that had been in the camp had numbered about a hundred. Due to this, and the number of pack animals, it was possible to believe the force was much larger. The total number of animals numbered about eighty-seven. The second item the Confederate determined was that the insurgents were heading toward the Rio Grande. Third, the Confederate pursuit was at least three days behind them. Therefore, the pursuit would be a hard ride over rough country if they were going to catch up with the quarry. This camp described by Williams is not the one located at the head of Turtle Creek. Williams describes the camp as being occupied for at least a month, confirms the insurgents had a base camp "deep in the mountains". It also confirms Cramer's comment, "At the well [or spring] of Guadeloupe we met 40 more men." In three days, the Confederates had only covered about thirty miles. The wagon would not be able to keep up and had to be sent back.[37] The Confederates spent the remainder of the day resting their horses and preparing field rations, consisting of bread, coffee, bacon, and sugar. The rations were carried in a sack swung across a saddle. The scout was divided into messes of five men each. Each mess had one sack. One small hunting party killed a small bear, but it was "so miserably poor as to be uneatable."[38]

Early in the morning of August 7th, the Confederates, now rested and with their field rations, were back in the saddle. From the head of the Southern Fork of the Guadalupe River, the scout traveled west-southwest until they reached the head of the Medina River. "A tremendously rough and hilly country, and we could only follow the trail in Indian file, till we struck the head of the Medina River." This was actually the head of the North Prong of the Medina River, located just inside Bandera County. "Here the country became rough, rolling prairie studded with timber, and we pushed on along the wide trail at a smart pace, till we called a short halt at midday." The trail is generally along current Texas Ranch Road 39 in southwestern Kerr County. W. J. Edwards' description of the route states from the Guadalupe South Fork and says the trail "went South west for a while and then North West." The trail dropped down into the West Fork of the Frio Canyon. Williams describes the route to as "desperate country to ride over, for we were will in the mountains, and frequently had to dismount and lead our horses down rocky slides." Toward evening the Confederates reached a large waterhole on the head of the West Prong of the Frio River, "perhaps the only one to be found for many miles … which showed the enemy had good guides." The Confederates watered their horses, filled their canteens and after a brief rest continued the pursuit. They continued well into the night guided by a full moon. Williams' description continues, "the full moon rose gloriously, and by her light we rode, and clambered, and slid till midnight, when we camped for a brief rest on a rough and narrow plateau, where there was a little grass for the horses but not water," said Williams. The Confederates traveled due west from the head of the West Prong of the Frio and crossed several canyons including the Short Prong of the Frio, above headquarters of the current Prade Ranch. The plateau is located west of Short Prong and northeast of the head of Long Kent Creek in Real County and northeast of the North Prong of Bull Head Creek. It is east of Texas Ranch Road 3235 and about five miles from where its pavement ends. [Map K, p 416]. They traveled about thirty miles on August 7th. Williams said, "From the elevation on which we stood we could see that the whole country to the southwest was on fire. It was a magnificent sight, probably caused by the Indians firing the dry cedar-brakes, which burnt like pitch pine."[39]

Before sunup the morning of August 8th, the Confederates were back in the saddle. They dropped down off the plateau into a major canyon. "About ten o'clock we struck the eastern branch of the Nueces River, where there being good water and grass we halted for breakfast and to graze our horses." Instead of being the Nueces River, this stream is Bull Head Creek, which heads just southwest of where the Confederates spent the night of the 7th. It travels in a southwest direction until it runs into the East Prong of the Nueces River at the current community of Vance. Because of its size and water flow, it is often mistaken for the head of one of the prongs of the Nueces River. The Confederates found Bull Head Creek, "running strongly, though only a few inches deep, between cliffs a hundred feet high, and over a bed of solid rock about the same number of feet in width ... the trail was easy ... to follow." The rocks above their heads were rich with untold wealth of honey, and the river full of fish.

Another account also tells of the abundance of fish and wild honey found during this part of the scout. It describes the air as fresh, the streams filled with perch and mountain trout, and many hives of honey that hung from the mountainsides. "As we rode along that afternoon, another trail came into the one we were following, showing the [insurgents] had been reinforced [by] another party," the four-man group of the Scott brothers, Howard Henderson, and William Hester.[40]

The Confederates pushed their mounts on. They knew they were gaining on their quarry. "The fugitives were not far ahead of us now, for at two o'clock, close to another branch of the Nueces, we came on their camp in which they had slept the previous night. The fires were still smoldering, and great chucks of half-cook beef were lying about." This establishes that the insurgents spent the night of August 7th where Bull Head Creek runs into the Nueces River. This also proves that the night of August 6th they spent the night at or near the head of Bull Head Creek. The trail ran south along the Nueces for a few miles to near the current town of Barksdale. It then turned southwest back into the hills. The spot it turned was west of the abandoned post of Camp Wood. The route behind Military Mountain, across Spring Creek to Dutch Creek. Henry Schwethelm lived in this area in the 1890s and passed this information to local residents. Williams continues, "We pushed on, the country getting worse and worse, and we generally on foot, as it was impossible to ride, till light failed us, and then halted for the night. There was some talk of marching again when the moon rose, but both men and horses were too done for that, and we had to rest till morning. There was no water for any of us." The Scout had traveled about another thirty miles on August 8th.[41]

The morning of August 9th found the Confederates in the saddle and riding hard well before sunup. "Famishing and parched with thirst, we struggled to the top ... of a ridge ... next morning, and then led our horses down the most precipitous descent ever attempted by mounted men." At midday, they reached a "water hole on some stream which held a little muddy, evil-smelling liquid, but a perfect godsend to both men and horses. Around this water hole the 'sign' of the fugitives was quite fresh." McRae knew he was getting close to the fleeing insurgents. He ordered an advance guard to move ahead of his main body. It was about 2:00 p. m. on the afternoon of August 9th.[42]

The advance guard moved forward. About 2:15 p. m., they came across two insurgents who were part of one of the hunting parties. They quickly captured and interrogated them. The captives refused to tell where their camp was. The Confederates offered to spare their lives if they would diverge the camp location. Again, the two insurgents refused. The Southerners then hanged them before continuing on the trail of the main party. Since beginning the research, the author has heard rumors of an insurgent being captured before the battle and hanged because he would not disclose the location of the insurgent's camp. What gave creditability to this story are a few lines in Ella Lonn's book, *Foreigners in the Confederacy*. Lonn's story says, "Shortly before daybreak the Confederates approached the camp and captured one of the other party. Through his life was offered him in return if he would lead them to the camp of his companions, he refused to betray them and was hanged." The problem with Lonn's story was it did match known facts. The Confederate advance guard had located the insurgent's camp. McRae and his officers made a recon of the camp well before any approach to the camp. So, there was no need to offer a captured insurgent his life in exchange for the location of the camp, the night before the battle. However, the rumor persisted during oral interviews.

The one thing that gave Lonn's story creditability was she used as her source Alex Brinkman, the most knowledgeable authority on the battle until his death in 1947. So, the only place Lonn would have gotten the story was from him. Brinkman must have had some reason to think the story was truthful to have passed it to Lonn. Based on other data obtained by the author, it is believed that one or two insurgents were captured on the afternoon of August 9th and offered their lives in exchange for the location of the camp. They refused and were hanged. A German-Texan who was stationed in Carolan's Company H, 32nd Regiment Texas Cavalry at Fort Clark also confirms insurgents were hanged at or near the battle site.[43]

About 2:30 p. m., the Confederate advance party located the insurgent's camp. They quickly returned to the main group and informed McRae. Williams tells how "We had only ridden on about two miles from this spot [the water hole] when our scouts came hastening back to report that our long stern chase was at an end." The advance guard had found the insurgents' camp. It was three miles ahead of them, on a small prairie surrounded by cedar-brakes, on the other side of the western branch of the Nueces. "The prairie, on our side of the stream, ran up in steep rocky cliffs, and from the top of these the scouts had overlook the camp." McRae also tells about his advance guard finding the insurgent's camp, "on the evening of the 9th instant, about 3 o'clock, my advance guard reported a camp in sight on the headwaters of the Western Fork of the Nueces River." McRae, "immediately diverged from the trail to the right, secreting my command in a canyon about 2½ miles from the enemy, and at once proceeded in company with Lieutenants Homsley, Lilly, Harbour, and Bigham, to make a careful reconnaissance of the position of the enemy's encampment." They returned thinking they had not been seen. McRae "proceeded to make my dispositions for an attack at daylight on the following morning." Williams fills in some details, "We then, being halted on the return of the scouts, [the officers] went forward to reconnoiter the position before forming the plan of attack." In about an hour, the officers returned and "orders were issued for a night attack. Then we moved about a quarter of a mile up a ravine running at right angles to the river, where we were securely hidden, and there off-saddled, and spread our blankets to await the coming fight with what patience we might."[44]

The insurgents arrived at the West Prong of the Nueces about 10:00 a.m. August 9th and established camp. Horses were watered and put out to graze. The insurgent's camp was spread out along a prairie. The individual messes started about 150 yards west of the main waterhole and ran west about 100 yards. Two man-hunting parties came and went. It was one of these hunting parties the Confederates captured and hanged. About 3:00 p. m. the hunting party of Franz Weiss and Heinrich Stieler saw three men on a hill studying the camp. Stieler was the youngest member of the insurgent group. having been born in 1845 in the Duchy of Anhalt. His father, Gottlieb, was a Freethinker and Forty-Eighter who arrived in Texas in the early 1850s and settled in the Comfort area in 1856. Heinrich Stieler was a member of the Comfort *Turnverein*. Weiss and Stieler knew the three men were not part of their group. They hurried back to camp and reported what they had seen, "We saw three soldiers on a hill," they warned, but "their words were in vain."

Sansom explains the incident in more detail: "About that time one of the hunting parties returned with the report that they had seen strangers whose evident desire for concealment appears suspicious. This intelligence created considerable commotion. But at that juncture another party of hunters returned to camp, and learning of the excitement quickly allayed it by the statement that they were the strangers the first party had seen—that just to see what the first party would do, they had first shown themselves and then made a pretence of concealing themselves. This statement at once turned the laugh on the first party; it was too good a joke not to be enjoyed, too reasonable an explanation of the first party's alarm not to be accepted as the true and only one. All uneasiness vanishing and calm instantly restored, not an effort was made to verify or prove false the story told by the first party. And while the supposed victims of the second party's joke were being unmercifully teased and made fun of, a third party of hunters came in and unwilling to acknowledge themselves victims of the same joke failed, until too late to do any good, to report that they had also seen strangers who acted so suspiciously that they had come back at once to camp to report the circumstance."[45]

From this point, myths replace many facts in the story. R. H. Williams and John W. Sansom have often been misquoted to establish some point a writer wants to make. For example, in a 1952 *San Antonio Express Magazine* article, Louis B. Engelke relates the insurgents seeing the strangers as, "before darkness set in, Capt. John W. Sansom and others sighted two horsemen across the valley. The horsemen raised their rifles, not at the refugees, but at the brush far away, and fired. Then, they rode in the direction in which they had fired. 'Listen to that echo,' a youthful refugee explained. 'Those men are hunting with new breechloaders. I wish I had one.' … 'Those are Duff's scouts,' Sansom replied. 'If they were real hunters, they wouldn't be so unfriendly, not in lonesome country like this.'" Sansom never made such a statement! Another myth is that Henry Schwethelm, "vigorously protested against this [delay], stating that they had been dallying along too easily; and furthermore, some of the members of his command had observed two horsemen on a distant hill apparently on a reconnaissance of some kind." As a result, "Schwethelm and two of his messmates threatened to abandon the troop and make their way to the Rio Grande ... Answering the remonstrances of Schwethelm and others, Tegener agreed to start the next morning, and promised them they would not unsaddle their horses until they were over the Rio Grande." The fact is, Schwethelm made no such comment. The myth developed in a 1924 article by Albert Schutze where Schwethelm's comment is taken out of context.[46]

A third myth is the claim that, "two men had visited us in camp, with the message that 'things could be arranged—it would be all right to go back.' Trusting in this, many went back ... That night 28 of the German troop gave up the trip and returned to Fredericksburg and vicinity, by taking a new route from that by which they came, and managed to evade anyone in pursuit." The fact is no insurgent left the camp prior to it coming under fire. This myth was first stated in the Hoffmann interview in the 1929 *Dallas Morning News.* No other Hoffmann account repeats this data, nor does any other source. When one first reads this article, it appears Hoffmann was saying something like 28 or 35 left the night before the battle. But that is not what either Schwethelm or Hoffmann said. What they said was "that night" 28 or 35 left. Now night begins after the sun goes down and before the sun comes up, or the night of August 9th—August 10th that 'night' was after 9:00 p.m. and before 7:00 a. m. Writers who claim some 28 or more left 'that night' and imply it was before the battle, point to Kuechler's letter as confirming this belief. But again, careful reading of Kuechler's comment tells us those that left did so after the first shots were fired. What Kuechler said was, "On August 9th, the fleeing band became **aware of their pursuers** [emphasis added] and some 30 of them abandoned the group and avoided a confrontation by hastily taking off in a different direction." Kuechler clearly says it was only after the insurgents became **aware of their pursuers** did any leave. They became aware at 3:00 a. m. the night of August 9th—10th. To compound the issue later in the 1929 interview the author – supposedly quoting Hoffmann – writes, "Before this, two men had visited us in camp, with the message that 'things could be arranged—it would be all right to go back.' Trusting in this, many went, although the time set in the Governor's proclamation had expired ten days before." The 1929 writer, like the 1924 writer, adds, "Some were caught and killed later on." The writer's narrative continues quoting Hoffmann as saying, "The rest of us said we would go on." The writer wants the reader to believe, and perhaps she believed, all this took place prior to the opening shots of the battle. But, both the 1924 and 1929 writers identified some of those that went back as later being killed. An analysis tells who were 'later captured and killed'. They had all been members of the fleeing group, some of whom were wounded in the battle. The night the men visited the insurgent camp was at Turtle Creek, not at the Nueces.[47]

All primary accounts state that a large number of insurgents left the battle site after the first rounds were fired at 3:00 a.m. and before the final Confederate assault. The number of insurgents taking part in the battle does not match numbers given by survivors, if the assumption is made that some 28 left before the first shots were fired. There was a total of 69 insurgents. If 28 left, that would only leave 41 remaining in the camp. Nineteen were killed and left at the battle site. That only leaves 22. The five 'American' left before the final assault. That leaves only 17. That is the exact number of survivors that gathered after the battle in the hills. So, whose accounts say insurgents left after 3:00 a.m. and the final Confederate assault? The accounts that tells us when insurgents left include Captain Cramer, who says, "The men who had joined us at the Guadeloupe, deserted their posts one by one. … of the 68 men we had only 32 were with us. Now the soldiers charged." Hoffmann says, "When the sun camp up over the mountains … there were about 30 of us left." Captain Kuechler says, "After the second attempt to take our camp nearly half of our men could not face the fight any longer and left us." Sansom says, "The camp was abandoned by able-boded defenders."

Siemering says, "Of the original 72, only 47 held their post." Schlickum says, "As dawn broke, Kuechler had only 35 men left." Henry Schwethelm says, "Some 25 man or so had left our camp before day—open—the fight commenced". And finally, the January 1866 Harper's *Weekly*, says, "Daylight, now at hand, disclosed that they were assailed by a body of a hundred strong. *Their own ranks had been thinned by desertion*" [emphasis added]. There was only twice when outsiders 'could have been' in the insurgent's camp. One is the night of August 9th as these two writers seem to want the reader to believe. The other was the gathering at Turtle Creek. If one wants the readers to believe two men caught up with the insurgents on the afternoon of August 9th, it has to be explained how they found the insurgents.

They certainly did not by following their trail. The Confederates were between the insurgents and anyone else following the trail. Some have speculated the two men could have been members of the pursuit force. If so, why would the two only tell a few of the insurgents leave? If two outside men showed up in the camp, some of *Des Organisator* leaders would have known about it. Yet, none of them say anything in their writings. The other time there were outsiders in the camp was at Turtle Creek where several men came and went.[48] It is the author's belief that, "Before this, two men had visited us in camp, with the message that 'things could be arranged—it would be all right to go back'. Trusting in this many went," **was at Turtle Creek**. Sansom tells us about 80 insurgents met and about 20 did not accompany the fleeing insurgents and that when the 1924 and 1929 writers tell about some leaving 'that night' it was those who 'left the camp early.'

The insurgent's camp settled down for the night. "There were duties to be performed, for by the time the sun was fairly down, the horses had all been caught and tied. The horses were tied up close by, they fed the loyal animals." A night guard of four men was posted and "passwords were set up." That evening "was one of those soft, Southern nights.

Bright stars set against a clear dark blue sky." "On this evening we were all quite cheerful, not having any omen that this place would be the grave for most of use." The insurgents "built flaming fires." Then came a feast on the game that had been killed during the day. They "conversed throughout the frugal meal." After that, some of the young men amused themselves by wrestling, turning summersaults, playing leapfrog, and light games. "We sang and declaimed until late in the night." Speech-making began, "some of the subjects of the speeches being 'American,' 'Citizenship,' and 'Civil War.'" Hoffmann recalled "the early part of the evening was devoted to declamation, recitations, patriotic speeches and *zum Schluss* (for a finale) we enjoyed some of the favorite German folk songs." One of these was an old German poem called 'Black Crows Are Still A-Flying.' "Everyone was in a cheerful mood and not an inclination of any danger was noted, no idea that a betrayal was pending," remembered one insurgent until, "sleep took over." So the quiet of the summer night settled on the carefree resting men."[49]

Not everyone in the camp was "carefree." John Sansom felt there might be reasons for concern. He called Major Tegener aside and asked, "Are you entirely satisfied, Major, that our boys saw no strangers around this evening?" Tegener answered, "Of course I am. Why do you ask?" Sansom replied, "Because, I fear they did see strangers, and if they did, it means harm to us." Sansom suggested, as he had never been placed on guard duty, that Major Tegener should have him pull guard duty that night. Tegener replied,

"There was no need for Sansom standing guard at any time; anyway, the guard detail had already been assigned." Sansom than asked he be called at two o'clock in the morning. Tegener agreed to do so. Sansom made another attempt to alert Tegener. He stated, "Major, you can if you will, get ready and leave here in thirty minutes. The moon is shining and the night air will give us cool traveling. Suppose you pull right out from here, and cross over into Mexico before halting again." Sansom stated that he said "much more ... to the same purpose, and it evidently set the Major to thinking seriously. At any rate he said he would confer with others of the party, and if they thought as I did, the march would be resumed as quickly as possible."[50] Major Tegener spoke with Captain Kuechler and Lieutenant Degener. Both, "expressed themselves as being convinced that no strangers were anywhere around, and as being in favor of remaining in their present camp until morning. That settled the question against an immediate start, and dropping the subject, all retired to their pallets to rest." August Hoffmann states he, "believed their officers bull-headed in refusing to push on to the border." The young patriots lay down upon the ground confident of security.[51]

Much criticism has been directed at Tegener for not heeding the warning of seeing 'strangers' and pushing on toward the Rio Grande. Sansom's account shows Tegener did not make the decision to stay alone. He consulted with other insurgent leaders and a joint decision was made. The other League leaders, Kuechler and Degener, share the responsibility for being surprised the next morning.

Back in the Confederate camp, Lieutenant McRae decided to make a dismounted attack, leaving his horses under guard at the staging area, reducing his assault force by the number of guards left to secure and care for the mounts. At the most, he had only about 90 men left for the attack. He did not have the necessary strength required to be reasonably confident of victory, which according to military science, is in an attack with a three-to-one ratio. Therefore, McRae needed a sound and well-executed tactical plan.[52]

McRae's plan was to move his force across the river under cover of darkness and by maneuver conduct a double envelopment of the enemy's position. This would force the insurgents to fight in two directions and enable the Confederate to mass grazing and enfilading fire onto the insurgent's position. By this tactic, McRae would concentrate his firepower and force the enemy to disperse their return fire. He planned to attack at first light before the insurgents were fully awake and organized. McRae's plan further called for the capture or killing of any insurgent guards or pickets so no alarm could be given. This would use the military principle of surprise. One major criticism almost every writer has leveled toward the Confederates is they 'ambushed' the insurgents and gave them no opportunity to surrender. The criticism goes on to claim the insurgents were caught by complete surprise and stood no chance to defend themselves and therefore, it was just murder. It would have been absolutely foolhardy for the Confederates to have announced their presence before the battle. The responsibility for being caught by surprise is totally that of the insurgent's leadership. They had several warnings of an enemy force in the area and took no action. Regarding any offer to the insurgents to surrender, Eduard Degener said, "He knew the men well, that they were the best shots in the country, that they would never have surrendered." As for being 'ambushed,' it is very rare for an ambush to be carried out. Webster's Dictionary defines an 'ambush' as, "a trap in which concealed persons lie in wait to attack" or for the enemy force to arrive in the killing zone. This is not what took place. McRae did not lie in wait. He maneuvered his forces into position, based on the insurgent's deployment. He successfully double-enveloped them. The insurgents were not caught by surprise; their two guards, Bauer and Beseler, gave a warning which cost them their lives, but enabled the insurgents to take action before the Confederate assault. Another major criticism leveled against the Confederates is they gave the insurgents no warning, allowing them to surrender, before opening fire. Given that the Confederates did not have the necessary ratio for victory, it would have been a serious mistake for McRae to let the enemy know of his presence before opening fire. It was critical for McRae to reduce the insurgent's strength. There is nothing wrong in a military commander using all the tools available to him to inflict the maximum casualties on an enemy while keeping his own to a minimum. As General George S. Patton is quoted as saying, "The goal is not to give you life for your country it is to make the other SOB give his life for his country." Military tactics call for at least a three to one ratio [3:1] for an attack to be successful. Even with these odds, other factors should favor the attacks. These include battlefield preparation [pre-planned fire on the objective], well-trained troops, surprise, and poor fortifications by the defender. Odds of 3:1 or less are acceptable only if there are other major advantages. A military analysis shows McRae did not have sufficient troops to ensure victory. After leaving men with the mounts his assault force numbered, at the most 91 men. The insurgents numbered 67. This was only a ratio of one point three five to one [1.35:1]. Very poor and little chance for success! For the Confederates to have any chance of success, the size of the enemy force had to be greatly reduced before a ground assault was made.[53]

McRae split his force into two almost equal wings. One he placed under the command of First Lieutenant Homsley, the other McRae kept under his personal command. Lieutenant Harbour and Bingham's detachments accompanied Homsley. Lieutenant Lilly's detachment accompanied McRae. Homsley's wing was to take up positions northwest and north of the camp. McRae's wing was to take up positions south and southwest of the camp. Both wings were to approach to within forty to fifty yards of the camp, but remain within the cedar-brakes. The signal would be a pistol shot by McRae. To assist in distinguishing his forces from the insurgents, McRae directed the men to take off their hats and tie a handkerchief around their head. The men lay down for a few hours of sleep.[54]

Shortly before midnight McRae awakened his force. The men checked arms and tied handkerchiefs around their heads. The two columns moved forward or generally south in single file. The moon was out, but covered by clouds. The two wings moved down the steep slopes toward the river. No one spoke a word. The only sound made was that of the men climbing over the rocks and boulders. They slid down a steep incline into the river. Fortunately, the water was only knee deep. It was running fairly strongly and they could tell from the moonlight that it was clear. The bottom of the river contained scattered boulders. Excitement was high. Many of the men paused for a quick drink of water before continuing. The two assault teams turned east and exited the river up a stony slope onto the prairie where the enemy's camp was located. McRae now divided his command.[55]

Lieutenant Holmsley's wing continued east until it was on line with the camp. It stopped, then turned south and moved toward the camp. The long parallel line stopped near the edge of the cedar brake about fifty yards from the enemy's camp. Homsley's position was north and northwest of the insurgents, just as planned. There is some evidence that Holmsley split his wing into two sub-groups. One, under Lieutenant Harbour was positioned on the northwest area of the insurgents' camp. The second, his own, was about fifty yards south southeast of Harbour.[56]

Lieutenant McRae's wing wheeled right, or south, from the river until it too was in line with the camp. There it moved single file to the east also just inside a cedar brake and along the bank of a small stream or depression. McRae's position was south and southwest of the camp, and like Homsley's it was in the correct position. It was shortly after three a. m. and according to McRae about an hour before daylight.[57]

The two wings were in position for the attack. The positioning of the two wings of the enveloping force was tactically sound. They were in the woods and occupied the most likely escape route for the insurgents, as well, as having them pinned into an 'S' or 'L' curve with high cliffs blocking the open side of the camp. Because of a slight rise in the terrain on which the insurgents were camped, the Confederate's position was so their fire would not spill over into each other's positions. They were well within pistol range of the insurgents. All that was needed was daylight and the signal to attack. Some of the Confederates lay down on blankets to get a few more hours of sleep.[58] [Map N, p 465]

Map K – Route to the Nueces River

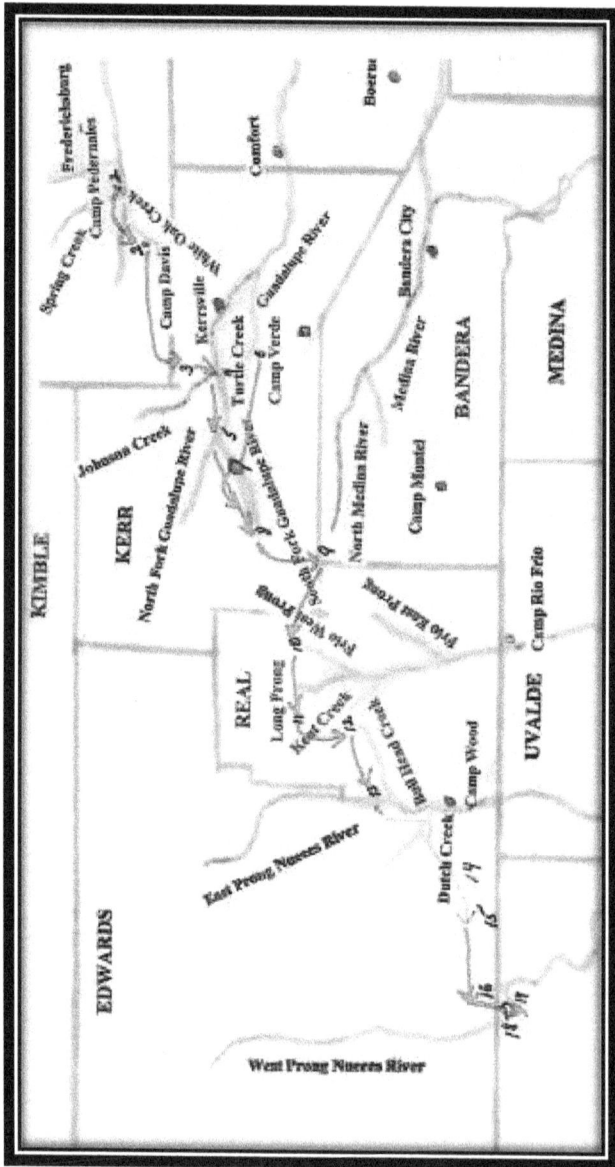

Endnotes -- To Fight or Flee

1. Degener's Letter of August 1, 1862; Sansom Battle *of Nueces*, p 3; Siemering's Germans During Civil War, June 8, 1923; Bennett *Kerr County*, p 5; "German Unionists in Texas", 'Harper's Weekly', New York, New York, January 20, 1866; Telephone Interview with Aleta Addicks Stahl, [Markwordt Family Member], October 21, 1997; Raley *Blackest Crime in Texas Warfare*; Cramer's Letter and Kuechler's Letter.

2. Siemering's Germans During Civil War, June 8, 1923; Bennett *Kerr County*, p 115; Weber *Die Deutsche Pioniere* p 13; Sansom *Battle of Nueces*, p 4; Letter, John W. Sansom, August 14, 1907, 1102 N. F. St. San Antonio, Texas to James T. DeShields, Esq.; Degener's Letter of August 1, 1862; Kuechler's letter and Cramer's Letter.

3. Degener's Letter of August 1, 1862; Raley *Blackest Crime in Texas Warfare* and Cramer's Letter.

4. Sansom *Battle of Nueces*, pp 3-4.

5. Barr *CMC*, Volume LXXIII, Number 2, (October 1969), p 256 and Degener's Letter of August 1, 1862.

6. Kuechler's letter; Sansom's *Battle of Nueces*, p 11; Schultz's *Was a Survivor of Nueces*, p 25; Davis' Proclamation; Raley Blackest *Crime in Texas Warfare*; Thorpe Historical *Friction*, p 78; Weber *Die Deutsche Pioniers*, p 25; Schlickum's Letter; Schellhase Journal; Williams' *Border Ruffians*; and *Harper's Weekly*, January 20, 1866.

7. Degener's Letter of August 1, 1862 and Sansom *Battle of Nueces*, p 4. It is Degener's Letter that established the earliest the two Degener brothers left home was August 1st. Sansom says he went to the assembly area with the two brothers. Heinrich Steves was born about 1834 in Prussia. He arrived in Texas with his family in 1849 on the *Neptune*. He was a member of Kuechler's December 1861 Company. Steves fled with the insurgent group and is believed was killed on the afternoon of August 9, 1862 generally north of the battle site.

   Fritz Schellhase was born in 1846 in Prussia. He arrived in Texas with his family in September 1853 on the *John Edward Grossie* and in Comfort in 1854. Sometime before 1913, he had his story written down by A. D. Stork. Fritz Schellhase died in 1928.

   Fritz Schellhase Journal" Dictated by Fritz Schellhase, to A. D. Stock, c1912, an unpublished manuscript; Eighth U. S. Census, 1860 Kerr County Texas Census, p 69; Geue *New Homes in New Land*, pp 125 and 137; Ransleben *100 Years*, pp 23-24, 94, 114 and 189; Bennett *Kerr County*, pp

50, 68, 71 and 113; Ninth U. S. Census, 1870 Kerr County Texas Census, p 75b and Boerne Area Historical Society *Gone But Not Forgotten,* Volume II, p 32.

One of Tegener's children was a female named Mary Augusta Texana who was born in November, 1859. The name of the second child is not known; the child died while Fritz Tegener was hiding in Mexico. A third child, Emelia Mexico was born on April 23, 1863. Tegener refused to accept this child as his. She was raised by her grandmother, Mary Benson Hart.

Eighth U. S. Census, 1860 Kerr County Texas Census, p 73; Case Number 61, Fritz Tegener vs. Susan E. Tegener, May 26, 1866, Kerr County District Court, Kerrville, Texas; Johnson County Texas Death Records for Amelia Mexico Tegener Bilbrey, Volume pp 4, 564; Ninth U. S. Census, 1870 Kerr County Texas Census, pp 138-139; Tenth U. S. Census, 1880 Gillespie County Texas Census, p 299; Twelfth U. S. Census, 1900 Comanche County Texas Census, E. D. 26, Sheet 4a; and Thirteenth U. S. Census, 1910 Tarrant County Texas Census, E. D. 90, Sheet 8a. An analysis of these sources shows Amelia/Emelia Tegener Bilbrey was born on April 23, 1863, which means she was conceived in July or August 1862 and thus the daughter of Fritz Tegener.

One "Heil" of a Family: Some Ancestors and Other Relatives of Louis Heil and Elizabeth Susan (Steves) Heil by Robert Allen Heil (Printed by Robert Allen Heil, (Comfort, Texas), n.d., pp 319-338 and *My Name is Nimitz* by Sister Joan of Arch, (Standard Printing Company, San Antonio, Texas), 1948, p 65.

8.   Julius Schlickum's Letter and Sansom *Battle of Nueces,* pp 3-4.

9.   Carl August Arlt was born on March 19, 1841 in Prussia. He arrived in Texas in 1853. He was conscripted into the 6th Field Artillery Battery on September 1, 1862. Arlt married Doris Katherine Pape on December 12, 1866 in Gillespie County. They were the parents of at least eleven children. Arlt died on March 19, 1927 in Gillespie County.

Frederick Fresenius 'may be the' Frederick Fresenius living in Gillespie County. He was born about 1826 in Prussia. The date he arrived in Texas is not known. Fresenius took the oath of allegiance to the Confederacy on February 23, 1862 while a member of Charles Nimitz's Gillespie Rifles. Frederick Fresenius was conscript in Company B, 3rd Regiment Texas Infantry in mid-too late 1862. He remained with the unit until the end of the war and received a parole on June 11, 1865 at San Antonio. The Gillespie Fresenius married Bertha [maiden name not known] about 1875. They were the parents of only one known child, Robert. Frederick Fresenius died about 1875, likely in Gillespie County.

Johann Adam Meurer was born on February 17, 1840 in Nassau. He arrived in Texas in 1845. He was conscripted into 6th Field Artillery Battery on September 1, 1862. He married Margaretha Kunz on November 21, 1867 in Gillespie County.

Sebastian Novian was born on November 16, 1841 in Nassau. The date he arrived in Texas is not known. He was conscripted into the 6th Field Artillery Battery on September 1, 1862. He married the widow A. W. Fiedler on August 21, 1866. After her death, he married Karolin Krause on April 22, 1867. They were the parents of thirteen children. Novian died on March 9, 1894 in Gillespie County.

Christian Probst was born about 1835 in Hanover. The names of his parents are not known. The date he arrived in Texas is not known. Probst was conscripted into Company E 1st Heavy Artillery.

Adolph Quintel was born on November 25, 1843, in Hanover. He arrived in Texas in 1846. He was conscripted into the 6th Field Artillery Battalion. He deserted and on December 3, 1863, he enlisted in Company C, First Regiment [Union] Texas Cavalry. Quintel married Maria Wendell on August 10, 1866 in Gillespie County. They were the parents of five children.

Frederick Rahe was born June 5, 1837 in Prussia. He arrived in Texas by 1853. Rahe was conscripted into the 6th Field Artillery Battalion on September 1, 1862. Frederick Rahe died on December 4, 1865 in Gillespie County.

Ernest Ludwig Theumann was born on June 12, 1833, in Hesse-Darmstadt. His father was very likely a Forty-Eighter and he was likely a *Turnverein*. He arrived in Texas in 1852. Theumann was conscripted onto the 4th Regiment Texas Infantry Regiment, Texas State Troops. He deserted and enrolled in Company H, First U. S. Infantry. Theumann married Maria Carolina Palm on September 24, 1865. They were the parents of seven children. Ernest L. Theumann died on January 20, 1920 in Colorado County.

Wilhelm Herman Toepperwein was born on February 2, 1837 in Burg bei Magdeburg, Saxony-Anhalt, Germany. He was a brother of Paul Johannes Toepperwein. The family arrived in Texas on 1850 and settled in Gillespie County. Wilhelm Herman Toepperwein married Amalie Augusta Luckenbach on May 27, 1860, in Gillespie County. He was conscripted in Company F, 32nd [aka] 36th] Regiment Texas Cavalry on August 12, 1862, at San Antonio. It appears he remained with the company until the end of the war. Wilhelm Herman Toepperwein died on November 7, 1916, in Kendall County.

Paul Johannes Toepperwein was born on February 14, 1844, in Neuruppin, Brandenburg, Germany. He was a brother of Wilhelm Herman

Toepperwein, above. The family arrived in Texas in 1850 and settled in Gillespie County. Toepperwein was conscripted in Company F, 32nd [aka 36th] Regiment Texas Cavalry on October 25, 1862, at Camp Bee in San Antonio, Texas. He deserted on February 24, 1864. Paul Johannes Toepperwein died on July 11, 1929, in Kendall County.

10. Alexander Brinkmann was born about 1837 in Prussia. He was a twin of Charles Brinkmann, below. They arrived in Texas in 1855 and settled in the Comfort area in 1858. He took the oath of allegiance to the Confederacy on September 30, 1861. Brinkmann was conscripted in Company B, 3rd Regiment Texas Infantry and likely served until the end of the war. He married Regina Wechsler on December 25, 1866, in Comal County. They were the parents of three children. Brinkmann died about 1890 in Travis County.

Charles Brinkmann was born about 1837 in Prussia. He was a twin of Alexander Brinkmann, above. They arrived in Texas in 1855 and settled in the Comfort area in 1858. Brinkmann was conscripted into Company B, 3rd Regiment Texas Infantry on September 1, 1862. He deserted on December 9, 1862 and on February 10, 1863, enlisted in Company C, First Regiment [Union]. Brinkmann was captured at the Battle of Las Rancho Rucias, near Brownsville on June 25, 1864. He died in the Chapel Hill, Texas, prisoner of war camp on October 15, 1864.

William F. Geissler was born about 1829 in Austria. The date he arrived in Texas is not known. He took the oath of allegiance to the Confederacy on August 4, 1861. He was conscripted in Company B, 3rd Regiment Texas Infantry in August 1862 and likely served until the end of the war. On April 28, 1866, Geissler married Elizabeth Mangold in Kendall County.

11. Kerr County Historical Commission *Kerr County Album*, p 14; Schutze *Was a Survivor of the Nueces Battle*, p 25, Sansom *Battle of Nueces*, p 3.

12. Sansom *Battle of Nueces*, pp 3-4; Tegener's Letter; Siemering's Germans During Civil War, June 3, 1923 and Kuechler's letter.

13. Letter Henry Schwethelm May 16, 1913, to his grandson, Otto. Kuechler's letter; Hoffmann's Letter of September 1, 1925; Raley *Blackest Crime in Texas Warfare*; Schutze *Was a Survivor of Nueces Battle*, pp 25-26; Schlickum's Letter; Cramer's Letter; Siemering's Germans During Civil War, June 8, 1923; Tegener's Letter and McRae's Report, p 615.

Henry Joseph Schwethelm was born on September 4, 1840, in Prussia. His father was a Freethinker and a Forty-Eighter. The family arrived in Texas about 1850 and in the Comfort area in 1854. Henry was a member of several ranger companies, to include G. H. Nelson, E. A. McFadier, and John W. Sansom. He was a member of Harbour's Home Guard Company in February

1861. After the Nueces Battle, he went directly to Mexico and on to New Orleans where he enlisted in Company A, First Regiment [Union] Texas Cavalry. He deserted the Union Army on June 24, 1864, after the Battle of Las Rusas. He returned to the Comfort area and hid out the remainder of the war. Governor E. J. Davis appointed him a Ranger captain in 1867. Henry Schwethelm moved to Edwards County by 1890 and lived near the town of Barksdale. He was an Edwards County Commissioner from 1890 until 1892. Henry Joseph Schwethelm died on August 16, 1924 in Kerr County. Kerr County Historical Commission *Kerr Album*, p 406; Ransleben *100 Years*, p 24; Eighth U. S. Census, 1860 Kerr County Texas Census, p 69; Bennett *Kerr County*, p 134; Henry Schwethelm's Union Service Records; Henry Schwethelm's Indian War Pension; Tenth U. S. Census, 1890 Special Schedule, Surviving Soldiers, Sailors, and Marines, and Widows, Etc. of Persons who served in the Army, Navy, and Marine Corps of the United States during the war of the rebellion; Edwards County Texas, E. D. 68, p 4; Schutze *Was a Survivor of Nueces Battle*, p 24; *A History of Edwards County* Edwards County History Book Committee (Anchor Publishing Company, San Angelo, Texas) 1984, p 22; Barr *CMC*, Volume LXXIII, (October 1969), p 252 and *Dallas Herald,* August 30, 1862.
14. "Gott mit Uns: The Texas German Confederates" by Egon Richard Tausch in 'Southern Partisan', Fall Issue 1995, p 36.

15. *Index to the First and Second Regiment Texas Cavalry Volunteers (Union),* Compiled by Jim and Doris Menke, 1998 for Medina County History Project, Hondo, Texas, and Glenn *Capt'n John,* 5.

16. Casper H. Sueltenfuss was born on April 15, 1844, in the Rhine District of Prussia. He arrived in Texas on the *Iris* from Düsseldorf in 1859. Sueltenfuss returned to Kendall County after the war and died in 1916. Kendall County Historical Commission *History of Kendall County*, p 220.

17. Siemering's Germans During Civil War, June 5, 1923; Weber *Die Deutsche Pioniere,* p 13; 1861 Kerr County Texas Tax Roll; Bennett *Kerr County*, pp 113 and 115; McRae's Casualty List; Charles Burgmann's Confederate Service Records, Record Group 109, National Archives, Washington, D. C. and *Escape from Reconstruction* by W. D. Nunn (Texas Christian University, Fort Worth, Texas), 1956, p 129.

18. Sansom *Battle of Nueces*, p 4.

19. Schutze *Was a Survivor of Nueces Battle*, p 26.

20. Kuechler's letter; Raley Battle *of Nueces*; Siemering's Germans During Civil War, June 8, 1923; Schultz *Was a Survivor of Nueces*, p 26; Degener's Letter of August 1, 1862, and McRae's Report, p 615.

21. Sansom *Battle of Nueces*, p 10; Schultz *Was a Survivor of Nueces Battle*, p 25 and McRae's Report, p 615.

22. Siemering's Germans During Civil War, June 8, 1923, and Cramer's Letter.

23. Kuechler's letter; Weber *Die Deutsche Pioniere*, p 13; Sansom *Battle of Nueces*, p 4; Raley *Blackest Crime in Texas Warfare*; Schultz *Was a Survivor of Nueces Battle*, p 25 and Schlickum's Letter.

24. Sansom *Battle of Nueces*, p 4 and Cramer's Letter.

25. Thomas J. Scott was born on June 25, 1836, in Illinois. He arrived in Texas by 1851 and settled in Bexar County before moving to the Hill County about 1861. He was likely a member of the Kerr Company of the Union Loyal League military battalion. After the battle, he went on to Mexico and joined the Union Army at New Orleans on October 27, 1862. He was mustered out of service on October 31, 1865 at San Antonio. Thomas J. Scott died on January 2, 1907, in Bexar County.

   Warren B. Scott was born about 1843 in Iowa. He likely arrived in Texas about 1851 and settled at Lavaca County. By the spring of 1861, he was living in Kerr County. He was a member of the Kerr Company of the League's Military Battalion. After the battle, he too went on to Mexico and on October 27, 1862, joined Company A, First Regiment [Union] Texas Cavalry. He was mustered out of service on October 31, 1865, at San Antonio. Warren B. Scott died on January 27, 1906 in Guadalupe County.

   William Hester was born in 1839 in Texas. His father died when he was about seven years old and his mother married Joel Casey on June 28, 1849, in Hunt County. It is believed the family lived near the Blanco and Hays County line in 1860. Hester was a member of the Kerr Company of the League's Military Battalion. After the battle, he too went on to Mexico and on October 27, 1862, enlisted in Company A, First [Union] Regiment Texas Cavalry. Hester was wounded near New Iberia, Louisiana on October 4, 1863. He was mustered out of service on October 31, 1865, at San Antonio. William Hester later owned a ranch that included the battle site. He died on February 6, 1912, in Uvalde County.

   Howard Henderson was the nephew of Sebird Henderson who was captured about August 5th in Gillespie County. Howard Henderson was born in 1842 in Tennessee. He arrived in Texas with his uncle Sebird in 1857. Howard Henderson enlisted in Company A, First Regiment [Union] on October 27, 1862, at New Orleans. He served until the end of the war, being mustered out on October 31, 1865, at San Antonio. He married Narcissa R. Turknett, whose father was killed in Kerr County in July 1864, on May 20, 1866. They were the parents of ten children. Howard Henderson died on November 18, 1908, in Kerr County.

Sansom *Battle of Nueces*, p 4; Cramer's Letter; Eighth U. S. Census, 1860
Bexar County Texas Census, p 47; Thomas J. Scott's Union Service
Records, National Archives, Washington, D. C.; Thomas J. Scott's Union
Pension Records, National Archives, Washington, D. C.; Ninth U. S.
Census, 1870 Kerr County Texas Census, p 147; Warren B. Scott's Union
Service Records, National Archives, Washington, D. C.; Warren B. Scott's
Union Pension Records, National Archives, Washington, D. C.; Howard
Henderson's Union Service Records, National Archives, Washington, D.
C.; Twelfth U. S. Census, 1900 Uvalde County Texas Census, E. D. 70,
Sheet 7a; William Hester's Union Service Records, National Archives,
Washington, D. C.; Obituary, William Hester from an unidentified Uvalde
newspaper, believed dated February 6, 1912. Copy provided by Helen
Hester Williams, Uvalde, Texas; Oral Interview with Helen Hester
Williams, a granddaughter of William Hester, on October 18, 1996: Kerr
County Historical Commission *Kerr Album*, pp 270-271 and 293 and
Watkins *Kerr County*, pp 39, 64, 70-71, 117, 133, and 279-280.

26. Cramer's Letter.

27. Raley *Blackest Crime in Texas Warfare*.

28. Cramer's Letter; Siemering's Germans During Civil War, June 8, 1923;
Kuechler's letter and Sansom *Battle of Nueces*, pp 4-5.

29. The two who joined Duff's Company were Robert N. Robinson and J. K.
Spears. They both enrolled on August 1st and remained with the company
until the end of the war. Robinson received a parole on August 12, 1865, at
San Antonio. Spears spent most of his service with the Quartermaster
Department. Banta's *27 Years*, pp 110-111; Compiled Service Records 33rd
Regiment Texas Cavalry, Record Group 109, National Archives,
Washington, D. C.; Hopkins Diary; Schellhase Journal; and Williams's
Border Ruffians, pp 237-238. R. H. Williams does not identify the guide by
name. He says, "One of the prisoners, an old soldier, and a friend of
[Duff's], had been released, and he was to act as our guide and betray his
friends, if possible, into our hands." This comment has supported the notion
that Bergmann was in the pay of Duff. It is more likely that Duff knew
Bergmann from the time he either was in the U. S. Army or a contractor.
The 1860 Bexar Census shows a Charles Bergmann, age 29, as a Teamster
working for the U. S. Army. However, this 1860 Charles Bergmann age
does not match the 39-year-old Charles Bergmann, shown on the Kerr June
1862 List of Males not in military service. Based on Williams' comment,
there very likely was a possibility that Duff and Bergmann knew each other
before August 1862, but there is also the possibility that Burgeman and Duff
knew each other as Burgeman worked for a stage line before the war.
Williams *Border Ruffians*, pp 237-238; Eighth U. S. Census, 1860 Bexar

County Texas Census, p 438; Bennett *Kerr County*, p 113 and Off the Record Oral Interview on July 16, 1997.

30. Eighth U. S. Census Kerr County Texas Census, p 73; Watkins *Kerr County*, p 6 and Benson family stories.

31. Williams *Border Ruffians*, pp 237-238.

32. Wisseman *Fredericksburg ... First Fifty Years*, p 4; Off The Record Interview of July 19, 1977; Mary Turner's oral presentation to the German-Texan Heritage Society at its September 5-7 Convention at Kerrville, Texas, Ransleben *100 Years*, p 121; Hopkins Diary; Schellhase Journal; Williams *Border Ruffians*, p 237-238; Letter, Michael E. Pilgrim, Textual Reference Division, National Archives, Washington, D. C., January 31, 1995, containing a copy of Lieutenant C. D. McRae's Casualty List, Record Group 109, E 22; Schultz *Was a Survivor of the Nueces Battle*, p 26 and Sansom *Battle of Nueces*, p 4.

33. Colin Dickson McRae was born on September 23, 1835, near Clarksville, in Red River County, Texas. McRae's father died in 1835. His mother married Ira S. Poor and by 1848, the family was living in Bexar County. McRae enrolled in Donelson's Company on September 25, 1861, and was appointed 1st lieutenant. He was wounded during the Nueces Battle. Two months after the battle, on October 12, 1862, McRae was promoted to captain. Colin D. McRae died on September 10, 1864, while in Confederate service. Camp Pedernales Post Returns for August and September 1862; McRae's Report, 615; Williams *Border Ruffians*, p 44; Colin D. McRae's Confederate Service Records, Record Group 109, National Archives, Washington, D. C.; *Biographical Sketch of Colin D. McRae,* by Douglas N. Travers, San Antonio, Texas, (an unpublished manuscript), 1997 and Eighth U. S. Census, 1860 Bexar County Texas Census, p 477.

34. McRae's Causality List; McRae's Report, p 614; Hopkins' diary and Williams *Border Ruffians,* pp 237- 238.

35. Williams *Border Ruffians*, pp 238-239 and Bennett *Kerr County*, p 39.

36. Williams *Border Ruffians*, pp 238-239.

37. William. J. Edwards was born on May 12, 1840 in Choctaw County, Mississippi. The family arrived in Texas in the mid-1840s. Edwards father died about 1845. His mother married Thomas W. Grayson and by 1847, they were living in Comal County. Edwards enrolled in Duff's Company on May 4, 1862 at San Antonio. He testified against Eduard Degener during the Confederate Military Commission. Edwards remained with Duff's Company until the end of the war, receiving a parole at San Antonio on

August 30, 1865. William J. Edwards date and location of death is not known.

Williams *Border Ruffians*, p 239; McRae's Report, p 614; Barr *CMC*, Volume LXXIII, (October 1967), 252-253; W. J. Edwards' Confederate Service Records, National Archives, Record Group 109, Washington, D. C.; *The Trail Drivers of Texas,* Compiled and edited by J. Marvin Hunter, (University of Texas at Austin, Press) 2000, pp 602-603; Siemering's Germans During Civil War, June 8, 1923, and Cramer's Letter.

38. Williams *Border Ruffians*, p 239.

39. Williams *Border Ruffians*, pp 239-241 and Barr *CMC*, Volume LXXIII, (October 1967), pp 252-253.

40. Williams Border *Ruffians*, p 240-243; San *Antonio Semi-Weekly News,* August 25, 1862 and Barr C*MC*, Volume LXXIII, (October 1967), pp 252-253.

41. Williams *Border Ruffians*, p 243; Oral Interview with L. J. Dean, a Nueces Canyon historian, March 9, 1998 and 1890 Union Veterans Census, Edwards County, p 4.

42. Williams *Border Ruffians*, pp 243-244 and McRae's Report, p 614.

43. Lonn *Foreigners in the Confederacy*, p 430 and *Neu Braunfelser Zeitung* August 25, 1862.

44. Williams *Border Ruffians*, pp 244-245; *Neu Braunfelser Zeitung* August 25, 1862; and McRae's Report, pp 614-615.

45. Schellhase Journal; Eighth U. S. Census, 1860 Kerr County Texas Census, p 70; Ransleben *100 Years*, pp 24, 94-95 and 202; Bennett *Kerr County*, pp 68-69 and 112 and Sansom *Battle of Nueces*, p 5.

46. Schultz Was *a Survivor of Nueces Battle*, p 26; San *Antonio Express, of* an unknown date in 1952 and *Kerrville Mountain Sun,* August 6, 1953.

47. Raley *Blackest Crime in Texas warfare*.

48. Sansom Battle *of Nueces*, pp 5-6; Schultz Was *a Survivor of Nueces Battle*, p 26; Schlickum's Letter; Cramer's Letter; Raley *Blackest Crime in Texas Warfare* and Siemering's Germans During Civil War, June 8, 1923.

49. Ibid.

50. Sansom *Nueces Battle*, p 6.

51. Sansom *Battle of Nueces,* p 6; Raley *Blackest Crime in Texas Warfare* and *Harper's Weekly*, January 20, 1866.

52. Ibid.

53. McRae's Report, pp 614-616.

54. McRae's Report, p 615; Barr *CMC*, Volume LXII, (October 1969), p 252 and *Combat Jump: The Young Men Who Led the Assault into Fortress Europe, July 1943* by Ed Ruggero (Harper Collins Publishers, New York, New York), 2003, p 120.

55. McRae's Report, pp 615-616 and William *Border Ruffians*, p 245.

56. Ibid.

57. Ibid.

CHAPTER 10 – The Battle

Inside the insurgent camp, John Sansom had not been awakened at two a.m. as he requested. About three a.m. he awoke. He was sure of the time because he looked at this watch. A few minutes later one of the guards, twenty-two-year-old Prussia native Leopold Bauer, son of a Forty-Eighter, Freethinker, and *Turnverein*, came and got Sansom. Bauer and Sansom moved toward the northwest side of the camp, Bauer was about twenty feet ahead of Sansom. The two went about sixty more yards to the end of a cedar brake. They walked straight into Lieutenant Harbour's position. Instead of attempting to capture Bauer and Sansom, as was the plan, Harbour fired, killing the Prussian instantly. Sansom says, "Bauer was killed first; he fell right in front of me." Sansom immediately returned fire, causing many of Harbour's group to rise "from their blankets and rushed pell-mell over a space of open ground to a part of their command which lay under the cedars some sixty yards east of the place where Bauer was killed."[1] [Map N, p 469]

The opening shots were fired just after the two Confederate wings had settled into their attack positions. McRae states, "Shortly after having secured our position a sentinel on his rounds came near the position of Lieutenant Homsley's division … He was shot dead." R. H. Williams states, "Hardly had we [gotten into position] when a rifle shot … rang out."[2]

Harbour and Sansom's opening shots alerted the insurgents. The element of surprise was lost! Henry

Schwethelm says, "About three o'clock in the morning [we] were aroused by shooting in the proximity of the camp ... [two guards were killed and] ... the other two guards thereupon came rushing into the camp. Every man was then ready to repulse the expected attack." In another account Schwethelm says, "Confederate Soldiers and the[y] shot into camp about Two [hours] before day [break]." August Siemering adds "It was three o'clock in the morning. Suddenly the camp was aroused by a series of close-by shots. The guards ran into the camp and reported they were attacked by a number of men and two comrades were killed. The darkness of the night did not allow them to recognize if they were whites or Indians." Julius Schlickum says at 3 a.m. in the morning the guard called out, "Wake up! We are betrayed!" At the very same moment came a cracking RATT TATT TATT and lighting flashes all around. "From all sides came a hail of bullets that whistled around their ears into the camp. All this was enforced by a terrible hollering and screaming, as if 1,000 Indians were possessed by the devil," relates Schlickum. Captain Ernest Cramer states, "About an hour before sunrise we heard one shot that awakened us. Immediately after came another and another. We leaped to our feet and were met with a volley of about 100 shots." Hoffmann says, "We had put out guards but by daybreak as the rest slept peacefully they heard shots and in a few minutes our camp was a battlefield." Sansom recollected later the shots "alarmed the camp, and fast and furious firing began between the contending parties." Confederate R. H. Williams says, "Instantly the camp was in a buzz, like a swarm of bees. Men ran hither and thither in great confusion." A *Harper's Weekly* article of January 20, 1866, which was based on insurgent accounts, states, "At 3 o'clock a. m. they were awakened by a volley of musketry. Two of their pickets, wounded, came running in and gave the alarm. A strong body of men, whether white or red men it was impossible to say, was advancing upon the camp."[3]

In the firefight that followed, the tragedy was multiplied by the fact that there were in-laws and family members on both sides. One example was Fritz Tegener's two brothers-in-law, William Thomas and John Washington Benson. Unknown to them they were actually exchanging gunfire on the camp's north side. In Homsley's wing on the north were several neighbors of insurgents.

While the Confederates had lost the element of surprise, it was nonetheless the insurgents who took the first casualties. In addition to Bauer, a second member of the camp was soon killed. Ernst Beseler, a second guard, was outside the insurgent camp and near Homsley's group when he was caught in the crossfire. Sansom "saw him fall and know [sic] that he fell fighting." Sansom says elsewhere, "Beseler was 50 yards to my right, I saw him fall." McRae reported Beseler'sdeath this way, "Another sentinel hailed us on the left, and shared the fate of the first." Other insurgents were wounded early in the fight. Captain Kuechler tells about four of those. "On the 10th, about half an hour before daylight, our guard near the spring was shot, and at the same time a volley was fired into camp; we suffered a severe loss, as Fritz Tegener was among the three severely wounded men, who had a great influence and would have done a great deal during the fight." Kuechler's statement about a guard near the spring being shot does not match Sansom's and other accounts that tell about Bauer being killed. The spot where Bauer was killed was on the far west side of the insurgent's camp, about 200 yards from 'the spring.' Beseler was closer to 'the spring,' but only by another 50 yards. Therefore, it seems that Kuechler is referring to another guard. There are two secondary sources that identify two other insurgent guards were killed. The first is the 1866 *Harper's Weekly* article. It says, "Two pickets Ernest Beseler and Louis Schierholz had been killed." The Schellhase Journal says, "Louis Boerner was on guard. He was one [of] the first ones that fell." Both Schierholz and Boerner are known to have been killed at or near the battle site. Captain Cramer confirms that four insurgents were wounded: "Leopold fell dead and four others were wounded." These wounded likely included August Luckenbach, Conrad Bock, and one of the Ruebsamen brothers. Sansom's says that only two additional insurgents were wounded.[4]

Lieutenant McRae's attack plan had been compromised, but not fatally. While losing the element of surprise, the Confederates had the advantage of facing a disorganized foe. Had the two wings launched an immediate attack, they would have likely carried the battle very quickly. Failing this, had both wings opened fire simultaneously, they likely would have inflicted heavy casualties on the insurgents. However, McRae's plan did not include a contingency for the battle to start before daylight. He wrote in his report, "It being still too dark for the attack, I ordered my men to hold quietly their positions until daylight. The enemy in the mean time were actively engaged preparing to resist us." Williams, in McRae's wing, stated, "If what had happened had been foreseen, and orders given to charge at once, no doubt we could have carried the camp with little loss. But no one knew what to do, and we on our side lay low, waiting for developments." This was a tactical error by McRae. Had he at least fired into the insurgent's camp he would have inflicted some casualties. Homsley's fire resulted in two insurgents killed and four wounded. If McRae's fire had been as effective, it would have resulted in mounting insurgent casualties. The insurgent force was down to 67 before the opening shots and down to 61 as a result of Homsley's fire alone.[5]

It seems Homsley's wing did attempt to quickly "carry the camp," but the insurgents were now fully awake and prepared to fight. Sansom says shortly after Beseler fell, "the Confederates made a charge upon the insurgents which was gallantly repulsed." This was followed by a "counter-charge made upon the Confederates," say Hoffmann. He further says that after Major Tegener was hit, Amil Schreiner took command, rallied the insurgents and led the counter-charge shouting, "*Laszt us unser Leben so teuer wie moglich verkaufen.* [Let us sell our lives as dearly as we can]." Schreiner was killed at the head of this attack. Captain Kuechler was now in full command of the north side. The insurgents fully recovered from their surprise, grabbed their weapons, and began firing volleys at Homsley's wing, "but without much execution in the darkness of the night. This was replied to by our people, and the firing became general on their side," says Williams.[6]

The insurgents took the opportunity to prepare defensive positions. Captain Kuechler says, "After this the enemy fell back and we prepared for the coming fight. Our position was not good and could not be improved." He felt they were too crowded together in a small cedar grove that had no undergrowth and insufficient shelter against the heavy fire of the rebels, "which outnumbered us a great deal." Siemering describes the attempt to improve a defensive position, "Immediately orders were given for a hasty defense. Flour sacks, saddles, dishes and whatever was present was used to build a kind of wall which would provide protection for at least part of the riflemen." Others stood behind trees, which were wide enough to provide some protection. Thus, they awaited the beginning of daylight.[7]

After this initial exchange of gunfire, a lull in fighting followed. The odds greatly favored the insurgents, even given the fact they were the only ones to have suffered casualties. McRae's force numbered about 91 men. The insurgent force numbered about 61 non-wounded. This was a ratio of only 1.5:1 in favor of the Confederates, not the neccessary 3:1. The element of surprise was lost. The insurgents had the opportunity to make some type of fortification and conduct a defense. It would be costly for the Confederates to make a deliberate attack. Sansom said there "Came a lull of an hour during which there was but an occasional shot fired." Captain Cramer describes this lull as, "Then all became quite followed a deadly stillness that was almost unbearable."[8]

During this lull the insurgents continued to fortify their positions and tried to decide their next move. "We had a consultation and decided to fortify ourselves as well as we could. That was quickly done." Cramer gave his orders to Lieutenant Hugo Degener, the older of the two Degener brothers, to cover the right wing or the southwest part of the camp. Cramer "was fully determined to stay and die at my post if need be." Hugo Degener and his men, including his brother Hilmar, Pablo Diaz and Albert Bruns, covered the southeast part of the camp. To Cramer, the chances of escape seemed impossible. The soldiers greatly outnumbered them and "had just as good position as we had and had at least 100 to 200 men available from Fort Clark."[9]

Thus far, the presence of McRae's wing was unknown to the insurgents. No Confederate casualties had occurred to this point. The Confederates were now content to wait for daylight or "developments" as Williams described the evolution of the battle. But, as McRae stated, "The enemy in the meantime were actively engaged preparing to resist us."[10]

Sansom was also busy, determining the size and locations of the Confederate force. He was caught outside the lines, roughly in front of or just to the west of Harbour's detachment. He first attempted to rejoin the camp, but the insurgents, believing he was a Confederate, fired on him and as he states, "came very near killing me, one ball passing through my clothing and grazing the skin about my stomach, and another cutting the flesh from a finger of my right hand." Sansom crawled away from the locations of both forces to decide what he should do next. He decided the greatest good he could do was make a reconnaissance of the Confederates position. Sansom was unaware of McRae's wing. He crept behind Homsley's wing and determined their numbers and deployments. He estimated there were about sixty men in the force. He then attempted to again rejoin the insurgents. He took a course, which he thought was well around any Confederates. However, unaware of McRae's wing, he walked into a group of them concealed in a thick grove of cedars, sixty yards southwest of the insurgent's camp. To his shock, he got so close he could have put his hands on them. He saw they were not wearing hats. Instead, they had handkerchiefs tied around their heads. He immediately took off his hat, and carrying it in his hand, backed away. He stated, "They saw me plainly, but I reckon supposed me to be a Confederate; at any rate, they did not take me to be a Unionist, and so let me go."[11]

Knowing there were Confederates southwest of the camp as well, Sansom turned back to the northwest. He then got down on his stomach and crawled toward the insurgent's camp. Shortly thereafter, he heard the click "of locks of guns about to be aimed at me." He called out, "Don't shoot, Sansom." There was no reply. He repeated the call. This time Captain Cramer answered, "Come on, come on, I came near shooting you." Sansom did and reached the relative safety of the insurgent camp.[12]

Sansom reported his observations to the insurgents at this location. He identified this group as headed by Captain Cramer and Lieutenant Simon, who were brothers-in-law. He advised them to, "abandon their present position and select one where they would not only be less exposed to the fire of enemy, but have a better chance to damage the enemy." Sansom then located Major Tegener, and found him "bleeding profusely," although still retaining overall command. He was lying on his pallet using his saddle as a breastwork. Sansom reported the results of his reconnaissance. He put the number of Confederates, "at one hundred and perhaps more ... well-armed." Sansom advised a prompt withdrawal. He felt Tegener favored the withdrawal.[13]

Sansom returned to his pallet. Close by were his four messmates: Hugo Degener; Hilmar Degener; twenty-three-year-old Prussia native Albert Bruns, son of a Forty-Eighter and a Comfort *Turnverein*; and a twenty-year-old Mexican, Pablo Diaz, who as a young boy in the late 1840s survived one Indian raid, but was taken prisoner and later rescued by Germans settlers. He had grown up in their households. According to Sansom, they were all "ready, if not anxious to continue fighting." They asked if he knew what were Major Tegener's plans. Sansom told them that he believed they would withdraw to a better position. He was not aware the insurgents had had a council of war and decided to stay and fight. "Withdraw!" exclaimed Lieutenant Hugo Degener. "Never! Our two guards have been killed, Major Tegener and two others of our comrades wounded, and if we leave here, they will get our horses, our rations, and all our equipage. I would rather fight here until every man of us is killed than to go anywhere else." Sansom replied, "Hugo, they out-number us greatly, and they have a much better position than we. For these reasons, we ought to withdraw. In the shuffle and excitement of going from one place to another, we may get their horses and equipments in exchange for those we may lose. I am in favor of retirement from our present position, so am going to carry my saddle with me and look for a safe route for our withdrawal." Sansom left. Hugo Degener's four messmates remained and were killed when the fight continued.[14]

The four men, who had joined the insurgents only two days before, Thomas Scott, W.B. Scott, Howard Henderson, and William Hester, were close by and heard the conversation. They immediately accompanied Sansom. The loss of another five insurgents reduced the number of non-wounded to about 56. The odds were now 1.6:1, still not enough for Confederate success. The five cautiously made their way to the spot where Bauer lay dead. Sansom was carrying his saddle. It is not clear if the other four had their saddles as well, but likely did. Sansom turned Bauer's body on its back and covered it with a Confedcrate blanket he found lying close by. Sansom and his small group located some horses. At the location where Bauer's body lay, they looked back at the insurgent's camp and saw no signs that Tegener had ordered a withdrawal. Sansom said, "Well, boys, they have not made up their minds to withdraw, but may later, so let's tie some of these horses back in the cedars where the boys can easily get them." The departure of the five Anglos caused consternation among the Germans. Siemering says, "During this time, five Americans, the only ones which were in the group ... fled the battle." Simon told McRae, "The five Americans ran at the first fire." Schlickum says, "Americans of the Union loyalists stole away from the camp."[15]

Morale or 'esprit de corps' in a military unit is one of those factors that cannot be measured. One of the things an attack does is to lower the enemy's morale or desire to fight. It is a military axiom that attacking inspires a soldier, it adds to his power, rouses his self-reliance, and confuses the enemy. The side attacked always overestimates the strength of its attacker. It was at this critical junction that this effect came into play against the insurgents' side. Fear, or a lost of morale, took over for many of the insurgents and a total of about twenty-three Germans left the camp, leaving at most 31 men to carry on the fight. The odds were now 3:1, enough for a Confederate victory by standard military calculations. [Maps N & P, pp 416-417].

Many insurgent descendants discount the fact that some of the insurgents left after the opening rounds were fired and before the Confederate ground assault. Instead, they argue that these were the 28 or so who left the 'night of August 9th'. The fact is, these 23 German's 'left early'. There are several references to 'those that left early.' Siemering says, "14 Germans fled the battle." Kuechler says, "about half of our men could not stand the fight any longer and left." Schlickum says, "As dawn broke Kuechler had only 35 men left." Cramer says, "The men … deserted their posts one by one … we had only 32 were with us." Sansom says the Confederates, "felt no assurance of victory until the camp was abandoned by ablebodied defenders." Schwethelm says about 25 insurgents left. Siemering provides a 'reason' for the early withdraw of the 23: "They did not have the courage to continue fighting in this death battle." "As dawn broke, [the insurgents] had only 35 men left," says Schlickum. [Appendix Z] This small group was determined to fight and die if necessary. They used the few trees for cover. "Each one had a six-shooter and a hunting gun." All of them were Germans; many had a wife and children at home. They were the elite of the German youth in West Texas: "All of them excellent hunters, sure shooters, raised at the frontier in the wild hills and woods, at home on the pathless prairie. Stout-hearted young men and all of great intelligence" explains Schlickum.[16]

Sansom's account says his five-man group had been somewhat successful gathering mounts by the beginning of the Confederate ground assault. They had, "tied a number of the horses to the trees … when day began to dawn, and the firing commenced." Other insurgents were also trying to get horses for their escape. Williams explains, "So far we hadn't fired a shot, and our presence was unsuspected; but now two of their picket-guards came running a few yards of our positions, driving some horses before them." It is not clear as to the identify of these two 'pickets.' Several insurgent accounts state there were four guards. Two, Bauer and Beseler, were killed in the opening exchange of gunfire. At which time, the "other two guards ... came rushing into our camp." There are other accounts that say two other guards were killed; Louis Boerner and Louis Schierholz. At this point of the battle, Williams describes there were no additional insurgent's guards outside the camp. These two men were likely trying to get out of camp when they ran into McRae's wing. Williams' group fired a shot without effect. Three more followed, and a heavy thud told them one had been hit and, later determined, killed. A Confederate ran out and brought in the dead man's arms, a Colt six-shooter and a Jaegar rifle. The insurgent was likely twenty-nine-year-old Louis Schieholz, a native of Hanover. This firing resulted in the start of the ground assault.[17]

As dawn broke, any doubts that it was Confederates who had fired into the camp, "were swept aside." The Southerners were on two sides of the insurgents. On the south side, they "were occupying the opposite brush" from which a dry creek ran up to a second mott of brush. Using this trail, the Confederates attempt to creep closer to the insurgent camp. From this dry creek bed, it was about fifty years of open ground before there was additional cover. It still was not fully daybreak. McRae's assault group crossed this open area in individual rushes, firing as they went. They all got across safety and, finding whatever cover they could, opened fire on the camp. The Confederates were now within thirty yards of the defenders. Williams' small group of six or seven "blazed away in great excitement and didn't do much execution." An insurgent described the action this way: The Confederates were armed with breechloaders, which were much more accurate. "At times, you could hear massed fire, shot after shot right behind each other, and a shot only after a long intermission." "The Confederates had a lot of ammo and were always ready to and easier to load than the Unionists' muzzle-loaders," he wrote.[18]

Now McRae ordered both wings forward. They advanced, "by a steady and slow advance upon their positions, firing as we advanced," McRae reported. An insurgent says, "The Rangers sent a hail of bullets into the camp of the insurgent loyalists and stormed the camp howling like that of Indians." The small group with Williams' crawled cautiously forward and, dodging behind trees, worked their way to the mott where the insurgents were located. "Bullets were whistling pretty thickly over their heads," explains Williams. "It was difficult to return the fire with much effect" but they, "kept up a galling fire on the defenders." The insurgents let the Confederates approach to about 50 feet and "with deadly aim they shot into their ranks." The fire was effective and some of the men close to Williams "suffered some loss through foolishly exposing themselves." Among the Confederates, nineteen-year-old Littleton Stringfield, a native of Limestone County and a member of Donelson's Company K, fell dead with a bullet in his head, and three others were "severely wounded." One was twenty-two-year-old Albert J. Elder from San Antonio and a member of Duff's Rangers, with a serious head wound. His twenty-eight-year-old brother, Robert, also from Duff's Rangers rushed to his side. Robert was hit in the right arm, breaking the bone. Others in McRae's wing were hit.

Twenty-two-year-old Benjamin Franklin Rosser, also from Donelson's Company K, went down with a bullet in his arm. Charles Burgmann was hit in both hips and a shoulder. McRae's wing was forced to withdraw back to the tree line.[19] [Map N, p 465] Holmsley's wing, on the north side of the insurgents was also having difficulty. Only the men from Taylor's Battalion had breechloaders. Harbour's men were armed with muzzle-loaders and pistols and therefore had to expose themselves to insurgent fire more often. As soon as he saw and heard McRae's wing begin attacking, Holmsley ordered his men forward. No sooner had they got up from their positions than they were hit by insurgent fire. Captain Kuechler had his men behind trees and other cover. He called out, "Don't expose yourself, hold your weapons at hand!" He had instructed his men to fire only when they had a target. The *Neu Braunfelser Zeitung* reported the insurgents fought like the 'devil.' John K. Morris, a twenty-one-year-old native of Mississippi from Taylor's Company B, fell mortally wounded. Twenty-four-year-old South Carolina native, Albert Wilborn of Davis' Company, fell wounded with a bullet in his right breast. Two members of Company A of Taylor's Battalion, twenty-three-year-old Wiley Williams and twenty-three-year-old Mississippi native John Welch, were hit and went down. The insurgents' fire stopped Homsley's assault just as it had McRae's. Even with their numbers reduced by half, the insurgents were giving a good account of themselves.[20] [Appendix AC, p 674]

From their concealed positions, the Confederates kept firing into the insurgent position. According to Williams, the insurgents on the south, "showed a bold front, and dared us to come on. They even threatened to charge out." Now the breechloaders played an important part. Captain Cramer had a bullet pass through his pants. Another tore away all of the front of his shirt. Neither touched his skin. William Tellgmann of Comfort, who had a brother fighting for the Confederates, received a very serious body wound. He "kept his place bravely and continued to fight." Gottlieb Vetterlein, a thirty-one-year-old native of the Duchy of Atenburg, was hit in the wrist. His thirty-year-old brother, Carl, came to his aid and bandaged the wound. Ferdinand Simon was hit in the ribs. Messmates Albert Bruns and Pablo Diaz went down with serious wounds. Franz Weiss was hit twice, once in the heel and one in the lower leg.[21]

The breechloaders also an impact on the northern side of the battle. "Don't shoot! Calm Blood!" cried Kuechler. Siemering described him as, "behind a tree cold-bloodedly giving orders, loading his gun and firing." Despite his best efforts, insurgents in his command continued to be hit. Twenty-one-year-old Henry Kammlah, a son of an *Adelsverein* settler who arrived in Gillespie County in 1846, was hit in the shoulder or upper arm. Thirty-one-year-old Karl Itz, also from Gillespie County and a native of Nassau, had a bullet hit a hunting knife he wore in his belt causing extensive internal injuries. Brothers, twenty-two-year-old Adolph Vater and twenty-five-year-old Freidrich "Fritz" Vater, natives of Prussia and sons of a Forty-Eighter, fell with life threatening wounds. Their older brother, William, went to their aid. A Confederate minnie ball hit him in the breast. Another struck him in his arm.[22]

The Confederate barrage went on for about forty-five minutes. McRae ordered both wings to charge again. The result was the same. In the south, McRae was wounded in the arm. Thirty-two-year-old Second Sergeant Emanuel Martin of Donelson's Company K received serious wounds in the right shoulder and in a knee. Two other members of Company K were hit. Forty-two-year-old First Sergeant John Hill was hit in the right shoulder and twenty-one-year-old trooper L. Dow Yarbrough went down with a wound in the hip. McRae's wing was stopped again.[23]

However, they had inflected heavy casualties on Cramer's men. Twenty-one-year-old Sisterdale Freethinker Adolphus Zoeller, a brother of a *Vierziger,* was hit in the chest or upper arm. Both Hilmar and Lieutenant Hugo Degener were seriously wounded. Hugo crawled to Captain Cramer's position to, "bid me farewell." Hilmar received a wound in the mouth. Henry Schwethelm had a bullet, "pass through his hat making a furrow through his long hair without touching his skin."[24]

In the north, Kuechler and his men were fighting grimly for their lives. He continued to encourage his men, "Don't use your powder carelessly. When you shoot, have a sure target." Their aim was good. Company A, of Taylor's Battalion, had three more men hit. Thomas J. Singleton, a twenty-two-year-old Texas native fell with a bullet wound. Latimore Edmonson, a twenty-two-year-old Mississippi native, was the next one to be hit. A third casualty at this time was twenty-seven-year-old Stephen L. Erwin, whose wounds would subsequently get him a medical discharge. Taylor's Company E had only one man wounded, twenty-four-year-old Henry W. Barker. One other trooper was hit, twenty-two-year-old William Thomas Benson of Davis Company, a native of Indiana, living in Kerr County and a brother-in-law of Fritz Tegener went down. These heavy causalities also stopped Homsley' second assault. Sansom watched this second attack as "twice during the engagement [the Confederates] reeled under the fire of the insurgents and staggered back to cover." Siemering says after the second attacks the Confederates "promptly were repelled because of a few well aimed shots". [25]

Holmsley's wing also inflicted heavy causalities on Kuechler's men. John G. Kallenberg, a twenty-year-old native of the Duchy of Saxe-Weimar and son of a Forty-Eighter, fell with a deadly wound. Forty-Eighter, thirty-two-year-old Henry Markwordt of Gillespie County, whose wife gave birth to a daughter only four days before, fell mortally wounded. Michael Weyrick, from Gillespie County a twenty-two-year-old native of Prussia, was also seriously wounded. Twenty-year-old Nassau native Heinrich Weyershausen, also from Gillespie County, was another insurgent seriously wounded. Christian Schaefer, an eighteen-year-old native of Nassau, was mortally wounded. Schaefer left a pregnant fiancée in Gillespie County; on February 16, 1863, she gave birth to a son.[26]

The fight had gone on for almost two hours now, "After the struggle had lasted an hour and a half all but twelve or fourteen of the [insurgents] were wounded and the others exhausted," says Siemering. Schwethelm adds, "out of the 40 members [left after those 'left early'] of the [insurgents] contingent 23 had been killed or wounded." In a letter dated May 16, 1913, Schwethelm says, "we only had 17 man [sic] left able to fight." An analysis of the battle to this point shows that there were only about twelve to twenty-three insurgents left who still could fight, which included some of the wounded. Siemering says, "There were only 14 to continue this unequal fight, whose outcome was no longer uncertain." On Cramer's south side, "We realized that we could not continue to hold our position. Moritz Weiss and I were the only ones that were not wounded." Kuechler and Cramer held a quick meeting and decided on a withdrawal plan. They would repel the next attack, but as soon as the Confederates fell back, they would withdraw while the enemy was still disorganized. They would head for a small creek on the northeast side of the battlefield and into the hills. The wounded who could not travel would cover the withdrawal. Schwethelm tells about this decision, "The 17 remaining, after a hasty consultation scattered in various directions."[27]

But some of the insurgents had already decided to withdraw. Kuechler claims that after the second assault about half of the defenders pulled out. Another says after the second assault some of the "Union men made good their retreat." Hoffmann says, "A group of us who had run out of ammunition by sunrise also realized how badly outnumbered we were. At this stage it was every man for himself." They had to cross an open area before they entered the small streambed. "We ran but we had to cross a clearing where several more were shot down," said Hoffmann. "There would have been more killed but the enemy kept on shooting on those that were dead or wounded."[28]

It was about 9 a.m. when the Confederates launched their third attack. Again, screaming like Indians, they raced forward. Again, the insurgents let them come within "24—30 feet" before opening fire. This time "the rebels entered our camp, but after a fierce conflict they had to give way and fall back." During all this time, twenty-year-old Wilhelm Klier alongside Kuechler, "stood like a man [and] fought fearless to the end." Cramer's south side held. Thirty-one-year-old Trooper Frost Allen, a native Texan of Duff's Company, went down with a shoulder wound that would eventually take his life. John Frank Robinson, a twenty-four-year-old Texas native from Uvalde County and another member of Duff's Company, was hit in the thigh. Kuechler's north side inflicted more casualties on Homsley's men. Twenty-six-year-old Missouri native William E. Poe of Taylor's Company E was hit in the head. His twenty-two-year-old brother, Henderson, rushed to his side; it was too late. William was dead. Both Confedcrate wings began to fall back for the third time.[29] [Map O, p 466]

After this third attack, Schlickum says "There were only seven men not wounded and eight wounded men who still could walk." Kuechler says, "This last attack reduced our fighting force to about six men who were still unhurt." "We knew now that the time had arrived to leave our camp.

The wounded who could walk left camp first and we followed guarding the rear." [Appendix AD][30]

The insurgent's withdrawal almost turned into a disaster. There are two points when a defending force takes the majority of their casualties. The first is in the opening fire when it is unprepared and caught by surprise. The second is when it tries to withdraw, forced to leave protected positions, and is exposed to enemy fire. This was the case here. Hoffmann does not tell us exactly how many casualties, they took, just "several more were shot down." These are included in the nineteen killed at the battle site. Williams explains "some of [the Confederates] were inclined to bolt, but were promptly rallied by Harbour." Lieutenant Harbour, of Holmsley's wing, saw the insurgents leaving. He did not retreat, but shouted out in a commanding voice, "They are giving way boys, come on charge." Kuechler recalled the lieutenant crying out, "Boys they are running; let us take the camp." Schlickum says Harbour said, "Forward! Charge! To hell with them! Kill them!"

Homsley's wing turned and charged the camp, followed by McRac's men. The wounded insurgents, including Wilhelm Tellgmann, the two Degener brothers and an unidentified insurgent who died in or near a campfire, opened fire on the attacking Confederates. They slowed the attack, but could not stop it. The Confederates at last carried the camp as the withdrawing "unionmen disappeared through the thick timber to the other side of the water" to "the protection of dense underbrush and lost themselves."[31]

One withdrawing insurgent looked back and saw a horrific scene: "19 young men lay dead or deathly wounded on the *Walstatt* [meaning symbolic of a holy sacrifice ground]. There lay the two sons of my old friend Degener of Sisterdale on the ground, the pride and joy of their parents. There lay Karl Beseler ... fearless and brave. There also was Paul, a Mexican who was purchased from the Indians as a child and raised as a German. ... My friend Tellgmann ... his body pierced from three bullets and still fighting."[32]

The Confederates claimed that the insurgents "fled, scattering in all directions through the many dense cedar-brakes in the immediate vicinity." A modern analysis reveals that there were up to 23 survivors in the camp at sunrise. There were 14 non-wounded and nine wounded. The non-wounded included: Jacob Kuechler, Moritz Weiss, Ernest Cramer, Henry Schwethelm, William Klier, Carl Graff, William Graff, Jacob Kusenberger, Philipp Henrich Weber, August Hoffmann, Mathias Pehl, August Duecker, Joseph Poetsch, and Carl Vetterlein. The wounded included: Gottlieb Vetterlein, Fritz Tegener, Franz Weiss, Ferdinand Simon, Henry Kammlah, William Vater, Karl Itz, Adolphus Zoeller, and an unidentified insurgent. R. H. Williams says, "The defenders by this time had lost heavily, and began to make off in small parties through the thick brush." He continues, saying they "pursued one [insurgent group] ... but soon lost them." Siemering disagrees and says, "It should be noted that the Confederates did not have the courage to follow them." Schwethelm agrees and says, "17 man went across the Nueces into the Cedar Break, they never followed us."[33]

Sansom claims he and the other four Anglos attempted to assist the insurgent's withdrawal. He says, "Our party of five moved closer to the enemy and somewhat to their rear, intending to attack from that direction, but when we pulled triggers, four of the guns snapped, mine only firing." He claims while the Scott boys, Henderson and Hester were cleaning the tubes of their rifles in an effort to clear them of bad powder, the Confederates made the final "determined charge upon the Unionists' camp. I fired as rapidly as I could at the Confederates, but my comrades could do nothing." When he saw Major Tegener and the survivors of his command leave the camp and the Confederates take possession if it, "I told the men with me to get horses." This acknowledgement by Sansom is revealing. The firing had been going on for over an hour, and more likely two hours. Why hadn't the four previously fired their weapons, at least in the opening exchange of gunfire? What took so long to clear the blockage? No Confederate, or insurgent's account says anything like fire was placed on the attacking North Wing, from the rear, during the ground assault. If any attacking force is receiving fire from its rear, if only from one pistol, they would have known it and done something to silence the fire. Sansom seems to be providing an excuse for the five not helping in the defense of the camp. This certainly gives credibility too later insurgents statements saying the Anglos ran at the first fire.[34]

As soon as the Scott brothers, Howard Henderson, and William Hester got their mounts they left the battle area. They reached the Rio Grande later that afternoon and crossed safely into Mexico. Sansom hurried back to where he had left his saddle near Bauer's body. He saddled a horse, using a Confederate blanket for a saddle pad and taking another blanket to use as a pallet. He mounted and started west from the battle site. He only got about two hundred and fifty yards when a four-man Confederate group challenged him. This group that fired on Sansom was very likely part of the detail that remained with the horses. They exchanged several shots. Sansom turned south away from the battle area. After a safe distance, he turned east and crossed the river. Here he saw four other insurgents, Henry Schwethelm, twenty-eight-year-old Nassau native, Jacob Kusenberger, and two men he identified as Graff. The insurgents mistook Sansom for a Confederate and did not stop. Sansom continued until he reached a cedar brake on a high bluff that overlooked the insurgents' camp. From here, he watched the Confederates as they, "stood and walked about the camp and its dead and wounded, and going to and from the river, their only place of securing water." Then, satisfied that he could do no more, he rode away. It was about ten o'clock Sunday morning, August 10, 1862.[35]

The last twenty-three insurgents fled from the battle site in three general groups. First was the ten-man August Hoffman group. Instead of a single group, these ten men likely made their way out of the camp in small groups of one or two. They included August Hoffmann, Frederick Weber,

Mathias Pehl, August Duecker, Karl Itz [who left alone], Carl and Gottlieb Vetterlein, Joseph Poetsch, Henry Kammlah, and an unidentified insurgent. Kammlah, Gottlieb Vetterlien, and the unidentified insurgent were wounded. Karl Itz was injured. Next was the four-man Henry Schwethelm group; Henry Schwethelm, Jacob Kusenberger, and Carl and William Graff. Third was the nine-man Ernest Cramer and Jacob Kuechler group. Their group included Jacob Kuechler, Ernest Cramer, Moritz Weiss, Ferdinand Simon, Fritz Tegener, Franz Weiss, Adolphus Zoeller, William Vater, and Wilhelm Klier.

The August Hoffmann group made it across an open area to the northeast side of the insurgents' camp and across the river. On the east side of the river, they found a small draw or canyon. They made their way up this draw. Karl Itz's internal injuries kept him from being able to run or walk very fast. He followed the group across the river, but instead of going up the draw, which he felt he would not make, he turned west and went up the north side of the riverbank for about two hundred years and found a hiding place. The unidentified member was also too weak to keep up. Instead of going up the draw, he turned west as Itz had done and went about 100 years to another small draw or canyon. He went up the draw about fifty yards where he found a hiding place.[36]

The Henry Schwethelm group left the camp heading due east until they came to the river at a point with a high cliff on the other side. "We of course had no horses and no money but we went." They turned south and went down the river for several hundred yards, where they crossed to the east side of the river and located a hiding place.[37]

Captains Cramer and Kuechler and their group crossed the river just to the northeast of the camp and like the Hoffmann group found the small draw, but they lost one member. William Vater was too weak to keep up. He reached the river, crossed it, and found a hiding place on the riverbank at the base of the high bluff. The remainder of the group continued up the first draw. Wilhelm Klier description says, "Bullets whirled from all direction when they fled across an open plain, but [Klier] miraculously escaped injury." The group "went forward up the valley very cautiously so as to leave no trace that could be followed." They continued up the draw for about a mile until they, "found an advantageous position in a cedar brake on the east side of the Nueces River."[38] [Map P, p 467].

The Confederates entered the camp about 9:30 a.m. The time the Confederates entered the camp is based on an analysis of all the primary and the Schlickum and Siemering accounts. Daylight did not break until almost 8:00 a.m. Sunrise on August 10th was about 7:00 a.m., but due to the high bluff just east of the camp it does not occur until around 8:00 a. m. The conditions inside the camp were "ghastly." Williams found one seriously wounded insurgent with, "little life in him, but unable to move [lying] across the camp fire." Williams pulled him out of the fire and put out the fire on his body. Shortly thereafter the man died. Williams says, "For a while there was plenty to do separating the wounded from the dead and dressing the hurts of the former as best we could, for we had no surgeon with us." McRae reported, "From the many signs of blood, I infer many of those escaping was seriously wounded." Williams adds, "Seeing there were plenty of willing helpers for our own poor fellows, some of the more humane of us did what we could to ease the sufferings of the wounded Germans. We bound up their wounds, and gave them water, and laid them as comfortably as we could in the shade." One of Lieutenant McRae's first actions was to send to Fort Clark for medical aid and transportation for the wounded. Williams says, "Immediately after the fight a couple of the boys were sent off, post hast, to Fort Clark … to fetch the surgeon stationed there."

Meanwhile, "By this time some of the boys had cooked Breakfast, for there was an abundance of provisions in the camp, and I fell to with them, with an appetite, having tasted nothing, except the bacon soup, for two days."[39] The Confederates captured eighty-three horses, the camp equipment, provisions for one hundred men for ten days, as well as thirty-three rifles, and thirteen six-shooters. The number of the number of horses captured help determine when the 23 German members of the group left.

There was a total of 69 insurgents when the battle started, which means each had one horse for a total of 69 horses. According to Eduard Degener, they had a pack animal for each third man. That adds another 21 pack animals, for a total of 90 animals. Henry Schwethelm, whose accounts says they "had some 80 horses," confirms this. This was before the four men on the trail joined them. Therefore, counting the four who joined on the trail, the total number of animals was about 90. After the battle, the five Anglos each left on a horse, this brings the number down to about 85. The Confederate captured 83 horses, which only leaves two horses unaccounted for. Now, if 23 left before the battle started, surely they would have taken their mounts. They would have no reason not to. From this analyses it is clear the 23 insurgents did not leave the camp before the opening rounds at 3:00 a. m. August 10th. When they left, after the opening shots and before the ground assaults, they did not take mounts. Like the number of horses captured, the number of weapons captured is very revealing. It seems safe to assume, regardless of when the 23 insurgents left, they would have taken some of their weapons. It can be established that Sansom and the four with him had their weapons. Likewise, it seems safe to assume that of the nine men in the Hoffmann group, the six unwounded men took their weapons. The same is true for the four-man Schwethelm group. This makes a total of 38 men who would have taken weapons. Of the nine-man Kuechler and Cramer group only three were not wounded, so for sake of discussion let's assume only those three took weapons. That means 41 insurgents left with weapons. We know that many of these had more than one weapon, but for the sake of this discussion let's say only the five Anglos had two weapons. That brings the total number of weapons taken out of the camp as 46. McRae reports 33 small arms and 13 pistols, for a total of 45 weapons. Forty-five added to the known 46 that left the camp means the insurgents had, as a minimal, 91 weapons or a total of 1.3 weapons per man. This establishes the Confederate claim and supported by the Schwethelm account that the insurgents were well-armed and not 'poorly armed' as claim by Sansom. However, as pointed out by Gregory Krauter, a Comfort historian, 'poorly armed' also could mean poor quality, not only quantity of weapons. The Confederates also captured what appeared to be a muster roll. Found on the body of one of the Degener brothers was a letter written by his father to friends in Germany.[40]

The number of Confederate and insurgent casualties has been a matter of controversy. There should be little dispute about the Confederate casualties. McRae's report is clear on the point. He says, "We had two killed on the field and 18 wounded." His casualty report however names two killed and nineteen wounded. Of the nineteen wounded, four subsequently died from their wounds. Thus, out of 96 Confederates, they suffered a total of six killed and fifteen wounded; a casualty rate of twenty percent or one out of every five members of the attacking force. In military science, with a casualty rate of this magnitude a unit would be classified as 'combat ineffective.' An example of these types of numbers being used in books and historical papers is Don Biggers' book *German Pioneers in Texas* where he says, "During the battle following Duff's attack, 12 Confederates were killed and 18 wounded, according to official reports." This is another example where writers have just changed what a source says to fit the point they want to make. McRae's Report, which is the only 'official report' says 2 killed and 18 wounded. Note how Biggers just changes 2 to 12! Another example is a 1943 master thesis at the University of Texas that says 47 Confederates were killed while at the same time claiming the insurgents were, "almost unarmed, [and a] defenseless band of Germans."[41]

The insurgent casualties are harder to determine. Examples of books that exaggerate the number of insurgent casualties include Fehrenbach's *Lone Star*, where he claims a total of 34 were killed. Biggers, in *German Pioneers* says the number was 28, pro-Confederate writers like William Banta who says in his book *Twenty-Seven Years* that 52 were killed. Gilbert Benjamin in his study, *The Germans In Texas*, gives the number as 32. The *Treue der Union* Moment in Comfort, Texas list nineteen names as *Gefallen am 10 August 1862 am Nueces* [Killed at the Nueces: August 10, 1862]. The problem arises because elsewhere on the monument nine names are listed as *Gefangen genommend und ermordet* [Captured and murdered]. Most writers have understood this to mean these additional nine were the ones captured **and killed at the Nueces on August 10, 1862**, or a total of twenty-eight. This is not correct. The nine additional names were ones 'later captured' or as Sansom says, "subsequent to August 10th, they were killed, and later their names added to the monument shaft." These nine were killed at various locations during the period of August 18 to August 23. The nineteen names listed as "Killed at the Nueces" **include those wounded, captured, and executed**.[42]

The military used a formula for estimating casualties. It is about fifteen percent of the military force will be killed or wounded. The fifteen percent is further broken down into a third killed or seriously wounded and likely to die from their wounds if not treated quickly. Two-thirds of the wounded will recover from their wounds. Of the third killed or seriously wounded, one-third will be killed and two-thirds will be seriously wounded. So therefore, if a 100-man force is involved in a heavy battle, 15 will be casualties. A third of this number, or five will be killed or seriously wounded and ten others wounded. Of the five killed or seriously wounded, one or two will be killed and two to five will be seriously wounded. Another ten will be wounded and recover.

Applying this formula to the Confederates, the numbers work out correctly. The Confederates had a total force of 95 of which 21 were casualties. The casualty rate is 22%. So, the 22% is used as the basic of the formula. A third of 22 is seven, which should be the amount of killed and seriously wounded. Of this number seven, a third is 2.4% or two killed and five dying from wounds. The actual numbers were two and four; very close to the formula.

Applying the formula to the insurgent's casualties, a third of the number 31 should be killed or seriously wounded. That works out to ten that should be killed or seriously wounded. Of these ten, three or four should be killed and six or seven should be seriously wounded. Based on the formula, only about ten should have been left at the battle site and the other 21 safely evacuated. So, applying the formula to the total insurgent's casualties shows a higher rate of killed or seriously wounded. However, applying the formula using the known number of 19 as the actual killed or seriously wounded, the results are a third of the 19 is six and should be the number killed and 13 the number dying from their wounds. One other outside factor is considered. That is two of the insurgents were killed before the battle started, which would increase the number killed from six to eight and reduce the number seriously wounded from thirteen to eleven. Apply these numbers to the known actual number, they are within the ratio. Eight insurgents can be accounted for as dying inside the camp during the battle, two killed before the attack, leaving nine seriously wounded in or near the camp. Of these nine, two died from their wounds; the man pulled from the fire and the one hiding alone in the small draw north of the camp. This leaves seven seriously wounded left in the camp.

The belief that more than nineteen were killed at the battle site is due in part to McRae's report in which he says, the insurgents, "left on the field 32 killed." The higher number of total insurgent casualties have become a standard number in written accounts. Like the Confederate casualties, the insurgent casualties are placed at anywhere from nineteen to well over fifty. The two official reports of the battle both give the number of insurgents killed as 32 or 35. These two reports were McRae's and the U. S. Consul General at Matamoros, Mexico. McRae stated the number was 33. Consul Pierce gave the number as 32. Both sources were insurgent survivors; McRae's was Ferdinand Simon who was captured a few days after the battle and Pierce's were the seven survivors who reached Matamoros. It can be established that about ten insurgents were killed. These included two hanged by Confederate scouts prior to the battle, Bauer and Beseler at the beginning of the battle, Schreiner killed in a counterattack, and two or three killed as described by Hoffmann. That provides a base number of at least seven killed. An 1865 newspaper account of recovering the bodies says that Pablo Diaz was killed in the battle and moves the number killed to at least eight. This leaves the number wounded left on the battlefield as eleven. Of these eleven, one escaped the battlefield and died in a small draw north of the battlefield. This reduces the number of wounded 'that could have been left' to ten. The one Williams describes as pulling from the fire, reduces the number to nine wounded that 'could have been left' and two that died of their wounds. It is known at least three wounded were left. These were the Degener brothers and Wilhelm Tellgmann. When the two Degener brothers died is not known, but likely after the Confederates took the camp. Thus, as a base, it can be established that about eight were killed, two died of wounds, and nine wounded left on the battlefield and captured, and ten of the nine dying from their wounds. One reason for this high number is the subsequent insurgents' belief that more of their number had been killed because they did not know the exact number that had 'left early.' They only knew of about fifteen. When the survivors came together after the battle they could account for only their seventeen, plus the fifteen who left early,' plus the five Anglos, for a total of thirty-seven. Therefore, it was their belief that all those not accounted for were dead. If they had a total of 69 to begin with and 37 accounted for, that left 32 unaccounted for and believed dead. In point of fact, the insurgents had ten killed, two died of their wounds, and seven seriously wounded were left on or near the battle site and twelve wounded who escaped from the battlefield, a total of 31.[43]

Seven is the correct number of wounded Unionists left on the battlefield. This number is determined by the adding the ten who we know were killed, plus the two who died of their wounds, leaving seven wounded left on the battlefield, and twelve wounded who safely escaped. Putting names to these numbers is at least only guesswork. The known dead included: Bauer, Beseler, Schreiner, and Diaz. The dead included: Heinrich Steves, Louis Boerner, and Louis Schieholz for a total of seven of the ten killed. The known wounded left on the battle site included Hilmar Degener, Hugo Degener, and Wilhelm Tellgmann for a total of three of the seven. This leaves the following unknown as to when killed: Fritz Behrens, Albert Bruns, John George Kallenberg, Heinrich Markwordt, Christian Schaefer, Adolph Vater, Friedrich Vater, Heinrich Weyershausen, and Michael Weyrich.[44]

Several accounts of the battle are told without any source documents or seemly concern for the truths. One is T. R. Fehrenbach's *Lone Star*. His account is almost pure fiction. He states, "McRae rode down upon the camp while the Germans lay sleeping. He surrounded it and opened fire indiscriminately. The result was massacre. Nineteen Germans were killed by gunfire, and six more trampled to death by McRae's cavalry." Thomas T. Smith's book *Fort Inge* seems to use *Lone Star* as its source and says the same; "Nineteen killed, nine wounded, and six run down by cavalry as they attempted to flee on foot." Even as recently as 1998, and despite the 'Nueces Encounter Symposium' in March, 1997, some writers still will state in national publications, "The six men paid a dear price for their valor; a group of mounted Confederates spurred into the camp and trampled them to death under their horses' hooves." McRae did not "ride down upon the camp." He made a dismounted attack. There was no indiscriminately fire. The fire was directed into an enemy camp. The battle was far from a massacre. The Confederates were very fortunate in winning, it was a costly victory, over a 20% casualty rate. Nineteen insurgents were **NOT** killed in battle. A total of nineteen were killed at the battle site, but this includes those wounded later executed [which was a massacre]. Certainly, no insurgents were trampled to death by cavalry.

Map L – Location of Artifacts

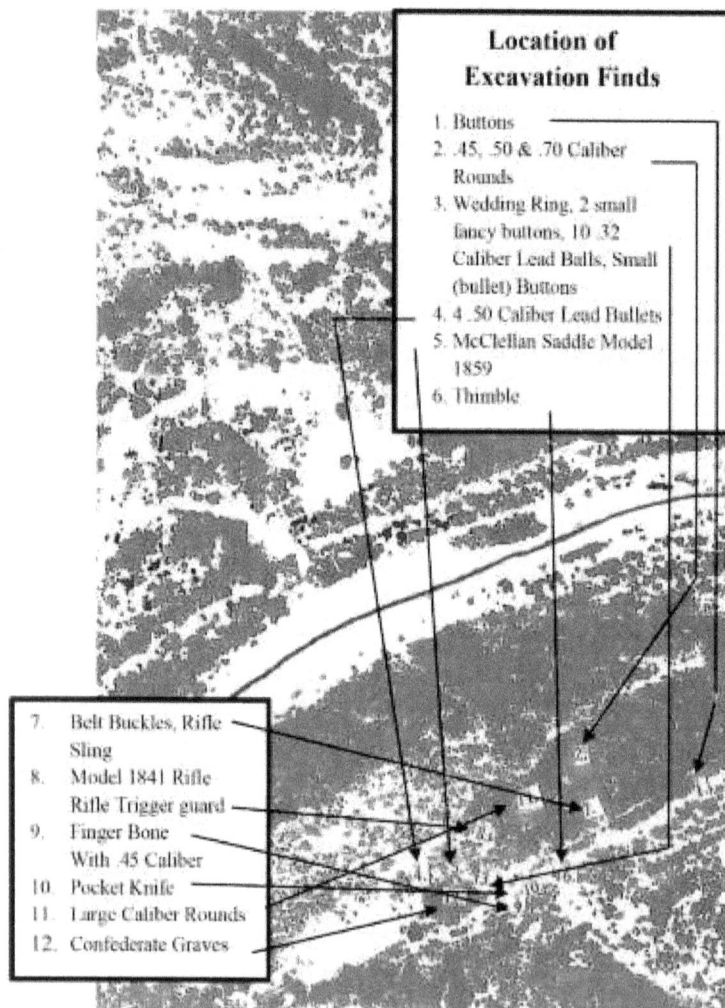

Location of Excavation Finds

1. Buttons
2. .45, .50 & .70 Caliber Rounds
3. Wedding Ring, 2 small fancy buttons, 10 .32 Caliber Lead Balls, Small (bullet) Buttons
4. 4 .50 Caliber Lead Bullets
5. McClellan Saddle Model 1859
6. Thimble
7. Belt Buckles, Rifle Sling
8. Model 1841 Rifle Rifle Trigger guard
9. Finger Bone With .45 Caliber
10. Pocket Knife
11. Large Caliber Rounds
12. Confederate Graves

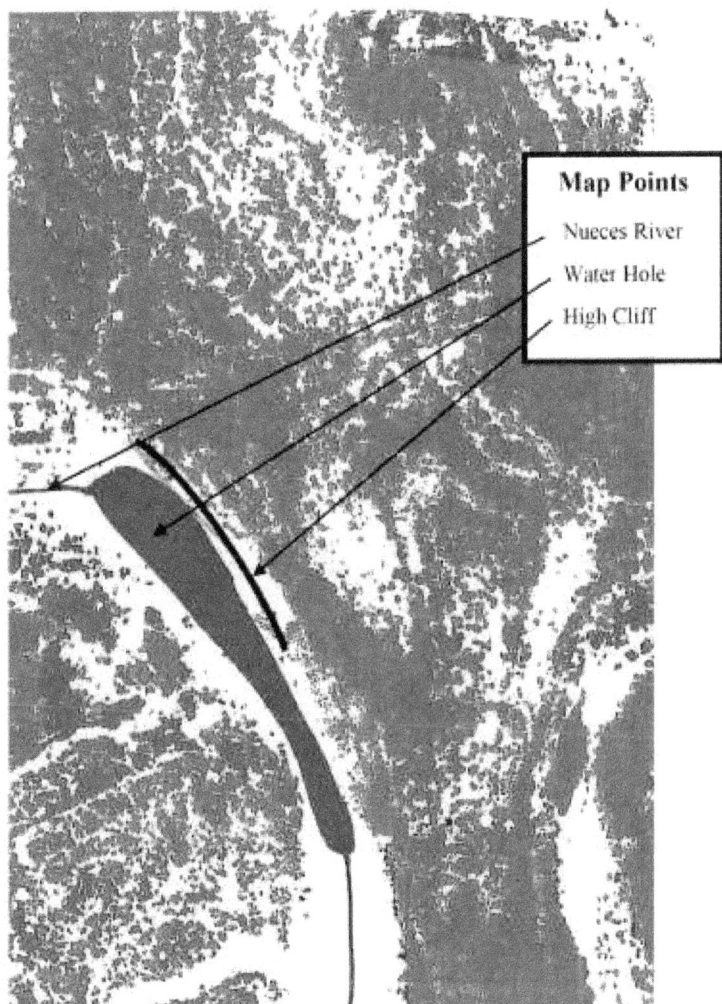

Map Points

Nueces River

Water Hole

High Cliff

Map M – Insurgent Camp at Nueces

Map N – Confederate Envelopment of Insurgent Camp

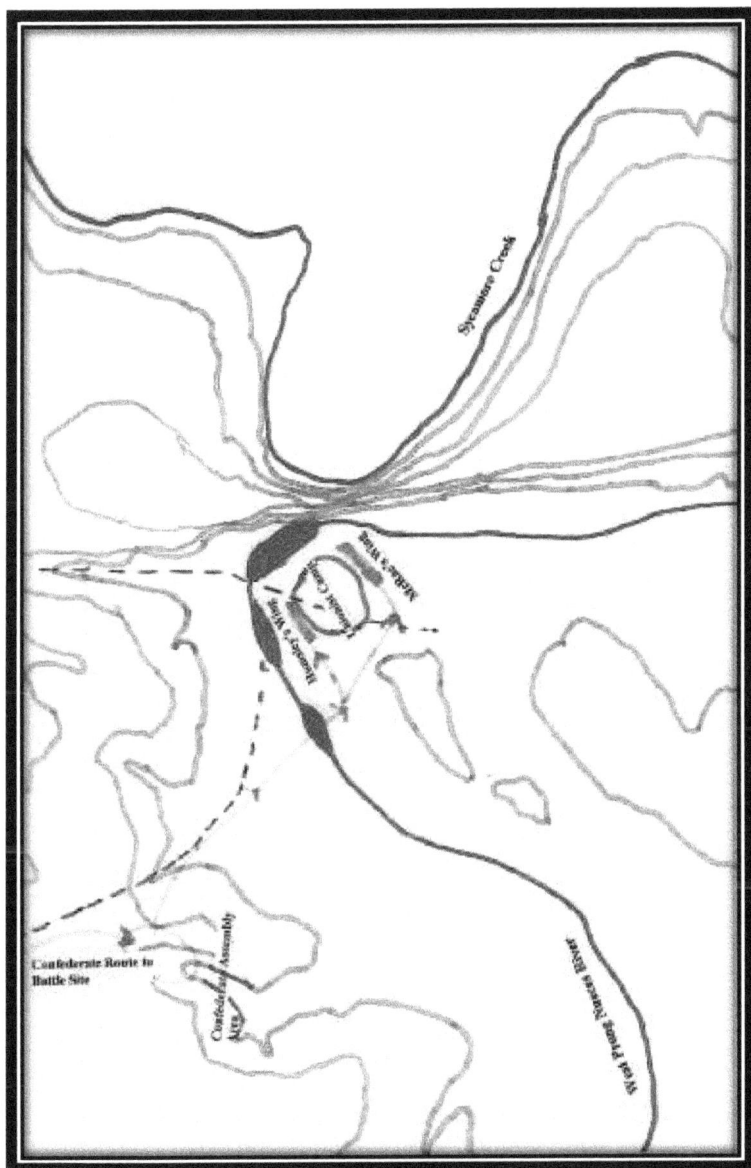

Map O – Movement of Insurgent Survivor Groups

Map P – Disposition of Forces

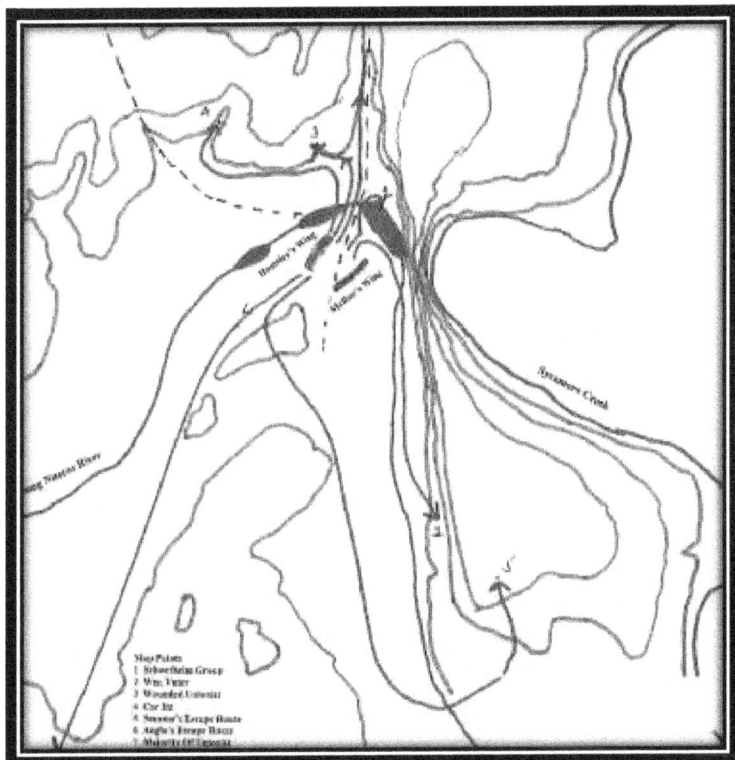

Endnotes - The Battle

1. Elghtlı U. S. Census, 1860 Kerr County Texas Census, 68h; Ransleben *100 Years*, p 24; Kerr County Marriage Records, p 8; *Bexar County, Texas Marriage Records; 1837 – 1866*, Compiled by Frances T. Ingmire, (Published by Frances T. Ingmire, St. Louis, Missouri), 1985, p 45 and Sansom *Battle of Nueces*, p 7.

2. McRae's Report, p 615 and Williams *Border Ruffians*, p 246.

3. Schultz *Was a Survivor of Nueces Battle*, p 26; Siemering's Germans During Civil War, June 5, 1923; Cramer's Letter; Schlickum's Letter of May 16th, 1913 to his grandson, Otto; Hoffmann's Letter; Sansom *Battle of Nueces*, p 7; Williams *Border Ruffians*, p 246; *Kinney County 1852 – 1977* Compiled by Kinney County Historical Commission (Published by Kinney

County Historical Society, Brackettville, Texas), 1977, p 36 and *Harper's Weekly,* January 20, 1866. An interesting item is included in Henry Schwethelm's interview. It says, "The unfortunate who had betrayed his comrades was killed in the melee whether by the Confederates or his friends is not known."

4. Sansom *Battle of Nueces,* p 7; Sansom's Letter of August 18, 1907; McRae's Report, p 615; Kuechler's letter; Cramer's Letter; *Harper's Weekly,* January 20, 1866; Letter, John W. Sansom, San Antonio, Texas August 18, 1909 to Mr. J. T. DeShields; Rivers, *Ranches, Railroads and recreation: A History of Kendall County, Texas* Compiled by Kendall County Historical Commission (Taylor Publishing Company, Dallas, Texas) 1984, pp 58-59 and Schellhase Journal.

5. McRae's Report, p 615 and Williams *Border Ruffians,* p 246.

6. Sansom Battle *of Nueces,* p 7; Raley Blackest *Crime in Texas Warfare*; Schlickum's Letter; Siemering's Germans During Civil War, June 5, 1923 and Williams *Border Ruffians,* pp 246247.

7. Kuechler's letter and Siemering's Germans During Civil War, June 5, 1923.

8. Sansom *Battle of Nueces,* and Cramer's Letter.

9. Cramer's Letter.

10. Williams *Border Ruffians,* pp 246-247 and McRae's Report, p 615.

11. Sansom *Battle of Nueces,* pp 7-8.

12. W. J. Edwards, a member of Duff's Company testified at the Confederate Military Commission that, "no one seemed to be in command [of the insurgent force] at this time." Sansom *Battle of Nueces,* p 8 and Barr *CMC,* Volume LXXIII (October 1969), p 252.

13. Sansom *Battle of Nueces,* p 8.

14. Albert Bruns was born about 1838 in Prussia. His date of arrival in Texas is not known. He arrived in the Comfort area in 1854. Albert Bruns was killed at the battle site.

Pablo Diaz was born about 1842 in Mexico. Indians captured him as a child. Germans rescued and raised him. The 1850 Comal Census shows him living in the Julius Dresel household. The 1860 Gillespie Census shows him living in the William Marschall household. He was a member of Kuechler's February 1862 Company. Pablo Diaz was killed at the battle site. Sansom *Battle of Nueces,* p 9; Eighth U. S. Census, 1860 Kerr County Texas Census, p 75b; Ransleben *100 Years*, p 24; Seventh U. S. Census, 1850 Comal County Texas Census, p 64; Eighth U. S. Census, 1860 Gillespie County Texas Census, p 23; Kuechler's MR and Sansom *Battle of Nueces*, pp 8-9.

15. Sansom Battle *of Nueces*, pp 8-9; Siemering's Germans During Civil War, June 5, 1923; McRae's Report, p 415 and Schlickum's Letter.

16. Siemering's Germans During Civil War, June 5, 1923; Kuechler's letter; Sansom *Battle of Nueces*, pp 7 and 11 and Schlickum's Letter.

17. Siemering's Germans During Civil War, June 5, 1923; Kuechler's letter; Schlickum's Letter; Cramer's Letter; Sansom's *Battle of Nueces*, pp 9 and 11; Siemering's Germans During Civil War, June 5, 1923; Williams *Border Ruffians*, p 247 and *Harper's Weekly*, January 20, 1866.

18. Siemering's Germans During Civil War, June 5, 1923 and Williams *Border Ruffians*, p 247.

19. Benjamin Franklin Rosser was born about 1838 in Alabama. The date he arrived in Texas is not known. By 1860, he was living in Bowie County. He enrolled in Company K, 2nd Regiment Texas Mounted Rifles on September 23, 1861. Rosser received a medical discharge on May 15, 1863. Rosser died in 1914 at Elk City in Beckham County Oklahoma.

Littleton Stringfield was born September 23, 1843 in Limestone County, Texas. On September 23, 1861, he enrolled in Company K, 2nd Regiment Texas Mounted Rifles. Family members removed his body and reburied in the Tehuacana Cemetery in Frio County, Texas.

Albert J. Elder was born in 1839 in Kentucky. The family arrived in Texas in the late 1840s and settled in Bexar County. Albert's father was a prominent member of the secessionist movement. His sister, Helen, was married to Judge Thomas Devine. Albert enrolled in Duff's Company on May 4, 1862 at San Antonio. He died of his wounds on August 17, 1862, at Fort Clark.

Robert G. Elder was born about 1835 in Kentucky. He enrolled in Duff's Company on May 4, 1862 at San Antonio. Robert G. Elder died on September 17, 1862, exactly a month after his brother, also at Fort Clark.

McRae's Casualty List; *San Antonio Weekly News,* August 25, 1962; Confederate Service Records of Littleton Stringfield, Record Group 109, National Archives, Washington, D. C.; *The Marked Gravesites of Citizens of the Republic of Texas: An Indexed Guide to Early Citizen's Burial Places,* p 436, Compiled by Daughters of the Republic of Texas, (Published by Daughters of Republic of Texas, Austin, Texas), 1986, p 50; Eighth U. S. Census, 1860 Bexar County Texas Census, p 412; *Neu Braunfelser Zeitung*

August 25, 1862 San Antonio Genealogical and Historical Society *Index Wills and Inventories of Bexar County, Texas*, p 28; Albert J. Elder's Confederate Service Records, Record Group 109, National Archives, Washington, D. C.; McRae's Casualty List; Robert G. Elder's Confederate Service Records, Records Group 109, National Archives, Washington, D.C.; McRae's Report, p 615; Williams *Border Ruffians*, p 247; and Confederate Service Records of Benjamin Rosser, Record Group 109, National Archives, Washington, D. C.

20. John K. Morris was born about 1841 in Mississippi. The date the family arrived in Texas is not known. He enrolled in 1ˢᵗ Regiment Texas Mounted Rifles in July 1861 at Belton and re-enlisted in Company B in May 1862. John K. Morris died of his wounds on September 1, 1862, at Fort Clark.

Albert L. Wilborn was born about 1828 in South Carolina. The date he arrived in Texas not known. He was a member of Harbour's February 1861 Home Guard Company. He enrolled in Davis' Company on March 4, 1862. After the battle, Wilborn re-enlisted in Hunter's Company on January 12, 1863. He received a medical discharge in May 1863. Albert L. Wilborn died on March 6, 1873, in Kerr County from complications of his wound.

Wiley Williams was born on September 3, 1838, in Arkansas. He enrolled in Company G, 1ˢᵗ Regiment Texas County Rifles in May 1861 at Camp Colorado. He re-enlisted in Company A, Taylor's Battalion in May 1862 and remained with the unit until it became Company G, 1ˢᵗ Regiment Texas Cavalry and served until the end of the war. Williams died on September 24, 1901, in San Saba County.

John Welch was born on March 11, 1839, in Alabama. The date he arrived in Texas is not known. He enrolled in Company G, 1ˢᵗ Regiment Mounted Rifles in May 1861 at Camp Colorado. Welch re-enlisted in Company A, Taylor's Battalion in May 1862 and remained with the unit until it became Company G, 1ˢᵗ Regiment Texas Cavalry. He transferred to Company E, 21ˢᵗ Regiment Texas Cavalry on December 2, 1863. His date and location of death are not known.

Neu Braunfelser Zeitung August 25, 1862; Schlickum's Letter; Siemering's German During Civil War, June 5, 1923; McRae's Casualty List; *San Antonio Weekly News,* August 25, 1962; Eighth U. S. Census, 1860 Bell County Texas Census, p 324b; John K. Morris' Confederate Service Records, Record Group 109, National Archives, Washington, D. C.; Wiley Williams' Confederate Service Records, Record Group 109, National Archives, Washington, D. C.; Ninth U. S. Census, 1870 Caldwell County Texas Census, p 287: Albert L. Wilborn's Confederate Service Records, Record Group 109, National Archives, Washington, D. C.; Eighth U. S. Census, 1860 Kerr County Texas Census, p 74b; Davis' MR; Hunter's MR:

Harbour's MR; Kerr Historical Commission *Kerr Album*, pp 458-459 and John Welch's Confederate Service Records, Record Group 109, National Archives, Washington, D. C.

21. The Vetterlien brothers were Charles and Gottlieb. They were born in the Duchy of Altenburg, Gottlieb about 1831 and Charles on January 22, 1832. They both fled to El Paso where they worked for the Union Army as teamsters. It is believed Gottlieb was the brother wounded and he died in the El Paso area. Charles returned to the Comfort area, where he died on July 9, 1914.

    Cramer's Letter; Letter from George H. Pettis to John W. Sansom, dated December 1, 1908, located in *A Hundred Years of Comfort in Texas*, Page p 119; Raley *Blackest Crime in Texas Warfare* and *Texas Burial Sites of Civil War Notables: A Biographical and Pictorial Field Guide* by James A. Mundie, Jr. et al, (Hill College Press, Hillsboro, Texas) 2002, p 292.

22. Karl Itz was born on October 26, 1831 in Nassau. The family arrived in Texas in 1853 and in Gillespie County by 1854. After the battle, Karl returned to Gillespie County and hid until 1863 at which time he enrolled in Schuetze's Company. Karl Itz died on January 31, 1908, in Gillespie County.

    Henry Kammlah II was born September 8, 1839 in Immensdorf, Braunschweiger. The family arrived in Texas on the *Gerhard Hermann* in 1845. Kammlah returned to Gillespie County after the battle and hid out the reminder of the war. Henry Kammlah II died January 26, 1923, in Gillespie County.

    Gillespie Historical Society *God's Hills*, Volume II, pp 45-48; Oral Interview by the author with Karl Biermann, a great-great-great-great-grandson of Karl Itz on November 9, 1997; Oral interview by the author with Leo Itz, a great-great grandson of Karl Itz, August 20, 1996; Gillespie Historical Society *God's Hills*, Volume I, p 73; Oral interview with Joe Kammlah, a great-great grandson of Henry Kammlah, April 26, 1996; Lindig Der *Friedhof Cemetery*, p 36 and Petition Louis Schuetze November 1, 1863 Fredericksburg, Texas to Texas Governor with muster roll of proposed company.

    Adolph Vater was born about 1839 in Prussia. He and his parents arrived in Texas about 1853. Member of the Luckenbach Bushwhackers. Adolph Vater was killed at the battle site.

    Friedrich [Fritz] Vater was born about 1837 in Prussia. He and his parents arrived in Texas about 1853. He married Christine Heubaum on July 15, 1858 in Gillespie County. She gave birth to a daughter who was baptized on August 7, 1862, just three days before her father was killed at the battle site.

Fritz's widow married George Heinrich Gombert on March 26, 1863, in Gillespie County.

William Vater is somewhat a mystery. He was not located on a Texas Census. He hid in the area after the battle and was rescued by a man from Uvalde. Sansom says he joined the Confederate Army, but no record of his has been found.

Sansom Battle *of Nueces*, p 13; Knopp *The Fredericksburg, Manuscripts*, p 27; Eighth U. S. Census, 1860 Gillespie County Texas Census, p 27; Luckenbach Bushwhackers; Gillespie County Marriage Records, Book 1, pp 278 and 392; Gold *Church Record Book of the Vereins-Kirche*, 57 and 65; and Gillespie County Probate Records, Volume C, Pages pp 229-230, 232-233, 279, 281-182, 322, 328-329, 331, 334, 356, 389-391 and 413.

23. John Hill was born about 1823 in Ireland. The date he arrived in Texas is not known. He enrolled in Donelson's Company K on September 23, 1861, and elected first sergeant. He was given a medical discharge on April 23, 1863. John Hill's died about 1890, likely in McMullen County.

     Emanuel Martin was born about 1829, location not known. He enrolled in Company K, 2nd Regiment Texas Mounted Rifles on September 23, 1861, at Brownsville and elected second sergeant. He died of his wounds on October 18, 1862, at San Antonio.

     L. Dow Yarboro [Yarbrough] was born in January 1840 in Texas. Yarboro enrolled in Donelson's Company on September 23, 1861. He transferred to Captain Tobin's Company F, 2nd Regiment Texas Cavalry on January 1, 1864, and remained with the unit until the end of the war. Dow Yarbrough died on January 5, 1907, in LaSalle County Texas.

     McRae's Casualty List; *San Antonio Semi-Weekly News,* August 25, 1862; Colin D. McRae's Confederate Service Records, Record Group 109, National Archives, Washington, D. C.; Emanuel Martin's Confederate Service Records, Record Group 109, National Archives, Washington, D. C.; John Hill's Confederate Service Records, Record Group 109, National Archives, Washington, D. C.; Dow Yarbrough's Confederate Service Records, Record Group 109, National Archives, Washington, D. C.; Dow Yarbrough's Confedcrate Pension Records, TSA, Austin, Texas and Nancy Yarbrough's Confederate Pension Records, TSA, Austin, Texas.

24. Adolphus Zoeller was born on December 1, 1839 in Hesse-Darmstadt. His date of arrived in Texas is not known. He settled in the Curry's Creek and Sisterdale area. After the battle, he returned home and recovered from his wounds. He later fled and enrolled in Company C, First Regiment [Union] Texas Cavalry on February 5, 1863, as a second lieutenant. On January 8, 1864, he was promoted to first lieutenant and to captain on July 7, 1864.

Zoeller was wound at Battle of Las Rucias on June 25, 1864. After the war, he returned to Kendall County was he was elected to the Texas legislature. Adolphus Zoeller died on September 18, 1909 in Kendall County Texas.

The fact one of the Degener brothers was wounded in the mouth is confirmed by W. J. Edwards' testimony before the Confederate Military Commission in October 1862. Edwards said, "... one man who was killed, shot in the mouth, was believed to be one of [Degener] sons."

Schultz *Was a Survivor of Nueces Battle*, p 26; Cramer's Letter; Oral Interview by the author with Wayne Meier, a great-great grandson of Adolphus Zoeller on April 26, 1995; Hoffmann's Letter; *A Family History of Hellwig Karl Ludwig Adolph Zoeller and Augusta Zoeller*, (Published by descendants), n.d. Copy provided by Margie Morgan, a great-great-granddaughter of Adolph Zoeller, Kerrville, Texas and Barr *CMC*, Volume LXXIII, (October 1969), p 252.

25. Latimore Edmonson was born about 1840 in Mississippi. The date he arrived in Texas is not known. By 1860, he was living in Brown County Texas. Enrolled in 1st Regiment Texas Mounted Rifles in May 1861 at Camp Colorado. Re-listed in Company A, Taylor's Battalion in May 1862. He remained with the unit when it became the 1st Regiment Texas Cavalry and served until the end of the war. Latimore Edmonson died on June 6, 1915, in Coleman County, Texas.

Thomas Jefferson Singleton was born March 20, 1840 near Houston, Texas. He enlisted in Company G, 1st Regiment Texas Mounted Rifles on May 12, 1861 and re-enlisted in Company A, Taylor's Battalion in May 1862. Singleton remained with the unit when it became Company G, 2nd Regiment Texas Cavalry and served until the end of the war. Thomas Jefferson Singleton died on August 16, 1919, in Tarrant County.

Stephen L. Erwin was born on September 3, 1835 in McMinn County, Tennessee. He enrolled in Company I, 1st Regiment Texas Mounted Rifles in May 1861. He re-enlisted in Company A, Taylor's 8th Battalion in May 1862. Stephen L. Erwin received a medical discharge on October 31, 1862 as a result of his wounds. He died on December 8, 1917 at Weatherford in Parker County.

Henry W. Barker was born in May 1839 in Texas. He enrolled in 1st Regiment Texas Mounted Rifles in May 1861 at Camp Colorado. He re-enlisted in Company E, Taylor's Battalion in May 1862 and remained with the unit until the end of the war. Barker died on July 13, 1917, at Mullin in Mills County.

William Thomas Benson was born on April 10, 1840 in Indiana. He had a brother, John Washington Benson, also in Davis's Company. The family arrived in Texas about 1854 and settled in Kerr County. He enrolled in

Davis' Company on March 4 1862. His sister, Susan, was married to Fritz Tegener, the major of the League's military battalion. Benson rejoined Davis' Company and served until the end of the war. He died on August 20, 1907 in Gillespie County.

McRae's Casualty List; *San Antonio Semi-Weekly News,* August 25, 1862; Stephen L. Erwin's Confederate Service Records, Record Group 109, National Archives, Washington, D. C.; Henry W. Barker's Confederate Service Records, Record Group 109, National Archives, Washington, D. C.; William Thomas Benson's Confederate Service Records, Record Group 109, National Archives, Washington, D. C.; William Thomas Benson's Confederate Pension Records, TSA, Austin, Texas; Nancy Isabellia Benson's Confederate Pension Records, TSA, Austin, Texas; Benson Family Bible Data Provided by John L. Mosley, Duncan Arizona; Sansom *Battle of Nueces,* p 11 and Siemering's German's During Civil War, June 5, 1923.

26. John George Kallenberg was born on September 6, 1842 in Duchy of Saxe-Weimar. His father, Johannes, was likely a Forty-Eighter. They arrived in Texas about 1853. John George Kallenberg was killed at the battle site.

Heinrich Markwordt was born about 1830 in Prussia. He and his father were likely Forty-Eighters. They arrived in Texas about 1853. Heinrich Markwordt was killed at the battle site. His widow married Carl Feller on August 19, 1867 in Gillespie County.

Christian Schaefer was born about 1844 in Nassau. Date arrived in Texas not known. He was killed at the battle site.

Heinrich Weyershausen was born about 1830 in Nassau. Date arrived in Texas not known. Heinrich was killed at the battle site.

Michael Weyrick was born about 1839 in Prussia. He arrived in Gillespie County in 1853 and settled in the Live Oak Community. He was killed at battle site.

Eighth U. S. Census, 1860 Gillespie County Texas Census, pp 16, 19, 23, 28 and 35; Kallenberg family data provided author by Kenneth O. Crenwelge, a Kallenberg descendants, Fredericksburg, Texas; Oral Interview by the author with Kenneth Crenwelge, May 7, 1997, Fredericksburg, Texas; Oral Interview with Clifton Stock October 1997; Telephone Interview with Aleta Addicks Stahl, October, 1997; Gold *Church Record Book of the Vereins-Kirche,* p 74 and Gillespie Historical Society *God's Hills,* Volume I, p 59.

27. Siemering's *Germans During Civil War,* June 8, 1923; Schultz's *Was a Survivor of Nueces Battle,* p 26 and Cramer's Letter. An analysis establishes

that 12 was the number of non-wounded that left the battle site after the third Confederate assault.

28. Kuechler's letter and Raley *Blackest Crime in Texas Warfare*.

29. Frost Allen was born on March 20, 1831 in Nacogdoches County Texas. On May 7, 1862, he joined Duff's Company at San Antonio. Allen never fully recovered from his wounds and died on January 28, 1880 in Nueces County.

John Frank Robinson was born about 1838 in Sabine County, Texas. The family arrived in Uvalde County in 1853. Indians killed his father, Henry, in 1861 in Uvalde County. John Frank enrolled in Duff's Company on May 4, 1862, at San Antonio. He recovered from his wound. On October 16, 1862, he was elected second lieutenant of Company B, Duff's 14th Battalion Texas Cavalry. He was promoted to captain on August 8, 1863, and remained with the unit until the end of the war. He was later the Uvalde County Judge. John F. Robinson died on February 20, 1917, in Uvalde County.

William E. Poe was born about 1836 in Missouri. The family arrived in Texas about 1860. Poe enrolled in 1st Regiment Texas Mounted Rifles in May 1861. He re-enlisted in Combs' Company E, Taylor's Battalion on July 6, 1862, at Fort Mason. He was buried at the battle site. At some point his family removed his body and reburied it at Johnson Station Cemetery in Tarrant County.

Schlickum's Letter; Cramer's Letter; McRae's Casualty List; *San Antonio Semi-Weekly News,* August 25, 1865; Frost Allen's Confederate Service Records, Record Group 109, National Archives, Washington, D. C.; Seventh U. S. Census, 1850 Nacogdoches Census, p 890; Eighth U. S. Census, 1860 Uvalde County Texas Census, pp 1b and 3; *Old Timers: Frontier Days in Uvalde Section of South West Texas* by Florence Fenley (State House Press, San Antonio, Texas), 1991, pp 65-70; John Frank Robinson's Confederate Service Records; John Frank Robinson's Confederate Pension Records, TSA, Austin, Texas; Mrs. M. P. Robinson's Confederate Pension Records, TSA, Austin, Texas; William E. Poe's Confederate Service Records, Record Group 109, National Archives, Washington, D.C.; Seventh U. S. Census, 1850 Jackson County Missouri Census, 256b; *Neu Braunfelser Zeitung* August 25, 1862; and *The Johan Frederick Gottlieb Striegler Family History: A Detailed Genealogy of the Descendants of Pioneer Citizens of Gillespie County, Texas*, Compiled, Edited and Published Under the direction of Selma Weyrich Striegler, Business Managers, January 1952, pp 77-78.

30. Schlickum's Letter; Kuechler's letter and Raley *Blackest Crime in Texas Warfare*.

Chapter 10 – The Battle

31. Williams *Border Ruffians*, p 248; Kuechler's letter; Schlickum's Letter; Siemering's *Germans During Civil War*, June 5, 1923; Raley *Blackest Crime in Texas Warfare* and Schultz *Was a Survivor of Nueces Battle*, p 30.

32. Schlickum's Letter.

33. McRae's Report, p 615; Williams Border Ruffians, pp 247-248 and Siemering's German During Civil War, June 8, 1923.

34. Sansom *Nueces Battle*, pp 9-10.

35. Jacob Kusenberger was born about 1834 in Nassau. He arrived in Gillespie County in the mid-1850s. He was a member of the Luckenbach Bushwhackers. Kusenberger fled on to Mexico after the battle and made his way to New Orleans where on October 27, 1862, he enlisted in Company A, First Regiment [Union] Texas Cavalry. He was mustered out of service on October 31, 1865, at San Antonio. Jacob Kusenberger died on December 18, 1898 in Gillespie County.

Sansom gives these two initials as A. and F. Graff. It is most likely they were William and Carl Graff. August Hoffmann identifies a Will Graff as a survivor. It is known the man named Graff who went to Mexico with Henry Schwethelm was Carl Graff. However, there was an August Graff listed on Kuechler's February 1862 muster roll, so an August Graff may have been in the Unionist group.

Sansom *Nueces Battle*, pp 10-11; Eighth U. S. Census, 1860 Gillespie County Texas Census, p 28; Jacob Kusenberger's Union Service Records, National Archives, Washington, D.C.; Jacob Kusenberger's Union Pension Records, National Archives, Washington, D. C.; *The Kusenberger Family,* by John and Dorothea Cotter (Published by John and Dorothea Cotter), 1991, copy provided author by Gregory J. Krauter, Comfort, Texas and Jean Kusenberger Kalina, Livingston, Texas; Schultz *Was a Survivor of Nueces Battle*, pp 26-29 Luckenbach Bushwhackers and Kuechler's MR.

36. Oral Interview with Leo Itz, Fredericksburg, Texas, August 21, 1996 and *Comfort News*, August 13, 1909.

37. Sansom *Nueces Battle*, p 10 and Schwethelm Letter.

38. Oral Interview with Werner Klier, Fredericksburg, April 1995; Striegler, Striegler Family History, pp 76-78; Hoffmann's Letter; Cramer's Letter and Kuechler's Letter.

39. Williams Border Ruffians, p 248; McRae's Report, p 615; Siemering's Germans During Civil War, June 8, 1923, and Schlickum's Letter.

476

40. McRae's Report, p 615; Schultz Was *a Survivor of Nueces Battle*, p 27; Williams *Border Ruffians*, p 248 and Barr *CMC*, Volume LXXIII, (October 1969), p 252.

41. McRae's Report, p 615; McRae's Casualty List; San *Antonio Semi-Weekly News,* August 25, 1862; Biggers *German Pioneers*, pp 59-60 and Curtis *History of Gillespie County*, p 59.

42. Sansom *Nueces Battle*, p 13.

43. McRae's Report, p 615; McRae's Casualty List; San *Antonio Semi-Weekly News,* August 25, 1862; Fehrenbach, *Lone Star*, p 364; Biggers *German Pioneers*, p 59; Banta *27 Years*, p 186; Benjamin *Germans In Texas*, p 110; Sansom *Battle of Nueces*, 3; *Freie Presse fuer Texas* September 1, 1865 and Letter, Leonard Pierce, U. S. Consul, Matamoros, Mexico to Hon. W. H. Seward, Secretary of State, Washington D.C., dated September 22, 1862, Dispatches from United States Consuls in Matamoros, 1826-1906, Volumes 7-9, January 1, 1858 – December 28, 1869.

44. Fehrenbach Lone *Star*, p 364; Fort *Inge: Sharps, Spurs, and Sabers On the Texas Frontier 1849 – 1869* by Thomas Tyree Smith, (Eakin Press, Austin, Texas), 1998, p 54 and Crosby *Wrong Side of the River*, p 54.

CHAPTER 11 – The Massacre

The Confederates treated the wounded as best as they could, "for we had no surgeon," explains Williams. Some of the troopers from Duff's Company "did what we could to ease the sufferings of the wounded Germans. We bound up their wounds, and gave them water, and laid them as comfortably as we could in the shade," said one of Duff's men. "How grateful they were!" R. H. Williams walked down to the edge of the river to see if any wounded insurgents were hiding there and needed assistance. He found none, but there were wounded insurgents close by. William Vater was hiding across from Williams along the riverbank. Two others, Karl Itz and the unknown insurgent, were hiding just to his north, also across the river.[1]

Williams found no one that needed assistance; he found plenty of scenic beauty close by the scene of so much bloodshed and human suffering. The natural beauty included a "cool spring bursting out below a great tree-shaded rock." Williams sat down to rest a few moments. It was a Sunday morning and he allowed his thoughts to turn to other times and places, to a far-away country church, where at that moment a simple service was being held, and those so dear to him were worshipping.[2] But Williams and the others had little time for rest. There was work to do; attend to the wounded, insurgents' horses and gather up equipment. Some horses had been badly wounded by stray shots during the battle. Williams stayed busy until about four o'clock when he went to where the wounded insurgents had been left. He was surprised to find them gone. He asked about them and was told they had been moved to a shaded area a short distance away. He went over to one of the wounded Confederates and gave him some cool water from the river. He heard shots. Williams at first though the Confederates were burying their dead with military honors, but realized this was not the case. He next thought some type of attack might be taking place. He seized his rifle and ran in the direction of the firing.[3] Williams met a fellow soldier coming from the direction of the firing. He said to Williams, "You needn't be in a hurry, it's all done; they shot the poor devils and finished them off." Williams replied, "It can't possibly be they have murdered the prisoners in cold blood!" The answer came, "Oh, yes; they're dead, sure enough—and a good job, too!"[4]

"It seems they were asked if they wouldn't like to be moved a little way off into better shade. The poor creatures willingly agreed, thanking their murderers for their kindness. They were carried way, but it was the shade and shadow of death, for a party of cowardly wretches went over and shot them in cold blood," Williams later recalled. He continues, "This was Mr. [Lilly's] work—the remorseless, treacherous villain!" Williams vowed to get even with Lilly if he ever got the chance. "Meanwhile, I denounced the bloody deed in as strong language as I could use, telling the perpetrator, to his face, what he was, and what every decent, honourable man would think of him as long as he lived." Williams watched Lilly, "handled his six-shooter, and looked as though he would like to use it on me; but the coward was afraid to shoot at live men, as I told him." Williams' comrades from Duff's Company backed him up and Lilly backed down. Despite subsequent claims that it was Lilly and Duff's men who executed the wounded insurgents, this statement by Williams that the other men from Duff's Company supported him and opposed Lilly in the showdown clearly establishes it was not Duff's men who killed the wounded.[5]

Why the wounded were executed is at the heart of the so-called "Nueces Massacre." Most writers have explained it using terms such as, "Duff had made it known he wanted no prisoners taken" or that McRae and Lilly were under some type of 'secret order' not to take any prisoners. An examination of the facts disproves both claims. If there had been orders to take no prisoners, then the wounded would have been immediately executed after the battle, not some seven or eight hours later. The reason had to be a deliberate decision made at the battle site. Of the five officers present, Lilly was the junior. Besides McRae, there were two other first lieutenants; Bigham and Homsley. The fifth officer, Harbour, was also senior to Lilly. Others have claimed that after McRae was wounded, Lilly took command, but this is not in accordance with military law and custom. There is no way the junior officer, on his own initiative would take on such a weighty responsibility. It is a military tradition that the junior office be given the most unpleasant duty.

The most likely reason the decision was made to kill the wounded was that McRae's force was overtaxed with wounded and captured equipment and did not have the necessary means to carry another seven seriously wounded men to Fort Clark. An examination of the numbers clarifies McRae's dilemma.

McRae's force originally numbered, at the most, 96 men. He had two killed and 19 wounded, leaving him with 75 effective men. Using Williams' data and cross-checking it with the service records of each wounded man, McRae's casualty list, and the list of causalities in the *San Antonio Semi-Weekly News*, it can be determined that at least eight Confederates had to be carried: Albert J. Elder, Robert G. Elder, John Frank Robinson, Charles Bergmann, Benjamin Rosser, John K. Morris, Dow Yarboro, and Emanuel Martin. McRae made the decision to use hand litters instead of travois or horse litters. Williams and many current day writers are critical of this decision. But when one examines the extremely rough terrain, much of which is covered by large boulders and cliffs and even today, is almost impossible to travel even with a four-wheel-drive vehicle, such a decision is rational. It took four men to carry each litter; so at least 32 men to carry the eight wounded. This reduced McRae's effective force to 42. If he added another seven litters to transport the wounded insurgents, that would have taken another 28 men to carry them. This would have further reduced his effective force to only 14 men; 14 men to relieve 60 men, carrying litters. The Confederates had 96 of their horses, plus 83 captured horses or mules for a total of 179 animals. This brings up another point; how were 14 men going to lead 178 animals, many loaded down with captured equipment? McRae still had the surviving insurgents to worry about. Even if he believed that 32 or 35 had been killed, that still would have meant there were about 30 insurgents hiding nearby. There was also the danger of Indians, as demonstrated by their fires burning the countryside and John Sansom saw Indians just south of the battle site. The Indians might take advantage of the situation.

What seems most likely is that McRae had a council of his officers to decide what to do. The decision was made to execute the wounded insurgents; many of whom the officers believed were going to die anyway. The junior officer, Lilly, was given the task. Schwethelm tells us that Lieutenant Lilly "called on volunteers to kill them and he got plenty of them."[6]

No matter what the circumstances which McRae faced, executing the wounded insurgents was, by today's terms a 'war crime.' Just who were the 'other men' who assisted in the murder of the wounded insurgents? The identity of all the executitoners is not known; Williams clearly names Lilly as one. Twenty-two-year-old William G'Laspie Wharton, of Davis' Company, was a second whose identity can be determined. In an 1895 letter to John Sansom, James M. Hunter, then the Edwards County judge, says one of the men was at the time living in Edwards County. Guido Ransleben did not identify the man when he reprinted the letter in his book, which contains a copy of Hunter's letter, but his identity can be determined from other sources. The Hunter letter says, "There is living in the county, today, a well-to-do ranchman whose head, likely yours and mine, is hoary with age, and who on that occasion was one of the Duff and Davis men that followed your party, and made the night attack, which brought on the battle you so well described. After the battle was over and the Union party vanquished he [name missing] was one of the detailed to murder the helpless wounded Unionist who lay upon the battleground. This may be the **reason why he is now allowed to carry a six-shooter**; [emphasis added] he no doubt fears that some relative of the murdered men might attempt to take his consecrated life." Historian Louis Leinweber of Harper, Texas recalls his grandfather telling the story of a man who lived in Kerr County as one of the men who killed the wounded insurgents. The Leinweber account also say **the man always carried a gun for protection** [emphasis added]. The Edwards/Kerr County ranchman identity is revealed in the book *Charles Schreiner: General Merchandise*. It says, "that upon getting up in the morning one of the first things **Bill Wharton** [emphasis added] did was to buckle his pistol and he would be more likely to be caught without his pants as his six-shooter. It was because that **he refused not to carry his pistol** [emphasis added] that the Kerr sheriff, hit upon the ingenious idea of deputizing Uncle Bill, and hence make his pistol packing legal." The book goes on, "The numerous German settlers dislike him, because, rumor had it, he was with Duff's command when it overtook the German Union sympathizers who were leaving Texas during the Civil War, and killed most of them on Nueces as they made their way into Mexico." The same sources credits Uncle Bill with "Having volunteered to finish off the wounded, which was done, but as he was never one to talk, much less brag, about his exploits this report goes unconfirmed."[7]

A third man is identified by Henry Schwethelm in a May 16th 1913 letter to his grandson, Otto. In it Schwethelm says, "After the fight Lieut Lilly had all of our wounded man [sic] killed. He called on volunteers to kill them and he got plenty of them. One of volunteers is living here in Kerr Co now his name is Alonzo Rees. He lives at Center Point. He is now a great churchman. I hope he will go to Hell."[8]

A fourth executor was likely twenty-six-year-old Prussian native Oscar Splittgerber. He was named in a July 1997 off the record interview. This account says after Dennis Kingston was killed, his widow swore at least ten Germans would pay with their lives. About October 1860, she married Oscar Splittgerber who worked for the Butterfield Stage Line and was a tanner for the Confederacy in Menard and Mason counties. Splittgerber joined the League using the alias Bargerum. According to this account, the Confederates captured him and, in order to save his life, he informed on the League's departure. He did so, not only to save his life, but also to fulfill his wife's oath. Another account says Oscar Splittgerber was "The Charles Bergman who, on order of a Confederate officer at the Battle of Nueces, executed the eleven Unionists" and Oscar was murdered in 1889 for his actions during the Civil War. An Oscar Splittgerber family story says after the "Civil War was over, the Splittgerbers and others were harassed by renegade northern soldiers because of their connection with the Confederacy" and the family fled to Mexico returning later. Five years later they returned to Texas and settled in west Menard County and later in Reeves County in West Texas.

It is a fact Oscar Splittgerber was killed on November 14, 1889, under mysterious circumstances. However, it is not believed Splittgerber used the alias Charles Bergmann and under that alias helped kill the wounded insurgents. It is very possible Splittgerber joined the League under an alias, but not that of Bergmann. The Charles Bergmann that was a guide for the pursuit force was hit three times during the battle; in both things and in an arm. With such wounds, it is not likely he was physically able to take part in the executions. One other incident supports the theory that Oscar was the *F. Bauman* that Alexander Brinkmann and his son-in-law, Guido Ransleben, say assisted the Confederates in murdering the wounded insurgents. There are two accounts of how the 'spy and murdered' died. One says he went to Mexico and killed by a "Seminole Indian Negro ... and thrown into the Rio Grande." A deathbed statement by Frederick Schultz[e], a Comfort insurgent, say much later the location of the 'spy' was learned. The Comfort insurgents drew straws to see who would go and kill him. Schultz[e] drew the short straw. He went to where the 'spy' was living and killed him. It is not known what became of Bergmann in Mexico. But Splittgerber returned to Texas, so he did not die at the hands of a "Seminole Indian Negro ... and thrown into the Rio Grande." Splittgerber's death occurred in 1889 in Reeves County. In 1888, Jacob Kuechler was a survivor of the railroad passing through West Texas. As such, he could have easily learned that Splittgerber was living there and passed the information to Comfort insurgents. Oscar Splittgerber's death under "mysterious circumstances" [one account says, "with a bullet hole in the back of his neck"] on November 14, 1889, in Reeves County closely matches Frederick Schultz[e]'s deathbed confession.

Williams may have been right on target in his judgment about why Lilly was not opposed to supervising the executions. Williams says Lilly was a "whitewashed Yankee," who until after secession, was a strong Union man. When the Confederates seemed likely to win the war he became the "hottest of hot Secessionists. The latter's [Lilly] chief motives, I believe was to prove his zeal and devotion to the Southern cause and by these base murders makes himself popular with the authorities in San Antonio." Two San Antonio newspaper articles seem to support Williams' claim that Lilly was a reluctant Confederate supporter. The *San Antonio Alamo Express*, a pro-Union newspaper before it was destroyed, referred to Lilly as "our enterprising energetic fellow-citizen, E. F. Lilly." Another says, "E. F. Lilly—our young Friend, who advocacy of the Union has heretofore been both decided and able, we have been gratified to learn, has within these few days passed his examination taken the oath to support the laws and Government of the Confederacy States and has been admitted to the Bar where we predict his success is sure."[10]

Williams was a private and not aware of the decision-making process to execute the wounded insurgents. By his own account, he was busy and not in the area where the officers were. In his own way, he adds support to the theory.

He excuses McRae by saying, "In justice of [Colin McRae], who was a brave and kindly man, I should mention that he was severely wounded in the fight, and had no knowledge of the crime." McRae was not that seriously wounded. He was still able to retain command, as established by his report of August 18th. Even if he was "seriously wounded," there were two other first lieutenants who would have assumed command.

Williams is surprisingly quiet about the other two first lieutenants, Bigham and Holmsley; not even mentioning them in connection with the executions. He also seems to remove any blame from Harbour whom he describes as "a rough but good sort of fellow and a 'number one' Indian fighter."[11] There are numerous other accounts of the executions, but they are based on what someone told someone who told someone else. One that is typical is an oral account told to U. S. Brigadier General David S. Stanley, while he was the commander of the Department of Texas in the 1880s. It relates how after the Confederate assaults, the insurgents surrendered. They were disarmed and Duff, "stood them up in one rank and shot down seventy of them — killed the wounded and marched back to the settlement, a hero." Stanley later included this account in his memoirs. General Stanley, like so many others, just accepted the insurgents' oral accounts without question. This resulted in him believing, "there was a failure of justice that this man Duff did not share in the retribution of Wirz." Captain Henry Wirz commanded the Confederate prisoner of war Camp Sumter at Andersonville, Georgia. He was tried and convicted of war crimes and executed on November 10, 1865.[12]

McRae explained the absence of any prisoners by saying, "They offered the most determined resistance and fought with desperation, asking no quarter whatever; hence I have no prisoners to report." However, there were reports of captured prisoners. When the detachment from Davis' Company returned to Camp Davis they explained why they had no prisoners. Sergeant D. P. Hopkins' diary entry shows how truthful he believed their story to be. "A number of prisoners were taken; but all got away, so the boys said; and I know they would not lie about a little thing like that." Both the San Antonio and Dallas newspapers reported that the insurgents had, "33 killed on the ground and many wounded of whom 3 are known to have since died." This seems to acknowledge that at lease three of the wounded were killed.[13]

The news regarding the killing of the wounded insurgents ultimately became general knowledge, especially among other Confederate units. Private Rudolf Coreth, a twenty-four-year-old Austrian native and member of Company F, 32nd Regiment Texas Cavalry, wrote his parents on August 26, 1862, telling them what he had heard about the battle, "There are reports almost daily about the conflict at Fort Clark [sic]." He told his parents, on August 24th a man from Carolan's Company arrived with information that the insurgents had suffered considerably more than the Confederate troops because of the effectiveness of their breechloaders. "I cannot give that report any credence though, because the man said that, after the fight was over, the soldiers dragged the wounded away from the camping ground and had them shot dead one by one, and that seems very improbable to me, really." Coreth added a hopeful sentence "Maybe it is put out by the Unionists in San Antonio."[14]

During his testimony at the Confederate Military Commission in San Antonio, Private W. J. Edwards, a member of the Confederate force, was asked about the battle and any prisoners and did the Confederates ask the insurgents to surrender? Edwards replied, "We did not [because], we supposed they would not surrender." He stated no orders were issued to 'refuse quarters'. Edwards simply said, "We killed 32 or 33 of them and took no prisoners – some escaped of them, some wounded as I supposed"[15]

Many claim after McRae was wound, Lilly, or Luck as they call him, took command and ordered nine wounded insurgents killed. Some have McRae riding down another seven or eight as they hid nearby. The most junior officer, which Lilly was, would not have taken command after McRae was wounded: one of the other first lieutenants would have. Neither did McRae, or any Confederates, ride down any insurgents.[16]

But it is not just pro-insurgent writers who fail to explain the facts. Some pro-Confederate writers try to explain the killed of wounded insurgents by saying, "Probably thinking that he was up against large armed force that was on its way to join in with the Federal First Cavalry under Davis, Colonel Duff gave no quarter to the enemy, and before the engagement was over the band of Germans was practically wiped out. Another excuse for the ruthless slaughter might be found in the fact that it was fought in darkness and Colonel Duff might have mistaken the Germans for a band of desperados. At this time this part of Texas was a hotbed of deserters, jayhawkers, and bad men of every type sought refuge in the fastness of the hills. And these banditti caused as much terror suffering in Fredericksburg and the other frontier towns as did the Comanches."[17]

There are literally hundreds of stories telling about the cruelty of the Confederates regarding the execution of the wounded. None have any basis in fact. One is by Henry Schwethelm. He says that not only were the wounded prisoners murdered, but that medical personal from Fort Clark wanted to treat the wounded insurgents but, "Duff's men said, 'Never mind, we will attend to them, and they thereupon went to the battleground and shot every wounded man in the head.'" This allegation is absurd. According to both Williams and McRae, the medical personnel did not arrive until the night of August 11th, well over 24 hours after the executions were carried out.[18] Some of the best examples of accounts of the killing of the wounded that are told without any basic for facts are in the writings of Lonn and Oaks. Lonn in *Foreigners in the Confederate* says, "The wounded, left in the hands of the attackers, were murdered in a barbarous manner a few hours later by the victors. Some dying men were dragged to trees and hung; one was thrown or fell on the camp fire when shot down, through just before he expired he was pulled form the fire by Williams."[19]

Another account is Stephen Oates, the editor of *Rip Ford's Texas* in which he says, "The soldiers charged in with pistols blazing. Again, Sansom sprinted through the gunfire, made it to the woods, and kept running until he collapsed on the banks of the Nueces some distance away. For a while he heard the swearing and muffled shouts of men, then tensed at the rapid crack of cavalry carbines. They were shooting the prisoners." As Colonel Potter of *M.A.S.H.* was fond of saying, "Horse hockey." Oaks' account is pure fiction.[20]

An article published in 1870 uses the testimony of Daniel Cleveland, the first Union mayor of San Antonio after the war; an emotional version and one used by many writers. It says, "The wounded, already in the hands of the insurgents [Confederates], were murdered in the most barbarous manner by bullets, bayonets, bowie knives, and hanging. Some who were actually dying, were dragged to trees and hung by the fiends. The commander of the butchers, Lieutenant Lilly, afterward boasted he killed several of wounded with his own hands, 'emptying two revolvers' in shooting them!" Again, to quote Colonel Potter; "Horse hockey!"[21]

Others claim "the Battle of Nueces was swift and decisive" and that McRae rode out across the terrain in search of the insurgents who had escaped in the fighting. He found eight hiding nearby and killed them." Another says, "The Dutch prisoners were marched up the canyon after the fight, and shot down in cold blood." One has the wounded insurgents being brought to Duff where he "ordered all of the wounded to be killed. Nine Unionists who were captured and brought in were shot immediately by Duff's orders." Yet another says, "Approximately thirty of Tegener's men were killed in the fighting, and twenty were wounded, captured, and then executed on the spot with bullet in the back of the head."

One of the most absurd accounts of the killings is a 'hand-written journal' that has been published and supposedly tells the 'way it really was' and based on a "partispant (sic) and an eye witness and was one of Sansoms men and another was one of Donelson's men." It says, "Then again in Agst about the 10ᵗʰ 1862 thare was a lot of Germans that did not want to take sides against the government of their adopted country so they held a meeting in Frederisbourge and thirty-three men agreed to leave the Confedacy and cross the Reo Granda and waite until the war was over … they secretly made preparations for their departure, collected about sixteen good wagons, good ox teams, loaded their wagons with provisions – flower, bacon and other supplies – and started for Mexico without any road … Duff orderd out Capt Donaldson with his full company with orders to pursue the trail and if possible to overtake and kill the last one of them … Two or three days after the train left Frederiksbourg, Capt. John Sansom with 15 or 20 companions left the vicinity of Currys Creek settlement in Blanco County to overtake and attach themselves to the German train. They overtaken the train late in the evening near Noesses Can at which place they all went into camp for the night. [During the] night and next morning Sansom urged the German organize and elect a leader, ureging the necessity of a commander in case of an attack by Indians or other enemy. The Germans held a council in their own language for some half hour, they supposing that Sansom being an American saught the command and they did not like to elect a Germans who new very little of American tactics over Sansom and his men, so they told him thare was no danger as they had journied so far without trouble. Sansom told them that he anticipated danger at any time while in Texas and that he wound not travil with them unless they organize." They crossed over the Noeses River and took up the line of march up a long holler or creek, traveled on until night began to set in and all went into camp. The Germans close beside a dense thicket of live oak saplins some 300 yards long, correled their wagons in front, having the thicket in their rear, always guarding against an attack by Indians and Sansom filed off to the left some six or seven hundred yards on another prong of the creek. After supper they put out their fire and went to sleep. About two hours before dawn the next morning they war arounsed by the fireing of guns.

Donaldson and his men had succeeded in crawling through the thicket and wainted until that hour, then opened a heavy fire on the. The Germans arose, frightened and excited and having no leader ran to and from only to be shot down like so many wolves, some fought bravely, the contest lasted until daylight when the Germans, seeing themselves surrounded, outnumbered, and overpowered, serendered."[23]

While this account of the battle is ridiculous, the account of the massacre and immediately aftermath is absurd. It says, "But Donaldson had orders to not receive a serender or take any prisoners, so kept up their bloody work until the last man was killed that hey could find, but during the contest three of the Germans had made their escape and finely after rambling around and dodging Duff's men for two weeks, cam into Frederiksbourg starved to death. Donaldsons men rounded up all the horses, cattle and oxen picked out a good tem and hitched to large wagon that was loaded with four thousand pounds of flour, piled up the other wagons with their effect, piled all the dead bodies on the pile after rifleing them, and set fire to pile, then moved off with the one wagon and the stock and arrived safe in camp. Duff confiscated the property and sold what they did not want to use themselves. Sansom and his men heard the firing commence and hastely got their ponies, saddled up, mounted and rode off up the side of the mountain near a mile away a watched the issued. When all was still and they saw Donaldson and his men move out of sight, they rode down to the camp and one of the men said in giving me the detailes, 'I never say such a sight before in my life, and I never want to again. Fifteen wagons with their loaded effects, yokes, saddles, blankets and the boddies of thirty men lay in one burning smokeing smoldering heap.' They gazed for a few minutes on the sickning sight, then turned and rode away. In a few days, they crossed the Rio Granda and was Safe." This is but one of many accounts of the battle and elated events that is supposedly told based on eyewitness's stories.[24]

These accounts can easily be discredited, but, they are no worst then many of the accounts made into mainstream books and other historical publications. These and other such accounts demonstrate how important it is not to accept secondary accounts, even if eyewitnesses supposedly tell them, unless checked against actual eyewitness accounts and primarily source documents. Examples of how many writers and historians have fallen into this trap are the ones where there is an eyewitness telling his story, many time in an interview, and information is added, which either he or the interviewer have heard or read from other accounts, and the data is accepted as correct. This incorrect information is thus accepted as factual because it came from someone who was any eyewitness to part of the events, but not the total event.

Some examples where this has happened are in the accounts of Sansom, Hoffmann, Schwethelm, Kuechler, and R. H. Williams. All these men were eyewitnesses to part of the events, but not to <u>all</u> events. Yet, the parts of their accounts that they did not personally see or take part in are accepted as factual. For over a hundred and fifty years such myths have grown and retold so often they are now accepted as facts.

A related decision regarding the slain insurgents resulted in almost as much opprobrium down on the heads of the Confederates as the execution of the wounded: they left the dead unburied. Williams says the insurgent dead were, "carried to where the murdered prisoners lay, and there left for prey to the buzzards and coyotes." Those who denounced this decision did so in the strongest terms. Don Biggers says, "Another barbarous feature of the Nueces Massacre was the fact that not one of the bodies was buried by the conquering hero, Duff. The dead were left for varmints to tear and eat the flesh, the bones to bleach." Guido Ransleben says: "The dead Unionists were left unburied on the battlefield; their comrades dared not return to perform that humane duty – their foes were too inhuman to do so. They were left as food for the wolf and the buzzard upon the soil they had glorified with their blood. And their bones lay bleaching in the sun, wept over by the rains from heaven, kissed by the winds – Nature's perpetual homage. They lay like warriors who had fallen asleep on the tented field–not entombed, but awaiting the bugle call of Freedom's resurrection day."[25]

As early as 1866, the families and descendants claimed they could not go to the battle site and bury their dead. An 1866 *Harper's Weekly* article states, "The bodies of the Union men who had been killed were left on the battlefield, and so long as the rebel flag floated over Texas no parent or friend ventured to afford them a resting place."[26]

While almost all religions place great emphasis on a 'proper' burial for the dead, to the Freethinkers it was almost an obsession paranoia. To the Freethinker, immortality was a myth because "no one believed in it." The dead lived on in their children, which was their only immortality. But, the body had to have a 'proper' burial. A formal funeral was a necessity. It was a last opportunity to say goodbye, for someone to tell the story of the deceased's life, a eulogy, to that person. Not to 'properly' bury the dead, or allow family and friends to bury, them were the ultimate insult.[27]

The seven or so executed insurgents brought the total number of bodies in the area to nineteen. With the temperature reaching the high 90s and into the 100s, this many bodies became a major problem. The two dead Confederates were buried in a long trench; both were later exhumed by their families and reburied in cemeteries close to their homes. The bodies of those insurgents killed in the battle were placed in a large pile with those executed. They were checked for documents or anything else of value. On the after noon of August 10th, the Confederates moved south of the insurgent's camp about five to six hundred yards and established a camp. The next day, August 11th, the Confederates were busy making litters for their own wounded, packing arms, ammunition, and other captured equipment. They planned to leave for Fort Clark the next morning, or on August 12th. "That night the doctor arrived, and was promptly at work," but expressed the feeling that several of the cases were very serious and would not live to see Fort Clark. The doctor was forty-seven-year-old Vermont native Edwin M. Downs of Bandera County, whom Bandera Unionists had threatened to hang in mid-1861 because of his support for the Secessionists. Now he was an assistant surgeon, hired as a contract surgeon for both the Frontier Regiment and several Confederate units.[28]

On the morning of August 12th, the Confederates set out for Fort Clark. Lieutenant McRae left a detachment of about twenty troopers in the general area of the battle site to continue searching for insurgent survivors. An insurgent stated, "They send out daily patrols and searched the whole area even though they came close to the main Unionist group they never discovered them." This was not quite correct for about August 13th, these scouts captured one of the wounded insurgents and took him to Fort Clark.[29]

Of the approximately twenty men from Duff's Company in the pursuit force, sixteen were all that were fit for duty. Lilly detailed them as litter bearers. These included, beside Williams, thirty-year-old Alsace native, Louis Ogi, twenty-two-year-old Mississippi native, W. J. Edwards, thirty-one-year-old Alabama native H. P. Ferrill, thirty-year-old Englishman William C. McDonna, and likely twenty-one-year-old Mississippi native Montalcon Woodward. Williams claimed these were the ones who had stood by him when he challenged Lilly about executing the insurgents. The plan was for the remainder of the Confederates to carry the weapons and other gear of the litter bearers. The only thing the litter bearers carried beside the litters was water-canteens. The trip would be broken down into five-mile intervals; at the end of each interval, says Williams, new litter bearers would take over. But based on the actual numbers another set of 32 men was just not available. McRae's pursuit force originally numbered about 96 men. It had suffered two killed and 19 wounded, thus reducing the force to about 75 men. Thirty-two men were detailed as litter bearers, reducing the force to about 43 men. Another approximately 20 men were left to scout the area for survivors, further reducing the force to only about 23 men, not enough for a complete change in litter bearers.[30]

R. H. Williams, for over a hundred years, has been the sole source historians have used to tell 'the Confederate side' of the 'Nueces Massacres.' Yet, a close examination of his book reveals it is full of exaggeration and incorrect data. He claims Lilly punished those who challenged him about the executions, and instead of relieving the 32 litter bearers, Williams says Lilly "lost his way" and never relieved them. But as can be seen from the numbers, not all the litter bearers could have been relieved. Williams also makes no mention of the other three officers, Homsley, Bigham, and Harbour. It is hard to believe they would have just stood by and allowed one group of litter bearers to go the entire distance.

Williams also claims that just as they were leaving the battle site four of the pack animals carrying forty captured weapons broke and ran away, resulting in the weapons and animals being lost, yet McRae makes no of mention of such an incident. Instead, he says the weapons were turned in at Fort Clark and the animals sent to San Antonio. A member of Davis' Company supports McRae's statement saying they captured eighty head of horses. Another thing which makes McRae's report more credible than Williams, is that if McRae reported he had captured eighty-three horses he had better be able to account for all of the captured animals. If some of the animals got loose and got away, he would have said so.[31]

There is no doubt that the trip to Fort Clark was a difficult one. It was after sunset before the Confederates linked up with the wagons and ambulances from Fort Clark. Thirty-nine-year-old English merchant David H. Brown provided the wagons. It was after ten p. m. on August 12th before they arrived at Clark. The wounded were taken to the post hospital. The Confederates rested a few days. Harbour's detachment returned to Camp Davis, arriving on August 22nd. The remainder of the pursuit force returned to San Antonio. Trooper H. P. Ferrill of Duff's Company remained at Fort Clark attending to the wounded. Four of the wounded died: Albert J. Elder of Duff's Company on August 17th and his brother Robert G. Elder on September 17th. Trooper John K. Morris of Taylor's Company B, Texas Cavalry died on September 1st. Trooper Emanuel Martin was transferred to San Antonio where he died on October 18, 1862.[32]

The twenty-man scout McRae left behind at or near the battle site produced results. About August 14th they captured a wounded insurgent, Ferdinand Simon. Simon was taken to Fort Clark for interrogation. He provided a great deal of information to Lieutenant McRae. Among the intelligence which he provided was the true size of the insurgent force; sixty-nine men consisting of "63 Germans, one Mexican, and five Americans." Simon claimed the "Americans ran at the first fire." He provided the name of the insurgent commander, Fritz Tegener. The capture and good treatment of Simon helps establish that there had been no "take no prisoners" orders, as well as disproving the claim that all insurgents captured "subsequently were put to death." Simon was transported to San Antonio to stand trial for his part in the insurgency. McRae included Simon's information in his report that he completed on August 18th. Also in his report, he expressed his thanks for the services of Dr. Downs and David H. Brown and closed with the comment, "My officers and men all behaved with the greatest coolness and gallantry".[33]

The number of wounded who died at Fort Clark was been another subject of great discussion. Williams claims five of the wounded died including his "poor comrade" Mont Woodward on August 13th. Williams goes into great detail describing Woodward during the trip to Fort Clark and how he suffered before dying. The only problem with this heart-rending story is Woodward was not among the wounded, much less dying at Fort Clark. He was killed on October 13, 1904, during a robbery in Arizona – nother example of Williams' incorrect data.[34]

Many of Duff's, Donelson's, and Taylor's troopers were given leave afterwards. Lieutenant James M. Holmsley took a sixty-day leave beginning August 28th. Lieutenant James S. Bigham also took a sixty-day leave beginning August 24th.[35] Four Confederates received medical discharges as a result of their wounds: Stephen L. Erwin of Company A, Taylor's 8th Battalion Texas Cavalry was discharged on October 31, 1862. Benjamin Rosser of Company K, 2nd Regiment Texas Mounted Rifles received his discharge on May 15, 1863. John Hill, also of Company K, 2nd Regiment Texas Mounted Rifles was discharged on April 23, 1863. Albert L. Wilborn of Davis' Company, the Frontier Regiment received his discharge on May 13, 1863.[36]

The story of the battle was quickly reported in newspapers around the state. For example, the *San Antonio Herald* of August 20, 1862, reported; By the El Paso mail this morning we learn that Capt. Duff's Company, has had a fight with the celebrated Mountain Guard, of that notorious scoundrel and Abolitionist, Jack Hamilton. The fight occurred a short distance above Fort Clark, and resulted in the entire defeat of the renegades; thirty-three were left dead on the field. One of Capt. Duff's men, Frank Robinson of Uvalde, was killed, and five wounded. There were about 100 men on our side, detachments from Duff's and Donelson's companies and Taylor's Battalion; all under the command of Lieut. McRae.[37]

Some accounts claim the battle was never reported and it was only after the war did news reach the population. An example is the book *Five Years in Texas: What you Did Not Hear During The War*. It says, "To the northwest of Austin, a hundred miles away, we heard the report that a serious fight had occurred between the State Militia and two or three hundred Indians, who had come down from the mountains to steal horses and cattle. That the Indians fought in ambush, and made many of the whites bite the dust. But when the lying spirit of the war was over, the truth came out that these Indians were a colony of German refugees fleeing from Texas persecution to Mexico. But few of them every reached there."[38]

The *San Antonio Semi-Weekly News* of August 25, 1862, also reported the battle. It corrected the misinformation J. Frank Robinson had been killed. It also provided many of the names of the Confederate casualties: We had the satisfaction of a meeting with one of Capt. Duff's Company who had been in pursuit of the enemy in the bush north of Fort Clark, and from him we obtained a list of the casualties of the battle in which Mr. Frank Robinson, of Uvalde was killed, but was not. There were two killed- W.E.

Poe and L. Stringfield—and sixteen wounded—W. Williams, Welch, J. Singleton, W. H. Barker, John Morris, S. Irvin, Charles Bergmann, Wm. H. Barker, R. Elder, Albert Elder, Benj. Rossey, (Rosser) John Hill, Dew Yarborough, Edmonson, and Lt. McRae. — This engagement has, we trust, broken up the foul nest in the Mountains, which has been [illegible] so many vile and cowardly enemies to our cause. More than half of them were left dead on the filed and the remainder-------------(?).[39]

McRae's casualty list shows a total of 19 wounded. He provides only the names of 12. He lists the other seven as names "not known". He does provide the units. Six were from the 8th Battalion and one from Davis's Company. This newspaper article provides the names of those seven not identified on McRae's list. The service records of the seven named in this article show they were on detached duty to Fredericksburg or a member of Davis' Company already there.

Three days later, on August 28th, the *San Antonio Herald* ran a follow-up story on the "The Battle with the Traitors": Since our last issue we have received more correct intelligence relative to the fight between our men and the traitors in the vicinity of Ft. Clark. Our losses were two killed and 18 wounded. The traitors lost 33 killed on the ground and many wounded of whom 3 are known to have since died. Their muster roll was founded, containing 69 names. There were five Americans [sic], conscious of their guilt fled at the first fire. The entire outfit of traitors, consisting of their horses, ammunition, guns, and provisions was captured by our men.[40]

McRae's report says nothing about capturing a muster roll. However, W. J. Edwards, a member of the Confederate force, testified that a muster roll was captured. The number 69 is the same number reported by McRae in his report and is the correct number. McRae says he obtained this information from "one of the party whom we fought, captured some four or five days subsequent to the fight." This prisoner was Ferdinand Simon.[41]

From the Southern side, there was nothing but official praise for the Confederate troopers and officers. McRae in his official report said, "My officers and men … seeming to vie with each other in deeds of daring chivalry. It would be invidious to attempt to draw any distinctions when all did their part most nobly and gloriously." General Bee, in his final report of October 21, 1862, praised Lieutenant McRae and the Confederate force with these words: "Lieutenant McRae and his command behaved with admirable coolness and bravery, and did their work most effectually."[42]

Endnotes – The Massacre

1. Williams *Border Ruffians*, p 248. It is not known why Captain Donelson did not send his surgeon with the pursuit force. Both the Compiled Service Records of the 2nd Regiment Texas Mounted Rifles and the Camp Pedernales Post Returns shown Surgeon George H. Doran present.

2. Ibid.

3. Williams *Border Ruffians*, pp 248-249. McRae's Report says nothing of any horses having to be killed. He says they captured 83 horses.

4. Williams *Border Ruffians*, p 249.

5. Williams *Border Ruffians*, pp 249-250.

6. Schwethelm's Letter.

7. Letter, James M. Hunter, County Judge, Edwards County, Texas November 17, 1895 to Captain J. W. Sansom, located in Guido Ransleben's *A Hundred Years of Comfort in Texas,* p 120; Telephone Interview with Louis Leinweber by the author on January 29, 1998; *Charles Schreiner: General Merchandise 1869-1969* by J. Evetts Halcy, (Published by Charles Schreiner Company, Kerrville, Texas), 1969, pp 40-42; and Schwethelm's Letter.

8. Schwethelm's Letter.

9. Off The Record Interview, July 19, 1997; Mary Turner's oral presentation to the German-Texas Heritage Society at its September 5-7 Convention at Kerrville, Texas; Menard County Historical Society *Menard County*; *Pecos: A History of the Pioneer West, Volume One* by Alton Hughes Seagraves, Texas (Pioneer Book Publishers) 1978, p 350; Volume I; Turner *Julius Theodor Splittgerber, Volume Two,* pp 212 and 215; McRae's Casualty List; Ransleben *100 Years*, p 121; Letter, Catherine Carrigan, August 1, 1991 to Gregory Krauter, Comfort, Texas, copy kindly provided author by Gregory Krauter; Letter Jacob Kuechler, Toyah, Pecos County [Texas] October 26, 1888 to Dearest Marie. Several individuals have transcribed the date of this letter as 1862. An example is in the book *Hermann Lungkwitz: Romantic Landscapist on the Texas Frontier* by James Patrick McGuire (University of Texas Press, Austin, Texas) 1983, p 20. This is incorrect. Toyal and Pecos County where not created until 1881 and 1883. Copy of letter provided by Anne Stewart.

10. Williams *Border Ruffians*, p 250; *San Antonio Weekly Ledger and Herald*, April 20, 1861 and *Alamo Express* March 29, 1861.

11. Williams *Border Ruffians*, pp 249-250.

12. *Personal Memories of Major General D. S. Stanley* by David S. Stanley, (Harvard University Press, Cambridge), c1917, p 234 and Long *Civil War Day By Day*, p 695.

13. McRae's Report, p 615; Hopkins' Diary: *San Antonio Herald*, August 28, 1862; and *Dallas Herald*, September 6, 1862.

14. Rudolf Coreth was born May 7, 1838, at Schloss Kahlsperg, Austria. The family arrived in Texas on October 3, 1846, and settled near New Braunfels. In November 1861, he enrolled in Company F, 32nd Regiment Texas Cavalry and served until the end of the war. He died on October 17, 1901. *Lone Star and Double Eagle* by Minetta Altgelt Goyne (Texas Christian University Press, Fort Worth, Texas), 1982, p 66.

15. Barr CMC, Volume LXXIII (October 1969), p 2.

16. *Fort Clark: The Lonely Sentinel: On Texas's Western Frontier* by Caleb Pirtle III and Michael F. Cusack, (Eakin Press, Austin, Texas) 1985, p 50.

17. *A Tale of Men Who Knew Not Fear* by Gertrude Harris (Alamo Printing, Company, San Antonio, Texas), 1935,

18. Schultz *Was a Survivor of Nueces Battle*, p 27; Williams *Border Ruffians*, pp 249-251 and McRae's Report, p 615.

19. Lonn *Foreigners in the Confederacy*, p 430.

20. *RIP Ford's Texas by John Salmon Ford* by Stephen B. Oaks, (University of Texas Press, Austin, Texas), 1963, pp 339-340.

21. *Sufferings of Texan Loyalist* by Benson J. Lossing, LL.D. in the *Pictorial Field Book of the Civil War*, Volume II, P 537, Reprint of 1870 Original by The John Hopkins University Press, Baltimore, Maryland.

22. Pirtle and Cusack *Fort Clark: The Lonely Sentinel On Texas's Western Frontier*, p 50; *Nueces Headwater Country: A Regional History* by Allan A. Stovall (Naylor Company, San Antonio, Texas) 1959, pp 94-95; *Massacre by White Men* by Ruel McDaniel in Frontier Times, August-September 1969, Volume 43, p 52; *The Dogs of War Unleashed: The Devil Concealed in Men Unchained* by Joe Baulch in West Texas Historical Association Volume LXXIII, 1997, p 133.

23. McDonald *Now You Hear My Horn*, pp 171-175.

24. Ibid.

25. Williams *Border Ruffians*, p 251; Biggers *German Pioneers*, p 60 and Ransleben *100 Years*, p 99.

26. *Harper's Weekly*, January 20, 1866.

27. Flack *A Yankee in Texas Hill County*, p 30 and *Houston Chronicle*, January 1, 1995.

28. Sansom Battle *of Nueces*, p 10; Williams Border *Ruffians*, 251; Glenn Capt'n *John*, pp 14, 21; Montague's Letter; Hunter *100 Years in Bandera*, p 59; McRae's Casualty List; Littleton Stringfield's Confederate Service Records and DRT *Gravesites of Citizens of the Republic of Texas*, p 50.

29. Sansom Battle *of Nueces*; Raley Blackest *Crime in Texas Warfare*; Schultz Was *a Survivor of Nueces Battle*; Kuechler's letter.

30. Montcalon [Mont] Woodward was born about 1841 in Mississippi. The family arrived in Texas about 1848. He enrolled in Duff's Company on May 4, 1862 at San Antonio.

    Williams Border *Ruffians*, p 251; Kuechler's letter; Siemering's Germans During Civil War, June 8, 1923; McRae's Report, p 615; Hunter *Trail Drivers of Texas, p* 682; Albert J. Elder's Confederate Service Records; Robert G. Elder's Confederate Service Records; John Frank Robinson's Confederate Service Records; Benjamin Rosser's Confederate Service Records; Dow Yarboro's Confederate Service Records; Emanuel Martin's Confederate Service Records; H. P. Ferrill's Confederate Service Records, Record Group 109, National Archives, Washington, D. C.; McRae's Casualty List.

31. Williams *Border Ruffians*, p 253: McRae's Report, p 615 and Letter, J. W. Seal, Camp Davis, September 8, 1862, 'Texana', Volume IV, Number 1, Spring 1966, pp 54-55

32. David H. Brown was born about 1823 in Coventry, England. He arrived in the United States in the late 1840s or early 1850s. On October 3, 1850, he enlisted in the First U. S. Infantry Regiment and was discharged on October 2, 1865 at Fort Duncan, Texas. At the time of the battle, he was a merchant, living at Fort Clark. Brown enrolled in Duff's Company B, 14th Battalion Texas Cavalry on October 1, 1862 and served until the end of the war. He died in August 1866 in San Antonio.

    Williams *Border Ruffians*, pp 253-254; McRae's Report, p 615; Series of personal letters from Jesse Rodriquez, great grandson of David H. Brown to Donald Swanson, Curator Emeritus, Old Guardhouse Museum, Fort Clark Springs, Texas. Copies provided author by Donald Swanson; Albert J. Elder's Confederate Service Records; Robert G. Elder's Confederate Service Records; John K. Morris' Confederate Service Records; and Emanuel Martin's Confederate Service Records. This date of August 22nd as the date Davis' men return to Camp Davis is extremely important. It was on this date that seven Unionist were hanged on Spring Creek in Gillespie County. Hopkins' diary.

33. McRae's Report, p 615. McRae's report refers to the five Americans' leaving early. It says, "5 Americans (the latter running [at] the first fire)." In one of his accounts August Siemering also refers to the five Americans leaving early. It says, "During this time, [the period after the opening shots and the Confederate assaults] 5 Americans, the only ones which were in the group … fled the battle." Siemering's Germans During Civil War, June 8, 1923.

34. Mont Woodard's Confederate Service Records, Record Group 109, National Archives, Washington, D. C.; Hunter *Trail Drivers*, pp 680-682 and Helen B. Woodard's Confederate Pension Records, TSA, Austin, Texas.

35. James Homsley's Confederate Service Records and James S. Bigham's Confederate Service Records.

36. Stephen L. Erwin's Confederate Service Records; Benjamin Rosser's Confederate Service Records; John Hill's Confederate Service Records and Albert L. Wilborn's Confederate Service Records.

37. *San Antonio Herald,* August 20, 1862.

38. *Five Years in Texas; or What You Did Not Hear During the War, by* Thomas North, Elm Street Printing Co., Cincinnati, Ohio), 1871, pp 192-193. Let's quote Colonel Potter again.

39. *San Antonio Semi-Weekly News,* August 25, 1862. This article names two men named Barker. While their initials [W. H.] are the same, the article seems to take care to show they are two separate men. It does this by using "W. H." in one case and "Wm. H." in the second case. The Compiled Service Records show a Henry W. Barker [instead of Wm. H.] as a member of Company E, 8th Battalion on detached service to Fredericksburg, so this is a match. McRae's Official Casualty List shows a second man from Davis' Company also wounded. This means the other W. H. Barker must be from Davis' Company. However, no 'W. H. Barker' is shown on Davis' muster rolls. There are two possible men with similar names. One is a H. A. Baker, the second is Wm. T. Benson. Benson family stories state he was wounded while in the war, but does not say when or where. Based on this fact, it is likely Wm. T. Benson was the 2nd man from Davis' Company wounded.

40. *San Antonio Herald,* August 28, 1862. Also see *Dallas Herald,* September 6, 1862.

41. McRae's Report, p 615.

42. McRae Report, pp 615-616 and Bee's Instructions.

CHAPTER 12 – Insurgent Survivors

The killing had not stopped. Recall that three different insurgent groups left the battle site. First, the approximately twenty-three, "who left early" including those who left after the opening rounds were fired and before the Confederate assaults. Next; John W. Sansom and his five-man group. Third; the survivors of the several Confederate assaults. This last group consisted of three sub-groups: the Hoffmann [Gillespie County] group, the Henry Schwethelm group, and the Ernest Cramer/Jacob Kuechler [Kendall County] group. The subsequent movements of these three groups must be examined in detail before the story is complete.

Those Who Left Early

This group left the battle site after the opening shots were fired, [about three a. m.] and before the beginning of the Confederate assaults [about seven a. m.]. This 23-man body was not in a single group. They left by ones, twos, and small sub-groups. The members of this 23-man group likely included Conrad Bock, Wilhelm Boerner, Theodore Bruckisch, Joseph Elstncr, Ernst Felsing, Herman Flick, Henry Hermann, Valentine Hohman,; Fritz Tays, August Luckenbach, Adolph and Louis Ruebsamen, Heinrich Stieler, Jacob Gold, Peter Jacoby, Carl Moellering, Michael Tatsch, Jacob Usener, Ludwig Usener, Peter Gold, John Hoerner, and Sylvester Kleck. They first went due east until they came to the Nueces River, south along the river until they reached the San Antonio and El Paso Road. Here they turned east and passed through or near the towns of Uvalde, D'Hanis, Quihi, New Fountain, and Castroville. From the Castroville area, they turned north either toward Boerne or up the Medina River Valley toward Comfort and Kerrville. It was on this third leg that they met disaster. Nine of them were captured and executed.[1][Map Q, p 538]

This group of twenty-three probably broke into smaller groups as they made their way east. The two Usener brothers, twenty-eight-year-old Jacob and thirty-one-year-old Ludwig, both natives of Prussia, were in one of the parties "that head homeward [and] some split up onto smaller groups." Why they split up is unclear, but a Usener descendant recalls, "Stories were that differences in opinion [arose] as how to elude the Confederates." The Usener brothers were in a group of twelve. Of these twelve, only three survived; the Usener brothers and one other. The name of the third man who survived is not known. The other nine were among those, "the Confederates overtook [and] hung in the western part of the [Gillespie] county." One of those hanged was Conrad Bock, a twenty-eight-year-old native of Prussia and resident of Gillespie County. He was probably wounded in the first exchange of gunfire at three a. m. the morning of August 10th. He joined the Usener brothers and the twelve men described in the Usener account. Before he left, he gave his coat and his father's military identification papers to one of the insurgents going back with another group, twenty-seven-year-old Peter Jacoby from Gillespie County.[2] Those captured and executed beside Conrad Bock were twenty-year-old Fritz Tays, a native of Hanover living near New Braunfels; twenty-one-year-old Nassau native Herman Flick, a brother-in-law of Conrad Bock, thirty-two-year-old Wilhelm F. Boerner, whose brother was killed near the battle site, seventeen-year-old Heinrich Stieler, twenty-two-year-old Prussia native Theodore Bruckisch, twenty-nine-year-old August Luckenbach, twenty-two year-old Adolph Ruebsamen and his nineteen-year-old brother Louis Ruebsamen, both natives of Nassau and members of the Luckenbach Bushwhackers.[3]

The specific circumstances these executions are not known, only bits and pieces of what happened. Conrad Bock and Fritz Tays were captured by a detachment of Duff's Partisan Rangers led by Second Lieutenant Richard Taylor on August 23rd near Boerne. Taylor's detachment hanged them on August 24th about a mile and a half north of Boerne near Cibolo Creek. The death of Bock and Tays is briefly mentioned by August Siemering. "Two [Fredericksburg insurgents] were hanged near Boerne, one of them wounded." After the war, a Kendall County Reconstruction Grand Jury Indicted Duff and Taylor for these executions. Neither of them were located and therefore never tried in court. In 1879, the district attorney requested the cases remain open. He stated he had information that Taylor was dead and Duff had fled the country. He desired the cases to remain open if either were ever captured. They remain open today. Of all the 'murders' Duff is accused of committing these are the only indictments.[4]

According to Luckenbach family stories, August was captured about eight miles north of San Antonio on the San

Antonio and Fredericksburg Road. He was hanged from a "huge live oak tree." His body was found and recognized by Fredericksburg friends. August Siemering says two other Fredericksburg insurgents were hanged. One was captured "at the Medio and a fourth one close [by]." The insurgents "captured close by" was August Luckenbach. The only other Fredericksburg man later captured and executed was Herman Flick.[5]

Siemering provides some information on the deaths of the Ruebsamen brothers: "Two brothers who were wounded when captured and were [shot without pity]." Siemering does not say when or where the Ruebsamen brothers were captured and executed.[6]

There is much more information on the execution of two other insurgents: Theodore Bruckisch and Heinrich Stieler. Again, Siemering provides some of the information: "Two of the [insurgents] had come within reach of their homes and were found thirsty and tired and were shot without pity." There are several family accounts of the death of Bruckisch and Stieler. All agree on several points, while on others the accounts are very different. They all agree that Bruckisch and Stieler were captured at different locations, but about the same time. Stieler was the first captured. He had just about convinced the Confederates he was on his way to join the Confederate Army and had lost his way. About that time Bruckisch was brought in. Under questioning Bruckisch admitted he and Stieler had been in the insurgent group. The result was that both were executed and their bodies left unburied.[7]

Fritz Schellhase recalls a story he heard. "They fell out with each other. One said, 'We have to go right.' The other said, 'We have to go left.' But Stieler knew that he was right. He went down the river until he got to the Kerrville Road and reached the home of Sidney Rees. He knew that he needed a travel permit and asked Rees, who was the Kerr County Clerk, to help him get one. Rees told him since he was not yet 18 years old, he did not need one. Stieler next stopped at Mrs. [Fritz] Tegener's house. As he came into her house, Mrs. Tegener said, 'That is my Fritz's gun, how did you get it?' Stieler replied, 'I traded with your Fritz.' She asked, 'Where is Fritz, did he come through?' Stieler replied, 'I don't know, me and Fritz were separated and were not together for some time.' At that time, a man named Lowrance arrived with Bruckisch. At the same time, a Confederate scout arrived. Lowrance turned Bruckisch over to them. The soldiers asked Bruckisch if Stieler were in the battle too? He said yes."[8]

Henry J. Schwethelm tells of the Stieler and Bruckisch deaths in a 1913 letter. He says, "Stieler and Thomas Bruckisch were taken prisoners in Kerrville by Starkey [James Starkey the Kerr County provost marshal] and turned [them] over to some of Captain Davis Company and 3 miles above Kerrville on Goat Creek they killed them shooting target at them." One of the men who executed Stieler and Bruckisch was J. M. Seal. In a letter written to friends he said, "I had the pleasure of shooting one of the poor devils a few days ago myself."[9]

There is a belief among some insurgent descendants that Bruckisch and Stieler were killed at the Nueces battle site and that Wilhelmina Urban Stieler, the mother of Heinrich Stieler, and Wilhelmina 'Minna' Stieler, his sister, recovered them from the battle site. This belief is incorrect. Bruckisch and Stieler were killed near 'Kerrville and their home.' The "scene of the slaughter" or "to the place of the murder" or to "the murder scene" are all correct terms. It is just the scene or place of murder was at Goat Creek and not the battle site.[10]

Upon learning of Heinrich Stieler's execution, his mother, thirty-nine-year-old Wilhelmina and her seventeen-year-old daughter, Minna, decided to go to the death site and retrieve the body as "no man dared attempt" it. Upon reaching the site, the mother and daughter "passed through the ranks of the soldiers who, learning of [their] mission, were entirely chivalrous, and did not molest or prevent [them] from carrying out [their] intention," as several of the stories claim. They found the bodies of Stieler and Theodore Bruckisch hanging from a limb with a sign around their necks with the word 'traitor' or something similar. They also discovered they were unable to either properly bury them or bring them to Comfort. They dug a small grave, put the bodies in it and covered it with brush and rocks to keep the animals from dragging them up. The two women returned home, but later recovered the bodies, taking them to Comfort and burying them. Three Comfort-area men likely helped them: Henry Heinen, Peter and Martin Ingenhuett. The two bodies are likely buried on the Stieler farm.[11]

The Sansom Group

John Sansom states that he and the other four split up. The four went on to Mexico and enlisted in the Union Army while Sansom decided to return home. The Scott brothers, William Hester, and Howard Henderson crossed the Rio Grande the afternoon of August 10, 1862. Hester later described the outcome of the battle and his trip to Mexico. He claimed he lost seven of his, "brave companions in the [battle and] traveling the dreary road to the Rio Grande." In Mexico, "perfectly destitute" and "without clothes or money," Hester made his way to Matamoros where he found refuge with the United States Consul, Leonard Pierce, Jr. Pierce provided Hester and several other insurgents with transportation to New Orleans where they enlisted in the Union Army.[12]

Sansom remained hidden on the high bluff overlooking the battle site until about ten o'clock in the morning of August 10th. Satisfied he could do no more he rode away, "dazed by a tragedy that had robbed of their lives, nineteen vigorous young men, and wounded more or less seriously six, whom I know of." Sansom followed the Nueces River down an old Indian trail about four miles when he came upon three Indians about a hundred yards ahead of him traveling on foot. Neither Sansom nor the Indians wanted a fight so Sansom continued his journey. He continued the night of August 10th. About daylight, he came upon part of Dix's Company K, the Frontier Regiment. Before he had a chance to fight, he was arrested. News of the battle had not reached them. Sansom obtain his release because, "I gave them such a strong talk they let me pass."[13]

Sansom turned east and headed toward Camp Montel, which was commanded by his brother-in-law, Benjamin F. Patton, about forty miles away. About two miles from Camp Montel, Sansom saw another group of Indians driving a herd of horses and mules. When he reached the camp, he reported seeing the Indians. Lieutenant Patton led a ten-man detachment after the Indians, killing seven. It consisted of Lieutenant Patton, Peter Adams, Charles Cole, John Cook, Jack Davenport, Demps Forrest, Jasper Kinchaloe, Dan Malone, T. F. Moore, and Ed Taylor. John Cook was wounded in the engagement. The Indians were overtaken near the Harper ranch in the Sabinal Canyon.[14] Sansom ate and rested at Patton's camp then returned to Curry's Creek, arriving on August 18th ahead of the news of the battle.

Sansom quickly fell under "the eyes of the Confederates." As soon as news of the battle reached the area, "the people were in a rage of excitement over the Nueces Massacre, and soldiers were after me and others." Sansom concealed himself in a large cedar brake, where one morning he awoke to find, "my companions there was a great panther and a Mexican hog. I killed both of them."

Sansom "scouted around among the hills twenty days, recruited nine men [who were hunting him] and went to Monterrey, Mexico," where he: "borrowed $100.00 from Governor Bedowery [sic], then went on to Matamoros, where we met some hundred or more American Refugees. Stopping with Mr. Pierce, the American Consul. Eight days later we went twenty-five miles to the mouth of the Rio Grande to a chartered schooner planned by the Consul, boarded it, got into a storm and almost wrecked, on to New Orleans, and began the formation of the first regiment of Texas U. S. Cavalry on the 27th day of October, 1862. I enlisted as a private on the 28th inst."[15]

Survivors of Confederate Assaults

The survivors of the Confederate assaults left the battle site in three small groups: the Hoffmann [Gillespie County] group, the Henry Schwethelm group, and the Ernest Cramer and Jacob Kuechler [Kendall County] group. The Hoffmann and Cramer/Kuechler groups left the camp in a northeasterly direction. They found a small draw or canyon and went up it where they found hiding places. The Schwethelm group followed the Nueces River south for a few hundred yards than turned east back into the mountains. Karl Itz made it to the Nueces River, but unlike the other members of Hoffmann's group, turned west, and hid along the north bank of the river. He watched other insurgents cross the river northeast of the camp and go up a steep canyon. Believing the Confederates would follow them and knowing that because of his injury that he could not travel far, he decided to remain hidden where he was. He remained for several hours until he felt it was safe to move. He than traveled north, up the Nueces River bed for about a mile, where he spent the night and decided what he should do next.[16]

Two other survivors could not keep up with the groups fleeing from the battle site. William Vater was one. He hid on the east side of the Nueces River, near the base of a huge cliff. The second unidentified survivor reached the Nueces River and like Itz turned west. He went only about a hundred yards when he turned into a small draw or canyon. He found a hiding place. This survivor was badly wounded and soon died.[17]

As soon as it was safe, some members of the Hoffmann group began looking for water and food, but none were found. Hoffmann described the conditions: "It was on the 10th of August and the heat was insufferable. No water and nothing to eat." While most of the group hid, Hoffmann and another survivor, twenty-two-year-old Nassau native Mathias Pehl of Gillespie County went to a "high bank" hoping to find water. None was found. When they returned to where the group had been hiding, they were gone. A quick search found they had moved to some brush along the riverbed. Another insurgent, twenty-one-year-old Prussia native William Graff (a brother of Carl Graff, who had fled with the Schwethelm group) was hiding in a nearby thicket. He joined the Hoffmann group. Graff had been observing the movements of the Confederates and watched where they placed their guards. He recommended they take their shoes off so not to make any noise walking on the rocks. The group did so and two at time slipped to the waterhole. On one of these trips, Hoffmann noticed a ripple in the water. When he looked again, he saw a man in the water under a willow bush with just his nose above the water. When the man saw Hoffmann, he raised his head out of the water just enough to see. He recognized Hoffmann and said, "Oh God, it is you."

It was William Vater, wounded in the arm and breast. Hoffmann and his group took him a short distance from the river and made him as comfortable as they could. They dried his clothes and made him a cedar branch bed. "There he lived for 3 weeks on prickly pears and bear grass. He later surrendered to a scout who took him to San Antonio." It is not known what happen to him. Hoffmann and his group moved about two miles where they found another water hole and other insurgent survivors.[18]

Another insurgent group of survivors was the Schwethelm group. As they were fleeing across the Nueces River, John Sansom rode up to them. Not knowing whether Sansom was an insurgent or Confederate, they scattered and William Graff became separated. The Hoffmann group shortly found him. The other members of the Schwethelm group went up and over a hill and reached Sycamore Creek. They followed it until they located other survivors.

The last group of survivors was the Cramer and Kuechler group. Captain Cramer states they carefully traveled up a valley attempting to leave no trace of their presence. After about two miles, they met part of Hoffmann's group. "We met 6 or 8 of those men who had deserted us. They were lost and did not know which way to turn. They begged to be allowed to join us and as bitterly as we felt towards them we still could not refuse them and leave them there." The survivors continued until noon without finding any water. Because of the condition of the wounded, they made a camp for them in a "sheltered place." Captain Jacob Kuechler and Moritz Weiss remained with them.

Captain Cramer and a companion, whom he later identified as Weber [this was twenty-seven-year-old Nassau native Philipp Heinrich Weber of Comfort] continued the search for water. They found none. About four o'clock in the afternoon they returned to where they had left the wounded and the remainder of the survivors. They were gone![19]

They were gone because Captain Jacob Kuechler had located a spring in Sycamore Creek, about a mile from the battle site and moved the wounded there. But for the time being Captain Cramer and Weber did not try to find them. Captain Cramer said he was, "too tired and weak to hunt for them. After resting for a while we proceeded to walk back to the battlefield. About two hours after dark we came to the water. As soon as we had satisfied our thirst we felt ourselves strong again. We emptied our powder horns and filled them with water. We meant then to continue our search for the wounded." Before they could begin their search, a group of about ten insurgents showed themselves. They also had been getting water when they heard Captain Cramer and Weber approach and had hidden. Captain Cramer states the Confederate, "soldiers were still not more than 150 steps away."[20]

Captain Cramer asked the men to help him find Captain Kuechler and the wounded and to take water to them. They used every article they had to carry water. Cramer and Weber decided to try to kill a deer for food. They agreed on a meeting place. Cramer and Weber "went about four or five miles hoping to find a deer at dawn." They were not successful. About ten o'clock on the morning of the 11th, they went to the agreed meeting place. Captain Cramer found all the group, except his brother-in-law Ferdinand Simon, who, "had become too weak to continue with them."

Cramer expressed his concern, saying: "I was almost frantic—was too anxious about him to be able to take any rest. So, dangerous as it was, I started out in the evening [of the 11th]. It was a bright moonlight night and I hunted for him throughout the whole night. I had no success—he was nowhere to be found. I felt convinced that he was dead. I came back to the camp in the morning [of the 12th] completely exhausted. I had had nothing to eat for three days and the anxiety and worry strain was enough to exhaust the strongest man."[21]

On August 11th, while Cramer was trying to find Simon, three insurgents left the area and headed to Mexico. This was the Schwethelm group; Carl Graff, Jacob Kusenberger, and Henry Schwethelm. They felt their best chance was to go on to Mexico, as it was only forty or fifty miles farther. They believed if they returned to the Hill Country they would likely be killed. On the morning of August 11th, without food, water, money, or mounts they started for Mexico. A second group, the Vetterlein brothers, also decided not to return to the Hill County. Instead, they departed for El Paso. With the absence of Simon and these five men, the group of survivors was now down to fifteen.[22]

On August 12th, the main Confederate force left the battle site. A Confederate scouting party of about twenty men remained in area for a day or so. About the 13th, they located Simon and took him to Fort Clark. All the

Confederates had now left the area. Ernest Cramer and Jacob Kuechler took this opportunity to return to the battle site. They were horrified at the sight that awaited them. Cramer says, "The sight was horrible beyond description. They had been striped one over the other in a large heap. Those who were still living when we were forced to leave them had been lined up and used as targets. Their faces and bodies completely riddled with bullets. It was heart-rending and I could not linger there." Captain Kuechler's stated, "The rebels left the battle ground on the 13th of August. Soon afterwards I visited our old camp, and stood pale and shuddering at the sight of the fate, which had befallen the poor wounded, who could not leave the camp with the rest of us us having every one been murdered and mutilated by a cruel foe."[23]

Kuechler also found a live insurgent near the battle site. "I met Mr. Vater, of our party, badly wounded; he arrived in camp a little while before me." He said, "'I have returned to camp to die near my dead comrades.'" William Vater had been hiding in a dogwood thicket near the spring. "It was impossible to take him to our camp; I, therefore, visited him twice a day to dress his wounds, give him food and water. During our stay, three scouts of 20 men each, were continually hunting us. I lost sight of Mr. Vater after the last scout had left and could not find him." This was the second wounded insurgent Kuechler 'lost' or otherwise did not assist in getting to safety; Ferdinand Simon who ended up a prisoner and was sentenced to death and William Vater, who "was rescued by a Union man from Uvalde."[24]

The man from Uvalde who rescued Vater was Callimense Beckett. He learned of the battle about three weeks after it occurred. Beckett formed a scouting party and went to the battle site. He found the dead, "scattered around, and all of the valuables were gone supposedly stolen, either by the soldiers or others. He found one of the German boys in the water hole near the camp with a broken leg. He was almost dead from hunger and his wounds. He had bandaged his broken limb as best he could by tearing off some of his clothes. He was brought into Brackettville and doctored."

Hoffmann says the wounded insurgent gave himself up at Fort Clark and was taken to San Antonio. It is not clear how Beckett learned of the battle. It may have been from newspaper accounts or stories from the Confederate soldiers as they passed through Uvalde. His account also tells of a cave, "Right across from the Dutch [sic] battleground containing the skeletons of ten or fifteen people, which skeletons have been there since long before the Civil War. It is the supposition that they were killed by Indians in older times."[25]

The surviving insurgents prepared to depart, returning to the Hill Country, or go to Mexico, or El Paso. Five of them decided it was best if they did not return to the Hill County, believing that only death awaited them there.

The Henry Schwethelm Group

This three-man group left the battle site area early on the morning of August 10th. Jacob Kusenberger and Carl Graff told Schwethelm if he went to Mexico, they would go with him. They traveled west about ten miles where they crossed the road going between Fort Clark and Fort Concho. Schwethelm knew this area and he knew of a man named Riodan who owned a ranch nearby.

This was Thomas Riordan, an interesting individual. His date and location of birth are not clear; He was either born in 1816 in Canada or in 1826 in Ireland. He served in the Mexican War and arrived in Texas with the U. S. Army after that war. The 1850 Bexar Census lists him as a sergeant major stationed at Fort Duncan. Riodan re-enlisted on March 1, 1860, for a five-year term. The 1860 Dawson Census shows him stationed at Fort Clark and an orderly sergeant in the First U. S. Artillery. It does not appear he was discharged when the Federal Army left Texas. It may be that he was left behind. Riordan purchased land near the San Antonio and El Paso road on Pinto Creek, about ten to twelve miles west of Fort Clark on August 12, 1861. On November 24, 1861, he married Ellen Melefont at Brackettville. In early 1864, Riordan organized a 53-man militia company for the Third Frontier District. When the Federal Army arrived back in Texas, he again is shown on U.S. Army records as an orderly sergeant. Riordan died on March 21, 1867, in current Kinney County. His wife died on May 23, 1905, in Kinney County.[26]

Schwethelm proposed he go to the ranch house and try to get provisions. Both Kusenberger and Graff wanted to go with him. Schwethelm convinced them three men in the area would likely arouse suspicion. The others agreed and hid while Schwethelm was gone. The ranch house was only about half a mile away. When he reached the house, he made up the story that he had come to that part of the country on one of the hay wagons from D'Hanis which provided Fort Clark with hay and he had gone deer hunting and got lost.[25]

Just as Schwethelm sat down to eat, two men rode up and entered the house. One was Dr. Downs from Fort Clark, the other was a Confederate from the battle scene. Schwethelm incorrectly identifies this Confederate as Captain Duff. Duff was still at Fort Martin Scott. Riordan asked the two from where they had come. The Confederate replied that he had come from a fight on the Nueces, and a courier had called Doctor Downs from Fort Clark to treat the wounded. When asked if the fight had been with Indians. The Confederate replied, "No-with a bunch of d- - d Abolitionists!" Schwethelm later said he had just taken a bite of food when he realized whom the visitors were. He almost choked and had to quickly drink a swallow of coffee. Hoping not to be identified as an insurgent, Schwethelm asked, "Well, did you get them all?" The Confederate replied, "Well, nearly all." He continued the conversation, hoping not to seem nervous. The two men stated several wagons were on their way to carry the wounded. Schwethelm asked if he could get a ride with them on to Fort Clark. The Confederate replied he did not think there would be enough room once they reached the wounded. Schwethelm stated he would travel on by foot and catch up with the hay wagons. The Confederate expressed the view to the Riordans that perhaps he should take Schwethelm with them as he might be one of the insurgents. Mrs. Riodan replied that she knew Schwethelm to be one of the hay men. Schwethelm finished eating. As soon as the two men left he also departed, "As soon as he was hidden from view from the house he started in a run to where his comrades were waiting, and appraised them of Confederates being near and they hastened away as fast as their legs could carry them."[28]

The trio continued, still without food or water. They decided to travel at night in hopes of evading any Confederates. Early the morning of August 12th, they reached a waterhole that had a dead cow lying at the edge. They remained here during the day of the 12th. The only food they had was some cactus pears that gave them a fever. Towards evening they heard horses approaching. They hid as two riders passed close by. As soon as it was dark, they left. On the morning of August 13th, they reached the banks of the Rio Grande just as the sun was coming up. Carl Graff did not know how to swim, so they built a small raft and crossed on it.[29]

After crossing the Rio Grande, the three came upon a Mexican goat herder who mistook them for Indians. He ran and notified the captain of a Mexican frontier company that was stationed about thirty miles north of Piedras Negras. The captain quickly gathered a twenty-five-man detachment and came to fight the 'Indians.' When the Mexicans arrived, Schwethelm explained who they were and why they were in Mexico.[30] The Mexicans put them on mounts and took the three to a nearby hut and gave them food and water. The three spent the night of August 13th at the Mexican frontier company camp. The next day they reached Piedras Negras, where they met Adolph Real, a thirty-seven-year-old native

of Prussia whose older brother Casper lived on Turtle Creek
in Kerr County. Real was preparing to return home with two
wagons of flour he had purchased. The three sold Carl
Graff's pistol to Real for $30.00. Real was only able to pay
half the amount. The other half he paid to Henry
Schwethelm's father when he returned to Texas.[31]

The trio traveled on to Monterrey and from there to
Matamoros where Leonard Pierce, the U. S Consul, took
them in. Pierce sent a number of letters to the U. S. Secretary
of State describing the condition of the Texas refugees when
they reached Mexico and pleading for funds. Pierce reported
the information the Scott brothers, Henderson, Hester, and
the Schwethelm group had provided about the battle:

"A party of Union Men numbering sixty-eight left the
County of [Gillespie] in Texas to cross to this city from
whence they intended to embark for the northern states when
about one hundred and forty miles from San Antonio they
were attacked by three companies of rebel soldiers to the
number of about two hundred. A desperate fight ensued,
which resulted in the massacre of nearly all of the Union
Men, but seven of them having reached here, to this date.
They left thirty-two dead on the spot and many wounded
who were afterwards killed. The fight took place about the
tenth of August. Most of the killed were Germans who have
relatives in the Western states. As yet I have been unable to
obtain a complete list of the names of those missing, but as
soon as I can do so, I shall transmit it to the Department."[32]

Pierce sent them to Matamoros and from there they
went on to New Orleans where they joined with other Texans

and enrolled in the Union Army. Schwethelm and Kusenberger were mustered into Company A, First Regiment [Union] Texas Cavalry on October 27, 1862. Graff enlisted in Company B of the same regiment on November 10, 1862.[33]

The Vetterlein Brothers

Gottlieb and Carl Vetterlein decided to make their way overland to El Paso. "For clothing, they wore cornsacks they had gathered on the trail of Sibley's retreat from New Mexico." In mid-December,1862, they arrived at the camp of First Lieutenant George H. Pettis, commanding a 78-man detachment from Company K, First Regiment [Union] California Infantry at Camp San Elizario, about 20 miles southeast of El Paso. Pettis, a twenty-eight-year-old native of Rhode Island, had arrived at San Elizario on December 7, 1862. It took the Vetterlein brothers over three and a half months to make the 450-mile trip. "They were a woebegone sight. One of them was wounded in the Nueces fight." It was likely Gottlieb who was "badly wounded in the wrist." The surgeon with Pettis' command treated Gottlieb who recovered. Both were given employment in the quartermaster department.[34]

Survivors Return to the Hill Country - The Karl Itz Ordeal

After hiding north of the battle site for a day or so, Karl Itz decided to try to get home. His injury prevented him from traveling more than a few miles each day. On August 11th or the 12th, he began his long trip home. He lived off the land, eating whatever he could find, and traveling only about five miles a day. He generally retraced the route taken by the insurgents as far as the West Frio River. Here he turned northeast toward the South Llano River in present Kimble County, and then towards the Spring Creek Settlement where he had friends. From here, he went to the Cherry Spring area in northern Gillespie County, where his family lived on land owned by his brother-in-law, Ludwig Evers. It is not clear how long it took Itz to reach home. One story says 'days', another says much longer, as much as 'months.' Based on a map analysis of the likely route, it was a trip of about 140 to 160 miles. If he traveled about 5 miles a day, it would have taken him about 30 days, which would mean he reached home around September 12, 1862. Knowing that if he were found it would result in not only his death but also repercussions for his family, he remained hidden "in the bush." The threat of repercussions was real. In late February 1863, State Troops or pro-Confederate bushwhackers killed two of his brothers, Jakob and Hermann, because they would not tell where he was hiding.[35]

Karl Itz's wife worked out a method to get food to him as he remained hidden in the mountains near his home. Each day he would take a different route to a different location where he could see his house. His wife would send the children out to play. One of the children would have a small bucket containing food. The child was not told what was in the bucket. After the children returned from "playing," Karl would come out to the food.[36]

Karl never slept in the same place two nights in a row, and never made any type of shelter. One night he woke up to find he was surrounded by a pack of wolves or coyotes. Believing he would either be killed by the pack or his location revealed, he decided to just remain still. Soon the pack left without creating any unusual noise. Indians also assisted Karl. Previously, he and his wife worked out a type of peace treaty with Comanche Indians. When the Indians were in the area, they would not raid the Itz's farm if the family left a bucket of Clabber [which is sour milk that has thickened or curdled] for them. It was because of this relationship when the Indians found Karl Itz hiding they gave him food.[37]

The Hill Country Survivors

The remaining survivors of the Confederate assaults departed the battle site on August 15th to return to the Hill Country. "About twenty survivors came together after the Battle and about fourteen wended [sic] their way back home." There were actually fifteen who returned home; Ernest Cramer, August Duecker, William Graff, August Hoffmann, Wilhelm Klier, Henry Kammlah, Jacob Kuechler, Fritz Lange, Mathis Pehl, Adolphus Zoeller, Fritz Tegener, Philip Heinrich Weber, Franz and Moritz Weiss, and Joseph Poetsch. In addition to these fifteen, another five left for either Mexico or El Paso; Henry Schwethelm, Jacob Kusenberger, Carl Graff, Carl Vetterlein, and Gottlieb Vetterlein. An interesting fact is 14 is the number reported to the Confederates in Gillespie County as trying to make their way home. They were in at least two groups. Captain Ernest Cramer headed the Comfort group and Captain Jacob Kuechler headed the Gillespie group.[38]

The Kuechler—Gillespie County Group

The Gillespie County Group consisted of August Duecker, William Graff, August Hoffmann, Wilhelm Klier, Henry Kammlah [who was wounded], Jacob Kuechler, Mathis Pehl, and Joseph Poetsch. Captain Jacob Kuechler's brief account of the return home says, "We go along very well under the circumstances, all of our wounded recovering and at length reached the settlements, after enduring many hardships." August Hoffmann provides additional data. He stated the group's main diet was prickly pears. Like the Schwethelm group they, "suffered and … were weak from fever, caused by too steady a diet of prickly pears!" They found some flour that the Confederates had lost or left, "We pounced on it and at the next water hole had a grand baking on a flat rock. It was a good thing we had matches." This did not make much of a meal as each man only, "got a piece about an inch by half an inch." On one occasion, they were able to kill a deer. As they neared the Guadalupe River, they got food from the settlers. Some was given to them; some they took without asking. At one farmhouse, they were given two large loaves of bread and a bucket of milk. Hoffmann said, "We didn't leave very much." On another occasion, near Johnson Fork, they found a patch of sweet potatoes and killed a hog. Even without salt or bread, it was cooked and they "fared pretty well." As they neared the Pedernales

River, they became worried about the Confederate cordon. They received warnings of Confederate patrols. Hoffmann tells of how they received one of these warnings; "We noticed a man coming toward us on horseback. He was riding fast, he wanted to warn us that Duff's and Davis' Company and others had their camp on the Pedernales, and others had their camp close to the Pedernales falls."[39]

Hoffmann and the other survivors had reason to be concerned about being discovered by state and Confederate forces. On August 15th, the authorities were alerted about the battle and told that fourteen insurgent survivors had escaped and were heading back to Fredericksburg. Camp Davis and Camp Pedernales were put "on a war footing" and scouts were sent out to look for the survivors.[40]

Hoffmann stated, the insurgents all got home and "slipped in safely." But one man became ill and had to have a doctor, which caused his downfall: "One man ate too much when he got home and became sick. He needed a doctor; the doctor was compelled to report the case. The sick man was arrested at once and had to confess all names who had come back with him, but he gave names who were not with him, search was made everywhere but none found. Our names were well-known, we were hunted everywhere, all relatives were questioned."[41]

The insurgent arrested was Wilhelm Klier. Confederates seized him because he was, "in the nine [only eight names are known; August Duecker, William Graff. August Hoffmann, Wilhelm Klier, Henry Kammlah, Jacob Kuechler, Mathis Pehl, and Joseph Poetsch] who finally reached their homes" on August 21st. "Half-starved and sick [Klier] arrived one evening at his brother-in-law's home [Julius Ransleben]." Forty-eight-year-old Ransleben, who was married to Klier's sister, Josephine, did not know what to do, "in his shock and confusion, went to town the following morning [August 22nd] to seek counsel and advise. "He went from one to the other, haunted by fear" not knowing "what to do. The result was, that the news about Wilhelm's return had spread like wildfire. The same day, in the evening, he was arrested, even though the poor fellow lay sick in bed."[42]

He was kept under arrest overnight. The morning of the 23rd came news Klier was to be taken to San Antonio. Julius Ransleben asked if he could transport him in his wagon. The Confederates agreed. This was fortunate for Klier because when they arrived at Boerne, Klier was turned over to a group of Confederates there. Ransleben watched over Klier throughout the night. The next morning a hanging detail arrived. Ransleben convinced the Confederates not to hang Klier, but Klier had to watch as two of his fellow insurgents, Tays and Bock, were hanged.[43] In San Antonio, Wilhelm Klier "was under strict security. A clever lawyer defended his case", explains Klier's brother-inlaw, Julius Schlickum. The lawyer was paid for by Klier's many friends who "loved and respected" him. Klier was finally set free, but only after he swore allegiance to the Confederacy. He was immediately drafted into Confederate military service.

"He no longer is in personal danger," wrote Schlickum.[44] Another insurgent in the Gillespie County group, Henry Kammlah, says he, "was wounded at the Battle of the Nueces. Returning home, he hid in the pastures near Fredericksburg, where his wife brought him food." August Duecker was also in the Gillespie County group. When he returned home, he went into hiding and spent many daylight hours in the attic of his log cabin. At night, he worked in the field and performed other duties. While he was hiding, Duecker's wife became seriously ill. A doctor was called whom suggested if Duecker was anywhere near he should be called to her bedside. The family did not trust the doctor and Duecker was not called. His wife died on January 1, 1863. After her burial, he visited her grave under cover of darkness. The four children were taken by various families and raised until the end of the war.[45]

According to local folklore, two locations where the survivors hid were near the Albert community in eastern Gillespie County and near the Luckenbach community in southeast Gillespie County. These were the same locations where the "Luckenbach Bushwhackers" hid after Duff's first visit. Traces of furnaces dug into the ground that were used for cooking could still be seen as late as the 1950s. The Albert location was on ranch property formerly owned by Albert Wilke, about six miles southwest of the Albert Settlement. There is a spring that has never gone dry and is so well hidden that even from a short distance it is hard to see. Next to, or just overhead, is a cliff for shelter and a small cave which provided a storage place for food and other supplies. The spring is now known as 'Bushwhacker Springs.' Friends and relatives of the insurgents secretly provided them food and supplies. The other location was along South Grape Creek, near Luckenbach and on the Friedrich Scharnhorst farm.[46]

Other oral family stories tell of how family members would warn those who were hiding of danger or if it was clear advice and all right for them to come pick up provisions. These signals would be if certain laundry was on the cloths line, or if a horse or horses were saddled or unsaddled. Other signals were if an ox were tied to a certain tree, or which ox team was used to plow, or how a lantern was displayed at night. There were several observation points the insurgents used. One was *Der Spitze Berg* [the pointed hill] west of Fredericksburg, which overlooked the Live Oak Creek area.[47]

The Comfort Group

The Comfort area group consisted of Fritz Lange, Major Fritz Tegener, Philip Heinrich Weber, Franz Weiss, Moritz Weiss, Adolphus Zoeller, led by Captain Ernest Cramer. On August 15th, "I started back to my home to get provisions," wrote Cramer. Three days later, on August 18th, they reached the "first habitation." On the 19th, they "arrived in the outskirts of Comfort." The group separated.[48]

The Weiss brothers contacted members of their family and hid along Cypress Creek, just northeast of town. Their family brought them food and attended to Franz's wounds. One of the survivors, likely Fritz Tegener, hid on the Lich farm on Cypress Creek. The Lich and Steves family took him food and water once or twice a week. Fritz Lange hid near his home near Verde Creek in southern Kerr County where his family provided him with food and shelter. His brother Frederick who was living near Dietert's Mill, hid Philipp Heinrich Weber. The three wounded insurgents, Fritz Tegener, Adolphus Zoeller and Franz Weiss, received medical attention from a local self-educated 'doktor,' forty-eight-year-old Prussia native, Adolf Pfeiffer who did not have a medical degree.[49]

Cramer tells what he did: "It was a terrible day [the day he reached home] for me. I stayed in the thicket about a half a mile from my home and dared not go to My Lottchen. I could not face her with the news I had to tell of Leopold's death and Simon's." Cramer did not visit his wife and family. The next day, August 20th, he went on to Boerne, "I went 18 miles further to the home of a friend to beg him to break the sad news to my sister-in-law and to the others. By the time that I reached my friend's home … I had a high fever."[50]

Boerne was not the safest place for an insurgent as Duff's Partisan Rangers were in the vicinity. They were being withdrawn from Fredericksburg and were on their way back to San Antonio. On August 23rd, a detachment of Duff's Company arrested two of the insurgent who had 'left early.' On August 24th, the two were hanged on Cibolo Creek, about a mile and half north of town. Cramer was not exaggerating the danger his life was in when he wrote, "It was too dangerous for me to attempt to go home. Every one was under suspicion. If anyone had even been suspected of giving aid to one of us they would have been taken and hanged. Relatives of those who had been with us were especially watched."[51]

As soon as he could travel, Cramer left Boerne. It was about the 26th of August when he went to Leon Springs. This area was not any safer. At Leon Springs, Cramer, "found the opportunity to speak with my sister-in-law. She had visited Simon in San Antonio and was very anxious about him. She begged me to be careful and for me to try to get to Mexico as I was the only one left for them to depend upon."[52]

Captain Cramer followed his sister-in-law's advice and left for Mexico. He got a horse from a friend and made his way to Mexico. "Through an earlier companion, Richard Brotze, I managed to secure a horse with the understanding that I leave immediately for the Rio Grande. He insisted on that and I could not blame him for had I been seen with his horse he would immediately have been hanged. I still had many dangers to face before I could get to the Mexican border and heard the whistle of many a bullet." Thirty-two-year-old Brotze was a second lieutenant in Company B [Kendall County], 3rd Regiment of the 31st Brigade District. On August 27, 1862, he was elected Kendall County sheriff.[53]

Fortunately for Captain Cramer, he made the trip to Mexico by himself and did not try to reestablish contact with other insurgents, such as Captain Kuechler. Cramer explained why he didn't. "[Kuechler] during this time had returned with the wounded. I did not have time to look for him." It was fortunate Cramer did not contact Kuechler and flee with him, as this second insurgent group was attacked on the Rio Grande and many were killed.[54]

Map Q – Routes of Surviving Insurgents

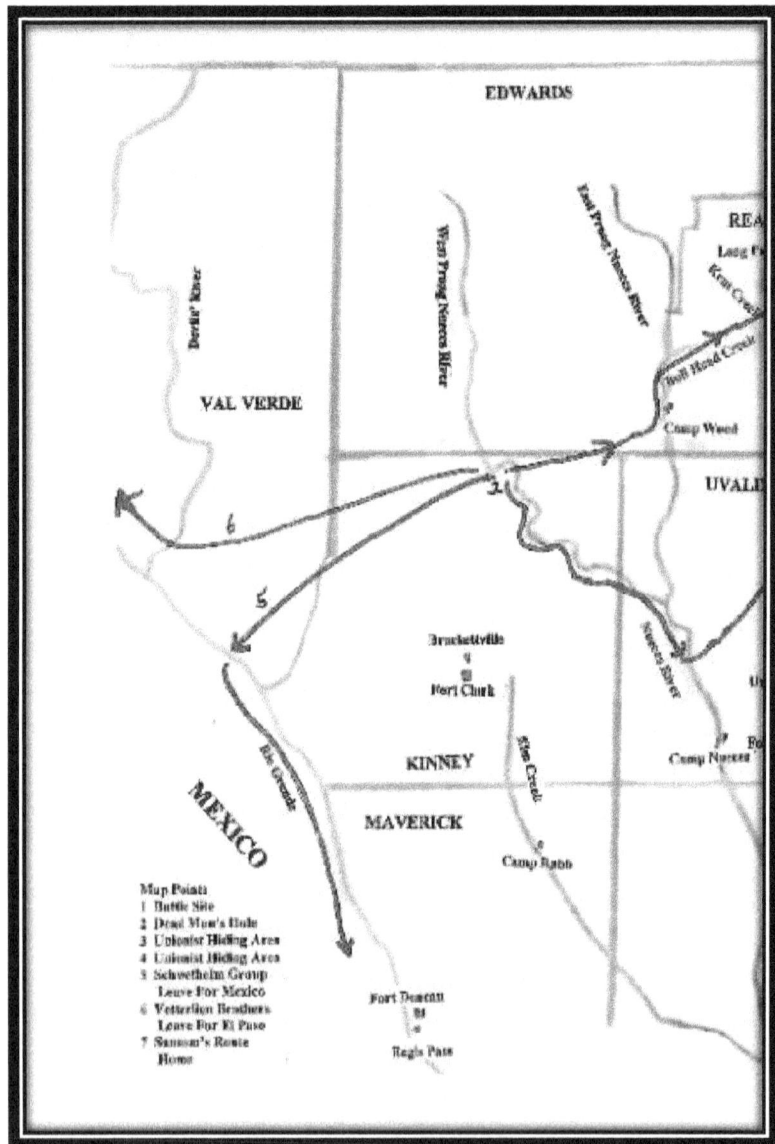

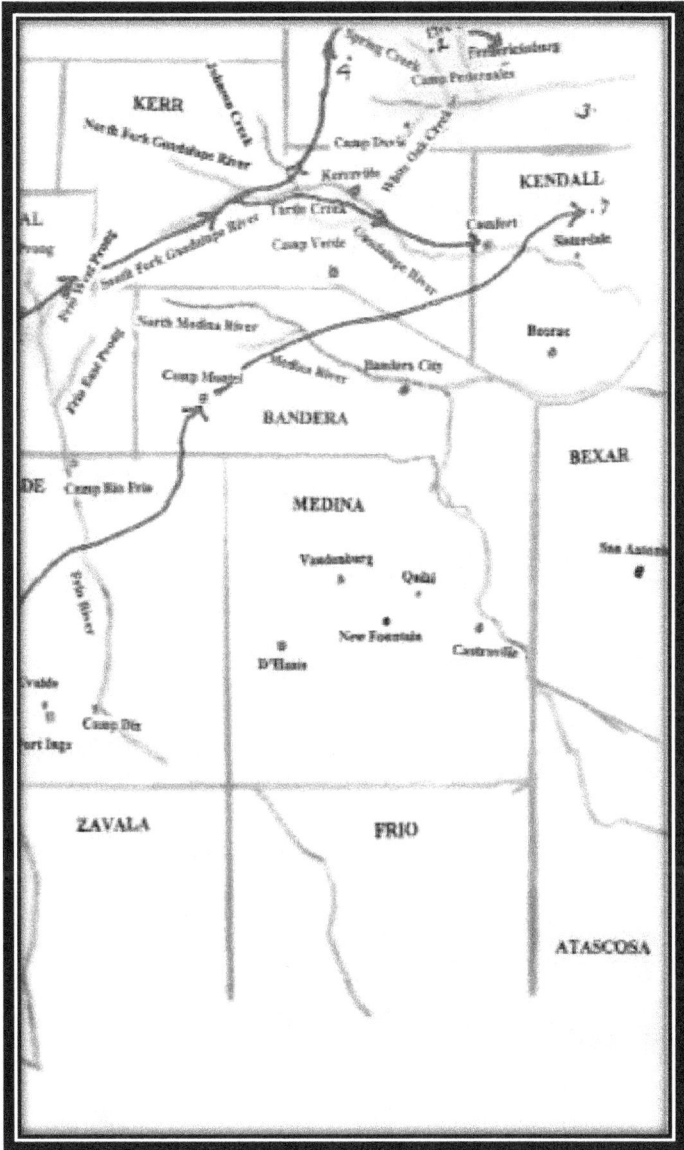

Chapter 12 – Insurgent Survivors

Endnotes -- Insurgent Survivors

1. The route the group took is an assumption by the author. No eyewitness or primary source describes a route. But there are clues. The Schwethelm account says they "returned to Fredericksburg and vicinity, by taking a new route from that by which they came." So, they either went back east, north or south of the original route. The shortest route would have been due east, but this was generally the route by which they came. It would have also taken them into Captain Montel's camps of Camp Rio Frio, on the Frio River, Camp Montel, on Seco Creek, and Camp Verde, in southern Kerr County. To have taken this route, without a guide, would have almost certainly resulted in them being captured. It is not likely they went north as they would have had to cross the dry Edwards Plateau. It seems safe to believe they went south. The belief they took this route is supported by the fact at least four, and maybe six, of the group were captured on this route. One was August Luckenbach. He was captured "just north of San Antonio.". Family member say it was about eight miles north of San Antonio. Another, likely Herman Flick, was captured on the Medio Creek in northwest Bexar County. Two others, Conrad Bock and Fritz Tays, were captured near Boerne. The other two were Heinrich Stieler and Theodore Bruckisch who "may have been captured near Kerrville" or "near home." (See Map Q p 538, for likely route and locations where captured). Schultz *Was a Survivor of Nueces Battle*, p 26; Newspaper Article titled "the Nueces Battle of 1862" by Hugo Weinheimer of Stonewall, Texas, located in the Vertical File under title "Local History", Pioneer Memorial Library, Fredericksburg, Texas; Oral Story and Interview with Dorothy Basse, a Luckenbach descendant on May 1, 1997; Siemering's Germans During Civil War, June 8, 1923; Diamond Jubilee Souvenir Book of Comfort, Texas, Commemorating the 75th Anniversary [of Nueces Battle], August 18, 1929; Folder, Centennial Observance "Battle of the Nueces" held at the Union Monument Comfort, Texas on Sunday August 12, 1862 and Kendall County District Court Records, Case Number 5, The State of Texas vs. James Duff and Richard Taylor.

2. Jacob Usener was born about 1834 in Nassau. Ludwig was born on January 20, 1831, in Prussia. The two brothers, along with their parents, arrived in Texas in August 1845 and were one of the first families to settle in Fredericksburg. Jacob Usener died on June 5, 1889, in Gillespie County. Ludwig died on June 28, 1930 in Gillespie County. The identity of this third man is not known. Raymond Usener, the grandson of Ludwig Usener, and the individual who recalls this family story states it 'may have been' a man named Koehler. The only Koehler shown on the 1860 Gillespie is a Franz Koehler, age 38, a farmer, born in Hanover. There is no other record of Franz Koehler being in the Unionist group. Undated manuscript by

540

Raymond E. Usener, 760 Treibs Road, Fredericksburg, Texas. Copy provided by Mr. Usener to author. Mr. Usener read this manuscript at the "Nueces Encounter Symposium" on March 22, 1997 at the Palace Theater, Fredericksburg, Texas; Siemering's Germans During Civil War, June 8, 1923; Kenn Knopp *Fredericksburg, Manuscripts*, p 14; Gillespie Historical Society *God's Hills*, Volume I, p 241; Lindig *Der Friedhof* Cemetery, p 87 and Oral family story told by Raymond Usener, grandson of Ludwig Usener, at meeting in Fredericksburg, October 4, 1997.

Conrad Andreas Christian Bock was born about 1834 in Prussia. The date he arrived in Texas is not known. He married Pauline Flick, the sister of Herman Flick, and another member of the Unionists group. on May 5, 1860, in Gillespie County.

Peter Jacoby was born on October 18, 1835, in Prussia. The date he arrived in Texas is unknown. He was a member of Braubach's Company, February-November 1861. Peter Jacoby died on January 28, 1930, in Gillespie County. Eighth U. S. Census, 1860 Gillespie County Texas Census, pp 16, 32 and 35; Gold *Church Record Book of the Vereins-Kirche*, pp 69, 107, 116, 134, 161 and 173 and *Comfort News,* August 16, 1912.

3.   Heinrich Friedrich [Fritz] Tays was born about 1842 in Hanover. The family arrived in Texas on the Weser in 1845. His father died shortly after arriving and his mother married Conrad Engelke. Fritz's brother was Johann Heinrich Tays who was a Union soldier. Herman Flick was born about 1839 in Nassau. The family arrived in Texas on the *Washington* in 1846. His sister, Pauline, married Conrad Bock, another member of the Unionist group, on May 5, 1860. Wilhelm F. Boerner was born about 1830 in Hanover. The date he arrived in Texas is not known. He and his family arrived in the Cypress Creek area, near Comfort, in 1855. He married Laura Steves, a sister of Heinrich Steves, also a member of the Unionist group on July 31, 1860 in Kerr County. Laura gave birth to a son, Henry, shortly after Wilhelm's death. She died on December 9, 1861. Emile Steves Voigt, a sister of Laura's, raised Henry. Heinrich Stieler was born about 1844 in Duchy of Anhalt. The date he arrived in Texas is not known. The family arrived in the Comfort area in 1856. His sister, Emilie, married Henry Schwethelm, another member of the Unionist group. Theodore Bruckisch was born about 1838 in Prussia. His father was likely a Freethinker and Forty-Eighter. Theodore arrived in Texas with his family on the *Friedrich Gross* in 1853. He was a second lieutenant in Company D [Kendall County], 3rd Regiment, 31st Brigade District. Bruckisch was a member of Harbour's Home Guard Company in February 1861. August Luckenbach was born about 1833 in Nassau. He along with brothers, Jacob and William, arrived in Texas in December 1845. He married Bertha Goebel about 1858. Member of Luckenbach Bushwhackers. Adolph Ruebsamen was born about 1840 in

Nassau. He was a brother of Louis. The family arrived in Texas in 1845 on the *Harriet.* Member of Luckenbach Bushwhackers. Louis Ruebsamen was born about 1843 in Nassau. He was a brother of Adolph. The family arrived in Texas in 1845 on the *Harriet.* Member of Luckenbach Bushwhackers. Eighth U. S. Census, 1860 Comal County Texas Census, pp 35,176, and 201; Geue *New Land Beckoned*, pp 93, 133 and 146; Comal County Marriage Records, Volume D, p 15; Gillespie County Marriage Records, Book 1, p 324; Eighth U. S. Census, 1860 Kerr County Texas Census, pp 70 and 75; Ransleben *100 Years*, pp 24, 25, and 66; Kerr County Marriage Records, Book A, p 40; Ransleben *100 Years*, p 24; Bennett *Kerr County*, pp 68-69 and 112; Geue *New Homes In New Land,* p 58; Records, 31st Brigade District, A.G.C., TSA, Austin, Texas; Eighth U. S. Census, 1860 Gillespie County Texas Census, pp 29, 30; Biographical Sketch of Jacob Luckenbach, n. d. in the possession of author; Gillespie Historical Society *God's Hills,* Volume II, pp 84-86; Gillespie County Marriage Records, Book 1, p 287 and Luckenbach Bushwhacker and Luckenbach Ranger List.

Richard Taylor is somewhat of a mystery because nothing certain can be found about him before or after the war. He is believed to be the R. C. Taylor shown on the 1860 Bexar Census as born in Scotland about 1829. It is known he was a first lieutenant in Company Number 3, Precinct Number 3, of the 1st Regiment, 30th Brigade District in early 1862. He enrolled in Duff's Company on May 4, 1862 and elected second lieutenant. He was promoted to first lieutenant on October 16, 1862 and to captain on January 12, 1863. He remained in command of Company A until he resigned his commission on September 5, 1864. Eighth U. S. Census, 1860 Bexar County Texas Census, p 391; Richard Taylor's Confederate Service Records, Record Group 109, National Archives, Washington, D. C. and Records 30th Brigade District, A.G.C., TSA, Austin, Texas.

4. Kendall County District Court Records, Case Number 5, The State of Texas vs. James Duff and Richard Taylor and Siemering's *Germans During Civil War*, June 8, 1923; Oral interview with Evelyn Woods, May 1997 and oral interview with Clara Garber, Fredericksburg, Texas, May 1997.

5. The location where Luckenbach was hanged is about a mile south of the USAA Complex on Fredericksburg Road in San Antonio. It is just south of a hill. There is an apartment complex with oak trees at the location. The Medio is the Medio Creek. It is an intermittent stream originating in eastern Medina County. It flows southeast into western Bexar County for about fifteen miles and empties into the Medina River southwest of San Antonio. Weinheimer *Nueces Battle*; Dorothy Basse Interview; and Siemering's Germans During Civil War, June 8, 1923.

6. There were only one set of brothers later captured and executed; the Ruebsamen brothers. The wounds occurred in the opening shots of the battle. This is the only account that says either of the Ruebsamen brothers were wounded. Siemering's Germans During Civil War, June 8, 1923.

7. Siemering's Germans During Civil War, June 8, 1923; Ransleben 100 *Years*, pp 95 and 118; Kerr County Historical Commission *Kerr County Album*, p 315; Stewart *Death on Nueces: Mina Stieler Stories*, pp 24-41; *"Treue Der Union": German Texan Women On The Civil War Homefront*, by Judith Dykes-Hoffmann, B. A. Master of Arts Thesis, Southwest Texas State University, San Marcos, Texas, 1966, pp 80-81; *Comfort Women in Comfort History* by Anne and Mike Stewart (Published by Anne and Mike Stewart, Comfort, Texas), 1993, pp 127-130; Schlickum's Letter; Notes taken from Autobiography of Hubert Heinen, Comfort, Texas 1872-1965, Exhibit 3—Genealogy—Stieler Family. Provided author by Anne Stewart of Comfort, Texas and Glenn *Capt'n John*, p. 20.

8. Mrs. Tegener was Susan Benson Tegener. She and Fritz Tegener were married in December 1858. She was likely one of the sources that told the Confederates the Unionist had fled as she had two brothers in Davis's Company. She and Tegener were divorced in May 1866. She married Charles Curren on December 6, 1869 in Kerr County. They left Kerr County and it appears they went to South Africa.

Lowrance was Daniel Boon Lowrance. He was born on April 1, 1829 in Tennessee. He married Leah Ann Thompson on February 24, 1855 in Alabama. He and his family arrived in Texas in 1859. The 1860 Kerr Census shows him as 31 years old, with a wife and two children, living between Kerrville and Camp Verde. He had signed the petition requesting Kuechler's February 1862 Company be disband. Family stories say he was shot during the later part of the Civil War and died on August 21, 1875, the thirteenth anniversary of the date he turned Stieler and Bruckisch in. Siemering's *Germans During Civil War*, June 8, 1923; Ransleben *100 Years*, pp 95, and 118; Schellhase Journal; Kerr Historical Commission *Kerr County Album*, p 315; Stewart *Death on Nueces Mina Stieler Stories*, pp 24-41; Hoffmann *German Women on the Texas Civil War Homefront*, pp 80-81; Schlickum's Letter; Notes taken from Autobiography of Hubert Heinen, Comfort, Texas 1872-1965, Exhibit 3—Genealogy—Stieler Family and Glenn *Capt'n John*, p 20.

9. Schwethelm Letter and J. W. Seal's Letter of September 8, 1862.

10. Stewart *Death on Nueces: Minna Stieler Stories*, pp 30 and 33 and Wisseman *Fredericksburg ... The First Fifty Years,* P 48. Also see Usener Family Story.

11. Wilhelmina Urban was born about 1823 in the Duchy of Anhalt. She married Gottlieb Stieler about 1842. She and her family arrived in the Comfort area in 1856. Family accounts state she suffered mental disorder as a result of her son's death and finding his body. She died about 1894. Wilhelmina 'Minna' Stieler was born on May 22, 1845. She married Henry Heinen on August 14, 1864 in Kerr County. She died on June 4, 1919. It is very difficult to believe two individuals traveled from Comfort to the battle site, almost 100 miles, over very rough terrain, with the danger of Indians, and returned with little supplies. It is much more logical for the two to have traveled from Comfort to a much nearer location. However, local folklore has them making a trip to the battle site.

The clues of the location of the deaths of Bruckisch and Stieler can be found in several accounts. First, in the very accounts that claim the two women went to the battle site. The accounts continue and tell the two women "passed through the ranks of the soldiers who, learning of [their] mission, were entirely chivalrous, and did not molest or prevent [them] from carrying out [their] intention." If the two bodies were at the battle site, there were no Texas or Confederate troops for the two women to "passed through the ranks of". Second, many of the Stieler family stories provide more data. Examples are: the Hubert Heinen's version, "there was added the tragedy of their oldest son Heinrich being murdered, near Kerrville on his way home from the 'Massacre of the Nueces'"; The Helen Ruth Stieler Crouch's version says, "When they arrived, [at the murder scene] they received a terrible shock. Instead of Henry Stieler and his friend, Theodor Bruckisch being killed in battle or executed in the group of prisoners taken by the Confederates unit, they found something much worse. Instead of dying with their companions in the Union Loyal League, Wilhelmine and her mother found the two teen-age men hanging from a tree, with a sign around their necks, SPIES."; The Tillie Reeh Heinen Lott's version, saying when they learned of the deaths the two women asked Balthasar Lick to accompany them, "Balthasar Lich hitched his horse team to the wagon and the two ladies, Mrs. Stieler and daughter Wilhelmine, went horseback. They packed some lunch and left early in the morning." These accounts all provide data that does not match what took place at the battle site, from "being murdered near Kerrville on his way home from the Nueces Massacre" to finding out the two were not killed in battle or executed in the group of prisoners, they were hung with a sign around their neck", to "packing some lunch and leaving early in the morning" clearly shows the death site of Bruckisch and Stieler was not at the battle site but much closer to Kerrville and home. Besides these printed versions there are others that state something like, "Heinrich Stieler took part in the 'Battle of Nueces'; escaped, killed on way home, near Kerrville."

When the Siemering and Schellhase data is added, a clearer picture emerges; Bruckisch and Stieler were killed near Kerrville on their way home from the Nueces Hoffmann *German Women on the Texas Civil War Homefront*, pp 8081; Ransleben *100 Years*, p 95; Stewart *Death on Nueces: Minna Stieler Stories*, pp 30, 32-37 and 41; Notes taken from Autobiography of Hubert Heinen, Comfort, Texas, Exhibit 3—Genealogy—Stieler Family. Copy provided by Anne Stewart of Comfort, Texas; and Personal Vignettes, "The Wilhelmine Urban Stieler Story" by Roberta Warren, an oral account presented at the 'Nueces Encounter – 1862, Battle or Massacre', Symposium on March 22, 1997. The likelihood that Henry Heinen, Peter and Martin Ingenhuett were the three Comfort area men that help bring the bodies home comes from the records of Provost Marshal James Starkey. He issued travel permits to these three men on August 29, 1862. Watkins *Kerr County*, p 116.

12. Ransleben *100 Years*, pp 111-112 and Thompson *Mexican Texans in the Union Army*, p 13.

13. Several accounts of the battle are told without any or little actual source documents or concern for the truth. One such example is in *Rip Ford's Texas*. It tells of Sansom returning to the battle site after the Confederates left. It states, "Hours later, when the Confederates had gone, Sansom, suffering from shock, wandered over the battleground talking to the corpses." John W. Sansom "The Bloody Battle of Nueces River" '*K. Lamaity's Harpoon*' Volume Six, Number Eight, October 1, 1908 and Oates *Rip Ford's Texas*, pp 339-340.

14. Harpoon; Montel's MRs; and Stovall *Breaks of The Balcones*, p 69. Benjamin F. Patton deserted the unit in March 1864 and on April 1st enlisted in Company I, First Regiment [Union] Texas Cavalry. Benjamin F. Patton's Confederate Service Records and Benjamin F. Patton's Union Service Records.

15. Sansom's Memoirs and Glenn *Capt'n John*, p 24.

16. Leo Itz's Interview.

17. Hoffmann's Letter and Comfort *News*, August 13, 1909. The name of this Unionist is not known, but he was one of the 19 listed on the Comfort Monument. He attempted to carve his name on the stock of his rifle. The body and rifle were found about 1867 by a Mexican. He buried the skeleton and gave the rifle to Henry Baylor of Uvalde. August Santleben procured it and made it a present to Captain Adolf Zoeller on August 10, 1909. The current location of the rifle is not known.

18. Mathias Pehl was born about 1839 in Nassau. He arrived in Texas with his family on the *Hercules* in 1845. He was a member of Braubach's Company from February 1861 until February 1862. He died about 1867 in Gillespie County.

William Graff was born about 1841 in Prussia, a brother of Carl Graff, also a member of the Unionist group. The date they arrived in Texas is not known. After the battle, William Graff returned to Gillespie County. He died on February 3, 1905 in Gillespie County.

Hoffmann's account identifies this man as Adolf Vater, which is almost certainly incorrect. The name Adolph Vater is shown on the Comfort Monument as killed at the battle site. Since Hoffmann states this man survived, it was not Adolph Vater. A second man named Vater, Fritz Vater, is also shown on the Comfort Monument as killed at the battle site. Kuechler later tells about finding a wounded man named Vater, who also survived. Both Adolph and Fritz Vater's estates were probated in Gillespie County, so it is certain they died at the battle site. Therefore, the Vater Hoffmann and Kuechler spoke with was William Vater. William Vater's date and location of birth or death are not known. He was not located on any 1850, 1860, or 1870 Texas Census. Hoffmann's Letter; Eighth U. S. Census, 1860 Gillespie County Texas Census, pp 13 and 27; Geue *New Land Beckoned*, p 127; Braubach's MRs; Gillespie Marriage Records, Book 1, pp 30 and 390; Gillespie County Probate Records, Book C, pp 466-467; Gold *Record Book of the Vereins-Kirche*, pp 68, 122, 137 and 152; Lindig *Der Friedhof Cemetery*, p 27; Kuechler's letter; and Gillespie County Probate Records, Volume C, pp 229-230, 232-233, 279, 281-284, 322, 328-329, 331334, 364, 367, 370, 372-373, 389-391 and 413.

19. These Unionists were likely the Vetterlein brothers, Frederick Weber, Henry Kammlah, August Duecker, Joseph Poetsch and the three members of the Schwethelm group. They were soon joined by August Hoffmann, Mathis Pehl, William Graff, and Wilhelm Klier. With the seven in the Cramer and Kuechler group, the survivors numbered about twenty. Cramer's account implies these were part of the group 'who left early', which they were not. They may have left before Cramer and Kuechler group, but left only shortly before them.

Frederick Weber was born about 1838 in Nassau. He arrived in Texas with his family on the *Washington* in 1845. He arrived in the Comfort area in 1854. Weber was a member of Kuechler's February 1862 Company. He returned to Comfort after the battle. Frederick Weber enrolled in Company A, First Regiment [Union] Texas Cavalry on December 22, 1863 at Brownsville. He was declared missing in action [MIA] after the Battle of Las Rucias on June 25, 1864 and never returned to his unit. Cramer's Letter: Eighth U. S. Census, 1860 Kerr County Texas Census, p 72; Geue *New*

Land Beckoned, p 151; Kuechler's MR; Menke *Index to First and Second Regiments [Union] Texas Cavalry* and Frederick Weber's Union Service Records, National Archives, Washington, D. C.

20. Kuechler's letter and Cramer's letter. This was likely the same water hole where the Hoffmann group was and where Hoffmann states, about twenty of the survivors "came together." August Siemering also tells about Unionist coming together. His account says 32 Union men, of whom 18 were wounded came together. This count of 32 seems to be much too high. There were 69 Unionists at the beginning of the battle. Nineteen were left at or near the battle site and five were with Sansom. This leaves 45. Of these, 23 'left camp early' and returned "home by a different route." This leaves only twenty-two. It is known Karl Itz, William Vater and the unknown Unionist did not rejoin. This leaves only nineteen. Hoffmann's Letter and Siemering's Germans During Civil War, June 8, 1923.

21. Cramer's Letter.

22. Schultz *Was a Survivor of the Nueces Battle*, p 26 and Pettis' Letter.

23. Cramer's letter and Kuechler's letter. These two eyewitness accounts of seeing the executed Unionist provide good data on how they died. Neither account says anything about seeing their hands or feet tied, nor do they imply any were hanged. The accounts describe seeing bodies killed by bullets. Cramer's statement about "their faces and bodies completely riddled with bullets" is further evidence that none were hanged. They were executed by gunfire as described by Williams.

24. Kuechler's letter.

25. Fenley *History of Uvalde*, pp 59-60 and Hoffmann's Letter.

26. Schultz *Was a Survivor of Nueces Battle*, pp 26-27. Thomas Riordan's U. S. Army Enlistment Records; Seventh U. S. Census, 1850 Bexar County Texas Census, p 285; Eighth U. S. Census, 1860 Dawson County Texas Census, p 390; Land Records Kinney County, Texas Branches and Acorns, Volume III, No. 4, June 1988, p 142; Wilbarger *Indian Depredations*, p 654; Sowell *Texas Indian Fighters*, p 483; Marriage Records of Uvalde County and Ellen Riordan's Mexican War Pension Records.

27. Schultz *Was a Survivor of Nueces Battle*, pp 26-27. An account told by the Kusenberger family states they found the Riordan house because they heard a roster crow. One of the men went to the house and asked the farm lady for food. Just at that time some Confederates rode up and the farm lady lied for the man, enabling them to get away. Letter, Frank Albrecht, Jr., Austin,

Texas, March 12, 1995, contained in series of documents entitled "The Kusenberger Family'. Copy provided by Gregory J. Krauter, Comfort, Texas.

28. Schultz *Was a Survivor of Nueces Battle*, pp 26-28.

29. Schultz Was *a Survivor of Nueces Battle*, pp 27-28; Letter, Frank Albrecht, Jr. March 12, 1995, Austin, Texas to Gregory J. Krauter and Letter, Frank Albrecht, Jr., Austin, Texas, August 11, 1999 to author.

30. Schultz *Was a Survivor of Nueces Battle*, p 28 and Albrecht's March 12, 1995 Letter. The Kusenberger family calls this part of the story "Graff on a raft."

31. Adolph Real was born on March 21, 1825 near Düsseldorf, Prussia. The family arrived in Texas about 1847. Real was a merchant living in Bexar County at this time. Adolph Real died on December 20, 1901 in Bexar County. Casper Real was born March 3, 1824, near Düsseldorf, Prussia. The family arrived in Texas in 1847. Casper married Emilie Schreiner, the sister of Amil, a member of the Unionist group and Charles, on November 5, 1853 in Bexar County. They were the parents of Julius Real who was a state senator and for whom Real County is named. Casper was a member of Harbour's Home Guard in February 1861 and in 1864 was a member of Farr's Company of the Third Frontier District. Casper Real died on August 1, 1893 in Kerr County. Geue *New Land Beckoned*, pp 130 and 139; Seventh U. S. Census, 1850 Bexar County Texas Census, p 275; Kerr Historical Commission *Kerr Album*, p 378; Harbour's MR; Farr's MR; Bexar County Marriage Records, Volume C, 127; Eighth U. S. Census, 1860 Kerr County Texas Census, p 74 and Schultz *Was a Survivor of Nueces Battle*, p 28.

32. Pierce's Letter of September 22, 1862.

33. Schultz Was *a Survivor of Nueces Battle* p 29; Schwethelm Letter Henry Schwethelm Union Service Records, National Archives, Washington, D. C., (hereafter cited as Henry Schwethelm's Union Service Records); Jacob Kusenberger's Union Service Records, National Archives, Washington, D. C. Carl (Charles) Graff's Union Service Records National Archives, Washington, D. C. This is an example of not being able to completely rely on data provided by someone who conducted an interview. It must be check against any official records that can be located. The Schwethelm account claims bad feelings developed toward Graff because Graff had a $20 gold piece and did not share the money with them. This was despite the fact they sold Graff's pistol and used the money for food and other items. The Schwethelm account says when Graff attempted to enlist in Company A, "they raised objections to his becoming a member of their company, and

succeeding in their efforts, Graff was mustered into another company which had no relationship to the First Texas Cavalry." Graff enlisted in Company B, First Regiment [Union] Texas Cavalry.

34. Ransleben *100 Years*, pp 59 and 89; Pettis' Letter and Raley *Blackest Crime in Texas Warfare*. George H. Pettis was a first lieutenant in Company K, First Regiment [Union] California Infantry. He commanded a 78-man detachment at Camp San Elizario, Texas that was located about 20 miles south of El Paso. Pettis and his detachment did not arrive at San Elizario until December 7, 1862. George H. Pettis was born about 1834 in Rhode Island. He died on January 28, 1909; George H. Pettis' Union Service Records, National Archives, Washington, D. C.; Webb *Handbook of Texas*, Volume II, pp 549-550' Eighth U. S. Census, 1860 San Francisco County California Census, p 59; George H. Pettis' Union Pension Records, National Archives, Washington, D. C.; OR, Series I, Volume 34, Part II, p 209; OR Series I, Volume 50, Part II, p 24; and *Records of California Men in the War Of The Rebellion—1861 To 1867"*, Revised and Compiled by Brigadier General Richard H. Orton, Adjutant General of California (California State Printing Service), 1890, pp 333-334.

35. Leo Itz Interview and Gillespie Historical Society *God's Hills*, Volume II, pp 46-48. The German word used by Leo Itz Interview to describe Itz and others who hid in the bush was *Buschwacker*, which means 'hidden in bush'.
36. Leo Itz Interview.

37. There is a brief account of Karl Itz's ordeal in *Pioneers in God's Hills*. It states, "After days of hunger and he had travel afoot, he made his way back to his parents' home. Lest he endanger them he soon left to live in hiding, first near friends in the Llano region, and then for eight months in the Cherry Springs area." Karl had a child and a nephew die of diphtheria while he was in hiding. This informal peace treaty with the Indians did not work for Karl Itz's brother-in-law, Heinrich Arhelger. Indians killed him on February 14, 1863 while Karl was in hiding. According to Leo Itz, the reason for Arhelger's death was his disrespect for the Indians. Gillespie Historical Society *God's Hills*, Volume II, pp 46-48 and Leo Itz Interview.

38. The date the Unionists departed can be established by Captain Cramer's letter. He says the fourth day after the battle, which would be the 14th, the Confederates, withdrew and "the next day", would be the 15th he started for home. Hoffmann's Letter and Cramer's Letter.

39. Kuechler's letter; Raley *Blackest Crime in Texas Warfare* and Hoffmann's Letter.

40. Hopkins' diary. The number of 14 is amazingly accurate. Did the Confederates have a source or was it just coincidence? The Hopkins' diary states 14 was the number of Unionist returning and Hoffmann's account states 15 were in fact the number returning. Schwethelm's letter says 14. This number does not count the 23 Unionists who 'left early.' These two accounts show somehow the Confederates acquired almost the exact number heading home after the battle. Simon was likely the source, but the Cramer account states he was separated from the group before the Schwethelm and Vetterlein groups left for Mexico and El Paso.

41. Raley *Blackest Crime in Texas Warfare* and Hoffmann's Letter.

42. Striegler *Johan Frederick Gottlieb Striegler Family History,* pp 77-78 and Schlickum's Letter. Hoffmann's and Schlickum's statements that there were nine Unionists who returned to the Fredericksburg area and Cramer's Letter that states 6 were in the Comfort group helps establishes there were a total of 15 who returned in the two groups, almost exactly the number the Confederates and Hoffmann reported. Julius J. L. S. Ransleben was born on August 17, 1814 in Prussia. He arrived in Texas on the *Mathilde* in 1846. He was of the first settlers in Fredericksburg. On October 18, 1852, he married Josephine Klier, Wilhelm's sister. Julius Ransleben died on October 9, 1897 in Gillespie County. Gillespie Historical Society *God's Hills*, Volume I, pp 169170.

43. Klier Story, pp 77-78; Schlickum's Letter and The State of Texas vs. James Duff and Richard Taylor.

44. Schlickum's Letter. No record of this trial has been found.

45. Gillespie Historical Society *God's Hills*, Volume I, p 73; Oral Interview with Amelia 'Mollie' Eckhardt Dennis, a descendant of August Duecker on September 5, 1997; *The Early History of Gillespie County,* by Ophelia Nielsen Weinheimer, Master of Arts Thesis, Southwest Texas State Teachers College, 1952, pp 58-59 and "Bushwhackers In Them Thar Hills" by Vernal Alberthal in *The Radio Post*, October 2, 1952.

46. Amelia 'Mollie' Eckhardt Dennis Interview; Weinheimer Early *History of Gillespie County* and Alberthal *Bush-whackers in Them Thar Hills.* Friedrich Scharnhorst was born about 1809 in Hanover. This story was passed down in the Scharnhorst family by Charles Scharnhorst, who was about 4 years old in 1862 and remembers the stories told by his father. Carl's daughter, Alma, passed the story on to family members and others. Eighth U. S. Census, 1860 Gillespie County Texas Census, p 28. One local folk story says that August Hoffmann and Heinrich (Henry)Rausch were among those that hid along South Grape Creek between the communities of

Luckenbach and Grapetown. "The Grapetown Legacy" by Cynthia Hohenberger 'The Texas Junior Historian', Volume 26 #1, (September 1965), p 8.

47. Usener family story.

48. Cramer's Letter and Kendall County Historical Commission *History of Kendall County*, p 235.

49. Oral Interview with Louis Weiss Haufler, a Weiss descendant on August 10, 1996; Family story of the Lick family of Comfort. Story provided to author by Gregory J. Krauter of Comfort; and "Comfort in Texas" by Walter Herbst in 'The Junior Historian', Texas State Historical Association, November 1945, Volume VI, No. 2, p 5. The family story states the Unionist hid on the Louis Lich farm. But, it was three years after the war before the Lick family purchased the farm. The farm that Balthasar Lich, the original Lich settler, purchased was next to the Henry Steves homestead. The story incorrectly identifies the Unionist as named Brinkmann. No record of a Brinkmann was located showing he was in the fleeing Unionist group. The only unaccounted Unionist returning to the Comfort area is Frederick Tegener. Adolph Pfeiffer was born about 1814 in Prussia. Date arrived in Texas not known. He arrived in the Comfort area in 1855 and settled in the Cypress Creek area. It is not known if he had a medical degree but was known as 'Doktor' He died on October 31, 1881. Eighth U. S. Census, 1860 Kerr County Texas Census, p 69 and Ransleben *100 Years*, pp 25, 56- 57.

50. Cramer's Letter.

51. State of Texas vs. James Duff and Richard Taylor and Cramer's Letter.

52. Cramer's Letter.

53. Cramer's Letter; Seventh U. S. Census, 1850 Guadalupe County Texas Census, p 295; Records, 31st Brigade, A.G.C., Austin, Texas and Kendall Historical Commission *A History of Kendall County*, p 9.

54. Cramer's Letter.

CHAPTER 13 – The Confederates Leave the Hill Country

At Camps Davis and Pedernales' Captains Davis and Donelson were continuing their mission of enforcing martial law and requiring the citizens to take the required oath of allegiance or else be "dealt with summarily" and taking such actions with "such vigorous measurers as … deemed necessary." With the departure of Lieutenant McRae and his men, Donelson's force was considerably reduced. He only had six officers and 150 men, plus the remainder of Davis's Company. There remained at Camp Pedernales one officer and 65 men from Duff's Company. The one officer was Second Lieutenant Richard Taylor. First Lieutenant James Sweet was on detached duty as a member of the Confederate Military Commission in San Antonio. At **NO** time does Captain James Duff's name appear on the Camp Pedernales Post Returns.

Duff's name is mentioned several times in Hopkins' diary: on July 31st the day Duff arrived, and on August 3rd where he says, "Sunday evening our captain [Captain Davis] called on Capt. James Duff. Duff camped about one mile below us." While Duff's Company was at Camp Pedernales, he was at Fort Martin Scott just outside of Fredericksburg, which is where he had his provost marshal headquarters. Therefore, it was Donelson not Duff who was enforcing martial law requirements.[1]

Complying with his orders, Donelson arrested several insurgents and their sympathizers. Many of the insurgents' families were also detained. No doubt farms and homes of some of these insurgents were destroyed as ordered by General

Bee. Several arrested insurgents were executed. Blame for their deaths and many others in the next three years has been placed on James Duff, virtually to the exclusion of every other Confederate involved in suppressing the Hill County insurrection. As a result, Duff has been dubbed "The Butcher of Fredericksburg." However, Duff is not entirely blameless; one Gillespie County death did happen while he was the provost marshal, although Duff was not likely aware of the death until after it occurred. This execution took place during the night of August 4th. R. H. Williams tells what he heard about the event as it took place **after** he left in the pursuit force. He is not a primary source. On August 4th, Monday, one man was brought in by a patrol, "accused of being a Northern sympathizer. Nothing could be proved against him, and that same night he was released, and given a pass," says Williams. "He went away happy enough no doubt, poor fellow, at his escape from such clutches. Next morning his body was found hanging in the woods near by, with the throat cut from ear to hear." Hopkins of Davis' Company says on August 5th, a Tuesday, "As our regular scouts came in today they found a man hanging right over the trail with his throat cut from ear to hear. Learned his name was [Howell], and hung over our trail by some of Duff's men, for two purposes. First, to leave the impression that our boys had done it; and second, we would have to bury him." R. A. Gibson, whose fences had been destroyed by the Union Loyal League, says a man named Howell was executed because he was "a proven abolitionist."

Another myth is that, "Over 150 men and boys were lynched during the tenure of James Duff in the Hill Country," that Duff conducted a reign of terror against innocent citizens of the area and he and his troops "had been on a lynching and burning spree in Unionist German communities." An analysis shows only three deaths took place while Duff was in the area: John Howell, Conrad Bock, and Fritz Tays. Captain Duff arrived in the area on July 31, 1862 and left, at the latest, on August 20, 1862, a total of 21 days. A total of 14 deaths can be determined to have taken place during the time Confederate troops were in the area, but except for these three, all were executed during the time Captain John Donelson, not James Duff, was the provost marshal.[2]

It was on August 16th that troopers at Camp Davis and Camp Pedernales learned of the Nueces Battle. The first report of casualties said that Texas and Confederate forces suffered one man killed while forty of the 'bushwhackers' had died. About 8 o'clock that evening the camps were advised to send out scouts to find fourteen insurgent survivors heading back to the Hill Country. The two camps were "placed on a war footing," which included putting pickets out to prevent a surprise attack. After the pickets and scouts were dispatched, not many men were left in the two camps. Camp Davis only had three men. By August 18th, the two camps settled back down into their normal routines.[3]

Duff and his company prepared to return to San Antonio. He visited Southern sympathizer Robert A. Gibson, who lived in the Spring Creek area. Duff advised Gibson to leave the area as he felt it was unsafe for him to remain. Fearing for his and his family's safety he left his 'stalk' in the field and moved about eighteen miles east of Fredericksburg to a spot near the

Blanco County line. Gibson was not safe here either. Again, his farm was raided. This caused him to take up arms and become a member of the *Haengerbande* [pro-Confederate bushwhackers]. In late 1863 and early 1864, Gibson was one of three major pro-Confederate leaders of the bushwhackers and committed several of the killings later blamed on James Duff. Gibson was likely a major source in identifying the Anglo members of the Union Loyal League. Several of these were executed including Sebird Henderson, Frank Scott, and Hiram Nelson.[4]

On August 20, 1862, Captain John Donelson replaced Duff as provost marshal at Fredericksburg, whereupon Duff and his company left for San Antonio. It is clear that Captain John Donelson was acting as the new provost marshal in Fredericksburg by August 21st. Duff was back in San Antonio by August 25, 1862: on that date he sent a letter to the Department of Texas from San Antonio requesting permission to recruit two additional companies of Partisan Rangers. Permission was granted and on January 12, 1863 the companies were organized into a battalion known as the 14th Battalion Texas Cavalry. Duff was promoted to a Confederate major on January 3, 1863, with date of rank as of November 22, 1863, and to lieutenant colonel on February 7, 1863. In the spring of 1863, the battalion was deployed to the Rio Grande. On May 2, 1863, it was increased to regimental size and renamed the 33rd Regiment, Texas Cavalry with Duff as colonel. He remained with the regiment until the end of the war. On May 10, 1865, as the War was ending an Austin newspaper had this to say about James Duff, "We had the pleasure of a call this morning from our old friend, Col Jas. Duff, of San Antonio. We are pleased to see him in fine health and spirits. Few men in Texas have made more sacrifices for the Confederacy than Col. Duff. After the commencement of the war he was engaged in a large and prosperous business, and having been a large contractor, for many years, for the old U. S. Army, he might still have been engaged in that capacity for the Confederate government, by which he might have realized an immense fortune, as many others have done; but after furnishing all the means within his control to support our cause, he closed up his business, and has now been for three years in active service in the field. Under ordinary circumstances we should not have made this statement, but in time like these, it is desirable we should know who are the patriots of the country." This glowing statement of Duff was reprinted in several Texas newspapers. First Lieutenant James Sweet was promoted to captain and placed in command of Company B. First Lieutenant Richard Taylor was promoted captain and placed in command of Company A. Second Lieutenant Edwin Lilly was promoted to first lieutenant in Company A.[5]

One of the major myths is Duff spent a long time in the Hill Country as provost marshal, during which he conducted a reign of terror. The fact is he spent less than 44 days in the area. The first trip was from May 30th to June 21st, a total of 23 days. His second trip was from July 31st to no later than August 20th, a total of at most 21 days. Another source that helps establish the fact is R. H. Williams. He says after the battle he returned to his ranch on the Frio River. By analyzing his account, it shows he returned to San Antonio on either August 22nd or 23rd.

Williams says, "I sent out to return to the command at San Antonio, where I arrived at the end of the week, and found my company encamped on the riverbank near the town." August 22nd was a Friday and the 23rd was a Saturday, the end of the week.[6]

Withdrawn about the same time as Duff's Company were the detachments from Taylor's 8th Battalion Texas Cavalry. New units replaced them. On August 25, 1862, a fifty-three-man detachment from Captain Josiah Taylor's Company G, 32nd Regiment Texas Cavalry commanded by twenty-seven-year-old First Lieutenant John R. Kelso, arrived as replacement for Duff's Company. Two new detachments from Taylor's Battalion arrived on August 28th. They were a seventeen-man detachment from Company B, under the command of Second Lieutenant George R. Kuykendall, a twenty-one-year-old Texan who had enrolled in Company B at Fort Clark in May 1861. A five-man detachment from Companies D and E, commanded by Second Lieutenant Young Collins, a thirty-three-year-old Alabaman who had enrolled in Company D in May, 1861, at Fort Clark, arrived at the same time. [Appendices AF, AG, and AH, pp 683-686]. Other changes were made. Second Lieutenant Leonardus S. Lawhorn, a twenty-nine-year-old Georgia native from Company K, 2nd Regiment Texas Mounted Rifles, was appointed acting commissary and subsistence officer responsible for receiving and disposing all confiscated insurgents' property. One trooper commented on confiscating property, "We have had a great deal of sport confiscating property of all kinds."[7]

Scouting continued. One member of newly arrived Company G of the 32nd Regiment described the area near Fredericksburg: "There is now a daily guard around Fredericksburg. The 'bushwhackers' or traitors are plentiful in this country but keep themselves hid, and they have selected a good country for the business." Two of these 'bushwhackers'; thirty-six-year-old Arkansas native Louis Nelson and thirty-one-year-old Reinhardt Reeh, a native of Nassau, hid in the hills near present-day Harper. Reeh's mother, Catharine, took food out to them.[8]

The stress was beginning to show in the relations between the Confederate troops and the citizens of the area. One state trooper observed some of those arrested and the general mood in Fredericksburg. On August 21st, he "went to town. Saw the conscripts. They had long faces and some of the good old German women had longer faces. Some of them sobbed as if their hearts would break. Oh, those dear old mothers, wives, and sweethearts." The trooper also got a sense of the general feelings of the citizens of the town. He went to the Nimitz Hotel for lunch. He sat down and ordered his meal. He was forced to wait even after others who had ordered after him received their food. He continued to wait while others were served. Still no lunch. Finally, he became angry and asked the waiter why he had not received his meal. The waiter pretended not to speak English, only German and the trooper never did get a meal.[9]

The departure of Duff on August 20th was a bad omen for the League's insurgents. Just two days after Duff left, state and Confederates forces went on a hanging spree, the same day that Davis' detachment returned from the battle. "Our big scout got in from the bushwhackers fight on Nueces River. The men on our side were made up from the different companies of our regiment and part of Duff's independent company, commanded by Major [John] Donelson," said one of Davis' sergeants. "There were two men killed and eighteen wounded on our side; and thirty bushwhackers killed. A number of prisoners were taken; but all got way, so the boys said; and I knew they would not lie about a little thing like that," reported D. P. Hopkins.[10]

The reasons for the hanging are not clear, but may have been in revenge for the deaths of the Confederates at the Nueces Battle. Whatever the reasons, that afternoon seven insurgent prisoners were taken from Camp Davis and Camp Pedernales and hanged from a large oak tree on Spring Creek. "We are having tough times in this neck of the woods these times. I counted seven men hang on one limb, cut down and thrown over the bluff into Spring Creek. One old gray-headed man, named Nelson, 80 years old, was among the number. When the body rose to the top of the water some women lifted him out in a sheet and buried him," wrote Hopkins in his diary. A second Confederate eyewitness reported seeing bodies just four days later, reporting only four not seven bodies. "The creeks in this vicinity are said to be full of dead men!! I witnessed a sight yesterday, which I never wish to see again in a civilized and enlightened country" recalled trooper Thomas C. Smith. He added, "in a water hole in Spring Creek (about two miles from camp) there are four human bodies lying on top of the water, thrown in and left to rot, and that too after they were hanged by the neck and dead." Another account says that the four were hanged "On Spring Creek near the present location of Harper." It continues saying "rocks were tied to their feet, and they were thrown into Spring Creek, now called 'Dead Men's Hole.'"

Smith continues his comments, "If they are traitors no doubt they deserved their reward but should have at least gave them a burial." A third trooper also wrote of the dead insurgents saying, "The tories in this part of the country is getting some what scarce that is to day those that are living. It would be a hard mater [sic] for me to give any thing like a definite account of those that have been shot or hung in this vicinity. They are lying and hanging all over the woods." These statements come from credible sources. They established that seven insurgents were hanged on August 22, 1862, but by the 26th families of four men's families arrived and buried the bodies. Not all accounts agree the deaths took place on August 22nd. The book *Kerr County* says it was "about the same time" as the battle, or about August 10, 1862. John W. Sansom, in a 1911 article, states the deaths took place "during the last days of July" 1862. James Starkey, the provost marshal for Kerr County, issued Hiram Nelson Travel Permit No. 10 on July 30, 1862, so it can be concluded that Nelson's death was after that date.

Four of these seven can be identified: Sebird Henderson, Hiram Nelson, Gus Tegener, and Frank Scott. There are questions as to the identity of the other three.[11] These four were the ones captured about August 3rd and kept at Camp Pedernales while the wives and children were sent to Fredericksburg. The Confederates, "took Frank, Sr. and old man Henderson-a man about 60 years old-with two more as prisoners, out to a camp at a water hole on the Divide about 35 miles west of Fredericksburg," and "took the women and children to Head-quarters in Fredericksburg where they were kept in a one-room hut; would not permit them to leave this hut unless they would ride out with them." Mary Scott's daughter recalled, "meanwhile [Mary, Frank Scott's wife] and the children had the measles. There were ten people in that one room. Mother later said she almost lost her mind with her three children sick and she had not heard a word in all that time about her husband."

Some of the state or Confederates troops attempted to help the wives, "Now, there were some good men in the bunch that had taken the oath or had joined the band to save their lives of them families. These men would help the women when they could do so without detection. They were sorry for the women and children and were kind to them as they could possibly be but they could not do very much for them."[12]

The wives of Henderson, Nelson, Scott, and Tegener learned of their husbands' deaths when, "One man named [William] Banta told mother that the Captain had ordered those four men killed as they would not take the oath, and that they killed them and threw their bodies into Dead Men's Hole." Either Captain Henry Davis or Captain John Donelson was the executioner as they were the only "captains" in the area. Davis was Banta's "captain" so it is likely he was the executioner. It was definitely not Duff, since he and his company left Gillespie County on August 20th. The executions did not take place until August 22nd.[13] Though the wives were quite ill with the measles, they asked permission to bury the bodies. They were told if, "they joined the Confederate troops," they would be allowed to bury the men. At least two of the wives/widows did "join the Confederate troops." Mary Jane Payton Tegener, who had married Gustav Tegener only on May 6th, married John Helm on October 5, 1862, just six weeks after Tegener was executed. While not a widow, eighteen-year-old Susan Eveline Benson Tegener married Frederick "Fritz" Schladoer on July 27, 1863. Schladoer was one of the men who likely provided information on the insurgents and later provided additional data to the pro-Confederate bushwhackers.[14]

Many of the men made lewd propositions to the women held at Fredericksburg. The wives told the men they would scald the next man who came after them. At dark, a man slipped up to a window and asked Mary Scott to, "go walking with me." The man attempted to climb into the hut through a window. Mary Scott threw hot water into his face. He fell out of the window howling. The guards made fun of the man and the women had no further problems with the troopers. The women were finally allowed to bury their husbands. They and their six children dug a big grave, taking turns and digging all night. "Wading into water up to their armpits in order to get the bodies of their loved ones, they placed them on a sheet, rolled them into the grave, said a prayer, and filled the grave." The bodies were so swollen and black they could not be identified except for Frank Scott who was missing a big toe. Shortly thereafter, the women were released. The four men were buried in a common grave in the Spring Creek Cemetery near Harper. There is a tombstone bears the inscription:

Sebird Henderson
Hiram Nelson
Gus Tegener
Frank Scott
1862
Hanged and Thrown in Spring Creek By Col. James Duff's
Confederate Regiment[15]

The names of the other three that were hanged are not easily identified. Subsequent writers incorrectly blamed Duff for these executions. "I know that J. M. Duff and his company of murders killed many of my neighbors and friends," wrote an insurgent. "My uncle and cousins, Schram Henderson, my wife's father and brother, Turknette, [sic] were murdered, my neighbors, Hiram Nelson, Frank Scott and his father, Parson Johnson and old man Scott were all butchered by Duff and his gang," says Howard Henderson. He continues, "Many others of my neighbors were put to death, their houses burned, and their wives and children taken to the camps of soldiers where, when not otherwise insulted, they were compelled to listen to the foul tongues that denounced their husbands and fathers." Sansom names a total of nine men who were, "put to death during the last days of July." His list includes several men killed much later and even one who was later a member of Duff's command. Sansom identifies them as, "Gustav Tegener, Young Turknett, Rev. Tom Scott, Frank Scott, Rev. Jim Johnson, Hiram Nelson, Warren Cass, Wm. Schultz, [and] Ephraim Henderson."[16]

From the various accounts, a total of fifteen names can be identified as being insurgents that Donelson and the Confederates allegedly executed near the end of July and early to mid-August 1862. These fifteen are Sebird Henderson, Schram Henderson [likely Sebird Henderson], Hiram Nelson; Ephraim Henderson [also likely Sebird Henderson], C. Frank Scott, Frank Scott's father [Benjamin Scott], 'old man Scott' [likely John Scott of Burnet County], Reverend Tom Scott, Gustav Tegener, Warren Cass, William Schultz, Philip Brandon Turknett, John S. C. Turknett, Jacob Turknett, and Parson/Reverend Jim Johnson. The only ones of these fifteen executed on or near August 22nd were Frank Scott, Hiram Nelson, Gustav Tegener, and Sebird Henderson; all executed by orders of Captains Davis or Donelson. Duff, or the Confederates under his control, executed none of the other eleven. Of those eleven, bushwhackers executed at least four in 1864, not in 1862. These four were Philip Brandon Turknett, John S. C. Turknett, Jacob Turknett, and Warren Cass. One was executed in Burnet County; 'old man Scott' – John Scott. Two of the fifteen were not killed: William Schultz and Frank Scott's father [Benjamin]. Three of the names are duplicates and are in fact Sebird Henderson. One cannot be identified, that being Reverend Tom Scott. So, who are the other bodies seen on the same limb on August 27, 1862? One 'could be' Reverend Jim Johnson. That still leaves two and perhaps three bodies not identified, since only four bodies were recovered from Spring Creek and buried. What happened to the other three bodies? If they were not buried than how were they disposed of? These three 'may have been' part of the nine insurgents who were 'later captured and executed.' One of these three was likely Wilhelm F. Boerner. The other two can not be identified.[17]

Captain Donelson kept his scouts out looking for insurgents, but it was not insurgents who were captured. Philip G. Temple, the lieutenant of the American Company, captured Donelson. After her husband's execution, Mary Scott stayed with her father-in-law, Ben Scott. The Confederates arrested Benjamin Scott and took him to their camp for questioning. There was fear he would be killed. Because of this concern, Lieutenant Temple sent word to Captain Donelson that if Benjamin Scott were harmed, he would kill Donelson in retaliation. Captain Donelson returned Scott to his home. Upon arriving at the Scott home, Donelson ordered Mary to fix him and his bodyguard's dinner. As Mary was preparing the meal, she saw Temple and several insurgents slipping toward the house. She became alarmed fearing a gun battle might start and either her children or other innocent people might be injured. Captain Donelson was setting with his back toward the door while his bodyguards were facing the door. Mary sent one of her daughters to give Donelson a table knife. By doing, so she put the girl between Donelson and the door. This prevented any likelihood Lieutenant Temple would open fire. Temple motioned Mary to get the child out of the way. She shook her head, "no".[18]

The Confederates had left most of their weapons on the horse saddles, which were in front of the house. Temple and his men slowly worked their way to the horses and began to collect the weapons. Mary moved to the door to block the

Confederates view. The insurgents got four weapons without any problem. As they attempted to get a fifth, the horse snorted and reared. The Confederates rushed outside past Mary. As they did, she grabbed two of their side arms. Without most of their weapons the Confederates offered little resistance. They quickly raised their hands over their heads and surrendered.[19]

Captain Donelson was scared and cursed his men for letting Lieutenant Temple capture them. Donelson begged Temple not to kill them. Temple threatened to kill them all. The captain promised to leave the Scotts alone and not brother them again. Temple and his men mounted the Confederate horses and made the Confederates trot ahead of them for about half a mile to where they had hidden their horses. Arriving they mounted their own mounts and gave the Confederates back their horses.

Temple told the Confederates to leave and said, "If you ever bother these people again, I'll kill you." As the Confederates rode away, Temple and his men fired over their heads. The relieved Confederates quickly made their way back to their camp. They did not keep their word. Instead, they ran off all of Mary Scott's livestock and burned her house. Mary Scott, her children, and father-in-law left Gillespie County and moved to Onion Creek in Travis County.[19]

Other insurgent families had fears of Confederate scouts and search parties. Wilhelmina Urban Stieler, the mother of Heinrich, one of the insurgents executed about August 25th, developed "mental illness" after finding her son's body and because of the many Confederate scouts arriving at her home looking for other insurgents. Wilhelmina developed such a paranoid fear of strangers for years afterwards that whenever one approached her home she cried out, "The Confederates are coming! The Confederates are coming to get me!" Another mother who "lost her mind" was Maria Degener, the mother of Hugo and Hilmar who were killed at the Nueces. She spent most of her remaining life "in black, lying on a bed, utterly crushed by the tragedy."[20]

Still another family victimized by the Confederate troops was the Casper Real family. Emilie Real Neill recalls hearing her grandfather, Arthur Real, telling stories that how as a young child he remembered the Confederate scouts searching for insurgents. One of the things he recalled was the sounds of the spurs and sabers of the mounted troops. He also recalled his mother getting up early in the morning and baking bread, some of which was taken to insurgents hiding nearby.[21]

One of Davis' State Troopers later told about this fear the local citizens had of the State Troops, "When we ride up to dutchmans house now they are the worst scared people you ever say. They don't know how to treat us. They invite us to lite [and] take something to eat or smoke or chew more expecially [sic] if they see an extra rope." This trooper also told about how well they were eating, "We are now living on the fat of the land. We have plenty of beef pork chicken turkey mutton and goat and c [sic] to eat Which [sic] as been confiscated here."[22]

For the most part, these scouts were very boring for the Confederates. A member of Company G, Wood's 32nd Regiment kept a diary that recorded some of them. On August 28th, an eight-man scout was sent out. On September 15th, another scout of eleven men was ordered out. Two scouts followed this on September 20th. One was a fifteen-man scout from the detachments of Taylor's Battalion. First Lieutenant Kelso from Company G, 32nd Regiment Texas Cavalry led the second consisting of eighteen troopers and five guides.[23]

Captain Donelson's reports reflect the lack of insurgent activity after the Nueces Battle. The insurgency was back to the beginning of Phase I. On September 8th, Donelson forwarded a report to General Bee in San Antonio that stated: "Since my last report nothing has occurred within this jurisdiction worthy of notice. Those disaffected citizens opposed to this Confederate government have generally yielded obedience by taking the oath of allegiance. Some twenty or thirty Unionist are still concealed in the cedar-brakes near this place. It is difficult to capture them as their friends and hiding places are ... [remaining is illegible]. Fifty men would now be sufficient to hold these counties in subjection. They would also be ... [illegible] ... in rendering assistance to the receiver while seizing and selling the confiscated property in this section a large amount of which is on hand. As before stated my opinion is confirmed that this population–principality German–will remain peaceable and not again commit an overt act of treason unless this State is invaded by our enemies in which event most of them will join the foe. Nearly every male citizen here is armed with a good six-shooter and rifle. Would it not be best to press these arms into the service before an opportunity could be afforded for having them [used] against us?"[24] Captain Donelson closed with these words: "Hoping the policy [of] Capt

Duff and myself have pursuit[ed] here has meet your approbation."[25]

Captain Josiah Taylor arrived in mid-September with the remainder of Company G, 32nd Regiment Texas Cavalry. He brought with him four men from Company B, Taylor's 8th Battalion. Taylor replaced Donelson as provost marshal on September 21, 1862. Captain Donelson and the remainder of his Company K, 2nd Regiment Texas Mounted Rifles returned to San Antonio. He was promoted to major on October 8, 1862. This left Captain Taylor with six officers and one hundred thirty-six men. The troopers included 65 men from Company G and 71 from Taylor's Battalion.[26]

Believing their men dead, many of the insurgent families started the process of rebuilding their lives. Two wives took legal action to prevent the loss of at least some of their property to the Confederates. Both filed affidavits claiming they owned separate property from that of their husbands. On September 12th, twenty-seven-year-old Louise Feuge Duecker, who had married August Duecker on June 11, 1854, in Gillespie County, filed documents claiming she owned a cow, valued at $18.00 and a sorrel mare, valued at $100. Forty-two-year-old Karoline Graff, who had married Carl Graff about 1851 in Prussia, filed documents on September 15, 1862, claiming she owned at the time of her marriage property valued at $160. The documents identified her husband as, "Charles Graff—said to be deceased." She did not learn of his fate until he returned at the end of the war.[27]

Donelson and Taylor's scouts kept the pressure on the hiding insurgents and forced some to make new attempts to reach Mexico. A group hiding between San Antonio and the Mexican border decided to try their luck. They included the three who had escaped from jail in San Antonio on July 19th: Julius Schlickum, Philip Braubach, and F. W. Doebbler. The prisoners had been kept in a jail that was connected to a firehouse. Six San Antonio Unionists assisted in the escape.

The prisoners had been provided with weapons. About 1:00 a. m., the three slipped through a half open door into the firehouse next door. They quickly made their way outside. Several of their friends were posted in nearby doorways. They made their way to the edge of the city where a Unionist hiding behind a mesquite bush challenged them. After exchanging countersigns, they mounted horses and in two hours fled about twenty-two miles west to a dry riverbed where there was a cave. Their friends told them where they would find food and other supplies. The other Unionists left with the horses.[28]

The three escaped men remained hidden near the cave in eastern Medina County for about six weeks. The nearest water was a spring about two miles away. A second source of water was about six miles away near the farm of a friend Wilhelm Huster, who supplied the three with groceries, bread, and salted meat. While in hiding, friends smuggled two survivors of the battle to their hiding place. Later a third survivor joined them.[29]

In mid-September, the decision was made to try and reach Mexico. Their friends again provided them with mounts. Julius Schlickum, a member of the group, states, "We detoured long miles around settlements and military post to avoid being seen, how, like Indians, we always had to cover our tracks." They reached the border near Eagle Pass about September 24th, out of food. The Rio Grande was very high and the current too swift to cross. For three consecutive days, they tried to cross, but the river remained too high and swift. On one attempted they ended up in the middle of one of Captain Thomas Rabb's camps but were able to get away before being caught. They returned to their hiding place. About October 2nd, they were able to purchased food, rest, and tend to their mounts.[30]

Friends on the Mexico side of the river realized where they were and sent a small boat. The six decided to cross at night, "right under the Stockade of Fort Duncan." At 1 a. m., they were safety across. However, they had to leave their horses on the Texas side. Schlickum hired a Mexican national to swim back across the river to retrieve their horses. The Mexican found the horses without a problem, but just as he was about to enter the water members of Captain Rabb's Company spotted him. He was arrested and taken to the provost marshal in Eagle Pass. As they entered the town, the Mexican found an opportunity to escape on horseback. With bullets flying past his head, he reached the Mexico side of the river safely. But that still left the insurgent horses on the Texas side. Schlickum and the others learned their horses were at a small ranch known as the William Dunavan's about nine miles north from Fort Duncan with only three men guarding them. On September 29[th,] they again attempted to get their horses.[31]

This time they built a small raft of dry wood. They bundled their cloths and weapons on the raft and at dusk crossed the river. Schlickum, Braubach, and two others were on the raft. The Mexican pulled it across the river. They made it safety across the 750-foot-wide waterway, dressed and retrieved their weapons. They approached the farm with caution; it was a clear, starry night, the moon went behind a cloud. They did not see any horses, and could not understand why there were no horses. Braubach and Schlickum crawled on their hands and knees closer to the building, taking care to keep upwind of several dogs. As they neared the buildings, they saw horses tied between the house and the barn. The two returned to their companions who were waiting behind some bushes about 50 feet away. As they were making plans to get their horses, the dogs started barking. The door of the house was flung open and Rabb's men came pouring out. It had been a trap![32]

Not only were Rabb's men coming out of the house, they had the little group surrounded. A barrage of fire came from all sides. The insurgents "pressed against the ground, wedged behind sheltering, low-growing shrubs." Rabb's men did not see them, but bullets flew over their heads. The insurgents returned fire, wounding one of the state troopers, twenty-three-year-old William Pelham an Alabama native. His friends "returned to the house, carrying [the] wounded soldier with them." All of a sudden it was quiet again.[33]

Schlickum looked around and found himself alone. He decided to try to get their horses. He crawled in a large circle around the buildings until he heard the hoof steps of horses. He saw a man leading two animals. Schlickum grabbed him from behind with one hand and with the other put his pistol to the man's head and said, "Don't move, or I'll shoot." The reply was in Spanish, "Oh, Don Julio don't kill me, it's me." It was the Mexican who in the confusion of the gunfight had cut the horses loose and brought them out. "Where are the missing horses?" asked Schlickum.

The Mexican replied, "There were only three and I brought them to the fence and since one horse didn't want to be led, I'm glad I got these two." Schlickum went back to the fence, found the third horse, and put a rope around his neck. He then swung onto the horse's back and followed the Mexican through the bushes. They were not followed. At the Rio Grande, they found the other three members of their party. They were surprised and couldn't believe their eyes when they saw the horses.[34]

The five men drove the horses into the river, launched their raft and soon were on the Mexican side of the river. When they examined the horses, they found the animals were not theirs, but as Schlickum said, "Well! It's not our fault." Three of the men mounted up and headed for Monterrey. Schlickum and Braubach remained in Piedras Negras, where in a few days Jacob Kuechler and Wilhelm Huster joined them.[35]

The skirmish between the insurgents and Rabb's Company made the San Antonio, Austin and New Braunfels newspapers. The Confederate version of the fight said: Twenty-two traitors who had left Texas for Mexico, to avoid the Conscript law, or to show their veneration for old Abe's government, recently re-crossed the Rio Grande, near Eagle Pass, and attacked a portion of Capt. Rabb's Company. Twenty of the renegades were killed, only two escaping to tell the tale. Rabb lost but two men.[36]

The October 15, 1862 issue of the *Austin State Gazette* reported: "We learn that Capt. Rabb's men of the State Troops some twelve days ago, were attacked by some thirty Dutchmen and bushwhackers who are fleeing from conscription. Rabb gave them a most royal drubbing leaving twenty-two of their number dead on the ground. The fight occurred near Eagle Pass on the Rio Grande. This fight and that of Duff's with Hamilton's followers will probably bring these miserable wretches to their sense."[37]

Captain Jacob Kuechler led another group of insurgents who decided to try again to reach Mexico. After hiding in Gillespie County for over six weeks, sixteen or seventeen fled toward Mexico. "In constant danger of being captured and murdered ... the survivors turned and made a second attempt to reach Mexico" says Kuechler. This second group included survivors of the battle, some of the twenty-three who 'left the camp early,' and others who were not in the original group. They left for Mexico about October 10th. They traveled the same general route as the August group. The men were well-armed and mounted and, "Of course they could not follow the regularly traveled road from San Antonio and Eagle Pass. They had to avoid all settlements for fear of being betrayed and hunted down," one account says. Another says, "They had to keep from being seen by other traveling like themselves, for the purpose of escaping death." When they reached the battle site they, "gathered the animal-scattered bones of their comrades and built a stone pyramid over them" and decorated the bones with green branches and wildflowers.[38]

Somehow the Confederates learned of this attempt. General Bee immediately sent a pursuit force to destroy them. Captain Stokely M. Holmes of Company K, 32nd Regiment

Texas Cavalry, commanded it. Captain Kuechler's group reached the Rio Grande on the evening of October 17th or the morning of 18th. There are conflicting stories to what they found upon reaching the river. Cramer says, "They had no trouble until they reached the Rio Grande. Just at the river they were attacked and under heavy fire [and] had to leave their horses and ammunition behind. They had to swim the Rio Grande." Kuechler and Huster "and 15 others crossed the river about 15 miles above [Piedras Negras] and were detected by the Rangers and shot at. They lost their horses and only 11 of them reached the free shore," explained Schlickum. Kuechler says they were overtaken, "before they could cross into Mexico." The Bonnet family stories say, "On reaching the Rio Grande they found the river well-guarded and decided to evade the guard by crossing at night." In this attempt, they found themselves suddenly right in the midst of soldiers. "Taking advantage of the confusion, [caused by] their sudden appearance ... they rushed on until they reached the edge of the water [where] a deadly fire was [opened] upon them. They found the river swollen and unfordable." Using the riverbank as a breastwork, they fired back and held their pursuers at bay until they could prepare for swimming across.[39]

While the insurgents were retreating toward the Mexican border, they were, "suddenly attacked by hostile forces and in the cruel contest that ensured more than one-third of [Kuechler's] command were slain," another account says. "In the fight, Captain Kuechler was severely wounded and escaped only after much suffering and hardship. The rest of them, with

Jacob Kuechler, happily reached the Mexican shore," wrote Cramer.[40]

It is a fact the group attempted to cross the Rio Grande River under heavy Confederate fire. The result was heavy insurgent casualties. Eight were killed or died later of wounds. Franz Weiss was hit. His brother Moritz went to his aid. Both drowned. Four others were killed while swimming the river. The four included twenty-two-year-old Joseph Elstner who had been a member of Kuechler's February, 1862, Company and in the August, 1862, group; Ernst or Edward Felsing, a thirty-three-year-old Forty-Eighter who was married to Caroline Schlickum, a sister of Julius Schlickum and who had also been a member of the August 1862 group; Lieutenant Valentine Hohmann of the Gillespie Company of the League's military battalion and a survivor of the August battle; and twenty-nine-year-old Henry Hermann who had also survived the August battle. Twenty-nine-year-old Johann Peter Bonnet, a native of Prussia, was hit below the shoulder. His brothers helped him made it across the river. "In crossing [the Rio Grande] the firing was increased and Peter Bonnet was shot through the body and had to be dragged along making it still more difficult to get across. After being carried down the river for several miles, they finally reached the opposite shore bare-headed and bare-footed, scantily clothed, without food or arms and one of them in a dying condition." After reaching the Mexican side of the Rio Grande, the Bonnet brothers "wandered about for two days without food except prickly pear before they came to a Mexican settlement." Peter Bonnet was taken to Piedras

Negras, Mexico, where he died from his wounds on March 12, 1863.[41]

Fritz Lange, a thirty-year-old survivor of the Nueces Battle from Comfort, was hit, but able to make to the Mexican side of the river. Local Mexicans from Piedras Negras also cared for him, but he died from his wounds in 1866. Captain Jacob Kuechler and Major Fritz Tegener both were hit. They recovered, but their wounds prevented them from enlisting in the Union Army.

Eighteen-year-old Sylvester Kleck, a member of the Luckenbach Bushwhackers, who had missed the August group, was also seriously wounded. He struggled to the Mexican side of the river and collapsed. He, like Peter Bonnet and Fritz Lange, was taken to Piedras Negras. There a fifteen-year-old Mexican girl, Juanita Sanchez, nursed him back to health. They were later married and became the parents of at least six children.[42]

After two attempts and the death of twenty-six men in his two groups, Captain Jacob Kuechler finally made it to Mexico. Afterwards, Cramer said, "Kuechler is now the only one of my intimate friends that is left. He is safe here." Kuechler joined Schlickum and others a few days later in Piedras Negras.[43]

On September 12, 1862, The Confederate War Department in Richmond, Virginia advised General Hebert, commander of the Department of Texas, that martial law in Texas was annulled. General Hebert was informed that the issue of martial law had been submitted to President Davis who disapproved it as an "unwarrantable assumption of authority and as containing abuses against even a proper administration of martial law." Hebert was further advised that, "military commanders have no authority to suspend the writ of habeas corpus. All proclamations of martial law by general officers and others assuming a power vested only in the President are hereby annulled."[44]

But Texas Governor Lubbock wanted some form of martial law continued. Upon hearing that martial law was to be abolished he wrote General Hebert expressing his concerns. His letter of September 26, 1862 states: "I trust it is not the intention of the Government or of the commanding general in this district to release the entire State from the operation of the law martial. If such a course is adopted I fear the consequences. It is useless to disguise the fact of there being many disloyal people in various localities of the State, whose vile tongues and bad example is held in check by the effect of martial law. I am also clearly of opinion that should martial law be abandoned in some of the localities, say those in which the heaviest amount of business is transacted, it will result in the most fatal depreciations of our currency, there being a great disposition on the part of many person to destroy the currency of the Confederate States. For these reasons, and many others that might be urged, I can but hope that Martial law will be kept in force in such localities as may be deemed necessary by you, or that some other plan be adopted by which the country will be kept quiet and our citizens required to remain at home and perform such duties as may be demanded of them by the Government of our choice."[45]

General Hebert had no choice but to annul martial law. On October 11, 1862, he notified Richmond that he had revoked martial law over the State of Texas and all orders based upon it. He explained the reasons why he declared martial law: "I had made [the declaration of martial law] because it had ecome an absolute military necessity, and at the request and petitions of the best citizens of the State and with the full consent and approbation of the Governor of the State." Hebert pointed out that a large portion of Texas citizens, especially the German element, was opposed to the Confederate government. He stressed that because of Texas' extensive seacoast, immense western and Indian frontier, civil authority by itself was unable to prevent "traitors" from communicating not only with the blockading forces but also with Union authorities in Mexico, and leaving the country only to congregate beyond Texas borders to "plot mischief" to the Confederate cause. In addition, large numbers of draft-age men were "running away to avoid the law" and that "the exercise of the law was as mild as could be under the circumstances." Hebert continued saying "There was no interference with the administration of civil law, and no suspension of the writ of habeas corpus, and persons arrested and tried had all facilities of defense as in ordinary trials under the common law." The General pointed out that even before he declared statewide martial law, Brigadier General H. P. Bee had found it necessary to declare martial law in the western portion of the State." Hebert closed saying, "I would, in conclusion, say that I consider martial law more necessary to day than ever in Texas. In this opinion both Governor Lubbock and General Bee fully concurs [sic]." The announcement of martial law repeal was published in local newspapers, beginning October 4, 1862. General Hebert was relieved of command in November 1862.[46]

The Confederate Military Commission in San Antonio stopped hearing cases. It is not clear what happened to all the cases it had already acted on. The case against Ferdinand Simon was transferred to the Confederate District in Austin and scheduled to start in early 1863, but records of the case are not on file today.[47]

By late October 1862, Camp Pedernales was closed. Company G, 32nd Regiment Texas Cavalry was transferred to Fort Clark and the detachments from Taylor's 8th Battalion Texas Cavalry returned to San Antonio. The defense of the area reverted to Captains Davis and Montel and the Frontier Regiment. It was not long before civil law enforcement broke down and vigilantes started operating.[48]

The organized insurgency was destroyed. Most of the young men of the 'hard-core' element were dead, either at the Nueces or the Rio Grande. Others made it safely to Mexico or into the ranks of the Union Army. The political leaders of *Des* Organisator took various actions to avoid capture. Eduard Degener, the head of the League, left Sisterdale after he was released from jail and moved to San Antonio. He was a member of the Texas Constitutional Conventions of 1866 and 1868-69 and elected to the U. S. Congress in March 1870. While a member of the Texas Constitutional Conventions led an effort to split Texas into two states, one of which would be West Texas. He went as far as writing a draft constitution for the new state. Degener died on September 11, 1890, in San Antonio.[49]

August Siemering enrolled in Van der Stucken's Company of Taylor's 8th Battalion on May 7, 1862, even before martial law was declared. He served as a second lieutenant until March 1, 1864, when he resigned. After his military service, he took a very low profile until the war was over, when again he became politically active. He served several terms as either Bexar County Chief Justice or as a Bexar County Court Judge. He wrote a book and several articles. In 1865, he established the German-language newspaper *Freier Presse fuer Texas*. He died on September 19, 1883 in San Antonio.[50]

Philip Braubach escaped from jail with Schlickum and made it safety to Mexico where he organized a company of guerrillas. On May 4, 1864, he enrolled in the First Regiment [Union] Texas Volunteer Cavalry where he was appointed captain of Company H and served until the end of the war. He married Louisa Schuetze, the daughter of Louis Schuetz, one of the original members of the Union Loyal League, on October 9, 1865. He died on June 30, 1888, of cancer of the bowels. In August 1890, former Confederate Brigadier General H. P. Bee assisted Louisa in obtaining a federal pension.[51]

The Confederate Military Commission banished Rudolf Radeleff from the Confederacy. He remained in Mexico until the war ended when he returned to Gillespie County. He served several terms as either a Justice of the Peace or Chief Justice. He died about 1878, likely in Gillespie County.[52] August Duecker escaped from jail and remained hidden the rest of the war. He married Wilhelmine Knetsch Lindemann on February 22, 1866. Wilhelmine helped raise his children by his first wife and together they had another two children. August Duecker died on September 13, 1913, in Gillespie County.[53]

Ferdinand Ohlenburger married Louise Schmidt on January 9, 1860 in Gillespie County. He, like Siemering, enlisted in Van der Stucken's Company on May 7, 1862. He transferred to William Krumbhaar's Artillery Battery on June 30, 1863. Ohlenburger deserted in January, 1864, and fled to Brownsville where he enrolled in Company G, First Regiment [Union] Texas Volunteer Cavalry. He became regimental sergeant major by the end of the war. On September 22, 1865, he was commissioned a second lieutenant in Company H, First U. S. [Colored] Cavalry. He was mustered out of service on February 4, 1866 at Brazos Santiago, Texas. He returned to Gillespie County where he was appointed sheriff on June 11, 1866. He moved first to Boerne in Kendall County, then to Colorado County and finally to San Antonio where he died on May 19, 1930, at the age of 95.[54]

Fritz Tegener remained in Mexico until the end of the war, at which time he returned to Kerr County. He found his wife, Susan E. Benson, had remarried several times and spent most of his money. He divorced her in 1865. Tegener was elected to the Texas House of Representatives in 1866 and again in 1870. He married Augusta Strunk on October 16, 1866, in Travis County. They were the parents of five children. After leaving the Texas House of Representatives, Tegener served as a Travis County Justice of the Peace. Frederick Tegener died in 1901 in Travis County.[55]

Louis Schuetze of Gillespie County remained one of the most vocal militant insurgents. He was the head of the Gillespie County pro-Union vigilantes. Upon learning of the Union invasion at Brownsville in early November 1863, he organized a 'home guard' Company for 'frontier protection.' The governor refused to call it into service. After the creation of the Frontier Districts, he organized another company and on January 27, 1864, it became part of the Third Frontier District. Many members of these two companies were pro-Union vigilantes. Pro-Confederate vigilantes hanged Schuetze on February 1, 1864. In March 1864, pro-Confederate bushwhackers killed at least three other members of the company in retaliation of their Unionists activities.[56]

Ernst Schwethelm of Comfort also kept a low profile until after the war. His health failed and he died about 1868 in Kerr County. Gottlieb Bauer escaped several pro-Confederate bushwhacker efforts to kill him. After the war, he was a stockraiser in Kendall County. His date of death is not known, but believed to be about 1885. Oskar von Roggenbucke remained near Comfort. After the loss of his two stepsons, Franz and Moritz Weiss, he toned down his anti-Confederate speech. He died in January 1883 at his home near Comfort.

John Abraham Staehely remained a merchant in New Braunfels during the war. It is not known what happen to him after the war, but it is believed he died about 1865.[57]

Dr. Ferdinand Charles von Herff continued his medical practice during the remainder of the Civil War gaining fame and financial success. It is rumored he helped many insurgents escape from the San Antonio jail as well to reach Mexico. He returned to Prussia in 1865 where he served in a military hospital. In December 1867, he and his family returned to San Antonio. "No citizen of San Antonio ever enjoyed a greater esteem among his fellow citizens than did Dr. Herff," wrote one newspaper. Dr. Ferdinand Herff died on May 18, 1912, at his home in San Antonio.[58]

Friedrich Wilhelm Doebbler was one of the two other men to escape jail with Julius Schlickum in July 1862 and fled to Mexico where he waited out the war. Doebbler returned to Gillespie County after the war and became a well-known journalist contributing many articles to local and national newspapers. He died about 1915 in Gillespie County. Frederick Lochte returned to Gillespie County after he was released from jail. He returned to his mercantile business. He died on July 19, 1867, in Gillespie County.[59]

Phillip Zoeller remained living near Sisterdale. In February 1864, he enrolled in W. E. Jones' 'home guard' company of the Third Frontier District and served until the end of the war. He and his wife Margarete Schneider, who he married on December 13, 1853, in Comal County, were the parents of eight children. Phillip Zoeller died in 1900 in Kendall County.[60] Jacob Kuechler remained in Mexico until the end of the war. He served as a member of the Constitutional Convention in 1868. In 1870, he was elected Texas Land Commissioner. He was defeated in his re-election bid in 1873. He served as chief surveyor for several railroads. Jacob Kuechler died on April 3, 1893, at Austin in Travis County, Texas.[61]

Confederate search parties at the battle site found a letter on the body of Pablo Diaz that implicated Julius Dresel as a member of *Des Organization*. Dresel was arrested and placed in jail at the same time as Eduard Degener. Unlike Degener, Dresel was not tried by the time martial law was annulled and he was released. In 1863, he moved to San Antonio and became a merchant. After the war, Dresel was appointed San Antonio Treasurer and wrote for *Der Freier Presse fuer Texas.* Julius Dresel moved to California in 1869 and took over the Dresel family business in the wine firm of Dresel and Company. He died in 1891, in Wiesbaden, Germany.[62]

While not original members of the *Des Organisator*, four other insurgents played major roles in the League's military operations. The commander of the Kendall Company, Ernest Cramer, remained in Mexico until the end of the war. He was Custom Inspector at Eagle Pass from 1868 until 1870. He returned to Comfort for a brief time then moved to Santa Clara, California by 1880. He settled in Hailey, Idaho, where he died about 1900. Henry Hartmann the commander of *Des Organisator*'s Kerr [American] Company fled to Mexico. He enlisted in Company A, First Regiment [Union] Texas Volunteer Cavalry on October 29, 1862. After the Union invasion of Brownsville, Texas, he was sent on 'secret recruiting duty' back to the Hill Country on January 20, 1864. Henry Hartmann's wife had meanwhile become involved with Sam Gibson, a pro-Confederate bushwhacker. The pro-Confedcrate bushwhackers learned of his return, likely from his wife, and he was executed in February or March 1864. The lieutenant of the Kerr Company, Philip G. Temple, also made it safely to Mexico. He enrolled in Company A, First Regiment [Union] Texas Volunteer Cavalry on October 27, 1862. He was appointed first lieutenant on November 6, 1862, and captain on September 18, 1863. Temple was mustered out of service on October 31, 1865 at San Antonio. He returned to Gillespie

County. Due to ill health, he returned to his family home in Ohio in 1896, where he died on February 28, 1903. Ferdinand Simon remained in jail at Austin until the end of the war. He returned to his home between Comfort and Boerne. His wife died in June 1878 and he died the next month, in July 1878.[63]

On the Confederate side, most of those involved in the Nueces Battle were rewarded for their participation. James Duff was authorized to raise a battalion and then a regiment, which he commanded until the end of the war. After the war, he fled to Mexico, then to Cuba, and returned to England. In 1877, the British financier James W. Barclay hired Duff to manage the Colorado Mortgage and Investment Company office in Denver Colorado. He became a well-respected businessman in Denver. Duff returned to London, England in 1885, where he died on April 16, 1900, from diabetes. John Donelson returned to his command near San Antonio in September 1862. He was promoted to major on October 8th. Donelson resigned on January 4, 1864, due to ill health and died at San Antonio in July 1864. Colin D. McRae recovered from his wounds and rejoined his command. He was promoted to captain on October 10, 1862, in command of Company K, 2nd Regiment Texas Mounted Rifles. He married Margaret D. Haw on December 1, 1862 in Bexar County. His wife gave birth to a son on November 3, 1863. In June 1864, the boy became ill and died. Colin McRae died of typhoid fever on September 10, 1864, while in service. Edwin Lilly was promoted to captain and appointed 33rd Texas Cavalry Regimental Quartermaster on February 24, 1863. He remained with Duff's command until the end of the war. After the war, he fled to Mexico. The date and location of his death are not known.

Hamilton Prioleau Bee was given command of a cavalry division in late 1864 in the Trans-Mississippi Department. Shortly before the end of the war he was promoted to major general. He fled to Mexico after the war, but returned in 1876. He died on October 2, 1897, in San Antonio, Texas. He was buried wrapped in the Confederate flag.[64]

The Forty-Eighters and Freethinkers did not realize what a Pandoras' box they had opened when they organized their insurgency. While they were no longer in active opposition to Confederate authority after October 1862, the results of their actions caused a situation that almost became as bad as that of "Bleeding Kansas." For the next two years, pro-Union and pro-Confederate Texans killed each other. It now became friend against friend and neighbor against neighbor. Frontier justice became a bloody way to settle grudges. For several generations of Hill County families look at each other with revenge on their minds and had "arguments that dragged on for decades after the Civil War was over."[65]

In February, 1863, the one-year enlistment term for the Frontier Regiment ended and new companies were organized. Captain Davis and about half of Company F, joined Duff's 14th Battalion Texas Cavalry as Company F and moved to San Antonio. Davis and his company remained with Duff's 33rd Regiment Texas Cavalry until the end of the war. By then, he was promoted to lieutenant colonel. Henry T. Davis moved to Orange County, Texas, where he died in 1918. The remainder of Davis' old company plus new recruits, reorganized the company and it became Company A, the Frontier Regiment on December 24, 1862, at Camp Davis. James M. Hunter of Gillespie County was elected captain. William Banta, also of Gillespie County was elected first lieutenant. Thirty-three-year-old New Hampshire-born Jeremiah M. Hays of Hays County and twenty-five-year-old Tennessee native Peter Osborn Alonzo Rees of Kerr County were elected second lieutenant.

Hunter's Company continued searching for insurgents, many of whom were executed. It is not clear if Captain Hunter and state officials authorized these searches. A secret group was organized and many members of Hunter's Company belonged. Its name was 'Friends of the Soldiers.' It was the major pro-Confederate vigilante group. The military leaders of this secret organization were William Banta and James P. Waldrip. Civilian leaders included Robert A. Gibson and William Paul.

Later this organization was known as the *Haengerbande* or Hanging Gang. It is very likely this group committed the hangings of August 22nd. Other known deaths committed by this group include: Hermann Itz and Jakob Itz, in February 1863; John Turknett and Jacob Turknett in January 1864; and Philip Bandon Turknett in July 1864. All these deaths have been blamed on James Duff. The tombstones of the Itz brothers reads, *Von Der Duff Bande Ermordet* [Murdered by Duff's Gang]. John and Jacob Turknett's tombstone reads, "Beat to death by bull whips by the men of Col. James M. Duff." The Philip Brandon Turknett's tombstone reads, "Murdered by Col. Duff's men." In additional to these five men, the four insurgents hanged on August 22nd tombstone reads, "Hanged and Thrown in Spring Creek by Col. James Duff's Confederate Regiment". All these deaths took place after James Duff left the Hill Country, but it remains **written in stone that Duff killed them**.[66]

Some of the insurgents tried to find a way to co-exist with the Secessionists. In April 1863, authorities allowed men to serve as teamsters in place of military service. August Hoffmann, Joseph Poetsch, Mathias Pehl, Henry Rausch, and August Duecker were among those who volunteered. Even then they were not safe. Captain Frank Van der Stucken arrested August Duecker and jailed him in Austin. Others fled to Mexico and many of them joined the First Regiment [Union] Texas Cavalry. Others waited out the war.[67]

In December 1863, the Texas legislature passed a law which exempted men from the draft who served in one of the three Frontier Districts. Many joined thereafter, but they were not entirely safe. Pro-Confederate bushwhackers killed at least seven who belonged to Gillespie County home guard units in the Third Frontier District. Likewise, pro-Union bushwhackers killed at least five Secessionists who belonged to the area home guard units.[68]

Many more men, on both sides, died during this 'Bushwhacker War' period from January, 1863, until late 1864, when order was finally restored. One's personal beliefs determined who was a bushwhacker or just a loyal law-abiding citizen taking actions to protect himself, his family, and friends. The best-known of these vigilantes was the 'Waldrip Gang' or *Haengerbande.* These vigilante raids continued long after the war ended. The 'Blanco County War,' the 'Burnet County War,' and the 'Mason County War' of the 1870s mainly resulted from bad feelings stirred up during this period. Deep emotional scars and bitter feelings remained with and between many Hill Country descendants, even today.[69]

An example of such scars and bad feeling is demonstrated by the last death related to the Nueces Battle. Oscar Splittgerber was one of the pro-Confederate men; it is believed that he assisted the Confederates, and likely to have helped killed the wounded after the Nueces Battle. According to family stories, fter the Civil War, Splittgerber was, "harassed by renegade northern soldiers because of [his] connection with the Confederacy." He and his family fled to Mexico. They returned to Menard County, Texas, in 1866, but still received death threats. "Things got so bad [the family] left Menard County" in the early1880s and moved to Reeves and Pecos counties in West Texas. After Jacob Kuechler visited Toyah, in Reeves County in 1888, the Comfort insurgents learned of Splittgerber's whereabouts. The insurgents drew straws to determine who would go and killed Splittgerber. Frederick Schultz drew the shot straw. On November 14, 1889, Oscar Splittgerber was killed with a shot in the back of the head near Toyah. After twenty-seven years, the killing ended.[70]

It appears *Der Organisator* continued to exist for a number of years. As late as July 20, 1865, Eduard Degener reported "our Organization remain in close contact with Governor ." The Nueces Battle will live on in history as the "Nueces Massacre".[71] Partisan feelings die hard — just like the men who held them 155 years ago.

Endnotes – The Confederates Leave the Hill Country

1. Camp Pedernales Post Returns for August and September 1862; Barr *CMC*, July, 1966 and Hopkins' Diary.

2. Williams *Border Ruffians*, pp 236, 258-259; Bee's Instructions; Biggers *German Pioneers*, p 58; Glenn *Capt'n John*, p 23; Fehrenbach *Lone Star, p* 377; Camp Pedernales Post Returns for August 1862; Hopkins' diary and Gibson's Statement; Comfort Handout; "Gillespie County in the Civil War" by Gerald R. Gold in 'The Junior Historian', Volume Number and date not known, p 30, copy located in Vertical File, Pioneer Memorial Library, Fredericksburg, Texas; "Comfort In The Country" by Michael D. Brockway, 'Texas Highways', Volume 42, No. 12, Dec. 1995, pp 5-6.

    The man named Howell that was executed was likely the man named by Duff in his June 1862 report and believed to be John T. Howell who at the time lived near North Grape Creek in western Blanco County. Duff's Report, p 786 and Moursund *Blanco County Families*, pp 218-219.

3. Hopkins' diary.

4. Gibson's Statement. Series of Indictments located in various criminal cases in District Clerks' Office Fredericksburg, Texas. These included Case Numbers 95, 96, 97, 100, 101, 102, 103, 104, 105,107 and 183; Sansom's Memoirs and Siemering's *Germans During Civil War*, June 12, 1923.

5. Camp Pedernales Post Returns for September 1862; Letter PM, August 21, 1862 and Camp Pedernales Post Returns for September 1862; Letter, Captain James Duff, Commanding Company of Texas Dragons, to Captain C. John Mason, Acting Assistant Adjutant General, Department of Texas, August 25, 1862. Copy located in James Duff's Confederate Service Records; and James M. Duff's Confederate Service Records. General Order Number 1, Headquarters Western Sub-District of Texas, San Antonio, January 3, 1863, Record Group 109, National Archives, Washington, D. C. and Compiled Service Records, 33rd Regiment Texas Cavalry, Record Group 109, National Archives, Washington, D. C; Duff's Report, pp784-786; Comfort Handout; Gold *Gillespie County in the Civil War*, p 30; Brockway *Comfort in Country;* and *Austin State Gazette,* May 10, 1865.

6. Williams *Border Ruffians*, p 258.

7. John Roebuck Kelso was born about 1835 in Tennessee or Mississippi. The family arrived in Texas about 1850 and settled in DeWitt County. He enrolled in Company G on April 3, 1862 and appointed first lieutenant. On March 27,

1864, he was promoted to captain and remained with the unit until the end of the war. John Kelso's data and location of death are not known.

Leonardus S. Lawhorn was born about 1833 in Georgia. The date he arrived in Texas is not known. He was a lawyer and lived in Karnes County. He enrolled in Company K, 2nd Regiment Texas Mounted Rifles on January 13, 1862 at Fort Brown. Lawhorn was promoted to first lieutenant on June 21, 1863. He resigned on March 1, 1864. Leonardus Lawhorn died about 1875 in Karnes County.

Smith *Here's Yer Mule*, p 19; Compiled Service Records, 32nd Regiment Texas Cavalry, Record Group 109, National Archives, Washington, D. C.; Compiled Service Records Taylor's 8th Battalion Texas Cavalry, Record Group 109, National Archives, Washington, D. C.; Eighth U. S. Census, 1860 DeWitt County Texas Census, p 498; "Kelso Family" DeWitt County Historical Commission, *The History of DeWitt County, Texas*, (Curtis Media Corporation, Dallas, Texas), 1991, p 525; John R. Kelso's Confederate Service Records, Record Group 109, National Archives, Washington, D. C.; Eighth U. S. Census, 1860 Bell County Texas Census, p 311; George R. Kuykendall's Confederate Service Records, Record Group 109, National Archives, Washington, D. C.; Ninth U. S. Census, 1870 Lamar County Texas Census, p 373; Young Collins Confederate Service Records, Record Group 109, National Archives, Washington, D. C.; L. S. Lawhorn's Confederate Service Records, Record Group 109, National Archives, Washington, D. C.; Ninth U. S. Census, 1870 Karnes County Texas Census, p 135 and J. W. Seal's Letter of September 8, 1862. A search of Gillespie County Deed Records shows only two traces of land confiscated and sold at public auction. Both were ordered sold by the Court of the Western District of Texas. The court appointed receiver in both cases was Thomas Moore. The first property was 640 acres on the mouth of Spring Creek and the mouth of While Oak Creek and belonged to Peter Hayden. It was ordered sold at the June 1863 Court Term and sold on January 14, 1864 to J. C. Rushing for $800. The second property was also 640 acres belonging to Frances Morris and Peter Hayden. It was sold on October 26, 1864 to Dr. William Keidel for $480. Gillespie County Deed Records, Volume H, pp 283 and 286.

8. There are two Louis Nelsons living in western Gillespie and Kerr Counties. One was Hiram Louis Nelson who was born on November 8, 1835 in Washington County, Illinois. He was a son of Hiram Nelson, known as 'old man Nelson.' It is not believed this is the Louis Nelson referred to in this account as he went to Mexico in 1860 to manage an import/export business. This Louis Nelson was later identified as one of the pro-Union bushwhackers. Hiram Louis Nelson died on May 8, 1922 in Kerr County. The other was Lewis Nelson born about 1826 in Arkansas. He and his family arrived in Texas about 1858.

Reinhardt Reeh was born on September 15, 1831 in Nassau. The date he arrived in Texas is not known. Reinhardt Reeh died on April 23, 1905 in Gillespie County.

Smith *Here's Yer Mule*, p 19; Harper Centennial Committee *Here's Harper*, p 2; Eighth U. S. Census, 1860 Gillespie County Texas Census, pp 4 and 6; Kerr Historical Commission *Kerr Album*, pp 121-122; Eighth U. S. Census, 1860 Gillespie County Texas Census, p 22; Tolman *Kerr Cemeteries*, p 354; *Texas Ranger Indian War* Pensions by Robert W. Stephens, (Nortex Press, Quanah, Texas), p 78 and St. Mary's Catholic Cemetery Records.

9. Hopkins' diary.

10. Ibid.

11. Hopkins' diary; Bennett *Kerr County*, pp 114 and 117; Sansom *German Citizens*. Duaine *Dead Men Wore Boots*, p 31; and Smith *Here's Yer Mule*, p 20. This 'Nelson' was Hiram Nelson, the individual who Duff had attempted to arrest in June, but who had "taken to the cedar breaks and escaped." Nelson was about 66 years old at the time of his death. *Wagons, Ho!: A History of Real County, Texas* Compiled by Marjorie Kellner (Curtis Media, Inc., Dallas, Texas), 1995, pp 576-577; Duff's Report, p 786; Eighth U. S. Census, 1860 Gillespie County Census, p 410 and Seventh U.S. Census, 1850 Jefferson County Illinois Census, p 362.

12. Harper *Mary Scott's Story*; Bennett Kerr *County, p 117; Letter*, Howard Henderson to J. W. Sansom, Ingram, Texas, October 16, 1908, located in Ransleben *100 Years*, pp 119-120 and Harper Centennial Committee *Here's Harper*, p 12.

13. Harper *Mary Scott's Story*.

14. Kerr County Marriage Records, Volume A, 52 and 54 and Statements Fritz Schladoer, March 26, 1864 and March 31, 1864, Camp Davis, Texas, Case Number 101, District Clerk's Office, Fredericksburg, Texas.

15. "The Massacre" by Hatty L. Sagebiel, Llano, Texas, an unpublished manuscript, 1990, provided author by Gregory Krauter, Comfort, Texas. Ms. Sagebiel is a granddaughter of Howard Henderson; Harper Centennial Committee *Here's Harper*, pp 12-13; "Henderson Cemetery Historical Marker Dedication" Program by Wanda Henderson on June 10, 1990 and *Here's Harper Two* by Harper Sesquicentennial Committee, (Nortex Press, Austin, Texas), 1986, p 327.

16. Howard Henderson's Letter; Sansom's Ledge located in Sansom's File DRT Library at the Alamo; Sansom *German Citizens* and Telephone Interview with Mrs. Temple Henderson, Mountain Home Texas on February 17, 1997.

17. Before Howard Henderson died in 1908, he provided Sansom with names he claimed Duff killed. Besides Henderson's letter to Sansom, the following note was located in Sansom's Ledger, "Mr. M. [H?] L. Henderson – Mountain Home, Kerr County Texas says that John Turknett 16 years of age was whipped to death by Duff and soldiers in 1862 – his father Turknett and son shot at 13 years of age and burned their home. Hiram Nelson, Frank Scott, Gustav Tegener were hung and houses burnt are men thrown into Spring Creek. All had families." A copy of Howard Henderson's Letter to Sansom is located on Pages 119-120 of Ransleben *100 Years*. The Turknett deaths took place in 1864, not 1862 as Henderson told Sansom. This is another example of where events Sansom was told are incorrect. The statement by Howard Henderson is not clear. It seems to be talking about three different men; one "my uncle"; one "my cousins"; and a Schram Henderson. Howard Henderson's uncle was Sebird Henderson. It is known he was killed on August 22nd but no record of "my cousins" or Schram Henderson has been located. On February 17, 1997, in a telephone interview with a member of the Henderson family the author was told they did not know whom Howard Henderson was referring to as "my cousins". They believed Schram Henderson was Sebird Henderson. Howard Henderson's wife's father was Philip Brandon Turknett. Jonas Harrison killed him on July 13, 1864. The identity of "and brother" is confusing. Was this his wife's father brother? Or was this his wife's brother? Two other Turknett men can be identified as being killed. One was a Jacob Turknett; an older man who could be Howard Henderson's wife's father's brother. Killed at the same time was John Turknett who was Henderson's wife's brother. Pro-Confederate bushwhackers killed both in January 1864. So, all three of the Turknett men were killed long after James Duff left the Hill Country and during the period known as the 'Bushwhacker War'.

Frank Scott's father was Benjamin Scott. The Scott family story says nothing about him being killed. Just the opposite is said. Benjamin Scott was alive after the Civil War. Howard Henderson could be correct and the Rev Johnson 'may be' the Reverend Jim Johnson who lived in the Mountain Home/Spring Creek area. He may also be the J. H. Johnston who signed the April 22, 1862, petition. There was a Jim Johnson who was the first men to settle on or near Johnson Fork in far western Kerr County. Howard Henderson draws a distinction between 'Frank Scott's father' and 'old man Scott'. John Koepke, a great grandson of Frank Scott and who provided the author with a copy of Mary Scott's Story, is unable to identify anyone who may be 'old man Scott'. 'Old man Scott' may be John Scott, the first chief justice of Burnet County. Judge Scott was killed early in the war, likely in 1862, and his body thrown into 'Deadman's Hole' in Burnet County. John Scott was born about 1801 in New York. He had made quite a bit of money in the California gold rush. In 1851, he settled on Oatmeal Creek where he planted the first orchard in Burnet County. His friends advised him to flee to Mexico early in the Civil War. He took $2,000 dollars with him and started for Mexico. Another Unionist, a man named McMasters, joined him. Just before crossing a ford on the Colorado River between Smithwick and Marble Falls, they were held up. The two were robbed and killed. Their bodies were taken several miles away and thrown into 'Dead Man's Hole' After the war, their remains were recovered.

Young Turknett is John S. C. Turknett, the younger brother of Howard Henderson's wife. As previously stated, he was killed in January 1864, long after Duff and the Confederates left the Hill Country. The Mary Scott family is unable to identify a Reverend Tom Scott. He may be the 'old man Scott' Howard Henderson mentions. According to the Scott family, Frank Scott had two younger brothers; Tom and Ike. The family history says these two fought for the Confederacy in Hood's Texas Brigade, which is not correct, they were also insurgents.

Warren Cass was born about 1827 in New York. The date he arrived in Texas is not known. He was living in Helena in Karnes County in 1855. He and his family moved to Uvalde County on December 16, 1855. They moved to Gillespie County in 1859. Warren Cass was a second lieutenant in Company D, 2nd Regiment [Gillespie County], 31st Brigade District. Pro-Confederates hanged him on March 4, 1864 in Gillespie County.

William Schultz was born about 1833 in Prussia. Arrived in Texas about 1855. He enrolled in Duff's Company E, 14th Battalion Texas Cavalry on November 1, 1863. It is believed he remained with the unit until the end of the war.

Case Number 183, District Clerk's Office, Fredericksburg, Texas; Testimony of John Larremore before Justice of the Peace, Charles Feller on April 2, 1864, located in Case Number 183, District Clerk's Office, Fredericksburg, Texas; Testimony of James W. Turknett, Spring Term 1870, Case Number 183, District Clerk's Office, Fredericksburg, Texas; Statement by John Larremore, April 2, 1864 before Charles Feller, J. P. in Gillespie County, A.G.C., TSA, Austin, Texas; Harper *Mary Scott's Story;* Bennett *Kerr County*, p 39; Petition, County of Kimbal [Kimble], April 22, 1862 to Governor F. R. Lubbock, Governor Lubbock's Papers, TSA, Austin, Texas; Debo *Burnet History*, Volume I, pp 3536; Oral Account of the hanging by Clifton Stork, Fredericksburg Historian, April, 1995; *Marble Falls The Picayune*, May 27, 1998; Eighth U. S. Census, 1860 Gillespie County Texas Census, p 26; *Life and Diary of Reading Black: A History of Early Uvalde* Arranged by Ike Moore (Printed by El Progreso Memorial Library, Uvalde, Texas), pp 85-88; "Uvalde County Government" El Progreso Club *Proud Heritage*, p 13; Case Number 101, District Clerk's Office, Fredericksburg, Texas; Eighth U. S. Census, 1860 Kerr County Texas Census, p 71; William Schultz's Confederate Service Records, Record Group 109, National Archives, Washington, D. C.; Letter, Ernest Altgelt, Comfort, May 21, 1863 to John Thurmund, Esq. Sec., S.A.M.A.A. Copy provided by Gregory J. Krauter, Comfort, Texas; Kellner, *Wagons Ho,* p 577 and English Translation of Address Commemorating The

50ᵗʰ Anniversary Of The Battle On The Nueces, August 1862, by Eduard Schmidt of Kerrville, August 10, 1912, p 9, Copy provided by Gregory J. Krauter, Comfort, Texas. For an example of bad feelings between citizens of nearby Burnet County see *Was Grossmutter Erzaehlt [Memoirs Of A Texas Pioneer Grandmother]* by Ottilie Fuchs Goeth (Eakin Press, Austin, Texas) 1982, pp 75-79.

18. The account incorrectly identifies the Confederate officer as James Duff. Since the incident took place after Frank Scott's death on August 22ⁿᵈ and Duff had left by the 20ᵗʰ, it could not have been Duff. Donelson was the provost marshal and troop commander at the time. Harper *Mary Scott Story*.

19. This story is similar to the one about Unionists surprising Captain Van der Stucken and capturing his horse and weapons described on Page 314. Harper *Mary Scott Story*.

20. Stewart *Death on the Nueces: Minna Stieler Stories*, pp 30, 32-37 and 41 and *A Boy's Civil War Story*, by Charles Nagel, (Eden Publishing Company, St. Louis, Missouri), 1934, p 249.

21. Oral Interviews with Emilie Real Neill, a great-granddaughter of Emilie Schreiner and Casper Real on August 20, 1997 and October 4, 1997.

22. J. W. Seal Letter of September 8, 1862.

23. Smith's *Here Yer's Mule*, pp 19 and 30. The eleven men on the September 15ᵗʰ scout were: John Adcock; Jess K. Brown; Russell Baker; Rousseau Baker; C. Gable; H. McHall; Fred Henneck; H. Korth; J. B. King; A. Kell; and J.W. Murray. The eighteen men on the September 20ᵗʰ scout were: First Lieutenant John R. Kelso; H. Menn; John M. Murray; W. C. Middleton; J. A. Middleton; C. W. Kort; C. J. Orman; W. J. Parker; George R. Gusworm; Louis von Roeder; Theodore Spies; Joe Taylor; H. F. Kelso; William Koehn; John York; William Gerhardt; Thomas C. Smith; and E. F Thie. The five guides were: a man named Man; a man named Hinke; E. W. Marlow; Anton Ott; and H. H. Ward.

24. Letter, Office of Provost Marshal, Fredericksburg, Texas, September 8, 1862 from Captain Donelson to Brigadier General H. P. Bee, San Antonio, Texas, contained in John Donelson's Confederate Service Records.

25. Ibid.

26. Compiled Service Records of Taylor's 8ᵗʰ Battalion Texas Cavalry, Record Group 109, National Archives, Washington, D. C.; John Donelson's Confederate Service Records; and Camp Pedernales Post Returns for September 1862. The six officers included: Captain Josiah Taylor; First Lieutenant John R. Kelso; Second Lieutenant William A. Adams; [all from

Company G]; Second Lieutenant George R. Kuykendall; and Second Lieutenant Young Collins; [of Taylor's Battalion]. Second Lieutenant L. S. Lawhorn, from Company K, 2nd Regiment Texas Mounted Rifles, was still the Receiver. The four men from Company B were: W. J. B. Johnston; James M. Lane; William B. Marshall and James Mosley.

27. Louise [Luise] Feuge Duecker was born in 1836 in Hanover. Her parents were Christiana and Christopher Feuge. Her family arrived in Texas on the *Mathilda* in 1846. She married August Duecker on June 11, 1854, in Gillespie County. She died on January 1, 1863.

   Karoline [maiden name not known] Graff was born about 1826 in Prussia. She married Carl Graff about 1851 in Prussia. They arrived in Texas on the *Weser* in 1859. Her date and location of death is not known.

   Gillespie County Deed Records, Volume H, pp 205-206; Seventh U. S. Census, 1850 Gillespie County Texas Census, p 312; Gillespie County Marriage Records, Book 1, p 19; Gold *Church Records of the Vereins-Kirche*, p 185; and Geue *New Homes in New Land*, p 92. Other families or friends had even harder legal actions to take, that of probating the estates of the deceased. The first probate filed was that of Basil Stewart, the individual whom the Unionists executed for being a 'spy'. His probate was filed on July 31, 1862, just weeks after his death. Louis Schieholz's was filed on October 27, 1862. Wilhelm F. Boerner's was filed on November 24, 1862; Henry Steves' Jr., was filed on January 26, 1863. August Duecker's was filed on March 14, 1863, even though he was not dead. August Vater filed his son Fritz's, on September 18, 1863, and his son Adolph's, on November 28, 1864. Conrad Bock's was filed on July 25, 1864 and August Luckenbach's on January 30, 1865. Herman Flick's was filed on January 27, 1865. Heinrich Weyershausen's was filed on February 7, 1866, and Hiram Nelson's was filed on February 26, 1866. Gillespie County Probate Records, Volume C, 195, pp 219-220, 229- 230, 256, 282-283, 294-295, 303, 372-273, and 378 and Kerr County Probate Records, Volume A, pp 59, 60 and 62,

28. Schlickum's Letter.

29. Ibid.

30. Ibid.

31. Schlickum's Letter and *Neu Braunfelser Zeitung* October 15, 1862.
32. Ibid.

33. Schlickum's Letter and *Neu Braunfelser Zeitung* October 15, 1862.

34. Ibid.

35. Ibid.

36. *San Antonio Herald*, October 11, 1862.

37. *Austin State Gazette*, October 15, 1862.

38. The names of all the members of this second group are not known. Very little has been written about this event. Seven of the group were killed and their names added to the Comfort Monument. These are: Joseph Elstner, Edward Felsing, Henry Hermann, Peter Bonnet, Valentine Hohmann, Moritz Weiss and Franz Weiss. Jacob Kuechler was in the group. Others included: Albert Beversdorf, August Beversdorf, Charles Bonnet, Daniel Bonnet, William Bonnet, Fritz Lange and, Sylvester Kleck. This makes a total of fifteen. Fritz Tegener may have been in the group, which would bring the total to sixteen. The Schlickum Letter says Wilhelm Huster was in the group, which would bring the known members to seventeen. Both the Schlickum and Kuechler accounts say there were seventeen in the group. Why the insurgents did not bury their fallen comrades as they passed the August battle site, is a mystery. They had plenty of time. One of the main criticism, the insurgents and their descendants level at the Confederates is they did not bury the bodies. Here is a group of their comrades not burying their own. This pyramid was still visible when the bodies were recovered on August 10, 1865. *Bonnet Brothers,* by F. W. Schweppe, an unpublished and an undated manuscript collected and preserved by the Edith Gray Library, Boerne, Texas. Copy provided by Ester Strange of Kerrville, Texas; Kelton's Article; Siemering's Germans During Civil War, June 8, 1923; Oral Interview with Ester Bonnet Strange, Fredericksburg and Kerrville, Texas, May 1997, July 1997 and October 1997; Oral Interview with Peter Kleck, a grandson of Sylvester Kleck on May 1, 1997 at the Nimitz Museum; Ransleben *100 Years,* pp 94-95; *The History of Eastern Kerr County, Texas,* by Gerald Witt, (Nortex Press, Austin, Texas), 1986, p 57 and Weber *Die Deutsche Pioniers,* p 14.

39. Sansom Battle *of Nueces,* p 13; Cramer's Letter; Schlickum's Letter; Weber Die *Deutsche Pioniers,* p 14; Schweppe *Bonnet Family Story* and "Requisition for Forage" October 21, 1862, Stokely Holmes' Confederate Service Records.

40. *In Memorial – Hon. Jacob Kuechler, Remarks* at The Funeral Services of Jacob Kuechler, April 4, 1893, by Rev. E.M. Wheelock, and Weber *Die Deutsche Pioniers,* p 14.

41. Joseph Elstner was born about 1840, location of birth not known. Believe he was part of the August 'Elsner' family that arrived in Texas on the *Adolphine* in 1851. Joseph Elstner arrived in the Comfort area in 1854. He was a member of Kuechler's February 1862 Company. Elstner was a member of the August fleeing group.

Ernst/Edward Felsing was born about 1829 in Hesse-Darmstadt. Likely he took part in the 1848 German Revolution. Date arrived in Texas not known. He arrived in the Comfort area in 1854. On January 6, 1861, he married Caroline Schlickum, a sister of Julius Schlickum, in Bexar County. Felsing was a member of the August fleeing group.

Henry Hermann was born about 1833 in Brunswick. Date arrived in Texas not known. It is believed he is the Henry Hermann shown on the 1860 Bexar Census as 27, a farmer. Hermann was a member of August fleeing group.

Johann Peter Bonnet was born at Charlettenberg, Prussia on February 8, 1833. The family immigrated to the United States in 1845 and settled in northern Bexar County. Peter Bonnet died from his wounds on March 12, 1863 at Piedras Negras, Mexico.

Webb *Die Deutsche Pioniers,* p 15; Sansom *Battle of Nueces,* p 13; Cramer's Letter; *The Bonnet Family From Chambons In The Dauphine,* by Dr. Jur Bernhard Boerner, [An English Translation] in 'Deutfsches Geschlechterburh', Volume 60, 1, 1928, copy provided by Esther Strange of Kerrville, Texas, a Bonnet descendant; Kerr County, p 111; Geue *New Homes In New Land,* p 66; Kuechler's MR; Eighth U. S. Census, 1860 Kerr County Texas Census, p 74; Bexar County Marriage Records, Volume D2, p 262; and Ransleben *100 Years,* pp 23, 94, 95 and 114; Eighth U.S. Census, 1860 Bexar County Texas Census, p 437; Schweppe Bonnet Family Story; *The Bremers and Their Kin in Germany and in Texas* by Robert R. Robinson, Jr., (Nortex Press, San Antonio, Texas), 1986, pp 219-220 and 749.

42. Fritz Lange was born June 20, 1832. His parents were Johanna Streuer and Heinrich Christian Ludwig Lange, a likely Forty-Eighter. The family arrived in Texas in 1851 and settled between Camp Verde and Zanensburg [Center Point]. He married Auguste Hasser on December 7, 1853, in Bexar County. They were the parents of three children: Auguste, Louise, and Emma. Fritz Lange was able to return home, but died from his wounds in 1866. His widow married Dr. George Zimmerman about 1867, likely in Bastrop County.

Sylvester Kleck was born on December 31, 1844 in Hohenzoeller, Prussia. His parents were Victoria [maiden name not known] and John Kleck. The family arrived in Texas on the *Andacia* from Harthausen in 1846. Member of Luckenbach Bushwhackers. He was wounded at Rio Grande on October 18, 1862, and spent several months recovering from his wounds. Juanita Sanchez, who he married, nursed him back to health. Sylvester Kleck died August 21, 1914, in Gillespie County.

Ransleben *100 Years,* pp 94-95 and 114; Witt *History of Eastern Kerr County,* p 157; Jacob Kuechler's Eulogy; Oral Interviews with Roland Hall, a great-great grandson of Fritz Tegener on August 10, 1996 and March 23, 1997; Bennett *Kerr County,* pp 71 and 135; Bexar County Marriages, Volume C, 135;

Eighth U. S. Census, 1860 Kerr County Texas Census, pp 74-74b; Ninth U. S. Census, 1870 Bastrop County Texas Census, p 424; Geue *New Land*, p 109; Lindig *Der Friedhof Cemetery*, p 43 and Peter Kleck's Interview.

43. Both of Kuechler's attempts to reach Mexico resulted in the Confederates surprising the two groups and death to a number of insurgents. Cramer's Letter and Schlickum's Letter.

44. Letter and General Order Number 66, Headquarters Confederate Army, Richmond, Virginia, September 12, 1862, contained in OR, Series I, Volume IX, pp 735-736.

45. Letter, Governor F. R. Lubbock to General P. O. Herbert, San Antonio, Texas, September 26, 1862, contained in OR, Series I, Volume LIII, pp 829-830.

46. Report, P. O. Hebert, Brigadier General, Provisional Army, Headquarters First District of Texas, San Antonio, October 11, 1862 to General S. Cooper, Adjutant General, Richmond, Virginia, contained in OR, Series I, Volume LII, pp 828-829; *San Antonio Weekly Herald,* October 4, 1862 and *San Antonio Semi-Weekly News,* October 27, 1862.

47. Barr *CMC*, October 1969 and *Austin Tri-Weekly Gazette* June 27, 1863.

48. Compiled Service Records of 32nd Regiment Texas Cavalry and 8th Battalion Texas Cavalry, Record Group 109, National Archives, Washington, D. C.

49. Webb *Handbook of Texas*, Volume I, p 482; Ron Tyler et al, eds., *New Handbook of Texas*, Volume 2, pp 562-563; and Knopp *German State in the New World*, p 168.

50. August Siemering Confederate Service Records and Webb *Handbook of Texas*, Volume II, p 609.

51. Schlickum's Letter; Santleben *Texas Pioneer,* p 35 and Louisa Braubach's Union Pension Records.

52. Barr *CMC*, October 1967, p 277; OR, Series II, Volume IV, p 863 and Wisseman *Fredericksburg ... First Fifty Years*, pp 54 and 56.

53. Comal County Marriage Records, Books B and D and Lindig *der Friedhof Cemetery*, p 16.

54. Ferdinand Ohlenburger's Company C, 8th Battalion Texas Cavalry Service Records; Ferdinand Ohlenburger's Company E, 1st Regiment Texas Cavalry Service Records; Ferdinand Ohlenburger's Krumbhaar's Artillery Battery Service Records, Ferdinand Ohlenburger's First Regiment [Union] Texas

Volunteer Cavalry Service Records; Ferdinand Ohlenburger's Company H, First Regiment U. S. [Colored] Cavalry; Wisseman *Fredericksburg ... First Fifty Years,* pp 53 and 56 and Louise Ohlenburger's Union Pension Records.

55. Roland Hall Interviews; "Personal Vignettes, Story of Fritz Tegener" Oral Presentation by Roland Hall at Nueces Symposium, Fredericksburg, Texas March 22, 1997; Kerr District Court Case # 61 and *The Texas House of Representatives: A Pictorial Roster 1846-1992,* Edited by Charles E. Spellmann (Texas House of Representatives, Austin, Texas), 1992, pp 12-14.

56. Gillespie Historical Society *God's Hills,* Volume I, pp 187-188; Schuetze's November 1863 Muster Roll; Schuetze's January 1864 Muster Roll and Gillespie District Court Case #51 and 101.

57. Ransleben *100 Years,* pp 24, 80, 102, 136 and 198; Watkins Kerr *County,* pp 68, 112, and 134; Tolman Kerr *County Cemeteries,* p 249; Freund *Dresel's Houston Journal;* pp xiv, xxv, xxvi, 121 and 125; Fischer *Marxists And Utopias in Texas,* p 86; Zucker *Forty-Eighters,* pp 289 and 331; Kaufmann *Germans in Civil War,* p 317 and Boerne Area Historical Society *Gone, But Not Forgotten,* Volume II, p 32.

58. *Early Texas Physicians 1830-1915* Edited by R. Maurice Hood, M. D., (State House Press, Austin, Texas), 1999, p. 182.

59. Schlickum's Letter; Cade *Mathilda Doebbler Gruen Story,* pp 156-188 and Lindig *Der Friedhof Cemetery,* p 53.

60. Jones' Muster Rolls; Comal County Marriage Records, p 20 and Boerne Area Historical Society *Gone, But Not Forgotten,* Volume I, p 29.

61. Webb Handbook *of Texas,* Volume I, p 975 and German *Artist on the Texas Frontier: Friedrich Richard Petri* by William W. Newcomb, Jr., (University of Texas Press, Austin, Texas), 1978 p 151.

62. *The Dresel Family* compiled by Clyde H. Porter, privately published at San Antonio, Texas 1952, p 14-C and 14-D.

63. Cramer's Letter; Kampoefner's Letter; Hartmann's Union Service Records; Gillespie Marriage Records, Book 1, pp. 358 and 395; Temple' Union Service Records; Temple's Union Pension Records; Simon Family Letters and Ferdinand Simon's Probate Records.

64. Duff's Confederate Service Records; Series of Letters from July 1996 to August 1996 containing research on James Duff, from Patricia A. Kemper, Golden, Colorado; Series of Letters from December 2001 to March 2002 containing research on James M. Duff from Joanne M. Gonsalves, Littleton, Colorado; Donelson's Confederate Service Records; San Antonio

Genealogical and Historical Society *Index Wills and Inventories of Bexar County*, p 26; McRae's Confederate Service Records; Bexar County Marriage Records, Volume D2, p 269; Travers *Biographical Sketch of Colin D. McRae*; Hunter *Texas Trail Drivers*, p 631; Lilly's Confederate Service Records; Nunn *Escape From Reconstruction*, 133; p Bee's Confederate Service Records; Webb *Handbook of Texas*, Volume I, pp 135-136 and Pease *They Came to San Antonio*, p 18.

65. "Eckstein family had its share of adventure" By Irene Van Winlde, *West Kerr Current*, July 6, 2002.

66. Henry Davis' Confederate Service Records; Compiled Service Records, 33rd Regiment Texas Cavalry; Ninth U. S. Census, 1870 Orange County Texas Census, p 218; Tenth U. S. Census, 1880 Orange County Texas Census, p 42b, Brinley *Orange County, Texas Cemetery Inscriptions*, p 78; and Muster Rolls, Captain James Hunter's Company, December 24, 1862, February 28, 1863, April 30, 1863, and October 31, 1863, TSA, Austin, Texas. Peter Osborn Alonzo Rees was born on September 6, 1837 in McNairy County, Tennessee. The Rees family was among the first settlers in Kerr County. He was the first Kerr County Clerk. Rees enrolled in Davis' Company on March 4, 1862 and remained with the unit until the end of the war by which time he was a captain and in command of the company. He served as Kerr County sheriff from 1867 to 1869 and again in 1876 and as a county commissioner from 1881 to 1897. Peter Rees died on January 26, 1919 in Kerr County. Kerr County Historical Society *Kerr Album*, pp 382-383; Peter Rees' Confederate Service Records, Record Group 109, National Archives, Washington, D. C.; Sansom's Memoirs; Testimony of John Banta before Theodore Bucholz, Chief Justice of Gillespie County, September 15, 1865, Case Number 101, District Clerk's Office, Gillespie County, Texas; Tombstone Inscriptions of Hermann and Jakob Itz, Fredericksburg City Cemetery; Sesquicentennial Committee *Here's Harper Two*, pp 327-328; Compiled Service Records of 14th Battalion Texas Cavalry and 33rd Regiment Texas Cavalry, Record Group 109, National Archives, Washington, D. C.; Eighth U. S. Census, 1860 Travis County Texas Census, p 276; Davis' MRs and Hunter's MRs.

67. Hoffmann's Letter; Frank Van der Stucken's Letter of December 19, 1863 and Menke's Index First and Second Regiment [Union] Texas Volunteer Cavalry.

68. *House Journal of the Texas Tenth Legislature Regular Session* Compiled by James M. Day, (Texas Library and Historical Commission), 1965, p 218; Day *Senate Journal of the Texas Tenth Legislature Regular Session*, pp 136 andd List of Individuals Killed During Bushwhacker War compiled by Wm. Paul Burrier, Sr.

69. *Bushwhackers and the Haengerbande—Texas Hill Country Germans and the Civil War,* by Karl Biermann, December 27, 1997. Copy in possession of author. For a discussion of the Mason County War see *The Mason County "Hoo Doo" War, 1874-1902* David Johnson (University of North Texas Press), Denton, Texas 2006).

70. Off the Record Interview, July 19, 1997; Menard Historical Commission Menard *County History*, p 602; Hughes *History of Pecos County*, p 350; Turner *Julius Splittgerber*, Volume II; Kuechler's letter of October 26, 1888 and Carrigan's Letter of August 1, 1991.

71. An example of the Nueces Battle still being called a 'Massacre' is a 2006 article in the *Fredericksburg Standard* of August 16, 2006, Section D, pp 7 and 10.

Appendices

A: Roster, W.T. Harbour's Minute Man Co., Feb. 27, 1861

Surname	Given Name	Age	Rank	Remarks
Harbour,	W. T.	27	Capt.	Cdr of Davis' Det in Pursuit Force
Rogers,	A. J.	49	1Lieut.	
Rosenthal,	Adolph	25	2Lieut.	
Hilliare,	Michael	31	1Sgt.	
Martin,	Robert	27	2Sgt.	
Ingenhuett,	Thomas	28	3Sgt.	Joined Union Army
Walker,	Donelson	45	4Sgt.	
Taylor,	Thurman T.	32	1Cpl.	Joined Union Army
Schwethelm	Henry	21	2Cpl.	Member Insurgent Group
Burney,	H. M.	34	3Cpl.	
Quinlan,	C. C.	27	4Cpl.	
Berger,	Louis	24	Pvt.	
Brinkmann,	Alex	24	Pvt.	
Brown,	John	24	Pvt.	
Bruckish,	Theodore	23	Pvt.	Member Insurgent Group
Burney,	R. H.	27	Pvt.	
Cramer,	Ernst	25		Member Insurgent Group
Crawford,	T. H.	38	Pvt.	
Dickson,	William	35	Pvt.	
Harrison,	Jonas	36	Pvt.	
Ingenhuett,	Peter	27	Pvt.	
Lane,	Samuel	25	Pvt.	
Lowrance,	Daniel	32	Pvt.	
Lowrance,	Miles	34	Pvt.	
Nelson,	A. B.	19	Pvt.	Joined Union Army

Ochse,	John E.	27	Pvt.	
Paul,	A. P.	36	Pvt.	
Real,	Casper	37	Pvt.	
Rees,	Alonzo	23	Pvt.	
Sherwood,	Thomas H.	38	Pvt.	
Schacker,	L.	29	Pvt.	
Schaefer,	Robert	28	Pvt.	
Schreiner,	Amil	21	Pvt.	Member Insurgent Group
Schreiner,	Charles	22	Pvt.	Likely Member of Insurgent Group
Stanford,	P. M.	50	Pvt.	
Starkey,	James M.	39	Pvt.	
Taylor,	James T.	19	Pvt.	Joined Union Army
Tullord,	A.	30	Pvt.	
Wilborn,	Albert L.	30	Pvt.	
Wharton,	William G.	20	Pvt.	

B: Roster, W.A. Blackwell's Minute Man Co., May 4, 1861

Surname	Given Name	Age	Rank	Remarks
Blackwell,	William A.	45	Capt.	
Tally,	J. C.	33	1Lieut.	Resigned Oct 1861 Replaced by Felps
Felps,	W. D.		1Lieut.	
Carson,	Joseph		2Lieut.	
Vaughn,	J. L.		1Sgt.	
Lindimen,	W [H]. C.	25	Cpl.	
Johnson,	S. A.		Cpl.	
Durham,	R.		Cpl.	
Hamilton,	John		Cpl.	
Alley,	J. N.		Pvt.	
Bishop,	W. H.		Pvt.	
Brewer,	L.		Pvt.	
Campbell,	E. B.		Pvt.	
Davis,	Milton		Pvt.	
Felps,	B.		Pvt.	
Felps,	Benjamin		Pvt.	
Felps,	T. C.		Pvt.	
Ferguson,	P. G.		Pvt.	
Gates,	A. V.	34	Pvt.	
Glenn,	J. B.		Pvt.	
Gray,	Samuel B.		Pvt.	
Jones,	W. E.		Pvt.	
Kelllam,	J. P.		Pvt.	
Kercheville,	A. J.	41	Pvt.	
Lindimen,	Adam	24	Pvt.	
Lindimen,	Edward		Pvt.	
Miller,	J. M.	35	Pvt.	
Nowlin,	J. C.	19	Pvt.	
Nowlin,	R.	19	Pvt.	
Palmer,	J. M.		Pvt.	

Robison,	Neill	50	Pvt.
Rogers,	Joseph		Pvt.
Silliman,	R.		Pvt.
Steele,	Charles		Pvt.
Steele,	Hardin		Pvt.
Trainer,	David		Pvt.
Trainer,	W. N.		Pvt.
Watson,	Benjamin		Pvt.
Watson,	J. M.		Pvt.
Williams,	W. B.		Pvt.
Zork,	Louis		Pvt.

C: G. Freeman's Company – Aug. 24, 1861

Surname	Given Name	Age	Rank	Remarks
Freeman,	George		Capt	
Cloudt,	Arthur		1Lieut	
Roberts,	Alexander		1Lieut	Not Able Bodied* Elected Nov 14, 1861
Hudson,	Richard		2Lieut	
Watson,	James		2Lieut	
Kent,	David B.		2Lieut	Elected Nov 14, 1861
Lewis,	Harrison		1Sgt	
Jones,	Wilson		2Sgt	
Smith,	Thomas M.		3Sgt	
Snow,	Moses M.		4Sgt	Pro-Union Bushwhacker Killed Jan, 1864
Wimberley,	Pleasant		1Cpl	
Moore,	Joseph W.		2Cpl	
Stayton,	John A.		2Cpl	
Smith,	James		3Cpl	
Crider,	Joseph		4Cpl	
King,	Edward A.		Bugler	Pro-Union Bushwhacker Not Able Bodied*
Cloudt,	Richard		Bugler	
Alberthal,	John A.		Pvt	
Alexander,	George		Pvt	
Arriola,	A		Pvt	
Arriola,	John		Pvt	
Arriola,	Massimo		Pvt	Not Able Bodied*
Bostick,	Charles W.		Pvt	
Coldwell,	James		Pvt	
Crider,	Daniel		Pvt	
Crownover,	Aaron		Pvt	
Cude,	John M.		Pvt	
Cude,	Solomon M.		Pvt	
Daniels,	George F.		Pvt	

Daniels,	William	Pvt	
Davis,	Robert	Pvt	
Edon,	Archibald, B.	Pvt	Pro-Union Bushwhacker
Gibson,	Joseph H.	Pvt	
Grey,	John	Pvt	
Hickson,	William J.	Pvt	
Hill,	William W.	Pvt	
Hinds,	Gering F.	Pvt	
Jackson,	John	Pvt	
Johnson,	Andrew Jackson	Pvt	
Jones,	James	Pvt	
Lackey,	Green B.	Pvt	
Lackey,	Henry L.	Pvt	
Lackey,	James	Pvt	
Lackey,	Jones	Pvt	
Lackey,	M. B.[V].	Pvt	
Lester,	Calvin	Pvt	
Lewis,	John	Pvt	
Lunday,	William	Pvt	Pro-Union Bushwhacker Killed Jan, 1864
McKeller,	Hector	Pvt	
Minor,	Martin	Pvt	
Moore,	John	Pvt	
Morris,	Spencer	Pvt	Not Able Bodied*
Pearson,	Joseph H.	Pvt	
Pearson,	Michael	Pvt	
Phillips,	William	Pvt	
Porter,	William	Pvt	Not Able Bodied*
Pruitt,	Christopher	Pvt	
Pruitt,	Solomon	Pvt	
Ray,	Job W.	Pvt	
Roberts,	Alexander	Pvt	
Roberts,	Jacob F.	Pvt	

Roberts,	John C.	Pvt	
Sharp,	William	Pvt	
Ship,	William	Pvt	
Smith,	Henry	Pvt	
Smith,	John H.	Pvt	
Smith,	Tell	Pvt	
Smith,	Walter S.	Pvt	
Snow,	William F.	Pvt	Pro-Union Bushwhacker Killed Jan, 1864
Stayton,	Charles M.	Pvt	Not Able Bodied*
Stayton,	David W.	Pvt	
Strickland,	Samuel	Pvt	
Stuckey,	Price	Pvt	
Waldrup,	Sullivan E.	Pvt	
Walker,	John A.	Pvt	
Walker,	John W.	Pvt	
Walker,	William	Pvt	
Watson,	Benjamin W.	Pvt	
Watson,	Jacob	Pvt	Over 50 Years Old
Westfall,	Willaim R.	Pvt	
White,	Simeon T.	Pvt	
Wiley,	Isaac	Pvt	
Wiley,	John W.	Pvt	Not Able Bodied*
Winkleman,	William E.	Pvt	
Wood,	Columbus P.	Pvt	
Wood,	Elihu	Pvt	
Wood,	George	Pvt	
Wood,	Thomas	Pvt	
Wood,	William P.	Pvt	

D: Roster, C. Nimitz's Home Guard Co., June-July 1861

Surname	Given Name	Age	Rank	Remarks
Nimitz,	Charles H.	32	Capt	
Radeleff,	J. Rudolph	31	1Lieut	Unionist Resigned in June 1861
Fresenius,	Frederick	34	1Lieut	Elected Jun 19, 1861
Weirich,	Charles	41	2Lieut	Not On Feb 1862 MR
Hunter,	John		2Lieut	Enrolled Mar 27, 1862
Krauskopf,	Engelbert	29	2Lieut	
Maier,	Anton	45	1Sgt	Not On Feb 1862 MR
Splittgerber,	Julius		1Sgt	Enrolled Mar 27, 1862
Schmidt,	Jacob	34	2Sgt	Not On Feb 1862 MR
Walch,	John	31	1Cpl	Not On Feb 1862 MR
Tatsch,	Jacob	30	2Cpl	
Arthelger,	August		Pvt	Not On Feb 1862 MR
Basse,	Oscar	16	Pvt	
Beckmann,	Henry		Pvt	Member By Feb 1862
Cameron	Edwin		Pvt	Member By Feb 1862
Cooley,	A. O.		Pvt	Enrolled Feb 23, 1862
Dennis,	N. M.	21	Pvt	
Doebbler,	F. W.	34	Pvt	Unionist Not On Feb 1862 MR
Doss,	John E.	21	Pvt	Enrolled Feb 23, 1862
Fensky,	James		Pvt	Not On Feb 1862 MR
Foster,	A. G.		Pvt	Enrolled Feb 23, 1862
Genteman,	Friedrich		Pvt	Not On Feb 1862 MR
Hahn,	Fr.		Pvt	Not On Feb 1862 MR
Hahne,	Conrad		Pvt	Not On Feb 1862 MR
Human,	Charles		Pvt	Enrolled Feb 23, 1862
Hunter,	James		Pvt	Member By Feb 1862
Jordan,	Fritz	20	Pvt	Not On Feb 1862 MR
Keidel,	Wilhelm	37	Pvt	
Koock,	Wilhelm	22	Pvt	
Kott,	August	38	Pvt	
Ludwig,	Daniel	26	Pvt	Not On Feb 1862 MR

Lungkwitz,	Adolph	39	Pvt	
Maier,	Edward	28	Pvt	Not On Feb 1862 MR
Markwordt,	Heinrich	30	Pvt	Unionist Not On Feb 1862 MR
Meinhardt,	Albert		Pvt	Enrolled Feb 23, 1862
Mengers,	Antone		Pvt	Enrolled Mar 27, 1862
Meyer,	Ludolph	26	Pvt	Not On Feb 1862 MR
Mietinger,	Wilhelm	30	Pvt	Not On Feb 1862 MR
Mueller,	Ottocar		Pvt	Member By Feb 1862
Ochs,	Henry	25	Pvt	Not On Feb 1862 MR
Pape,	Oscar		Pvt	Not On Feb 1862 MR
Peter,	George	26	Pvt	Not On Feb 1862 MR
Ransleben,	Julius		Pvt	Member By Feb 1862
Riedel,	F.	42	Pvt	Not On Feb 1862 MR
Schneer,	William		Pvt	Member By Feb 1862
Schmidt,	Peter	42	Pvt	Not On Feb 1862 MR
Schwarz,	Charles	21	Pvt	Not On Feb 1862 MR
Siemering,	August	30	Pvt	Unionist
Smith,	Thomas		Pvt	Member By Feb 1862
Speier,	John	22	Pvt	Not On Feb 1862 MR
Staus,	E,		Pvt	Member By Feb 1862
Stoffers,	Christian	19	Pvt	Not On Feb 1862 MR
Van der Stucken,	Emil		Pvt	Member By Feb 1862
Van der Stucken,	Frank	29	Pvt	
Wahrmund,	Louis		Pvt	Member By Feb 1862
Wahrmund,	William	37	Pvt	
Wathersdorff,	Albert		Pvt	
Walter,	Henry		Pvt	
Walter,	John		Pvt	
Weber,	John	33	Pvt	
Weiss,	Adolph		Pvt	Member By Feb 1862
Weiss,	Louis		Pvt	
Wight,			Pvt	Member By Feb 1862

Wilke,	Frederich	31	Pvt	Not On Feb 1862 MR
Wilke,	Henry	38	Pvt	Not On Feb 1862 MR
Wrede,	Frederick	39	Pvt	

E: Roster, P. Braubach's Home Guard Co., February 1871 – February 1862

Surname	Given Name	Age	Rank	Remarks
Braubach,	Philip	31	Capt.	Arrested by Duff in Jun 1862
Wahrmund,	Carl	29	Lieut.	
Burg,	Peter	40	Lieut.	Killed by Haengerband in Mar 1864
Peterman,	Franz	35	Sgt.	
Hohmann	Valentin	33	Sgt.	Member Insurgent Group
Kirchler,	Henry	29	Sgt	Killed by Haengerband in Mar 1864
Schneer,	William	37	Cpl.	
Weinheimer,	Anton	33	Cpl.	
Wahrmund,	Emil	32	Cpl.	
Basse,	Charles	22	Pvt.	
Becker,	Franz	30	Pvt.	
Becker,	John	33	Pvt.	
Bender,	Conrad	30	Pvt.	
Betz,	Reynold	22	Pvt.	
Blank,	John	36	Pvt.	Killed by Haengerband in Mar 1864
Duecker,	August	32	Pvt.	Member Insurgent Group
Feller,	William	30	Pvt.	Killed by Haengerband in Mar 1864
Feuge,	Christoph	22	Pvt.	
Gold,	Peter	21	Pvt.	Member Insurgent Group
Graff,	Carl	30	Pvt.	Member Insurgent Group
Jacoby,	Peter	25	Pvt.	Member Insurgent Group
Juenke,	William	21	Pvt.	
Klier,	William	26	Pvt.	Member Insurgent Group
Leyendecker,	Jacob	28	Pvt.	Joined Union Army
Meckel,	Bernhard	31	Pvt.	
Mosel,	Peter	29	Pvt.	
Pehl,	Mathias	21	Pvt.	Member Insurgent Group

Sauer,	Fritz	23	Pvt.	
Schildknecht	August	34	Pvt.	
Schildknecht	Adolph	43	Pvt.	
Schmidt,	Jacob	30	Pvt.	
Schmidt,	Ludwig	29	Pvt.	
Schmidt,	Lorenz	35	Pvt.	
Schmidt,	Mathias	50	Pvt.	
Stiehler,	Julius	30	Pvt.	
Tatsch,	Michael	26	Pvt.	Member Insurgent Group
Tatsch,	Peter	22	Pvt.	Member Insurgent Group
Wahrmund,	Louis	39	Pvt.	Joined Union Army
Weinheimer,	George	36	Pvt.	
Weinheimer,	John	28	Pvt.	

F: Organization, 31st Brigade, Texas State Troops, March-August 1862

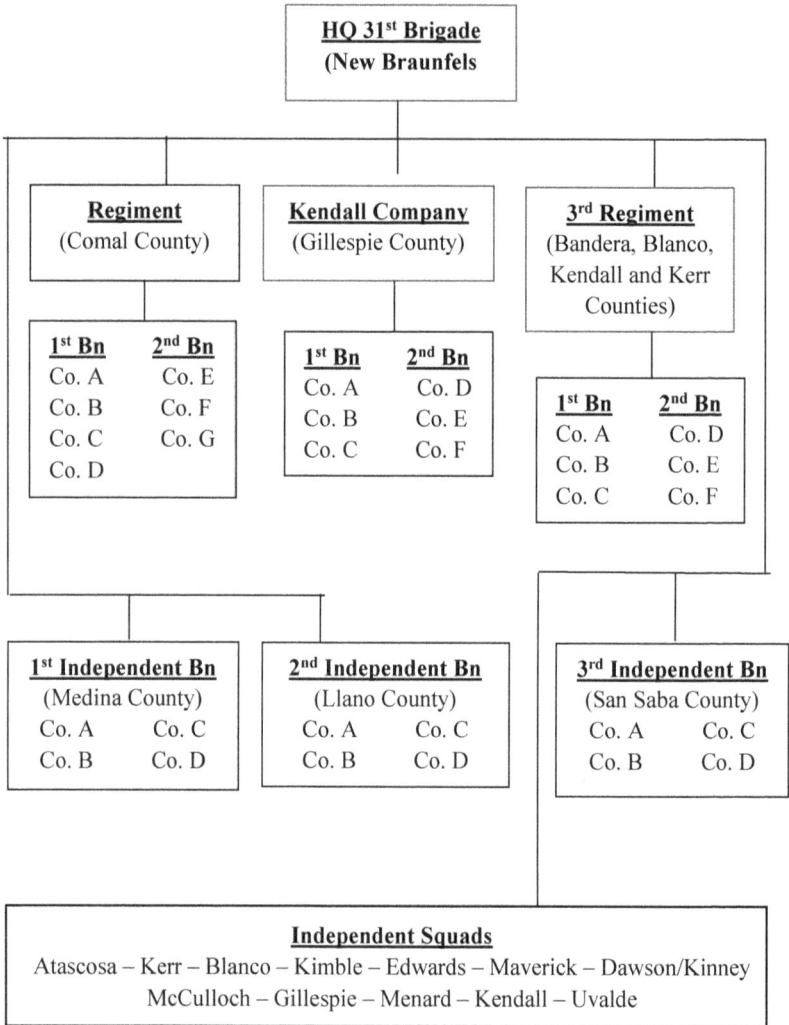

HQ 31st Brigade
(New Braunfels

Regiment (Comal County)	**Kendall Company** (Gillespie County)	**3rd Regiment** (Bandera, Blanco, Kendall and Kerr Counties)

1st Bn	**2nd Bn**
Co. A	Co. E
Co. B	Co. F
Co. C	Co. G
Co. D	

1st Bn	**2nd Bn**
Co. A	Co. D
Co. B	Co. E
Co. C	Co. F

1st Bn	**2nd Bn**
Co. A	Co. D
Co. B	Co. E
Co. C	Co. F

1st Independent Bn (Medina County)	**2nd Independent Bn** (Llano County)	**3rd Independent Bn** (San Saba County)
Co. A Co. C	Co. A Co. C	Co. A Co. C
Co. B Co. D	Co. B Co. D	Co. B Co. D

Independent Squads
Atascosa – Kerr – Blanco – Kimble – Edwards – Maverick – Dawson/Kinney
McCulloch – Gillespie – Menard – Kendall – Uvalde

G: Organization, 2nd Regiment 31st Brigade District, Texas State Troops, March-August 1862

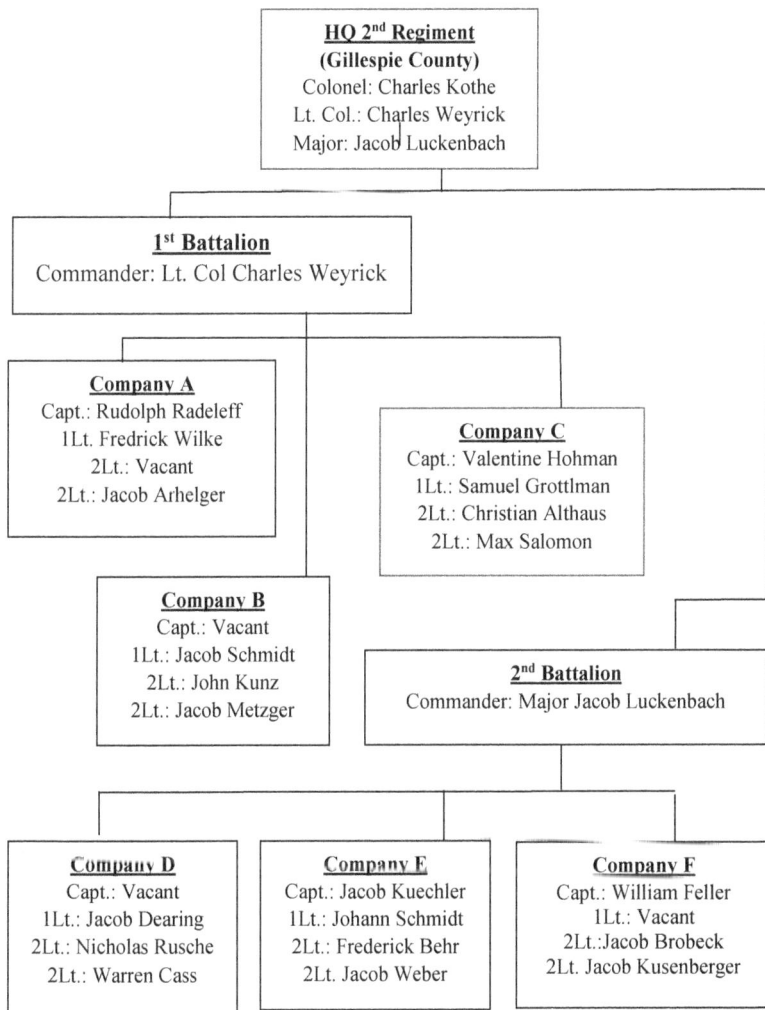

HQ 2nd Regiment
(Gillespie County)
Colonel: Charles Kothe
Lt. Col.: Charles Weyrick
Major: Jacob Luckenbach

1st Battalion
Commander: Lt. Col Charles Weyrick

Company A
Capt.: Rudolph Radeleff
1Lt. Fredrick Wilke
2Lt.: Vacant
2Lt.: Jacob Arhelger

Company C
Capt.: Valentine Hohman
1Lt.: Samuel Grottlman
2Lt.: Christian Althaus
2Lt.: Max Salomon

Company B
Capt.: Vacant
1Lt.: Jacob Schmidt
2Lt.: John Kunz
2Lt.: Jacob Metzger

2nd Battalion
Commander: Major Jacob Luckenbach

Company D
Capt.: Vacant
1Lt.: Jacob Dearing
2Lt.: Nicholas Rusche
2Lt.: Warren Cass

Company E
Capt.: Jacob Kuechler
1Lt.: Johann Schmidt
2Lt.: Frederick Behr
2Lt. Jacob Weber

Company F
Capt.: William Feller
1Lt.: Vacant
2Lt.:Jacob Brobeck
2Lt. Jacob Kusenberger

H: Organization, 3rd Regiment, 31st Brigade District, Texas State Troops, March-August 1862

HQ 3rd Regiment
(Bandera, Blanco, Kendall and Kerr Counties)
Colonel: Frederick Tegener
Lt. Col.: Julius Schlickum

1st Battalion
Commander: Lt. Col. Julius Schlickum

Company A
(Blanco County)
Capt.: A.J. Kercheville
Lt.: John Carson
Lt.: Thomas Carson
Lt.: Isaac Tanner

Company C
(Kendall County)
Capt.: Ottomar Labbardt
1Lt.: Louis von Breitenbauch
2Lt.: Ernest Beselar
2Lt.: Edgar von Westphalen

Company B
(Kendall County)
Capt.: Vacant
1Lt.: Frederick Lenz
2Lt.: Richard Brotze
1Lt.: Jacob Metzger

2nd Battalion
Commander: Major Ernest Cramer

Company D
(Kendall County)
Capt.: Michael Lindner
1Lt.: Ernest Schwethelm
2Lt.: William Schlickum
2Lt.: Theodore Bruckish

Company E
(Kerr County)
Capt.: Thomas Saner
1Lt.: Johann Schmidt
2Lt.: Edward Steves
2Lt. Thomas Selph

Company F
(Bandera County)
Capt.: Braden Mitchell
1Lt.: Robert Ballentyne
2Lt.: George Hay
2Lt. Robert L. Perkins

I: Organization, Medina County Independent Battalion, 31st Brigade District, Texas State Troops

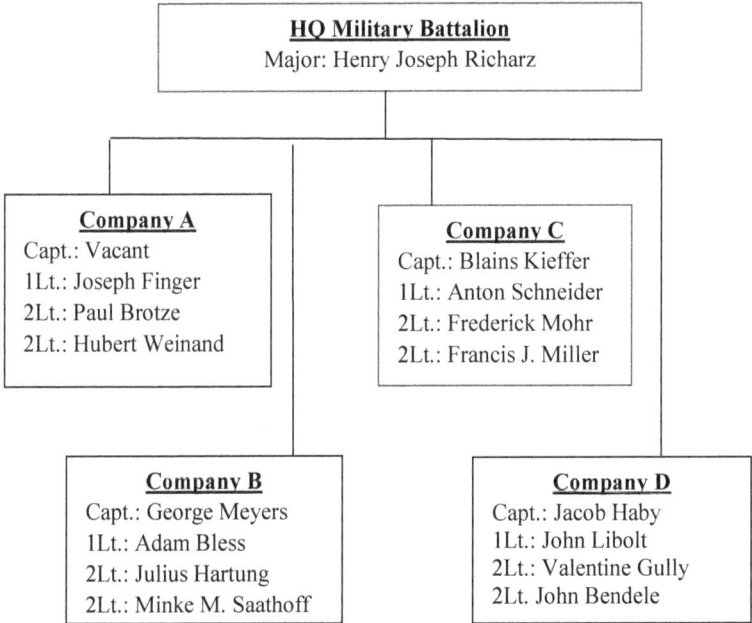

HQ Military Battalion
Major: Henry Joseph Richarz

Company A
Capt.: Vacant
1Lt.: Joseph Finger
2Lt.: Paul Brotze
2Lt.: Hubert Weinand

Company C
Capt.: Blains Kieffer
1Lt.: Anton Schneider
2Lt.: Frederick Mohr
2Lt.: Francis J. Miller

Company B
Capt.: George Meyers
1Lt.: Adam Bless
2Lt.: Julius Hartung
2Lt.: Minke M. Saathoff

Company D
Capt.: Jacob Haby
1Lt.: John Libolt
2Lt.: Valentine Gully
2Lt. John Bendele

J: Roster, C. de Montel's Co., Frontier Regiment, June, 1862

Surname	Given Name	Age	Rank	Remarks
de Montel,	Charles	49	Capt	
McCall	Thomas T.	30	1Lieut	
Gates	Amos V.	37	2Lieut	
Patton,	Benjamin F.	25	2Lieut	Deserted Joined Union Army
Lawhorn,	John	40	1Sgt	
Hill,	G. W.		2Sgt	
Harper,	Robert	23	3Sgt	
Cooper,	J. W.	30	4Sgt	
Croke,	James	37	5Sgt	
Malone,	Daniel	18	1Cpl	
Pafford,	Thomas	26	2Cpl	
Bandy,	Thomas	42	3Cpl	
Davis,	R. W.		4Cpl	
Ganyon,	W.		Bugler	
de Montel, Jr.	Charles	14	Bugler	
Kriesle	John	33	Blacksmith	Discharged Oct 23, 1862
Brown,	James		Farrier	
Adams,	Peter T.	27	Pvt	
Allen,	Hugh		Pvt	
Ballintyne,	William	38	Pvt	
Bandy,	John J.	17	Pvt	Discharged Oct 16, 1862
Bates,	E. A.	41	Pvt	
Bates,	Felix	21	Pvt	
Billharz,	Joseph	29	Pvt	Enlisted Jul 7, 1862
Bird,	Charles	35	Pvt	
Bowles,	David C.	23	Pvt	
Bowles,	G.		Pvt	
Brewer,	Leo	36	Pvt	
Brown,	John	25	Pvt	

Burney,	William C.	34	Pvt	
Bushnell,	Lester	25	Pvt	
Carnaham,	D. S.	25	Pvt	
Casner,	Francis M.	20	Pvt	
Casner,	Martin V.	20	Pvt	
Chipman,	Heber N.	20	Pvt	
Cole,	William A.	32	Pvt	
Conrad,	Peter	24	Pvt	
Cook,	John C.		Pvt	
Cook,	Thomas	22	Pvt	
Cooper,	T. A.		Pvt	
Croke,	James	37	Pvt	Promoted to 5Sgt
Cude,	R. D.	21	Pvt	
Davenport,	Andrew Jackson	19	Pvt	
Davenport,	William H.	21	Pvt	
Dodd,	Curtis		Pvt	
Dodd,	James C.	27	Pvt	Enlisted Aug 20, 1862
Dodd,	William		Pvt	
Dolch,	Louis	29	Pvt	
Duncan,	Benjamin	28	Pvt	Enlisted Jul 7, 1862
Dutcher,	Charles L.		Pvt	
Forrest,	Demps	33	Pvt	Enlisted Aug 20, 1862
Gray,	John D.		Pvt	Discharge Jul 8, 1862
Graylin,	August	24	Pvt	
Green,	John		Pvt	
Griner,	Nolbe J.	21	Pvt	
Haller,	Paul	30	Pvt	
Hamilton,	John W.	20	Pvt	
Harr,	Jephta	32	Pvt	
Heath,	Lewis	20	Pvt	
Hegg,	Frederick	36	Pvt	Deserted Joined Union Army

Hicks,	Fablan L.	34	Pvt	Enlisted Jul 7, 1862
Horton,	Thomas		Pvt	
Johnson,	Oscar	20	Pvt	
Kapper,	Frank	32	Pvt	Enlisted Jul 4, 1862
Kennedy,	John	25	Pvt	
Kent,	David B.	45	Pvt	
Lamon,	John	46	Pvt	
Lindemann,	Adam A.	25	Pvt	
Lindemann,	Edward D.	25	Pvt	
Lindemann,	H. C.	20	Pvt	
Lindsay,	Andrew		Pvt	Enlisted, Jul 21, 1862
Lundey,	M. R.	21	Pvt	Pro-Union Bushwhacker
Lundey,	William		Pvt	Discharged May 15, 1862 Pro-Union Bushwhacker
McKinney,	Thabert G. [V]	46	Pvt	
McKinney,	Thomas J.	30	Pvt	
Manning,	Thomas H.	36	Pvt	
Manning,	William	35	Pvt	
Messer,	Thomas C.	21	Pvt	
Miller,	John W.	21	Pvt	
Moore,	T. J.	25	Pvt	
Norris,	John		Pvt	
Nowlin,	Richard W.	21	Pvt	
Nowlin,	Samuel	24	Pvt	
Onion,	Joseph	19	Pvt	
Owens,	William P.	21	Pvt	
Patterson,	Daniel J.	23	Pvt	
Patton,	Pickens,	40	Pvt	Deserted Joined Union Army
Peden,	H. D.	24	Pvt	
Pingenot,	Celestin	23	Pvt	
Pulliam,	B. A.	21	Pvt	
Rackley,	Wilson	42	Pvt	

Reichertzer,	Thomas		Pvt	
Reinhardt,	John	30	Pvt	Enlisted Apr 1, 1862
Reynolds,	W. M.	38	Pvt	
Runnels,	W.		Pvt	
Sanders,	Robert	21	Pvt	
Sansom,	Joseph N.	22	Pvt	Deserted Joined Union Army
Schuckhard,	Friedrich F.		Pvt	
Schwarz,	Reinhard		Pvt	
Sharp,	George W.	19	Pvt	
Short,	Thomas	32	Pvt	Deserted Joined Union Army
Shults	John		Pvt	
Smith,	Henry R.	37	Pvt	
Smith,	R. B.	25	Pvt	
Stanford,	C. C. [Kit]	33	Pvt	
Stayton,	David W.	26	Pvt	
Stayton,	John A.	22	Pvt	
Taylor,	N. A.		Pvt	Enlisted in CSA
Taylor,	Thomas E.	32	Pvt	Enlisted Apr 7, 1862
Thompson,	Hiram G.	16	Pvt	
Tilley,	Gibson E.	35	Pvt	
Tomberlin,	Allen	20	Pvt	
Tondre,	Francis	28	Pvt	
Walker,	John	19	Pvt	
Ward,	H. L.		Pvt	
Watson,	Benjamin H.	49	Pvt	
Watson,	James	18	Pvt	Enlisted Jun 29, 1862
Westfall,	S. C.	17	Pvt	
Westfall,	William R.	22	Pvt	
Wight,	Lamoni L.	23	Pvt	
Wilkins,	Dwight G.	17	Pvt	
Wood,	Columbus P.	45	Pvt	Killed by Pro-Union Bushwhackers

Wood,	William P.	23	Pvt
Wyley,	Isaak	20	Pvt
Zimmermann,	Alexander	26	Pvt
Zorger,	Peter	22	Pvt

K: Roster, J. Keuchler's Co., February, 1862

Surname	Given Name	Age	Rank	Remarks
Kuechler,	Jacob	39	Capt	Member Insurgent Group
Braubach,	Philip	35	1Lieut	Arrested By Duff
Weiss,	Moritz	24	2Lieut	Member Insurgent Group
Degener,	Hugo	20	2Lieut	Member Insurgent Group
Bucholz,	Theodore	30	1Sgt	Later Member of Davis' Company
Duecker,	August	33	2Sgt	Member Insurgent Group
Sauer,	Friedrich	25	3Sgt	Later Member of Davis' Company
Boerner,	Louis	28	4Sgt	Member Insurgent Group
Becker,	Johann	34	1Cpl	
Sauer,	Gottlieb	22	2Cpl	Joined Union Army
Nixon,	A. II	21	3Cpl	Later Member of Davis' Company
Reagan,	Josiah	26	4Cpl	
Renick,	Jacob	38	Farrier	Joined Union Army
Behrend,	Heinrich	23	Pvt	Later Member of Van der Stucken's Company
Beseler,	Ernst	20	Pvt	Member Insurgent Group
Brinkmann,	Carl	25	Pvt	Joined Union Army
Blucher,	Adam	21	Pvt	Later Member of Davis' Company
Boerner,	Heinrich	37	Pvt	Member Insurgent Group
Cloudt,	Arthur	45	Pvt	Later Member of Davis' Company
Coffey,	Benjamin	18	Pvt	
Degener,	Hilmar	22	Pvt	Member Insurgent Group
Diaz,	Pablo	22	Pvt	Member Unionist Group
Dengel,	Joseph	29	Pvt	Later Member of Davis' Company
Elstner,	Joseph	21	Pvt	Member Insurgent Group
Feuge,	Christian	23	Pvt	
Fritz,	Casper	20	Pvt	Joined Union Army
Gold,	Jacob, Sr.	20	Pvt	Member Insurgent Group
Gold,	Peter, Sr.	22	Pvt	Member Insurgent Group

Gold,	Peter, Jr.	22	Pvt	Later Member of Van der Stucken's Company
Freeman	E. R.	42	Pvt	
Geissler,	W. T.	33	Pvt	
Graff,	August		Pvt	Member Insurgent Group
Herbst,	Carl	33	Pvt	Later Member of Confederate Army
Hartbeck,	R.		Pvt	Joined Union Army
Heumann,	Heinrich		Pvt	
Heinen,	Anton	22	Pvt	Joined Union Army
Hartzberg,	John		Pvt	
Juenke,	Wilhelm	23	Pvt	
Jung,	Frank	25	Pvt	Later Member of Van der Stucken's Company
Jung,	John	25	Pvt	
Kapp,	Alfred	26	Pvt	Joined Confederate Army
Karder,	Carl	17	Pvt	
Kallenberg,	George	19	Pvt	Member Insurgent Group
Klier,	Wilhelm	27	Pvt	Member Insurgent Group
Kuehne,	Wilhelm	22	Pvt	
Klein,	John	29	Pvt	Joined Union Army
Klein,	John Adam	29	Pvt	Joined Union Army
Leyendecker,	Joseph	32	Pvt	Joined Union Army
Lott,	Anton	39	Pvt	
Niemeyer,	J.		Pvt	
Nieubert,	Otto		Pvt	
Ottmers,	George	18	Pvt	Later Member of Van der Stucken's Company
Pfieffer,	Adolph	49	Pvt	
Quitzow?	Albert		Pvt	
Rische,	Ulrich	24	Pvt	
Reigner,	John	31	Pvt	
Schierholz,	Louis	32	Pvt	Member Insurgent Group
Schladoer,	Frederick	21	Pvt	
Schaefer,	Christian	18	Pvt	Member Insurgent Group

Schneider,	Jacob	18	Pvt	Later Member of Van der Stucken's Company
Schwartz,	Carl	21	Pvt	Later Member of Van der Stucken's Company
Siemering,	August	34	Pvt	Later Member of Van der Stucken's Company
Steves,	Heinrich	27	Pvt	Member of Insurgent Group
Striegler,	Arthur	21	Pvt	Later Member of Davis' Company
Striegler,	Olfert	23	Pvt	Later Member of Davis' Company
Striegler,	Owe	19	Pvt	Later Member of Davis' Company
Telegmann,	Wilhelm	31	Pvt	Member Insurgent Group
Tatsch,	Peter	19	Pvt	Joined Union Army
Usener,	Ludwig	20	Pvt	Member Insurgent Group
Vetterlein,	Gottlieb	30	Pvt	Member Insurgent Group
Walter,	John	20	Pvt	Later Member of Van der Stucken's Company
Weber,	Friedrich	28	Pvt	Member Insurgent Group
Weiner,	Joseph		Pvt	
Weiss,	Franz	19	Pvt	Member Insurgent Group
Westphal,	Edgar	42	Pvt	

L: Roster, H. T. Davis' Co., August 1862

Surname	Given Name	Age	Rank	Remarks
Davis	Henry T.	28	Capt	
Hunter,	James M.	33	1Lieut	
Harbour,	William T.	29	2Lieut	
Williamson,	James L.	39	2Lieut	
Hopkins,	Desmond P.	24	1Sgt	
Dennis,	Nathan M.	33	2Sgt	
Rees,	Daniel Adolphus	30	3Sgt	
Cloudt,	Arthur	46	4Sgt	
Miller,	Hugh	30	5Sgt	
Neil,	Ambrose H. S.	27	1Cpl	
Clopton,	Hoggitt	21	2Cpl	
Basse,	Oscar	19	3Cpl	
Ezell,	Benjamin F.	22	4Cpl	
Tadlock,	Sevies	24	Bugler	
Striegler,	Owe	19	Bugler	
Weiss,	Louis	33	Farrier	
Alexander,	C		Pvt	Enlisted Jul 1862
Anderson,	J. A.		Pvt	
Bagley,	David Ross	20	Pvt	
Bagley,	Marbeau Larmar	16	Pvt	
Bagley,	William M.	22	Pvt	
Baker,	Horatro N.		Pvt	
Baker,	John		Pvt	
Baker,	Leonard		Pvt	
Banta,	Jacob R.	27	Pvt	Pro-Confederate Bushwhacker
Banta,	John W.	29	Pvt	Pro-Confederate Bushwhacker
Banta,	William	35	Pvt	Pro-Confederate Bushwhacker

Bawson,	Sewart M.		Pvt	
Baylor,	M. B.		Pvt	
Benson,	John Washington	20	Pvt	Pro-Confederate Bushwhacker
Benson,	William Thomas	22	Pvt	Wounded Nueces Battle Pro-Confed.
Blevins,	Leroy S.	21	Pvt	Pro-Confederate Bushwhacker
Blucher,	Adam		Pvt	
Braden,	James H.		Pvt	
Brown,	Jacob		Pvt	
Brown,	Joshua David	46	Pvt	
Brown,	Rufus E.	47	Pvt	
Brown,	William H.	30	Pvt	
Bryant,	Milton R.	22	Pvt	
Buchanan,	James H.		Pvt	
Buchanan,	Sidney S.		Pvt	
Bucholz,	Theodore	30	Pvt	
Bulter,	John		Pvt	
Colbert,	John	38	Pvt	
Colbert,	Winthrop S.	28	Pvt	
Corbell	Tillman A.	30	Pvt	Enlisted Jul 1862
Cowan	Samuel F.	29	Pvt	Pro-Confederate Bushwhacker
Dengel,	Joseph		Pvt	Member Kuechler's Company
Doss,	Willis	19	Pvt	Enlisted Jun 1862
Durham	Marcellus	22	Pvt	Pro-Confederate Bushwhacker
Earnest,	Joseph William	18	Pvt	
Evans,	John Wood	26	Pvt	
Evans,	Thomas A.	28	Pvt	
Farr,	David H.	42	Pvt	
Frieschmeyer,	Frances		Pvt	
Gatlin,	William L.		Pvt	
Gibson,	Samuel L.	18	Pvt	

Goss,	Joshua Spencer	24	Pvt	Enlisted Jun 1862
Hall,	Richard G.	25	Pvt	Deserted Joined Union Army
Hampton,	Jordan E.		Pvt	
Haud,	J. A. [B]		Pvt	
Hayes	Jeremiah M.	33	Pvt	Enlisted Jun 1862
Hester,	Newton	25	Pvt	Deserted Joined Union Army
Hewitt,	James A.		Pvt	Enlisted Jul 1862
Hill,	Elijah J.	19	Pvt	
Hill,	Eli T.		Pvt	
Hopkins,	C. A.		Pvt	
Imhof,	John Peter	34	Pvt	Enlisted Jul 1862
Jackson,	Aaron W.	22	Pvt	
Jackson,	Nathan	28	Pvt	
Kell,	Scott	26	Pvt	
Lamb,	Hiram	22	Pvt	
Lane,	Samuel Jr.	26	Pvt	
Long,	William	61	Pvt	
Lowrance,	John Shelby	22	Pvt	
Lowrance,	Miles Abernathy	35	Pvt	Enlisted Jun 1862
McCann	Samuel Houston	16	Pvt	
McClusky,	Michael	18	Pvt	
McDaniel,	Silas		Pvt	
McDonald,	Benjamin F.	16	Pvt	
McDonald,	Robert L.	17	Pvt	
McDonald,	Shepard		Pvt	
McDonald,	Zachariah	21	Pvt	
Mayes,	Hollen	26	Pvt	Enlisted Jun 1862
Mayes,	John	22	Pvt	
Meyers,	Freeman		Pvt	
Moore,	William	21	Pvt	
Mosel,	John Peter	30	Pvt	Member Kuechler's Company

Neal,	James M.	18	Pvt	Pro-Confederate Bushwhacker
Neil,	Ambros H.	27	Pvt	
Neil,	Thomas S. D.	23	Pvt	
Nixon,	Andrew Jackson II	22	Pvt	Member Kuechler's Company
Oatman,	Marshall C.		Pvt	
Ottenhause,	Henry	22	Pvt	
Overstreet,	William H.	32	Pvt	
Paul,	Andrew Jackson	36	Pvt	
Phillips,	George P.	43	Pvt	
Picket,	John R.		Pvt	
Rees,	Peter Osborn A.	24	Pvt	
Rhodes,	Andrew Jackson	30	Pvt	
Riley,	David Crocket		Pvt	
Riley,	James Bowie	35	Pvt	
Rodgers,	Jesse W.		Pvt	
Rowe,	Enoch		Pvt	
Sauer,	Frederick	24	Pvt	
Schreider	Christian		Pvt	
Scott,	William A.		Pvt	
Seal,	John W.		Pvt	
Smith,	Henry	19	Pvt	
Smith,	Napolion	21	Pvt	
Smith,	Thomas J.	30	Pvt	Pro-Confederate Bushwhacker
Stoke,	James C.	39	Pvt	
Striegler,	Arthur	22	Pvt	
Striegler,	Olfort	23	Pvt	
Tanner,	Samuel T.	21	Pvt	Pro-Confederate Bushwhacker
Taylor,	James Thurman	29	Pvt	Deserted Joined Union Army

Thomas,	John S.		Pvt	
Thompson,	William Delbert	26	Pvt	
Vaughn,	Jesse H.	22	Pvt	
Wallace,	John	23	Pvt	
Wharton,	William G.	31	Pvt	
White,	Richard R.	28	Pvt	
Whitley,	Josiah	25	Pvt	
Whitley,	Elisha		Pvt	
Wilburn,	Albert L.	34	Pvt	Wounded Nueces Battle
Williams,	William A.	30	Pvt	Pro-Confederate Bushwhacker
Williamson,	John Roland	25	Pvt	
Williamson,	Newton Cannon	22	Pvt	
Williamson,	William T.	30	Pvt	
Wilson,	John T.		Pvt	
Winters,	Willis		Pvt	
Wittens,	Benson M.		Pvt	
Witters,	William A.		Pvt	
Wood,	John	56	Pvt	

M: Organization of *Des Organizator* Military Battalion, Organization of Military Arm, Union Loyal League

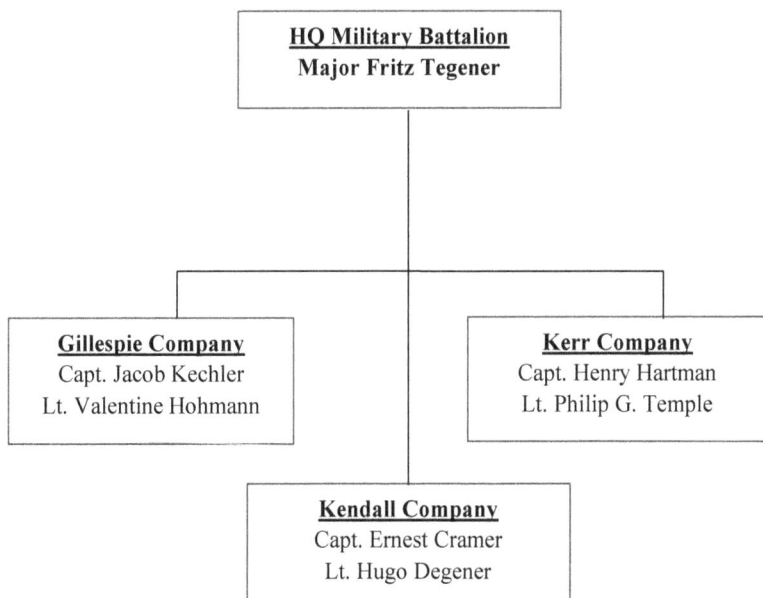

HQ Military Battalion
Major Fritz Tegener

Gillespie Company
Capt. Jacob Kechler
Lt. Valentine Hohmann

Kerr Company
Capt. Henry Hartman
Lt. Philip G. Temple

Kendall Company
Capt. Ernest Cramer
Lt. Hugo Degener

N: Roster, Likely Members Gillespie Unionist Company, March – July 1862

Surname	Given Name	Age	Rank	Remarks
Kuechler,	Jacob	37	Capt	Kuechler's Co. Unionist Group Oct Group
Hohmann,	Valentine	31	Lieut	Insurgent Group - Oct Group KIA Rio Grande
Ahrens,	Christian	20	Pvt	Joined Union Army
Beck, Jr.	Phillip	25	Pvt	Luckenbach Bushwhackers Insurgent Group - Joined Union Army
Becker,	Franz	37	Pvt	Braubach's Co
Becker,	Johann	34	Pvt	Kuechler's Co
Behrends,	Johann Heinrich	23	Pvt	Kuechler's Co
Behrens,	Frederick [Fritz]		Pvt	Insurgent Group - KIA Nueces
Bender,	Conrad	27	Pvt	
Betz,	Reinhardt	20	Pvt	Braubach's Co
Blank,	John	35	Pvt	Killed by pro-Conf Bushwhackers 64
Bock,	Conrad		Pvt	Luckenbach Bushwhackers Insurgent Group – Later Captured and Executed
Bratherrig/ Bratherich	Henry	21	Pvt	Joined Union Army
Braubach,	Phillip	33	Pvt	Braubach & Kuechler's Co
Bucholz,	Theodore	28	Pvt	Joined Union Army Braubach's Co
Burg,	Peter	40	Pvt	Braubach's Co - Killed by pro-Confederate Bushwhackers 1864
Cramm,	Henry	23	Pvt	Joined Union Army
Dambach,	Johann	26	Pvt	
Dambach,	Ludwig Frederick	27	Pvt	

Dengel/Dangel,	Joseph		Pvt	Kuechler's Co
Diaz,	Pablo	18	Pvt	Kuechler's Co - Insurgent Group - KIA Nueces
Duecker,	August	34	Pvt	Kuechler's Co - Insurgent Group
Ebers,	August	28	Pvt	Joined Union Army
Feller,	Carl	38	Pvt	
Feller,	Wilhelm	32	Pvt	Killed by pro-Conf Bushwhackers 64
Feuge,	Christoph	22	Pvt	Braubach & Kuechler's Co
Feuge,	Franz	19	Pvt	
Flick,	Herman	22	Pvt	Insurgent Group - Later Captured andExecuted
Fritz,	Casper	18	Pvt	Kuechler's Co - Joined Union Army
Gold,	Jacob Jr.	20	Pvt	Kuechler's Co
Gold,	Peter Jr.	22	Pvt	Braubach & Kuechler's Co
Graff,	August	26	Pvt	Kuechler's Co
Graff,	Carl William	31	Pvt	Braubach's Co - Insurgent Group Joined Union Army
Graff,	William	21	Pvt	Insurgent Group
Hartbeck,	Rudolph		Pvt	Kuechler's Co - Joined Union Army
Hartzberg,	John		Pvt	Kuechler's Co
Heimann,	Ludwig Heinrich	20	Pvt	
Hcumann,	Heinrick		Pvt	Kuechler's Co
Hoffmann,	August	19	Pvt	Insurgent Group
Itz,	Hermann	18	Pvt	Killed by pro-Conf Bushwhackers 63
Itz,	Jakob	20	Pvt	Killed by pro-Conf Bushwhackers 63
Itz,	Karl	28	Pvt	Insurgent Group
Jacoby	Peter	26	Pvt	Insurgent Group
Juenke,	Wilhelm	23	Pvt	Braubach & Kuechler's Co
Jung	Franz		Pvt	Kuechler's Co
Jung,	John	25	Pvt	Kuechler's Co
Kallenberg,	John George	19	Pvt	Kuechler's Co

Kammlah,	Henry II	22	Pvt	Insurgent Group - WIA Neuces
Kirchner,	Henry	28	Pvt	Braubach's Co - Killed by Pro-Conf Bushwhackers 64
Kirchner,	Richard	21	Pvt	
Kleck,	Syverster	17	Pvt	Luckenbach Bushwhackers Oct Group - Wounded Rio Grande
Klein,	Adam	29	Pvt	
Klein,	John	29	Pvt	Joined Union Army
Klein,	John Adams	27	Pvt	Joined Union Army
Kleir,	Wilhelm F. Eduard	27	Pvt	Kuechler's Co - Insurgent Group
Kuehne,	Wilhelm	22	Pvt	Kuechler's Co
Kusenberger,	Chrisian	35	Pvt	
Kusenberger,	Jacob	28	Pvt	Luckenbach Bushwhackers Insurgent Group - Joined Union Army
Langehenning,	Henry	31	Pvt	Joined Union Army
Leyendecker,	Franz Joseph	32	Pvt	Kuechler's Co - Joined Union Army
Lochte, Jr.	H. Frederick	28	Pvt	
Lott/Loth,	Anton	39	Pvt	Kuechler's Co
Luckenbach,	August	29	Pvt	Luckenbach Bushwhackers Insurgent Group - Later Captured and Executed
Markwordt,	Heinrich	32	Pvt	Insurgent Group - KIA Nueces
Meckel,	Bernhard	31	Pvt	Braubach's Co
Meckel,	Conrad	35	Pvt	
Meurer,	Peter	24	Pvt	Joined Union Army
Moellering,	Carl	22	Pvt	Unionist Group
Mosel,	John Peter	29	Pvt	Braubach's Co
Neuber/Neubert,	Otto Phillip	40	Pvt	Kuechler's Co
Niemeyer,	Gustav		Pvt	Kuechler's Co
Ohlenburger,	Ferdinand	27	Pvt	Joined Union Army

Ottmers,	George H. Gottfried	18	Pvt	Kuechler's Co
Pehl,	Mathias	22	Pvt	Braubach's Co - Insurgent Group
Petermann,	Franz	36	Pvt	Braubach's Co
Poetsch,	Christian	19	Pvt	Luckenbach Bushwhackers
Poetsch,	Joseph	22	Pvt	Luckenbach Bushwhackers Insurgent Group
Quitzow?,	Albert		Pvt	
Radcliff/Radeleff,	J. Rudolph	33	Pvt	Arrested by Confederates
Rausch,	Henry	19	Pvt	Insurgent Group
Reagan,	Josia		Pvt	Kuechler's Co
Reeh,	Christian	24	Pvt	Joined Union Army
Reeh,	Reinhardt	29	Pvt	
Remick,	Jacob		Pvt	Joined Union Army
Reigner,	John	31	Pvt	Kuechler's Co
Ruebsamen,	Adolph	22	Pvt	Insurgent Group - Later Captured and Executed
Ruebsamen,	Louis	19	Pvt	Insurgent Group - Later Captured and Executed
Sauer,	John F.	25	Pvt	Braubach & Kuechler's Co.
Schaefer,	Christian	18	Pvt	Luckenbach Bushwhackers - Insurgent Group KIA Nueces Insurgent Group KIA Nueces
Schildknecht,	August	36	Pvt	Braubach's Co
Schmidt,	Jacob	30	Pvt	Braubach's Co
Schmidt,	Lorenz	36	Pvt	Braubach's Co
Schmidt,	Ludwig	32	Pvt	Braubach's Co
Schneider,	Jacob	18	Pvt	Kuechler's Co
Schneer,	Friedrich Wilhelm	38	Pvt	Braubach's Co
Schoenewolf, Jr.	August	26	Pvt	Luckenbach Bushwhackers Insurgent Group
Schuetze,	Louis	42	Pvt	Killed by pro-Conf Bushwhackers 64
Siemering,	August	32	Pvt	Kuechler's Co

Stieler,	Julius	30	Pvt	Braubach's Co
Tatsch,	Michael	29	Pvt	Insurgent Group
Tatsch,	Peter	19	Pvt	Kuechler's Co - Joined Union Army
Usener,	Jacob August	28	Pvt	Insurgent Group
Usener,	Ludwig Wilhelm	18	Pvt	Insurgent Group
Vater,	Adolph	23	Pvt	Insurgent Group - KIA Nueces
Vater,	August	21	Pvt	
Vater,	Friedrich		Pvt	Insurgent Group - KIA Nueces
Vater,	William		Pvt	Insurgent Group - WIA Neuces
Weiner,	Joseph		Pvt	Kuechler's Co
Weinheimer,	Anton	34	Pvt	Braubach's Co
Weinheimer,	George	37	Pvt	Braubach's Co
Weinheimer,	John	29	Pvt	Braubach's Co
Weyershausen,	Heinrich	32	Pvt	Luckenbach Bushwhackers Insurgent Group - KIA Nueces
Weyrich,	Michael	23	Pvt	Insurgent Group - KIA Nueces

O: Roster, Likely Members Kendal County Unionist Company, March – July 1862

Surname	Given Name	Age	Rank	Remarks
Cramer,	Ernest	24	Capt	Insurgent Group
Degener,	Hugo	20	Lieut	Kuechler's Co Insurgent KIA
Aschman,	Friedrick	40	Pvt	
Bauer,	Leopold	23	Pvt	Insurgent Group - KIA Nueces
Bechstadt,	Ernest Theodore	23	Pvt	Joined Union Army
Becker,	Peter	36	Pvt	
Below,	Max	22	Pvt	
Berger,	Ferdinand	30	Pvt	
Berger,	John	33	Pvt	Joined Union Army
Berger,	Louis	27	Pvt	
Beseler,	Charles Philip	21	Pvt	Joined Union Army
Beseler,	Ernst	20	Pvt	Kuechler's Co - Insurgent KIA Nueces
Beversdorff,	Albert H.	30	Pvt	Oct Group - Joined Union Army
Beversdorff,	August	39	Pvt	Oct Group - Joined Union Army
Boerner,	Louis	29	Pvt	Kuechler's Co - Insurgent Group Recovered Bodies Aug '65
Boerner,	Wilhelm F.	34	Pvt	Insurgent Later Captured and Executed
Bonnet,	Heinrich Daniel	27	Pvt	Oct Group - Joined Union Army
Bonnet,	Johann Charles	34	Pvt	Oct Group - Joined Union Army
Bonnet,	John Peter	29	Pvt	Oct Group - KIA Rio Grande
Brinkmann,	Alexander	25	Pvt	
Brinkmann,	Charles	25	Pvt	Joined Union Army
Brinkmann,	H. Otto	20	Pvt	

Brunkish,	Charles	27	Pvt	
Brukish,	Theodore	23	Pvt	Insurgent Later Captured and Executed
Brunkish,	Wilhelm	23	Pvt	
Bruns,	Albert	24	Pvt	Unionist Group KIA Nueces
Burgmann,	Charles	36?	Pvt	Alleged 'Spy'
Degener,	Hilmar	22	Pvt	Kuechler's Co - Insurgent Group KIA Nueces
Diener,	V. [Fritz]	??	Pvt	
Elstner,	Joseph	21	Pvt	Kuechler's Co - Insurgent Group Oct Group KIA Rio Grande
Fabra,	Julius	35	Pvt	
Faltin,	August	32	Pvt	
Felsing,	Ernst/Eduard	34	Pvt	Insurgent - Oct Group KIA Rio Grande
Flack,	Christoph	35	Pvt	
Geissler,	William T. [F]	31	Pvt	Kuechler's Co.
Haerter,	Constantine	43	Pvt	
Harms,	Frederic [Fritz]	20	Pvt	Joined Union Army
Harms,	Hermann	35	Pvt	Joined Union Army
Heiligmann,	Henry	29	Pvt	
Heinen,	Anton	22	Pvt	
Heinen,	Heinrich Hubert	23	Pvt	Joined Union Army
Heinen,	Heinrich Joseph	18	Pvt	Joined Union Army
Heinen,	Peter	20	Pvt	
Heinen,	Theodore	23	Pvt	
Hermann,	Heinrich/Henry	29	Pvt	Insurgent - Oct Group KIA Rio Grande
Heuermann,	William	34	Pvt	
Hoerner,	Johann Charles	31	Pvt	
Ingenhuett,	Martin	27	Pvt	
Ingenhuett,	Peter Joseph	29	Pvt	
Ingenhuett,	Thomas	28	Pvt	Joined Union Army

Jonas,	August W.	20	Pvt	Joined Union Army
Jonas,	Hermann	27	Pvt	Joined Union Army
Karger,	Charles	17	Pvt	
Kempt.	H. L.	27	Pvt	Joined Union Army
Lamm,	Franz Joseph	36	Pvt	
Lange,	Frederick [Fritz]	30	Pvt	Insurgent Oct Group Wounded Rio Grande DOW
Lange, Jr.	Heinrich C. Ludwig	23	Pvt	Oct Group
Lich,	Balthasar	24	Pvt	
Lieck,	August	36	Pvt	
Marquart,	Christian Friedrich	27	Pvt	
Meckel,	Daniel	27	Pvt	Joined Union Army
Oberwetter,	Peter Heinrich	32	Pvt	
Perner,	John Fredrick	34	Pvt	
Pfieffer,	Robert	30	Pvt	
Pfeuffer,	Vitus	33	Pvt	
Rhodius,	Christoph	36	Pvt	Group Recovered Bodies Aug '65
Richter,	Henry I.	35	Pvt	
Rische,	Ulrich		Pvt	Kuechler's Company.
Rochau,	Henry	29	Pvt	Joined Union Army
Rosenthal,	Adolph	36	Pvt	
Sanger,	Fritz	34	Pvt	
Sanger,	H. C.	24	Pvt	
Sauer,	Friedrich [Fritz]	27	Pvt	Group Recovered Bodies Aug 65
Sauer,	Gottlieb	24	Pvt	Joined Union Army
Schaefer,	Robert	32	Pvt	Group Recovered Bodies Aug 65
Schieholz,	Louis	29	Pvt	Kuechler's Co - Insurgent KIA Nueces
Schilling,	Ernest Frederick	26	Pvt	

Schladoer,	Frederick [Fritz]	23	Pvt	Kuechler's Co
Schreiner,	Amil	23	Pvt	Insurgent Group KIA Nueces
Schultz[e] ,	Frederick [Fritz]	35	Pvt	Joined Union Army
Schultz[e] ,	William	39	Pvt	
Schwethelm,	Henry Joseph	22	Pvt	Insurgent Joined Union Army
Seewald,	Charles	24	Pvt	
Seidensticker,	Henry	32	Pvt	
Serger,	Emil	31	Pvt	Group Recovered Bodies Aug 65
Simon,	Ferdinand	36	Pvt	Insurgent WIA and Captured
Spenrath,	Franz	29	Pvt	
Steves,	Edward	32	Pvt	Group Recovered Bodies Aug 65
Steves,	Heinrich	28	Pvt	Insurgent Group KIA Nueces
Stieler,	Heinrich	18	Pvt	Insurgent Later Captured and Executed
Strohecker,	William Louis	27	Pvt	
Sueltenfuss,	Casper H.	18	Pvt	Joined Union Army
Tays,	Heinrich Friedrich "Fritz"	20	Pvt	Insurgent Group Later Captured and Executed
Tays,	Johann Heinrich	22	Pvt	Joined Union Army
Tegener,	Gustav	32	Pvt	Executed Aug 22, 62
Tellgmann,	Wilhelm	31	Pvt	Insurgent Group KIA Nueces
Vetterlien,	Charles Frederick	30	Pvt	Insurgent Group
Vetterlien,	Gottlieb	31	Pvt	Insurgent Group - WIA Nueces
Weber,	Frederick [Fritz]	24	Pvt	Joined Union Army
Weber,	Philipp Heinrich	27	Pvt	Insurgent Group
Weiss,	Franz	19	Pvt	Insurgent WIA Nueces Oct Group KIA Rio Grande

Weiss,	Moritz	26	Pvt	Insurgent - Oct Group KIA Rio Grande
Wittbold,	Henry	25	Pvt	
Zoeller,	Adolphus	22	Pvt	Insurgent Group - WIA Nueces Joined Union Army

P: Roster, Likely Members Kerr County (American) Unionist Company, March – July 1852

Surname	Given Name	Age	Rank	Remarks
Hartmann,	Henry		Capt	Joined Union Army - Killed by pro-Conf Bushwhackers 1864
Temple,	Phillip G.		Lieut	Joined Union Army
Arnold,	Frederick	29	Pvt	Joined Union Army
Baxter,	Jacob	31	Pvt	
Cass,	Warren	35	Pvt	Killed by pro-Conf Bushwhackers 64
Clements,	Adam		Pvt	
Clements,	Bird		Pvt	
Coffey,	Benjamin	16	Pvt	
Eastwood,	Joseph Frank	20	Pvt	Joined Union Army
Edon,	Archibald B.		Pvt	
Fairchild, Sr.	Hez Warden [Amos]		Pvt	Killed by pro-Conf Bushwhackers 64
Harris,	Frederick		Pvt	Joined Union Army
Henderson,	Casper W.	29	Pvt	
Henderson,	Howard	21	Pvt	Insurgent Group - Joined Union Army
Henderson,	John W.	32	Pvt	
Hester,	Newton	25	Pvt	Joined Union Army
Hester,	William	21	Pvt	Insurgent Group - Joined Union Army
Holden,	J. S.		Pvt	
Howell,	John T.		Pvt	Killed by Donelson's Men Aug 62
Johnson,	James	20	Pvt	Joined Union Army
Johnson,	William		Pvt	
Joy,	John	29	Pvt	
Joy,	Richard	26	Pvt	
Joy, Jr.	William [Tobe] T.	18	Pvt	

King,	Edward A.		Pvt	
Lacey,	A. B.	19	Pvt	
Lacey,	Asa Phelps	32	Pvt	
Lacey,	John Bunyon	32	Pvt	
Lacey,	Joshua Collins	20	Pvt	
Lacey,	James		Pvt	Killed by pro-Conf Bushwhackers
Lamb,	Albert	19	Pvt	
Locke,	William Jackson	34	Pvt	Joined Union Army
Lundy,	M. R.	21	Pvt	Killed by pro-Conf Bushwhackers 64
Lundy,	William		Pvt	Killed by pro-Conf Bushwhackers 64
Neel,	Xavier	38	Pvt	Killed by pro-Conf Bushwhackers 64
Nelson,	Allen B.	20	Pvt	Joined Union Army
Nelson,	Lewis	36	Pvt	
Nelson,	Hiram Louis	28	Pvt	
Nelson,	William	25	Pvt	
New,	John W.	32	Pvt	
Prescott,	Aaron	25	Pvt	
Scott,	Charles Franklin	26	Pvt	Executed Aug 22, 1862
Scott,	Thomas J.	26	Pvt	Insurgent Group - Joined Union Army
Scott,	Warren B.	19	Pvt	Insurgent Group - Joined Union Army
Snow,	Moses Moran	31	Pvt	Killed by pro-Conf Bushwhackers 64
Snow,	William F.		Pvt	Killed by pro-Conf Bushwhackers 64
Starr,	Andrew		Pvt	
Starr,	Charles		Pvt	
Starr,	Jesse	16	Pvt	Joined Union Army - Killed by pro-Conf Bushwhackers 64
Strong,	John	26	Pvt	Joined Union Army

Taylor,	Lewis	25	Pvt	
Taylor,	Morcy	23	Pvt	
Taylor,	Thurman Thompson	33	Pvt	Joined Union Army
Turknett,	John S. C.	17	Pvt	Killed by pro-Conf Bushwhackers 63
Turknett,	Phillip Brandon	40	Pvt	Killed July 1864

Q: Roster, F. Van Der Stucken's Co., August 1862

Surname	Given Name	Age	Rank	Remarks
Van der Stucken,	Frank	31	Capt	
Max,	George	35	1Lieut	
Heffler,	Hugo	24	2Lieut	
Siemering,	August	32	2Lieut	Member Kuechler's Co.
O'Neill,	Arthur J.	26	1Sgt	
Wahrmund,	Carl	30	2Sgt	
			3Sgt	
Schmidt,	Ludwig	34	4Sgt	
Durst,	John	33	1Cpl	
Johnson,	William	29	2Cpl	
Coleman,	August	24	3Cpl	
Kordzick,	Henry	25	4Cpl	
Landt,	Peter	28	Farrier	
Pfister,	Michael	47	Bugler	
Arhelger,	August	27	Pvt	
Bandje,	Fritz	19	Pvt	
Basse,	Carl	22	Pvt	
Bauer,	Jacob	22	Pvt	
Behrens,	Henry	22	Pvt	
Behrens,	William	25	Pvt	
Bender,	Conrad[John]	33	Pvt	
Bernhard,	Moritz	33	Pvt	
Bickenbach,	William	34	Pvt	
Billings,	Riley	21	Pvt	
Burrer,	Gottlieb	31	Pvt	
Cramm,	Henry	30	Pvt	Deserted Joined Union Army
Dambach,	Frederick	23	Pvt	
Dietz,	Henry	23	Pvt	
Durst,	Jacob	22	Pvt	
Ellebracht,	William	25	Pvt	
Fesistel,	Franz	37	Pvt	
Feuge,	Henry	19	Pvt	

Frischmeyer,	Franz	26	Pvt	
Gaedeke,	Christopher	27	Pvt	Enlisted Jun 1862 @ SA
Gold,	Jacob Sr.	20	Pvt	
Gold,	Peter Sr.	22	Pvt	
Grobe,	Henry	27	Pvt	Enlisted Jun 1862 @ SA Joined Union Army
Hahn,	Conrad	22	Pvt	
Hahn,	Frederick	28	Pvt	
Hasse,	Henry	28	Pvt	
Henke,	Carl	24	Pvt	
Henrich,	Jacob	21	Pvt	
Hilt,	Carl	23	Pvt	
Hoerster,	Fritz	20	Pvt	
Houker,	John	25	Pvt	Enlisted Jun 1862 @ SA
Jung,	Frank	24	Pvt	Member Kuechler's Co
Keller,	Adam	18	Pvt	
Kordzick,	Julius	22	Pvt	
Kott,	Hermann	22	Pvt	
Kunemann,	Henry	18	Pvt	
Lehne,	William	28	Pvt	
Metzger,	Joseph	22	Pvt	
Millering,	Henry	19	Pvt	
Nalte,	Otto	22	Pvt	Enlisted Jun 1862 @ SA
Nickel,	William	22	Pvt	Deserted Joined Union Army
Ohlenburger,	Ferdinand	27	Pvt	Deserted Joined Union Army
Otte,	Fritz	21	Pvt	
Ottmars,	George	18	Pvt	
Peter,	George	28	Pvt	
Peter,	Phillipp	22	Pvt	
Pfannensteel,	August	23	Pvt	Deserted Joined Union Army
Pfannensteel,	William	25	Pvt	Enlisted Jun 1862 @ SA Joined Union Army
Ressman,	Christian	27	Pvt	

Roesing,	Ernest G.	27	Pvt	
Roos,	Henry	23	Pvt	
Rosenbach,	John	19	Pvt	
Sagebiel,	August	24	Pvt	
Scheeler,	Henry	27	Pvt	Enlisted Jun 1862 @ SA
Schloeter,	Henry	21	Pvt	
Schneider,	Christian	25	Pvt	Enlisted Jun 1862 @ SA
Schneider,	Jacob	18	Pvt	Member Kuechler's Co.
Schuch,	Peter	18	Pvt	
Schueler,	August	47	Pvt	
Schumacher,	Gottlieb	18	Pvt	
Schwartz,	Charles	22	Pvt	Member Kuechler's Co.
Stockman,	Hardy	40	Pvt	Enlisted Jun 1862 @ SA
Stockman,	Hirman [Henry]	28	Pvt	
Stockman,	William	18	Pvt	Deserted Joined Union Army
Stoffers,	Christian	21	Pvt	
Stoffers,	Fritz	24	Pvt	
Strackbien,	Henry	20	Pvt	
Van der Stucken,	Emile	22	Pvt	
Van der Stucken,	Julius	35	Pvt	
Wahrmund,	William	18	Pvt	
Walter,	John	21	Pvt	Member Kuechler's Co.
Walter,	William	23	Pvt	
Wathersdorf,	Albert	29	Pvt	
Weber,	Peter	20	Pvt	
Wiedeman,	William	21	Pvt	
Wight,	Lehi L.	28	Pvt	
Wight,	Levi L.	25	Pvt	
Zimmer,	Adam	23	Pvt	

R: Roster, J. Duff's Co., June – August 1862

Surname	Given Name	Age	Rank	Remarks
Duff,	James M.	34	Capt	
Sweet,	James	44	1Lieut	Det Ser Military Commission @ SA
Taylor,	Richard C.	33	2Lieut	
Lilly,	Edwin F.	22	2Lieut	Leader of Pursuit Force Det
Field,	Simon	33	1Sgt	
Luckie,	William F.	30	2Sgt	Det Ser Jun 28, 1862
Newton,	Henry M.	27	3Sgt	
Horner,	George	27	4Sgt	
Abat,	L. Emile		5Sgt	Discharged Aug 20, 1862
Caldwell,	George	29	1Cpl	Det Ser Jun 27, 1862
Pue,	Edward B.	25	2Cpl	
			3Cpl	
Robinson,	John Frank	24	4Cpl	WIA Nueces Battle
Murphy,	Daniel		Bugler	
Smoot,	J. D.	30	Bugler	
Coleman,	S. T.		Farrier	
Schmitt,	Adam	35	1Cook	
Lang,	Albert		2Cook	
Gastring,	Edward	20	3Cook	
Adam,	J. M.	32	Pvt	Det Ser Jun 1862
Adam,	W. J.	23	Pvt	
Alexander,	J. S.	29	Pvt	Det Ser May 1862
Allen,	Frost		Pvt	WIA Nueces Battle
Anderson,	W. W.	24	Pvt	
Bacon,	F.	27	Pvt	Med Discharged Aug 1862
Bell,	Jessup M	27	Pvt	Det Ser Jul 1862
Bennett,	George H.	17	Pvt	Det Ser Aug 29, 1862
Binns,	John Frank	49	Pvt	Sick Jul-Aug 1862
Boehn,	Heinrich Wilhelm		Pvt	
Burney,	Robert Hamilton Jr.	31	Pvt	

650

Campbell,	James	23	Pvt	
Cassiano,	Ignacio	33	Pvt	
Childers,	John W.	39	Pvt	
Coleman,	J. C.		Pvt	
Crawford,	R. F.	24	Pvt	
Dashiel,	David H.	18	Pvt	
Davenport,	Samuel H.	19	Pvt	
Deats,	Thomas A.	32	Pvt	
Dixon,	E. G.	18	Pvt	
Edwards,	William J.	22	Pvt	
Elder,	Albert J.	27	Pvt	WIA Nueces Battle Died Aug 17, 1862 @ Ft. Clark
Elder,	Robert G.	27	Pvt	WIA Nueces Battle Died Sep 17, 1862 @ Ft. Clark
Eilliott,	William Henry	23	Pvt	Det Ser May 1862
Evans,	John W.	26	Pvt	
Faulconer,	R. F.		Pvt	
Ferrill,	H. P.	28	Pvt	
Gallagham,	Edward	27	Pvt	
Garcia,	Jesus	21	Pvt	
Gazley,	Alfred F.	20	Pvt	
Gazley,	William	30	Pvt	
Gravis,	F. C.	25	Pvt	
Hall,	William F.		Pvt	
Hummel,	Carl H.	43	Pvt	Med Discharged Aug 1862
Johns,	Joscph Depay	30	Pvt	
Johnson,	John H.		Pvt	
Johnson,	William H.	23	Pvt	
Kelly,	Charles C.	29	Pvt	
King,	Adolph		Pvt	
Losoyo,	Cavero	31	Pvt	
Luckie,	Samuel B.	21	Pvt	
Lytton,	J. L.		Pvt	

McDaniel,	Thomas Marley	23	Pvt	
McDonna,	William C.	30	Pvt	
Mullin,	F. J.	25	Pvt	
Muncey,	W. H.	35	Pvt	
Nash,	O. F.		Pvt	Transfer fr Allen's Regt Aug 16, 1862
Noonan,	Nelson	23	Pvt	
Ogi,	Louis	29	Pvt	
Palmer,	William L.	29	Pvt	
Pancost,	Josiah E.	18	Pvt	
Powell,	John		Pvt	Transfer to 2d Regt T.M.R. Jul 18, 62
Pue,	Samuel B.	24	Pvt	
Quinlan,	C. C.	29	Pvt	
Reaver,	Alfred		Pvt	Det Ser Aug 29, 1862
Redus,	George	24	Pvt	
Redus,	John	28	Pvt	
Reed,	P. L.	22	Pvt	
Robinson,	Robert N.		Pvt	Enlisted Aug 1, 1862 @ Fred'burg
Sanford	A. W.	41	Pvt	
Schneider,	Louis	25	Pvt	
Serna,	Ignacio F.	20	Pvt	
Smith,	Benjamin Franklin	21	Pvt	Transfer fr 2d Regt T.M.R. Jul 18, 62
Spears,	James K.	22	Pvt	Enlisted Aug 1, 1862 @ Fred'burg
Steel,	Elisha Asbury	23	Pvt	
Surratt,	Isaac D.		Pvt	
Sweet,	Alexander	22	Pvt	
Trimble,	John G.	34	Pvt	Transfer fr 2d Regt T.M.R. Jul 3, 62
Trimble,	W. F.	23	Pvt	
Truman,	A. J.	32	Pvt	

Vinton,	John R.	19	Pvt
Wehraham,	Ernst	33	Pvt
Williams,	Robert Hamilton	31	Pvt
Woodard,	Montealon F.	21	Pvt
Yturri y Castrillo	Manuel	21	Pvt

S: Organization of Benton's Task Force

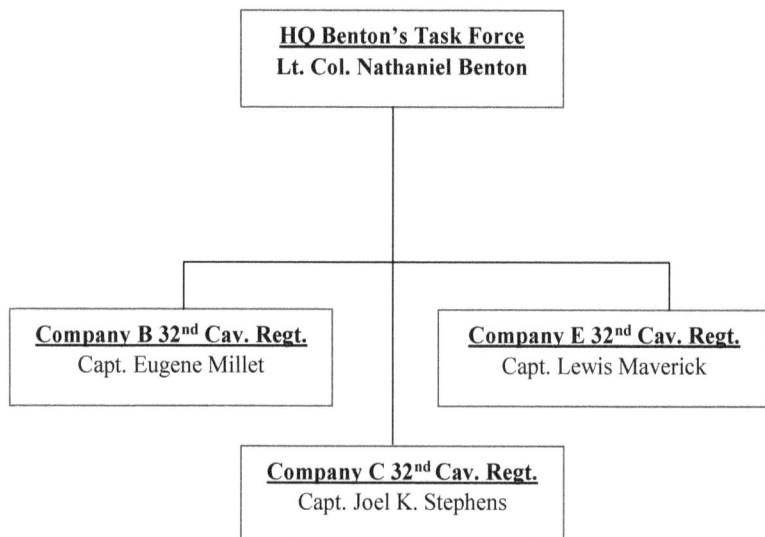

```
┌─────────────────────────────┐
│   HQ Benton's Task Force     │
│   Lt. Col. Nathaniel Benton  │
└─────────────────────────────┘
```

```
┌──────────────────────────────┐      ┌──────────────────────────────┐
│  Company B 32nd Cav. Regt.    │      │  Company E 32nd Cav. Regt.    │
│     Capt. Eugene Millet       │      │     Capt. Lewis Maverick      │
└──────────────────────────────┘      └──────────────────────────────┘
```

```
┌──────────────────────────────┐
│  Company C 32nd Cav. Regt.    │
│    Capt. Joel K. Stephens     │
└──────────────────────────────┘
```

T: Organization of Donelson's Task Force

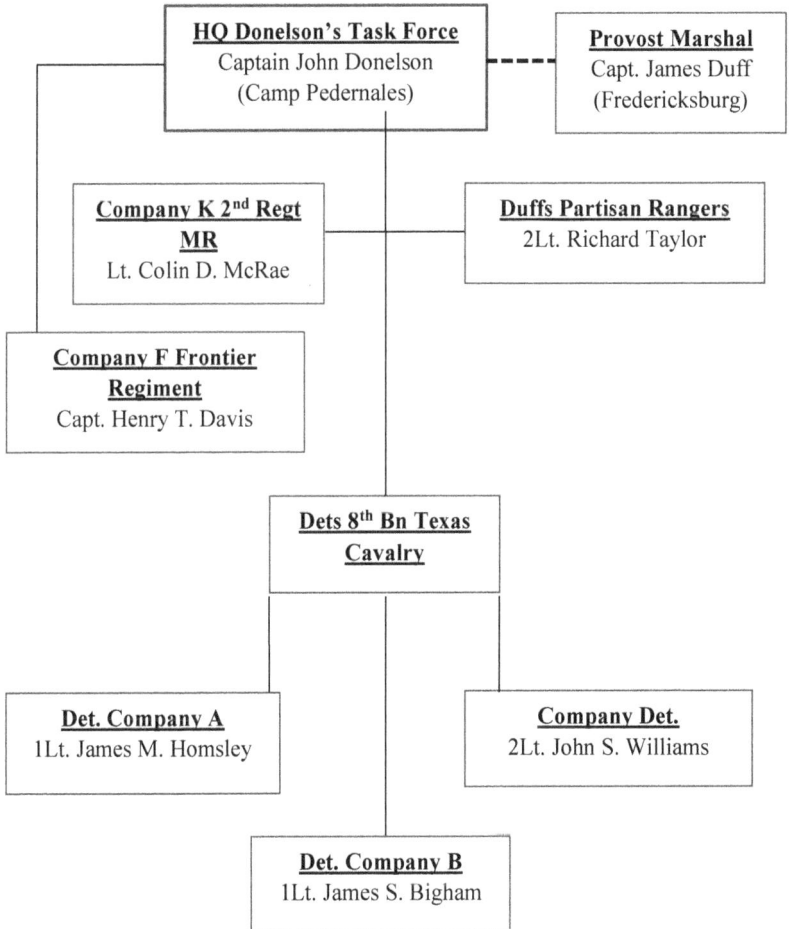

```
┌─────────────────────────┐        ┌─────────────────────────┐
│ HQ Donelson's Task Force │ ─ ─ ─  │     Provost Marshal      │
│  Captain John Donelson   │        │     Capt. James Duff     │
│     (Camp Pedernales)    │        │     (Fredericksburg)     │
└─────────────────────────┘        └─────────────────────────┘
```

Company K 2nd Regt MR
Lt. Colin D. McRae

Duffs Partisan Rangers
2Lt. Richard Taylor

Company F Frontier Regiment
Capt. Henry T. Davis

Dets 8th Bn Texas Cavalry

Det. Company A
1Lt. James M. Homsley

Company Det.
2Lt. John S. Williams

Det. Company B
1Lt. James S. Bigham

U: Roster, J. Donelson's Co. K, 2nd Regiment Texas Mounted Rifles, August 1862

Surname	Given Name	Age	Rank	Remarks
Donelson,	John	33	Capt	
McRae,	Colin D.	27	1Lieut	Wounded Nueces Battle
Lawhorn,	Leonardus S.	31	2Lieut	
Williams,	Robert	31	2Lieut	KIA Jun 21, 1863
Hill,	John	40	1Sgt	Wounded Nueces Battle
Martin,	Emanuel		2Sgt	WIA Nueces Battle Died SA Oct 18, 1862
Harrison,	Ira A.		3Sgt	
Richardson,	Samuel		4Sgt	
Winters,	Samuel		5Sgt	
Cude,	Alfred J.		1Cpl	
Waltrink,	Frances		2Cpl	
Walker,	Charles W.		3Cpl	
Wittiman,	Leonard		4Cpl	
O'Conner,	Charles		Bugler	
Doran,	George H.		Surgeon	
Antonio,	Dick		Pvt	Absent sick
Avant,	James	22	Pvt	
Beard,	John		Pvt	
Bell,	Samuel		Pvt	
Bennett,	L. C.		Pvt	
Bolton,	J. A.		Pvt	
Bolton,	Joseph E.		Pvt	
Butler,	Albert B.	18	Pvt	
Butler,	Marion		Pvt	
Carro,	Joseph M.		Pvt	
Cave,	Milton		Pvt	
Chambers,	John H.		Pvt	
Coons,	M. C.		Pvt	
Cravey,	George H.		Pvt	

Crouch,	Wiley		Pvt
Crouch,	William F.		Pvt
Cude,	W.		Pvt
Cude,	Willis F.	18	Pvt
Curry,	John L.		Pvt
Dockery,	Alfred J.		Pvt
Dockery,	Louis K.	20	Pvt
Dolan,	Thomas		Pvt
Duff,	Marion J.	31	Pvt
Edwards,	Cullen W.		Pvt
Edwards,	John D.		Pvt
Ervay,	Charles		Pvt
Frazer,	George W.		Pvt
Green,	Casper		Pvt
Green,	Charles		Pvt
Harrison,	James		Pvt
Humphries,	Thomas		Pvt
Inglish,	William		Pvt
Jones,	Calvin	23	Pvt
Kendrick,	J. W.		Pvt
Keys,	James O.		Pvt
Keys,	John W.		Pvt
Lansford,	Josiah R.		Pvt
Leazear,	James Franklin	18	Pvt
Leazear,	John Alexander	25	Pvt
Leazear,	William	22	Pvt
McCauley,	William		Pvt
Mangum,	Jasper Warren	28	Pvt
Mestez,	Lassero		Pvt
Moore,	William S.		Pvt
Morrissey,	Morris		Pvt

Odum,	Richard		Pvt	
Payne,	Henry		Pvt	
Pell,	Frederick		Pvt	
Pettitt,	William M.		Pvt	
Powell,	Floyd	22	Pvt	
Richardson.	John		Pvt	
Ricks,	Edward		Pvt	
Ricks,	Richard		Pvt	
Ricks,	Willey		Pvt	
Rosser,	Benjamin	24	Pvt	Wounded Nueces Battle
Smyth,	Patrick		Pvt	
Stringfield,	Littleton	19	Pvt	Killed Nueces Battle
Tullions,	Richard		Pvt	
Van Pelt,	William		Pvt	
Walker,	Goldsmith		Pvt	
Walker,	Samuel		Pvt	
Watkins,	Benjamin Franklin	18	Pvt	
Watkins,	Milton P.		Pvt	
Williams,	Joel G.		Pvt	
Williams,	John		Pvt	
Williams,	Joseph H.		Pvt	
Winters,	Daniel T.	20	Pvt	
Winters,	Marion		Pvt	Died Sep 27, 1862 @ SA
Yarbrough,	David	20	Pvt	
Yarbrough,	L. Dow	21	Pvt	Wounded Nueces Battle

V: Roster, Detachment Co. A, 8th Batallion Texas Cavalry, Fredericksburg, Texas, July – August 1862

Surname	Given Name	Age	Rank	Remarks
Homsley,	James M.	24	1Lieut	
Whatley,	William	26	3Cpl	
Alexander,	Elias	21	Pvt	
Beard,	Coleman Campbell	30	Pvt	
Bloucher,	George W.	24	Pvt	
Brawley,	James W.	19	Pvt	
Brawley,	John F.	28	Pvt	
Brawley,	William M.	24	Pvt	
Carnes,	Andrew Jackson	20	Pvt	
Carter,	Madison James	24	Pvt	
Cowan,	Samuel C.	26	Pvt	
Cowan,	William C.	33	Pvt	
Deane,	John Darnell	20	Pvt	
Eddy,	Lynch	28	Pvt	
Edmonson,	Lattimore	22	Pvt	Wounded Nueces Battle
Erwin,	John T.	19	Pvt	
Erwin,	Stephen L.	26	Pvt	
Erwin,	William A.	22	Pvt	
Huddleston,	William A.	24	Pvt	
Kemp,	William	24	Pvt	
Kreymour,	Constantine	20	Pvt	
Mercer,	John	22	Pvt	
Montgomery,	Addison W.	25	Pvt	
Neal,	Robert	21	Pvt	
Shaw,	William C.	25	Pvt	
Shirley,	George W.	24	Pvt	

Singleton,	Thomas Jefferson	22	Pvt	Wounded Nueces Battle
Stinnett,	James	19	Pvt	
Stratton,	Absalom J.	23	Pvt	
Swinney,	Robert	20	Pvt	
Thorpe,	George Jones	22	Pvt	
Titus,	Melzor	25	Pvt	
Walsh,	William R.	34	Pvt	
Welch,	John T.	23	Pvt	Wounded Nueces Battle
White,	Thomas B.	24	Pvt	
Williams,	Wiley	22	Pvt	Wounded Nueces Battle

W: Roster, Detachment B, Taylor's 8th Cavalry Battalion

Surname	Given Name	Age	Rank	Remarks
Bigham,	James S.	24	1Lieut	
Bellamy,	Asa F.	22	Pvt	
Bigham,	William N.	28	Pvt	
Elliott,	James A.	23	Pvt	
Hughes,	Calvin L.	19	Pvt	
Johnston,	W. J. G.	22	Pvt	
Marshall,	William Berry	24	Pvt	
Moore,	Daniel E.	24	Pvt	
Morris,	Edward J.	28	Pvt	
Morris,	John K.	19	Pvt	Wounded Nueces Battle Died Sep 1, 1862 @ Ft. Clark
Pace,	Edward D.	19	Pvt	
Polk,	James	20	Pvt	
Walker,	Jerry	21	Pvt	
Warren,	George W.	24	Pvt	
Williams,	James Madison	22	Pvt	
Williams,	William A.	20	Pvt	
Young,	Hasting	18	Pvt	

X: Roster, Composit Detachment, Taylor's 8th Cavalry Battalion

Surname	Given Name	Age	Rank	Remarks
Williams	John S.	22	2Lieut	Co. E
Fancher	Erwin S.	25	Sgt.	Co. E
Ford	Eli M.	24	Sgt.	Co. E
Ewing,	Thaddens C.	23	3Cpl.	Co. D
Allen	Joseph	22	Pvt	Co. D
Bacon,	James. M.	26	Pvt	Co. D
Barker,	Henry W.	23	Pvt	Co. E - Wounded Nueces Battle
Brown	Clinton P.	23	Pvt	Co. E
Coynor	Robert	29	Pvt	Co. E.
Ewing,	John K.	19	Pvt	Co. D.
Ford,	William W.	22	Pvt	Co. E
Foreman	George W.	18	Pvt	Co. E
Foreman	William W.	21	Pvt	Co. E
Frazier	John S.	23	Pvt	Co. E
Garland	Thomas L.	23	Pvt	Co. E
Gibson,	Thomas L.	23	Pvt	Co. E
Goodwin,	George W.	20	Pvt	Co. E
Gracy,	Emory A.	25	Pvt	Co. E
Grubbs	Benjamin	25	Pvt	Co. D
Hancock,	Joseph P.	28	Pvt.	Co. D
Hardigree	John T.	23	Pvt.	Co. E
Hobbs	Isaac M.	23	Pvt.	Co. D
House,	Bartlett	22	Pvt.	Co. E
Johnson,	William H.	19	Pvt.	Co. E
McCanless	James M.	42	Pvt.	Co. E
McCuistion	Bedford Lane	21	Pvt.	Co. D
McGasson	George M.	24	Pvt.	Co. D

Morris,	Gibson O.	25	Pvt.	Co. E
Parrish,	John D.	22	Pvt.	Co. E
Patton,	John	22	Pvt.	Co. D
Pickard,	George N.	28	Pvt.	Co. D
Poe,	Henderson	20	Pvt.	Co. E
Poe,	William E.	26	Pvt.	Co. E Killed Nueces battle
Rhodes,	D. B. F.	23	Pvt.	Co. E
Roy	John C.	22	Pvt.	Co. E
Sanders,	Lloyd P.	23	Pvt.	Co. E
Slavin,	George Alexander	24	Pvt.	Co. D
Smart,	Samuel G.	29	Pvt.	Co. D
Stephens,	Andrew	24	Pvt.	Co. E
Stephens,	William D.	22	Pvt.	Co. E
Stiles,	Samuel	24	Pvt.	Co. E
Stockman,	Hardy	40	Pvt.	Co. C
Stockman,	Hirman	28	Pvt.	Co. C
Strother	William D.	23	Pvt.	Co. D
Thompson	John W.	24	Pvt.	Co. D

Y: Kerr County Deserters

Surname	Given Name	Age	Rank	Remarks
Boerner,	Louis	28	Pvt	Militia Unit
Boerner,	Wilhelm	34	Pvt	Militia Unit
Boxter,	Jacob	31	Pvt	Militia Unit
Elstner,	Joseph	21	Pvt	Militia Unit
Felsing,	Ernst/Edward	34	Pvt	Militia Unit
Hasse,	Herman		Pvt	Militia Unit
Heinen,	Anton	22	Pvt	Militia Unit
Heinen,	Ferdinand	17	Pvt	Militia Unit
Heinen,	Henry	22	Pvt	Militia Unit
Heinen,	Theodore	22	Pvt	Militia Unit
Henderson,	Howard	21	Pvt	Militia Unit
Kell,	Scott	26	Pvt	Confederate Army
Kempt,	H. L.	27	Pvt	Militia Unit
Lamb,	Albert	17	Pvt	Militia Unit
Lane,	Paul		Pvt	Confederate Army
Nelson,	Allen	20	Pvt	Militia Unit
Nelson,	Louis	27	Pvt	Militia Unit
Sanger,	Fritz	30	Pvt	Militia Unit
Sanger,	Henry		Pvt	Confederate Army
Sauer,	Gottlieb	24	Pvt	Militia Unit
Schreiner,	Amil	22	Pvt	Militia Unit
Schwethelm,	Henry	22	Pvt	Militia Unit
Schieholz,	Louis	22	Pvt	Militia Unit
Scott,	Charles Franklin	35	Pvt	Militia Unit
Steves,	Heinrich	28	Pvt	Militia Unit
Stieler,	Heinrich	17	Pvt	Militia Unit
Strong,	John	26	Pvt	Militia Unit
Tegener,	Fritz	29	Col	Militia Unit
Tegener,	Gustav	35	Pvt	Militia Unit
Vetterline,	Charles	30	Pvt	Militia Unit
Vetterline,	Gottlieb	32	Pvt	Militia Unit

Z: Known and Likely Fleeing Insurgents

		Likely Left Camp Early	Likely In Camp at Final Assault	Killed At/Near Battle Site	Later Captured and Killed	In Oct Group	Killed at Rio Grande
KNOWN MEMBERS							
<u>Surname</u>	<u>Given Name</u>						
Bauer,	Leopold		X	X			
Behrens,	Fritz		X	X			
Beseler,	Ernst		X	X			
Bock,	Conrad	Xw			X		
Boerner,	Louis		X	X			
Boerner,	Wilhelm F.	X			X		
Bruckish,	Theodore	X			X		
Bruns,	Albert		X	X			
Cramer,	Ernest		X				
Degener,	Hilmar		Xw	X			
Degener,	Hugo		Xw	X			
Diaz,	Pablo		X	X			
Duecker,	August		X				
Elstner,	Joseph	X				X	X
Felsing,	Ernst	X				X	X
Flick,	Herman	X			X		
Graff,	Carl		X				
Graff,	William		X				
Henderson,	Howard	X					
Hermann,	Henry	X				X	X
Hester,	William	X					
Hoffmann,	August		X				

Hohmann,	Valentine	X				X	X
Itz,	Karl		Xw				
Kallenber,	John Georg		X	X			
Kammlah,	Henry II		Xw				
Klier,	William		X				
Kuechler,	Jacob		X			Xw	
Kusenberger,	Jacob		X				
Luckenbach,	August	Xw			X		
Markwordt,	Heinrich		X	X			
Pehl,	Mathias		X				
Ruebsamen,	Adolph	Xw?			X		
Ruebsamen,	Louis	Xw?			X		
Sansom,	John W.	X					
Schaefer,	Christian		X	X			
Schieholz,	Louis		X	X			
Schreiner,	Amil		X	X			
Schwethelm,	Henry		X				
Scott,	Thomas J.	X					
Scott,	Warren B.	X					
Simon,	Ferdinand		Xw				
Steves,	Heinrich		X	X			
Stieler,	Heinrich						
Tays,	Fritz	X			X		
Tegener,	Fritz		Xw				
Tellegmann,	William		Xw	X			
Vater,	Adolph		X	X			
Vater,	Fritz		X	X			
Vater,	William		Xw				
Vetterlein,	Carl		X				

Vetterlein,	Gottlieb		Xw?				
Weber,	Philipp Heinrich		X				
Weiss,	Franz		Xw			X	X
Weiss,	Moritz		X			X	X
Weyershausen,	Heinrich		X	X			
Weyrich,	Michael		X	X			
Zoller,	Adolphus	—	Xw__	—	—	—	—
Sub-Total		18	40	19	9	10	6

VERY LIKELY IN GROUP

Jacoby,	Peter	X					
Moellering,	Carl	X					
Poetsch,	Joseph		X				
Tatsch,	Michael	X					
Usener,	Jacob	X					
Usener,	Ludwig	X__					
Sub-Total		5	1	0	0	0	0

LIKELY IN GROUP

Gold,	Jacob	X					
Gold,	Peter	X					
Hoerner,	John	X					
Kleck,	Syvester	X				Xw	
Schreiner	Charles	X__					
Sub-Total		5	0	0	0	1	0
TOTAL		69					

MAY HAVE BEEN IN GROUP							
Beck,	Phillip	X?					
Diener,	V.	X?					
Doebbler,	Richard	X?					
Graff,	A. [August]		X?				
Graff,	F.?		X?				
Poetsch,	Christian	X?					
Poetsch,	Joseph	X?					
Rausch,	Henry??	X?					
Schoenewolf,	August	X?					
Weber,	Frederick	X?_					
Sub-Total		8	2	0	0	0	0
TOTAL		36	43	19	9	11	6
OTHER MEMBERS - IN OCTOBER GROUP							
Beversdorf,	Albert					X	
Beversdorf,	August					X	
Bonnet,	Charle					X	
Bonnet,	Daniel					X	
Bonnet,	Peter					Xw	X
Bonnet,	William					X	
Lange,	Fritz					Xw	X
GRAND TOTAL		36	43	19	9	18	8

Believe Actual Numbers in August Group ARE							
		28	41	19	9		
Believe Actual Numbers in October Group ARE							
						18	8

AA: Known and Likely Members of Pursuit Force
Duff's Company Partisan Rangers

Surname	Given Name	Remarks
	Known	
Lilly,	Edwin F.	Command of Det.
Allen,	Frost	WIA
Elder,	Albert J.	DOW
Elder,	Robert G.	DOW
Edwards,	William J.	Testified At CMC
Ferrill,	H. P.	Attended Wounded
McDonna	William C.	Named by Williams
Ogi,	Louis	Named by Williams
Redus	John	Names by Slave
Robinson,	John Frank	WIA
Williams,	Robert H.	Wrote Book
Woodward,	Montalcon	Named by Williams
Burgmann,	Charles	WIA
	Likely	
Bargerum?/Bauman,	F.	

Not Able to Determine
Other Members

Davis' Company, Frontier Regt.

Known

Benson,	John Washington	
Benson,	William Thomas	WIA
Harbour,	William T.	Command of Det.
Wharton,	William G'Laspie	
Wilborn,	Albert L.	WIA

Likely

Blevins,	Leroy S.	
Blucher,	Adam	
Farr,	David H.	
Hampton,	Jordan E.	
Hester,	Newton	
Hill,	Eli T.	
Jackson,	Aaron W.	
Jackson,	Nathan	
Lowreance,	John S.	
Moore,	William H.	
McCann,	Sam Houston	
Myers,	Freeman	
Paul,	Andrew J.	
Phillips,	George P.	
Rees,	Daniel A.	
Rees,	Peter O. A.	
Rogers,	Jesse W.	
Rowe,	Enoch	
Wallace,	John	
Whittey,	Elisha	
Williamson,	John	
Williamson,	William A.	

Co. K, 2nd Regt T.M.R.
Known

Hill,	John	
Martin,	Emanuel	DOW
McRae,	Colin D.	Command of Det.
Rosser,	Benjamin	WIA
Stringfield,	Littleton	KIA
Yarboro,	Dow	WIA

Likely

Pell,	Frederick
Pettitt,	William
Powell,	Floyd
Richardson,	John
Ricks,	Edward
Ricks,	Richard
Ricks,	Wiley
Shannon,	John T.
Smyth,	Patrick
Tullons,	Richard
Van Pelt,	William F.
Walker,	Goldsmith
Walker,	Samuel
Watkins,	Benjamin F.
Watkins,	Milton P.
Williams,	Joel G.
Williams,	John
Williams,	Joseph H.
Winter,	Daniel
Yarboro,	Dow

Co. A, 8th Bn.
Known

Edmonson,	Latimore	WIA
Erwin,	Stephen L.	WIA
Homsley,	James M.	Command of Det.
Singleton,	Thomas J.	WIA
Welch,	John T.	WIA
Williams,	Wiley	WIA

Likely

Not Able To Determine Other Members

Co. B, 8th Bn.
Known

Bigham,	James S.	Command of Det.
Morris,	John K.	DOW

Likely
Not Able To Determine
Other Members

Composit Det, 8th Bn.
Known

Baker,	Henry W.	WIA
Poe,	Henderson	
Poe,	William	KIA

Likely
Not Able To Determine Other Members

AB: Known and Likely Insurgent Casualties

Killed At/Near Battle Site

Surname	Given Name	Type of Wound	County
Bauer,	Leopold		Kendall
Behrens,	Fritz		
Beseler,	Ernest		Kendall
Boerner,	Louis		Kendall
Bruns,	Albert		
Degener,	Hilmar	Mouth	Kendall
Degener,	Hugo		Kendall
Diaz,	Pablo		
Kallenberg,	John George		Gillespie
Markwordt,	Heinrich		Gillespie
Schaefer,	Christian		Gillespie
Schieholz,	Louis		Kendall
Schreiner,	Amil		Kendall
Steves,	Heinrich		Kendall
Tellegmann,	William	Body Wound	Kendall
Vater,	Adolph		Gillespie
Vater,	Fritz		Gillespie
Weyershausen,	Heinrich		Gillespie
Weyrich,	Michael		Gillespie

Wounded 'Left Early'

Bock,	Conrad	Gillespie
Luckenbach,	August	Gillespie
Ruebsamen,	Adolph	Gillespie
Ruebsamen,	Louis	Gillespie

Wounded 'Left Late'

Itz,	Karl		Gillespie
Kammlah,	Henry II	Shoulder	Gillespie
Simon,	Ferdinand	Ribs and Leg	Kendall
Tegener,	Fritz		Kendall
Vetterlein,	Gottlieb	Wrist	Kendall
Weiss,	Franz	Heel and Leg	Kendall
Zoller,	Adolphus	Chest	Kendall

Wounded Left at Battle Site

Vater,	William	Breast	Gillespie

AC: Insurgents Remaining at Battle Site During Confederate Assault

At/Near Battle Site

Surname	Given Name
Bauer,	Leopold
Behrens,	Fritz
Beseler,	Ernst
Boerner,	Louis
Bruns,	Albert
Degener,	Hilmar
Degener,	Hugo
Diaz,	Pablo
Kallenberg,	John George
Markwordt,	Heinrich
Schaefer,	Christian
Schieholz,	Louis
Schreiner,	Amil
Stevens,	Heinrich
Tellegmann,	William
Vater,	Adolph
Vater,	Fritz
Weyershausen,	Heinrich
Weyrich,	Michael

Schwethelm Group

Graff,	Carl
Graff,	William
Kusenberger,	Jacob
Schwethelm,	Henry

Vetterlein Brothers

Vetterlein,	Carl
Vetterlein,	Gottlieb

Hoffmann Group

Duecker,	August
Hoffmann,	August
Itz,	Karl
Kammlah,	Henry
Pehl,	Mathias
Poetsch,	Joseph
Weber,	Frederick

Cramer and Kuechler Group

Cramer,	Ernest
Klier,	Wilhelm
Kuechler,	Jacob
Simon,	Ferdinand
Tegener,	Fritz
Vater,	William
Weiss,	Franz
Weiss,	Moritz
Zoeller,	Adolphus

AD: Confederate Casualties

Killed:	L. Stringfield	Captain Donelson's Co
W. E. Poe	Taylor's Battalion

Wounded:	Lt. C. D. McRae Comdg	Flesh wound in the arm
Albert Elder	Capt Duff's Co dangerously in head.
Robert Elder	Capt Duff's Co Right arm broken
Gregory	Capt Duff's Co Right arm severely.
Frank Robinson	Severely in thigh.
John Hill	Capt Donelson's Co Right shoulder severely.
Wm. Martin	Capt Donelson's Co Right shoulder broken severely
Ben Rosser	Capt Donelson's Co Left arm severely
Dow Yarboro	Capt Donelson's Co Flesh wound in hip
Wilborn	Capt Davis Co Right breast severely
Name not known	Capt Davis' Co
Morris	Taylor's Battalion
Name not known	Taylor's Battalion
Name not known	Taylor's Battalion
Name not known	Taylor's Battalion
Name not known	Taylor's Battalion
Name not known	Taylor's Battalion

Name not known	Taylor's Battalion
	Vol. Guide
Charles Bergmann	Wounded three places, both thights and arm

True Copy
E. F. Gray
Major and A.A.A. Gen

AE: Timetable of Insurgents Withdrawal

Time	Surname	Event	Number of Original Group Left in Camp	Combat Effective	Killed at or Near Battle Site
Aug 9			69	69	
2:00 p.m.	Two Unionist	Hanged	- 2	- 2	+02
	Total		67	67	
Aug 10					
3:00 a.m.	Bauer	Killed		-1	+01
3:03 a.m.	Beseler	Killed		-1	+01
	Total		67	65	04
3:03 a.m.	Tegener	Wounded		-1	
3:03-4:00	Luckenbach	Wounded		-1	
	Bock	Wounded		-1	
	Ruebsamen	Wounded		-1	
	Ruebsamen	Wounded		-1	
	Total		67	60	
3:03-4:00	Schreiner	Killed		-1	+01
	Total		67	59	05
4:30-5:30	Sansom	Leave	- 1	- 1	
	Scott	Camp	- 1	- 1	
	Henderson		- 1	- 1	
	Scott		- 1	- 1	
	Hester		- 1	- 1	
	Total		62	54	
6:30 a.m.	Pickett-Schieholz	Killed		- 1	+01
6:30	Total		62	53	06

6:30	Twenty-Three	Leave Camp	<u>-23</u>	<u>-19*</u>	
			39	34	
7:00-9:00		Confederate			
		Assault			
	Itz	Injured		- 1	
	Simon	Wounded		- 1	
	Zoeller	Wounded		- 1	
	Weiss	Wounded		- 1	
	Kammlah	Wounded		- 1	
	Vater	Wounded		- 1	
	Vetterlein	Wounded		<u>- 1</u>	
	Total		39	25	
	Degener	Wounded		- 1	
	Telgmann,	Wounded		- 1	
	Degener	Wounded		<u>- 1</u>	
	Total		39	24	
	Behrens	Killed		- 1	+01
	Bruns,	Killed		- 1	+01
	Kalllenberger	Killed		- 1	+01
	Diaz	Killed		- 1	+01
	Markwordt	Killed		- 1	+01
	Schaefer	Killed		- 1	+01
	Weyershausen	Killed		- 1	+01
	Weyrich	Killed		- 1	+01
	Vater	Killed		- 1	+01
	Vater	Killed		<u>- 1</u>	<u>+01</u>
	Total		39	14	16
ca 9:00	Schwethelm	Leave Camp	- 1	- 1	
	Kusenberger	Leave Camp	- 1	- 1	
	Graff	Leave Camp	- 1	- 1	

	Hoffmann	Leave Camp	- 1	- 1	
	Decker	Leave Camp	- 1	- 1	
	Weber	Leave Camp	- 1	- 1	
	Poetsch	Leave Camp	- 1	- 1	
	Pehl	Leave Camp	- 1	- 1	
	Graff	Leave Camp	- 1	- 1	
	Total		30	05	
	Vetterlein	Leave Camp	- 1		
	Vetterlein	Leave Camp	- 1	- 1	
	Total		28	04	
	Kammlah	Leave Camp	- 1		
	Itz	Leave Camp	- 1		
	Tegener	Leave Camp	- 1		
	Vater	Leave Camp	- 1		
	Simon	Leave Camp	- 1		
	Weiss	Leave Camp	- 1		
	Zoeller	Leave Camp	- 1		
	Weiss	Leave Camp	- 1	- 1	
	Klier	Leave Camp	- 1	- 1	
	Kuechler	Leave Camp	- 1	- 1	
	Cramer	Leave Camp	- 1	- 1	
			17	0	
4:00 p.m.	Degener	Murdered			+01
	Telgmann	Murdered			+01
	Degener	Murdered			+01
					19

AF: Roster, Composit Detachment Talor's 8th Cavalry Battalion Sent to Fredericksburg in Late August, 1862

Surname	Given Name	Rank	Age	Company
Colins,	Young	2Lieut	32	D
Clements,	Joseph L.	Pvt	31	E
Erwin,	William L.	Pvt	27	E
Sevier,	Frances A.	Pvt	18	E
Sevier,	John T.	Pvt	32	E
Kuykendall,	George R.	2Lieut	21	B
Bishop,	Samuel H.	3Cpl	25	B
Bishop,	Orlando L.	2Bugler	27	B
Bell,	William H.	Pvt	25	B
Bigham,	William N.	Pvt	28	B
Cabler,	Gastron D.	Pvt	19	B
Christian,	James C. J.	Pvt	19	B
Kelley,	John G.	Pvt	25	B
Kuykendall,	Robert	Pvt	18	B
Marshall,	James C.	Pvt	22	B
Moore,	Daniel E.	Pvt	24	B
Northcut,	Elisha L.	Pvt	20	B
Pullin,	Samuel J.	Pvt	23	B
Rich,	Robert	Pvt	22	B
Roberts,	Wade G.	Pvt	25	B
Scott,	Robert M.	Pvt	24	B
Whitney,	Allen J.	Pvt	25	B

AG: Roster, Company G, 32nd Regiment Texas Cavalry Sent to Fredericksburg on August 23, 1862

Surname	Given Name	Age	Rank	Remarks
Kelso,	John	28	1Lieut	
Adams,	William A.	34	2Lieut	
Porterfield,	Mathrew	43	2Sgt	
Smith,	Thomas C.	18	3Sgt	
Epperson,	Samuel	36	4Sgt	
Robertson,	Joseph	32	5Sgt	
Von Roeder,	Joachim [John]	25	1Cpl	
Lampley,	Benjamin	48	2Cpl	
York,	Jonathan	37	4Cpl	
Thie,	Charles F.	19	Bugler	
Kelso,	Henry F.	34	Farrier	
Aaron,	Martin	23	Pvt	
Adcock,	John	23	Pvt	
Baker,	Rosseau	22	Pvt	
Baker,	Russell F.	19	Pvt	Enlisted Camp Perdernales Oct 30th
Barber,	Benjamin F.		Pvt	
Bluntzer,	Urban	30	Pvt	
Brown,	Jesse K.	28	Pvt	
Ferrell,	William W.	34	Pvt	
Forbes,	Harvey P.	33	Pvt	
Gaebler,	Charles W.	24	Pvt	
Gerhardt,	William	23	Pvt	
Gill,	Richard M.	20	Pvt	
Hall,	Hiram M. C.	22	Pvt	
Hall,	Param T.	34	Pvt	
Hardy,	James T.	31	Pvt	
Hays,	James H.	31	Pvt	
Henning,	Frederick	28	Pvt	
Kell,	Abraham	30	Pvt	

King,	John B.	33	Pvt
Koehn.		25	Pvt
Korth,		22	Pvt
Marlow,		38	Pvt
Menn,		22	Pvt
Menn,		21	Pvt
Middleton,		21	Pvt
Middleton,		23	Pvt
Minke,		24	Pvt
Murray,	James H.	21	Pvt
Murray,	John M.	23	Pvt
Nann,	Charles W.	25	Pvt
Ormond,	Charles J.	24	Pvt
Ott,	Anton	38	Pvt
Parker,	Willey J.	30	Pvt
Price,	Edwin	54	Pvt
Russworm,	George R.	23	Pvt
Spies,	Theodore	23	Pvt
Taylor,	Joseph	29	Pvt
Taylor,	William A.	36	Pvt
Tinsley,	Bountain G.	30	Pvt
Von Roeder,	Lois	18	Pvt
Webb,	William H.	23	Pvt
Wilson,	Edward F.	25	Pvt
Word,	Hugh H.	28	Pvt
Word,	Mark	26	Pvt

AH: Roster, Composite Detachment, Taylor's 8ᵗʰ Cavalry Batallion Sent to Fredericksburg on August 23, 1862

Surname	Given Name	Age	Rank	Remarks
Kuykendall,	George R.	21	2Lieut	Co. B.
Bishop,	Samuel H.	25	3Cpl	Co. B.
Bishop,	Orlando L.	27	2Bugler	Co. B.
Bell,	Thomas J.	20	Pvt	Co. B.
Cabler,	Gastron D.	19	Pvt	Co. B.
Christian,	James C. J.	19	Pvt	Co. B.
Howard,	Orville	24	Pvt	Co. A.
Kelley,	John G.	25	Pvt	Co. B.
Kuykendall,	Robert	18	Pvt	Co. B.
Mankins,	Samuel J.	19	Pvt	Co. A.
Mankins,	William A.	21	Pvt	Co. A.
Marshall,	James C. J.	22	Pvt	Co. B.
Pullin,	Samuel J.	22	Pvt	Co. B.
Rich,	Robert	22	Pvt	Co. B.
Scott,	Robert M.	24	Pvt	Co. B.
Whitney,	Allen J.	25	Pvt	Co. B.
Wright,	Franklin	24	Pvt	Co. A.
Wright,	Leory	18	Pvt	Co. A.

AI: Comfort Area Men Who Took Confederate Oath of Allegiance at Fredericksburg on Sepember 18th, 1862

Surname	Given Name	Age and Date Arrived		Country	Date Applied	Sources
Stieler,	Gottfried	38	1854	Anhalt	Jul 01, 1861	KN-2-92
Lange,	Ludwig	23	1854	Hanover	Jul 08, 1861	KN-2-92
Faltin,	August	31	1856	Prussia	Jul 01, 1861	KN-2-93
Schultz,	Ferdinand	34	1846	Prussia	May 20, 1861	KN-2-93-94
Geisster,	William F.	30	1854	Austria	May 20, 1861	KN-2-94
Schultz,	William E.	38	1855	Prussia	May 20, 1861	KN-2-95
Boerner,	Louis	27	1855	Hanover	Aug 19, 1861	KN-2-95-96
Schwethelm,	Henry	21	1850	Prussia	Aug 26, 1861	KN-2-96
Stieler,	G.	44	1851	Anhalt	Sep 30, 1861	KN-2-97
Serger,	Emil	30	1854	Prussia	Sep 30, 1861	KN-2-97-98
Hanisch,	Paul	32	1858	Prussia	Sep 30, 1861	KN-2-98
Heinen,	Anton	21	1854	Prussia	Sep 30, 1861	KN-2-99
Bruckisch,	Charles	26	1853	Prussia	Sep 30, 1861	KN-2-99-100
Bohnert,	Anton	32	1855	Prussia	Sep 30, 1861	KN-2-100
Brinkmann,	Alexander	24	1853	Prussia	Sep 30, 1861	KN-2-101
Hillke,	Charles			Unk	Sep 30, 1861	KN-2-101-102
Dietert,	Fredrerick, Jr	25	1856	Prussia	Sep 30, 1861	KN-2-102
Lieck,	Edward	27		Prussia	Sep 30, 1861	KN 2 103
Strohecker,	Louis	29	1853	Prussia	Sep 30, 1861	KN-2-103 104
Dietert,	Frederick, Sr.	60	1859	Prussia	Dec 30, 1861	KN-2-104
Schmidt,	Charles	53	1854	Prussia	Dec 30, 1861	KN-2-104
Maetz,	C. C.	61	1854	Prussia	Dec 30, 1861	KN-2-105
Heinen,	C. W.	45	1858	Prussia	Dec 30, 1861	KN-2-105
Heinen,	H. H.	23	1860	Prussia	Dec 30, 1861	KN-2-105

Stecher,	John	23		Prussia	Dec 30, 1861	KN-2-106
Karger,	John	46	1860	Bavaria	Jan 06, 1862	KN-2-107
Harms,	Hermann	25	1855	Prussia	Jan 27, 1862	KN-2-108
Schierholz,	Louis	29	1855	Hanover	Feb 17, 1862	KN-2-109

AJ: Roster, Company B. Taylor's 8ᵗʰ Texas Cavalry Battalion Sent to Fredericksburg on September 18, 1862

Surname	Given Name	Rank	Age	Company
Johnston,	W. J. B.	Pvt	22	B
Lane,	James M.	Pvt	25	B
Marshall,	William B.	Pvt	24	B
Mosley,	James	Pvt	23	B

BIBLIOGRAPHY
Documents

Bylaws of Castroville Castle "K. G. C. [Knights of the Golden Circle], Approved August 16, 1861", Castroville Public Library.

Casualty List, Confederate Forces, Battle of Nueces, Record Group 109, National Archives, Washington, D. C.

Civil Records, Bandera County, Texas, County Clerk's Office, Bandera County Courthouse, Bandera, Texas.

Civil Records, Bexar County, Texas, County Clerk's Office, Bexar County Courthouse, San Antonio, Texas.

Civil Records, Blanco County, Texas, County Clerk's Office, Blanco County Courthouse, Johnson City, Texas.

Civil Records, Comal County, Texas, County Clerk's Office, Comal County Courthouse, New Braunfels, Texas.

Civil Records, Gillespie County, Texas, County Clerk's Office, Gillespie County Courthouse, Fredericksburg, Texas.

Civil Records, Hays County, Texas, County Clerk's Office, Hays County Courthouse, San Marcos, Texas.

Civil Records, Johnson County, Texas, County Clerk's Office, Johnson County Courthouse, Cleburne, Texas.

Civil Records, Kendall County, Texas, County Clerk's Office, Kendall County Courthouse, Boerne, Texas.

Civil Records, Kerr County, Texas, County Clerk's Office, Kerr County Courthouse, Kerrville, Texas.

Civil Records, Medina County, Texas, County Clerk's Office, Medina County Courthouse, Hondo, Texas.

Civil Records, Uvalde County, Texas, County Clerk's Office, Uvalde County Courthouse, Uvalde, Texas.

Civil Records, Young County, Texas, County Clerk's Office, Young County Courthouse, Graham, Texas.

Clark, Governor Edward, Papers, Texas State Archives, Austin, Texas.

Constitution of the "Committee of Safety" of Medina County. n. d., c1861.

Criminal Records, Gillespie County, Texas, District Clerk's Office, Gillespie County Court- house, Fredericksburg, Texas.

Criminal Records, Kendall County, Texas, District Clerk's Office, Kendall County Courthouse, Boerne, Texas.

Criminal Records, Kerr County, Texas, District Clerk's Office, Kerr County Courthouse, Kerrville, Texas.

Compiled Service Records of Confederate Veterans From the State of Texas, Record Group 109, National Archives, Washington, D. C.

Compiled Service Records of Union Veterans From The State of Texas, National Archives, Washington, D. C.

Court Martial File GG 49, Duff, James M., National Archives, Washington, D. C.

Dispatches From United States Consuls in Matamoros, Mexico 1826-1906, Volumes 7-9, January 1, 1858 – December 28, 1869, National Archives, Washington, D. C.

Dispatches From United States Consuls in Monterrey, Mexico 1849-1906 Volume 1, November 15, 1849 – December 9, 1869, National Archives, Washington, D. C.

Duff, James M., Enlistment Papers U. S. Army, National Archives, Washington, D. C.

General and Special Orders, Headquarters Sub-Military District of Rio Grande, May-Oct 1862, Entry 105, Record Group 109, War Department Collection of Confederate Records, National Archives, Washington D. C.

Hamilton, Governor A. J., Papers, Texas State Archives, Austin, Texas.

Lubbock, Governor Francis Richard, Papers, Texas State Archives, Austin, Texas.

Minutes of Meetings, Gillespie County Rifles, February 23, 1862, and March 29, 1862, with copy of Gillespie County Rifles Resolution, District Clerk's Office, Fredericksburg, Texas.

Muster Roll, Captain William C. Adams' Company, October 31, 1861, Texas State Archives, Austin, Texas.

Muster Roll, Captain William A. Blackwell's Company, May 4, 1861, Texas State Archives, Austin, Texas.

Muster Rolls, Captain Philip Braubach's Company, February 7, 1861, Mary 25, 1861, August 25, 1861, November 25, 1861 and February 25, 1862, Texas State Archives, Austin, Texas.

Muster Rolls, Captain Henry T. Davis's Company, March 5, 1862 and February 7, 1863, Texas State Archives, Austin, Texas.

Muster Rolls, Captain Charles de Montel's Company, February 17, 1862; June 30, 1862; and February 9, 1863, Texas State Archives, Austin, Texas.

Muster Roll, Captain Jacob Dearing's Company, October 1, 1864, Texas State Archives, Austin, Texas.

Muster Rolls, Captain John J. Dix's Company, March 30, 1862, June 30, 1862, and December 30, 1862.

Muster Roll, Captain James Duff's Company, January 31, 1862, Texas State Archives, Austin, Texas.

Muster Rolls, First Lieutenant D. H. Farr's Company, February 13, 1864 and March 25, 1865, Texas State Archives, Austin, Texas.

Muster Roll, Captain George Freeman's Company, March 1, 1861 and November 18, 1861, Texas State Archives, Austin, Texas.

Muster Roll, Captain George Haby's Company, October 12, 1863, Texas State Archives, Austin, Texas.

Muster Roll, Captain George Hancock's Company, February 25, 1862, Texas State Archives, Austin, Texas.

Muster Roll, Captain William T. Harbour's Company March 5, 1861, Texas State Archives, Austin, Texas.

Muster Rolls, Captain James M. Hunter's Company, December 24, 1862, February 28, 1863, April 30, 1863 and October 31, 1863 Texas State Archives, Austin, Texas.

Muster Roll, Captain W. E. Jones' Company, March 1, 1864, Texas State Archives, Austin, Texas.

Muster Roll, Captain Engelbert Krauskopf's Company, for month of May 1864, Texas State Archives, Austin, Texas.

Muster Roll, Captain Jacob Kuechler's Company, February 12, 1862, Texas State Archives, Austin, Texas.

Muster Rolls, Captain John Lawhon's Company, August 31, 1863, December 31, 1863 and February 29, 1864, Texas State Archives, Austin, Texas.

Muster Roll, Captain George Mayer's Company, August 31, 1862 Texas State Archives, Austin, Texas.

Muster Roll, Captain Charles H. Nimitz's Company, July 31, 1861, Texas State Archives, Austin, Texas.

Muster Roll, Captain James Paul's Company, March 4, 1861, Texas State Archives, Austin, Texas.

Muster Rolls, Captain Thomas Rabb's Company, August 7, 1861, September 30, 1861, July 25, 1862 and June 30, 1863, Texas State Archives, Austin, Texas.

Muster Roll, First Lieutenant Thomas Riordan's Company, September 21, 1863, Texas State Archives, Austin, Texas.

Muster Rolls, Captain George Robbins' Company, March 31, 1864 and January 18, 1865, Texas State Archives, Austin, Texas.

Muster Rolls, Captain Louis Schuetze's Company, November 1, 1863, and January 27, 1864, Texas State Archives, Austin, Texas.

Muster Rolls, Captain R. H. Williams' Company, November 23, 1863, and April 31, 1864, Texas State Archives, Austin, Texas.

Petition, County of Kimbal [Kimble] April 22, 1862, to Governor Lubbock, Governor Lubbock's Papers, Texas State Archives, Austin, Texas.

Register Of Enlistment in U. S. Army 1798-1914 – July 1853October 1854, Record Group 94, Microfilm Copy No. 233, Roll 24, Volume 50, Page 116, Dennis Kingston's U. S. Army Service Records, National Archives, Washington, D. C.

Register Of Enlistment in U. S. Army 1798-1914 – July 1826October 1850, Record Group 94, Microfilm Copy No. 233, Roll 22, Volume 49, Page 65, James M. Duff's U. S. Army Service Records, National Archives, Washington, D. C.

Special Orders Number 9, Headquarters 7th Military Department, Fort Smith, October 29, 1849, contained in letter dated June 12, 1997, from National Archives, Washington D. C.

Texas Adjutant General, Record Group 401, Adjutant General's Correspondence, Texas State Archives, Austin, Texas.

Texas Governor's Proclamation Book 1861-1863, Texas State Archives, Austin, Texas.

Texas Historical Records Survey, "Inventory of the County Archives of Texas, Gillespie County No. 867."

Texas Pardon Files, Texas State Archives, Austin, Texas.

U. S. Census Bureau, Seventh Census of the U. S. (1850), Schedule L., Free Inhabitants in various Texas Counties.

U. S. Census Bureau, Seventh Census of the U. S. (1850), Schedule I., Free Inhabitants in Jefferson County, Illinois.

U. S. Census Bureau, Eighth Census of the U. S. (1860), Schedule L., Free Inhabitants in various Texas Counties.

U. S. Census Bureau, Eighth Census of the U. S. (1860), Schedule 2, Slave Inhabitants in various Texas Counties.

U. S. Census Bureau, Eighth Census of the U. S. (1860), Schedule 3, Persons Who Died During The Year Ending 1st of June 1860, In The Counties of Blanco and Gillespie, State of Texas.

U. S. Census Bureau, Eighth Census of the U. S. (1860), Schedule 4, Production of Agriculture in Blanco County, Texas.

U. S. Census Bureau, Ninth Census of the U. S. (1870), Schedule L., Inhabitants in various Texas Counties.

U. S. Census Bureau, Tenth Census of the U. S. (1880), Schedule L., Inhabitants in various Texas Counties.

U. S. Census Bureau, Eleventh Census of the U. S. (1890) Special Schedule, Surviving Soldiers, Sailors, And Marines, And Widows, Etc. for various Texas counties.

U. S. Census Bureau, Twelfth Census of the U. S. (1900), Schedule L., Inhabitants in various Texas Counties.

U. S. Census Bureau, Thirteenth Census of the U. S. (1910), Schedule L., Inhabitants in various Texas Counties.

War of the Rebellion: A Compilation of the Official Records of the Union and Confederate Armies, U. S. Government Printing Office, 1884, 128 Parts in 70 Series.

Books

Adams, Ephraim Douglass, *British Diplomatic Correspondence Concerning The Republic of Texas—1838 – 1846* Austin, Texas: The Texas State Historical Association, c1918.

Arc, Sister Joan of, *My Name Is Nimitz,* San Antonio, Texas: Standard Printing Company, 1948.

Asprey, Robert B., *War in the Shadows: The Guerrilla in History,* Garden City, New York: Two Volumes, Doubleday and Company, 1975.

Bailey, Anne J., *Enemies of the Country: New Perspectives on Unionists in the Civil War*, Athens, Georgia: University of Georgia Press, 2001.

Bandera County History Book Committee, *History of Bandera County, Texas* Dallas Texas: Curtis Media Corporation, c1987.

Banta, William and Caldwell, J. W., *Twenty-Seven Years on the Texas Frontier* Council Hill, Oklahoma: Published by L. G. Parks, 1893.

Bartlett, John Russell, *Personal Narrative of Explorations and Incidents in Texas, New Mexico, California Sonora and Chihuahua 1850 – 1853,* Two Volumes, Chicago, Illinois: The Rio Grande Press Inc., 1965.

Baum, Dale, *The Shattering of Texas Unionist: Politics in the Lone Star State During the Civil War Era,* Baton Rouge, Louisiana: Louisiana State University Press, 1998.

Beckwith, Paul, *Creoles of St. Louis,* St. Louis, Missouri: Nixon-Jones Printing Company, 1893.

Benjamin, Gilbert Giddings, *The Germans In Texas: A Study in Immigration,* Philadelphia, Pennsylvania: German Amertran Annual, Volume VII, 1909.

Bennett, Bob, *Kerr County Texas 1856 – 1956,* San Antonio, Texas: Naylor Company, 1956.

Biggers, Don H., *German Pioneers in Texas,* Fredericksburg, Texas: Fredericksburg Publishing Company, Inc., 1983.

Biesele, Rudolph Leopold, *The History of the German Settlements in Texas 1831 – 1861,* Austin, Texas: Von Boeckmann-Jones Company, 1930.

Bitton, Davis, Editor, *Reminiscences and Civil War Letters of Levi Lamoni Wight: Life in a Mormon Splinter Colony on the Texas Frontier,* Salt Lake City, Utah: University of Utah Press, 1970.

Blanco County News, *Heritage of Blanco County, Texas,* Hurst, Texas: Curtis Media Corporation, 1987.

Blaufarb, Douglas S., *The Counter-Insurgency Era: U. S. Doctrine and Performance 1950 to the Present*, New York, New York: The Free Press, 1977.

Boerne Area Historical Preservation Society, *Gone – But Not Forgotten: A Survey of Cemeteries in Boerne and Surrounding Areas, Volume I,* Boerne, Texas: Published by Boerne Area Historical Preservation Society, 1983.

Boerne Area Historical Preservation Society, *Gone—But Not Forgotten: A Survey of Cemeteries in Boerne and Surrounding Areas, Volume II,* Boerne, Texas: Published by Boerne Area Historical Preservation Society, 1985.

Bowen, J. J., *The Exodus of Federal Forces From Texas 1861,* Texas: Eakin Press, 1986.

Brancaforte, Charlotte L., Ed., *The German Forty-Eighters in the United States,* New York, New York: Peter Lang Publishing Company Inc., 1989.

Branda, Eldton S., Ed, *The Handbook of Texas: A Supplement,* Austin, Texas: Texas State Historical Association, 1976.

Breitenkamp, Edward C., Translated, *The Cypress and Other Writings of a German Pioneer in Texas by Hermann Seele,* Austin, Texas: University of Texas Press, 1979

Brinley, Lorine, *Orange County, Texas Cemetery Inscriptions,* St. Louis, Missouri: Published by Frances T. Ingmire, 1985.

Brown, Harry James, *Letters From A Texas Sheep Ranch,* Urbana, Illinois: Illinois Press, 1959.

Brown, Mydcna and Kilborn, Helen Ione, *Cemetery Records City Cemetery No. 4, Confederate Section Bexar County San Antonio, Texas,* San Antonio, Texas: Published by Helen Ione Kilborn, 1988.

Bruncken, Ernest, *German Political Refugees in the United States during the Period from 1815 – 1860,* Milwaukee, Wisconsin: Deutsch-Amerikanische Geschichtsblaetter, 1904.

Buenger, Walter L., *Secession and the Union in Texas,* Austin, Texas: University of Texas Press, 1984.

Bibliography

Bynum, Victoria, *The Free State Of Jones: Mississippi's Longest Civil War*, Chapel Hill, North Carolina: The University of North Carolina Press, 2001.

Cade, Winifred S., Ed., *I Think Back: Being The Memoirs of Grandma Gruen-Mathilda Doebbler Gruen Wagner,* San Antonio, Texas: Published by Winifred Cade, 1937.

Castro Colonies Heritage Association, *History of Medina County, Texas,* Castroville, Texas: Published by Castro Colonies Heritage Association, n. d.

Clark, L. D. Ed., *Civil War Recollections of James Lemuel Clark and the Great Hanging at Gainesville, Texas in October 1861,* Plano, Texas: Republic of Texas Press, 1997.

Clavell, James, Ed., *Sun Tzu: The Art of War,* New York, New York: Delacorte Press, 1983.

Comal County Genealogy Society, *Comal County Census 1870,* New Braunfels, Texas: Published by Comal County Genealogy Society, 1992.

Comal County Genealogy Society, *Index to Naturalization Papers March 1847 thru January 18, 1927 Comal County, Texas,* New Braunfels, Texas: Published by Comal County Genealogy Society, 1989.

Comal County Genealogy Society, *Marriage Records of Comal County, Texas 1846 – 1864,* New Braunfels, Texas: Published by Comal County Genealogy Society, 1992.

Comal County Genealogy Society, *Marriage Records of Comal County, Texas 1864 – 1877,* Published by Comal County Genealogy Society, New Braunfels, Texas, n. d.

Comal County Genealogy Society, *Lest We Forget: Cemeteries of Comal County, Texas and Surrounding Areas Excluding New Braunfels,* New Braunfels, Texas: Published by Comal County Genealogy Society, 1989.

Conzen, Michael P., *The Clash of Utopias: Sisterdale and the Six-Sided Struggle for the Texas Hill County*, New York, New York: Rowman and Littlefield Publishers, Inc., 1980.

Copeland, Fayette, *Kendall of the Picayune: Being His Adventures In New Orleans, On The Texan Santa Fe Expedition, In The Mexican War, And The*

Colonization Of The Texas Frontier, Norman, Oklahoma: University of Oklahoma Press, 1943.

Cotter, John and Dorothea, *The Kusenberger Family,* Livingston, Texas: Published by John and Dorothea Cotter, 1991.

Crook, Cornelia English, *Henry Castro: A Study of Early Colonization In Texas,* San Antonio, Texas: St. Mary's University Press, 1988.

Crouch, Carrie J., *A History of Young County, Texas,* Austin, Texas: Texas State Historical Association, 1956.

Daughters of The Republic of Texas, Compiled By, *The Marked Gravesites of Citizens of the Republic of Texas: An Indexed Guide to 436 Early Citizen's Burial Place,* Austin, Texas: Published by Daughters of The Republic of Texas, 1986.

Day, James M., Compiler, *Texas House and Senate Journals of the Ninth Legislature of State of Texas: November 4, 1861 – January 14, 1862,* Austin, Texas: Texas Library and Historical Commission, 1964.

Day, James M., Compiler, *Texas House and Senate Journals of the Ninth Legislature of State of Texas: November 3, 1863 – December 16, 1863,* Austin, Texas: Texas Library and Historical Commission, 1965.

Debo, Darrell, *Burnet County History – A Pioneer History 1847 – 1979,* Burnet, Texas: Volume I, Eakin Press, 1979.

Denson, Mrs. Billy Burnes and Mrs. Howard Graves, Compilers *Bandera County Cemetery Records,* Bandera, Texas: Bandera County Historical Survey Committee, n. d.

DeWitt, County Historical Commission, *The History of DeWitt County, Texas* Dallas, Texas: Curtis Media Corporation, 1991.

Duaine, Carol L., *The Dead Men Wore Boots: An Account Of The 32nd Texas Volunteer Cavalry, CSA 1862 – 1865,* Austin, Texas: San Felipe Press, 1966.

Duden, Gottfried, *Bericht ber eine Reise nach westlichen Staaten Nord Amerika's und einen mehrjahrigen Aufenthalt am Missouri,* Switzerland, Published by Gottfried Duden, 1829.

Edwards County History Book Committee, *A History of Edwards County Texas,* San Angelo, Texas: Anchor Publishing Company, 1984.

Edwards, Walter F. Ed., *The Story of Fredericksburg, Its Past, Present, Points of Interest and Annual Events,* Fredericksburg, Texas: Published by Fredericksburg Chamber of Commerce, n.d.

Edwards, Walter F. *Tales Of Old Fredericksburg,* Fredericksburg, Texas: Gargoyle Press, 1974.

Engle, Stephen D. *Yankee Dutchman: The Life of Franz Sigel,* Baton Rouge, Louisiana, 1993.

Fehrenbach, T. R. et al, *The San Antonio Story—a pictorial and entertaining commentary on the growth and development of San Antonio, Texas,* Tulsa, Oklahoma: Published by Continental Heritage, Inc., 1978.

Fenley, Florence, *Old Timers: Frontier Days in Uvalde Section of South West Texas*, Austin, Texas: State House Press, 1991.

Fenley, Florence, *Old Timers of Southwest Texas*, Uvalde, Texas: Published by Hornby Press, 1957.

Feuge, Robert Lamar, Ph.D, *Christoph Feuge: A German Pioneer's Story*, Coral Springs, Florida: Llumina Press, 2009.

Flach, Vera, *A Yankee in German America: Texas Hill Country,* San Antonio, Texas: The Naylor Company, 1973.

Fischer, Ernest G., *Marxists And Utopias In Texas: The Lone Star State's Pioneer Flirtation With Socialism-Communism,* Austin, Texas: Eakin Press, 1980.

Forbes, Hugh, *Manual for the Patriotic Volunteer on Active Service in Regular and Irregular War: Being The Art and Science of Obtaining and Maintaining Independence*, New York: De Witt and Davenport, 1855. Reprint by AstroLogos Books, New York, New York 2007.

Francis, May E., *The Hermit Of The Cavern,* San Antonio, Texas: The Naylor Company, 1932.

Fremantle, Lt. Col, Arthur J. L., *Three Months In The Southern States,* Lincoln, Nebraska: University of Nebraska Press, 1991.

Freund, Max, Translated and Edited, *Gustav Dresel's Houston Journal: Adventures in North America and Texas, 1837 – 1841,* Austin, Texas: University of Texas Press, 1954.

Gallaway, B. P., Ed., *Texas, The Dark Corner of the Confederacy – Contemporary Accounts of the Lone Star State in the Civil War,* Lincoln, Nebraska: University of Nebraska Press, 1994.

Gammel, P. N., Compiled, *The Laws of Texas 1822 – 1897,* Volume V, Austin, Texas: Gammel Book Company, 1898.

Genealogical Society of Kendall County, Texas, *Genealogical Abstractions From Kendall County, Texas Probate Records: 28 April 1862 – 10 December 1900*, Boerne, Texas: Published by Genealogical Society of Kendall County, Texas, 1986.

Geue, Chester William and Ethel Hander Geue, Compiled and Ed., *A New Land Beckoned: German Immigration to Texas, 1844 – 1847,* Waco, Texas: Texian Press, 1966.

Geue, Chester W., Translated, *Sketches of Life in the United States of North America and Texas as observed by Friedrich W. von Wrede,* Waco, Texas: Texian Press, 1970.

Geue, Ethel Hander, *New Homes In a New Land: German Immigration to Texas 1847 – 1861,* Baltimore, Maryland: Genealogical Publishing Company, Inc., 1982.

Giap, General Vo Nguyen, *Banner of People's War, the Party's Military Line,* New York, New York: Praeger Publisher, 1970.

Gillespie County Historical Society, *Pioneers in God's Hills: A History of Fredericksburg and Gillespie County People and Events,* Austin, Texas: Eakin Press, 1960.

Gillespie County Historical Society, *Pioneers in God's Hills: A History of Fredericksburg and Gillespie County People and Events, Volume II,* Fredericksburg, Texas: Gillespie County Historical Society, 1974.

Gillett, Sergeant James B., *Fugitives from Justice: The Notebook of Texas Ranger Sergeant James B. Gillett,* Austin, Texas: State House Press, 1997.

Gish, Theodore, Translated, *The Diary of Hermann Seele and Seele's Sketches from Texas: Pioneer, Civil and Cultural Leader, German-Texan Writer,* Austin, Texas: German-Texan Heritage Society, 1995.

Gish, Theodore, *Eagle in the New World: German Immigration to Texas and American,* College Station, Texas: Texas A and M University Press, 1986.

Glenn, Frankie Davis, *Capt'n John: Story of a Texas Ranger,* Austin, Texas: Nortex Press, 1991.

Glenn, Frankie Davis, Frontier Series, *John William Sansom's Battle of the Nueces,* Boerne, Texas: Published by Frankie Davis Glenn, 1997.

Goeth, Ottilie Fuchs, *Was Grossmutter Erzaehlt [Memoirs Of A Texas Grandmother],* Austin, Texas: Eakin Press, 1982.

Gold, Ella A., Ed., *Kirchen-Buch: Church Record Book of the Vereins-Kirche 1849 – 1870,* Fredericksburg, Texas: Gillespie County Historical Society, Inc., 1986.

Goyne, Minetta Altgelt, *A Life among the Texas Flora: Ferdinand Lindheimer's Letters to George Engelmann,* College Station, Texas: Texas A and M University Press, 1991.

Goyne, Minetta Altgelt, *Lone Star and Double Eagle,* Fort Worth, Texas: Texas Christian University Press, 1982.

Graves, Mrs. Howard, *Veterans From Bandera County Of All Wars,* Bandera, Texas: Bandera Printing Company, 1978.

Greer, James K., *Buck Barry Texas Ranger and Frontiersman,* Lincoln, Nebraska: University of Nebraska Press, 1978.

Griffith, Samuel B., Translated, *Mao Tes-Tung on Guerrilla Warfare,* New York, New York: Praeger Publishers, 1961.

Guenther, Irma Goeth, Translator, Research and Additions, *Memoirs Of A Texas Pioneer Grandmother (Was Grossmutter Erzaehlt) 1805 – 1915 by Otillie Fuchs Goeth,* Austin, Texas: Eakin Press, 1982.

Gurasich, Marj, *A House Divided*, Fort Worth, Texas Christian University Press, 1994.

Halley, J. Evetts, *Charles Schreiner General Merchandise 1869 – 1969: The Story of a Country Store,* Kerrville, Texas: Published by Charles Schreiner Company, 1969.

Hamerow, Theodore S., *The Two Worlds of the Forty-Eighters*, New York, New York: Peter Long Publishers, 1989.

Hardin, Earl S., Jr., *The Ranger Companies Of Bandera County,* Bandera, Texas: Published by Earl S. Hardin, Jr., 1995.

Harper Centennial Committee, *Here's Harper 1863 – 1963,* Fredericksburg, Texas: Radio Post, Inc., 1963.

Harper Sesquicentennial Committee, *Here's Harper Two,* Austin, Texas: Nortex Press, 1986.

Harris, Gertrude, *A Tale of Men Who Knew Not Fear,* San Antonio, Texas: Alamo Printing Company, 1935.

Hass, Oscar, *History of New Braunfels and Comal County, Texas 1844 – 1946,* San Marcos, Texas: Steck, 1968.

Hawgood, John, *The Tragedy of German-America: The Germans in the United States of America during the Nineteenth Century—and After,* G. P. Putnam's Sons, New York, New York: 1940.

Hays County Historical Commission, *Clear Springs and Limestone Ledges,* San Marcos, Texas: Published by Hays County Historical Commission, 1986.

Hecke, J. Valentine, *Reise durch die Vereinigten Staaten von Nord-Amerika in den Jahren 1818 und 1819*, Berlin, Germany: 1820.

Heil, Robert Allen, *One "Heil" Of A Family – Some Ancestors and Other Relatives of Louis Heil and Elizabeth Susan (Steves) Heil,* Comfort, Texas: Published by Robert Allen Heil, n. d.

Helbich, Wolfgang and Walter D. Kamphoefner, *Deutsche im Amerikanischen Burgerkrieg: Briefe von Front und farm 1861 – 1865*, Munchen, Germany: Published by Ferdinand Schoningh, 2002.

Henley, Carol, *Medina County, Texas Cemeteries, Volume II*, St. Louis, Missouri: Published by Frances T. Ingmire, 1985.

Hood, R. Maurice M. C., Edited, *Early Texas Physicians 1830 – 1915,* Austin, Texas: State House Press, 1999

Hover, Herman D., *Fourteen Presidents Before Washington,* New York, New York: Dodd, Mead and Company, 1985.

Hughes, Alton, *Pecos: A History of the Pioneer West*, Seagraves, Texas: Pioneer Book Publishers, Volume I, 1978.

Hughes, Alton, *Pecos: A History of the Pioneer West*, Seagraves, Texas: Pioneer Book Publishers, Volume II, 1982.

Hunter, J. Marvin, *A Brief History of Bandera County,* Bandera, Texas: Published by Bandera Printing Company, 1949.

Hunter, J. Marvin, *The Lyman Wight Colony in Texas: Came to Bandera in 1854,* Bandera, Texas: The Bandera Bulletin, n.d.

Hunter, J. Marvin, *100 Years in Bandera 1853 – 1953,* Bandera, Texas: Published by Bandera Printing Company, 1953.

Hunter, J. Marvin, Compiled and Edited, *The Trail Drivers of Texas,* Austin, Texas: University of Texas Press, 1985.

Hurst, Kathryn Adam, Transcriber, *1870 Federal Census For Kendall County, Texas,* Boerne, Texas: Published by Genealogical Society of Kendall County, 1997.

Ingmire, Frances T., *Bexar County, Texas Marriage Records, 1837 – 1866*, St. Louis, Missouri: Published by Frances T. Ingmire, 1985.

Inscoe, John C. and Robert C. Kenzer, Ed., *Enemies of the Country: New Perspectives on Unionists in the Civil War South,* Athens, Georgia: University of Georgia Press, 2001.

Isaak, Alan C., *Scope and Methods of Political Science,* Homewood, Illinois: The Dorsey Press, 1969.

Jacobson, Lucy Miller and Mildred Bloys Nored, *Jeff Davis County, Texas,* Fort Davis, Texas: Published by the Fort Davis Historical Society, 1993.

Jenkins, Frank D., *The Turknett Family: Descendants of Jacob c1757 – 1816*, Lubbock, Texas: Published by Christine Know Wood, 1981.

Johnson, David, *The Mason County "Hoo Doo" War, 1874 – 1902*, Denton, Texas: North Texas State University Press, 2006.

Jordan, Gilbert J., *Yesterday in the Texas Hill Country,* Texas A and M. College Station, Texas: University Press, 1979.

Jordan, Terry G., *German Seed in Texas Soil: Immigrant Farmers in Nineteenth Century Texas,* Austin, Texas: University of Texas Press, 1966.

Kaufmann, Wilhelm, *The Germans In The American Civil War,* Carlisle, Pennsylvania: John Kallmann, Publishers, 1999.

Kearney, James C., *Nassau Plantation: The Evolution of a Texas German Slave Plantation*, Denton, Texas: North Texas Press, 2010.

Kellner, Marjorie, Compiler, *Wagons, Ho! – A History of Real County, Texas,* Fort Worth, Texas: Curtis Media Corporation, 1995.

Kelsey, Michael, et al compilers, *Miscellaneous Texas Newspaper Abstracts – Deaths,* Volume 2, Bowie, Maryland: Heritage Books, Inc., 1997.

Kendall County Historical Commission, *Rivers, Ranches, Railroads and Recreation: A History of Kendall County, Texas,* Dallas, Texas: Taylor Publishing Company, 1984.

Kerr County Historical Commission, *Kerr County Album,* Dallas, Texas: Taylor Publishing Company, n.d.

King, Irene Marschall, *John O. Meusebach: German Colonizer in Texas* Austin, Texas: University of Texas Press, 1967.

Kinney County Historical Society, *Kinney County 125 Years of Growth 1852 – 1977,* Brackettville, Texas: Published by Kinney County Historical Society, 1977.

Knopp, Kenn, *The Fredericksburg, Texas Manuscripts,* Fredericksburg, Texas: Published by Kenn Knopp, 1997.

Koerner, Gustav, *Das Deutsche Element in den Vereinigten Staaten von Nord-Amerika*, Cincinnati, Ohio: 1880.

Ledbetter, Barbara A. Neal, *Fort Belknap Frontier Sage: Indians, Negroes and Anglo-Americans on the Texas Border,* Graham, Texas: Published by Barbara Ledbetter, 1982

Ledecus, Eduard, *Reise durch die mexikanischen Provinzen Tumalipas, Cohahuila und Texas im Jahre 1834,* Leipzig, Germany, 1837.

Levine, Bruce, *The Spirit of 1848: German Immigrants, Labor Conflict, and the Coming of the Civil War,* Urbana, Illinois: University of Illinois Press, 1992.

Lich, Glen E. and Dona B. Reeves, Ed., *German Culture In Texas: A Free Earth; Essays from the 1978 Southwest Symposium,* Boston, Massachusetts: Twayne Publishers, 1980.

Lich, Glen E., *Goethe on the Guadalupe,* Boston, Massachusetts: Twayne Publishers, 1980.

Lich, Glen E., *The German Texas,* San Antonio, Texas: The University of Texas Institute of Texan Cultures at San Antonio, 1981 and 1996.

Lindig, Armand, Compiled, *Record of Der Friedhof Cemetery,* Fredericksburg, Texas: Published by City Cemetery Assoc., n. d.

Lochte, Emile W., *A History Of Frederick (Fritz) Lochte And His Descendants and A History of Henry C. Lochte,* New Orleans, Louisiana: Published by Lochte Descendants, n.d.

Loeher, Franz, *Geschichte und Zustaende der Deutschen in Amerika,* Cincinnati, Ohio: 1847.

Long, E. B., *The Civil War Day By Day,* Garden City, New York: Doubleday and Company, 1971.

Lonn, Ella, *Foreigners in the Confederacy,* Chapel Hill, North Carolina: The University of North Carolina Press, 1940.

Lowe, Richard, Edited, *A Texas Cavalry Officer's Civil War: The Diary and Letters of James C. Bates,* Baton Rouge, Louisiana: Louisiana State University Press, 1999.

Luttwak, Edward and Stuart L. Koehl, *The Dictionary of Modern War,* New York, New York: Gramercy Books, 1991.

Mainland, William F. Translated, *William Tell*, Chicago, Illinois: The University of Chicago Press, 1972.

Marten, James, *Texas Divided, Loyalty and Dissent in the Lone Star State 1856 – 1874*, Lexington, Kentucky: University Press of Kentucky, 1990.

Martin, Bernard, *Strange Vigour: A Biography of Sun Yat-Sen*, Port Washington, New York: Kennikat Press, 1944.

Mason County Historical Commission, *Mason County Historical Book*, Mason, Texas: Published by Mason County Historical Commission, 1986.

Mason County Historical Commission, *Mason County Historical Book, Supplement II*, Mason, Texas: Published by Mason County Historical Commission, 1994.

Mauldin, Lynn C., *Kerr County, Texas Marriage Records 1856 – 1899*, Kerrville, Texas: Published by Lynn C. Mauldin, n. d.

McCaslin, Richard B., *Tainted Breese – The Great Hanging at Gainesville, Texas 1862*, Baton Rouge, Louisiana: Louisiana State University Press, 1995.

McDonald, Archie, Edited, *Hurrah for Texas: The Diary of Adolphus Sterne 1838 – 1851*, Austin, Texas: Eakin Press, 1986.

McDowell, Catherine W., Ed., *Now You Hear My Horn: The Journal of James Wilson Nichols 1820 – 1887*, Austin, Texas: University of Texas Press, 1967.

McGowen, Stanley S., *Horse Sweat and Powder Smoke: The First Texas Cavalry in the Civil War*, College Station, Texas: Texas A and M University Press, 1999.

McGuire, James Patrick, *Hermann Lungkwitz: Romantic Landscapist on the Texas Frontier*, Austin, Texas: University of Texas Press, 1983.

Menard County Historical Society, Ed., *The Menard County History: An Anthology*, San Angelo, Texas: Anchor Publishing Company, 1982.

Menke, Jim and Doris, *1850, 1860, 1870 Federal Census's Medina County, Texas*. Hondo, Texas: Medina County History Series, 1998.

Menke, Jim and Doris, *Index to the First and Second Regiment Texas Cavalry Volunteers (Union),* San Antonio, Texas: Published by Jim and Doris Menke, 1998.

Menke, Jim and Doris, *Seventh-Eighth-Ninth Census of the United States, Medina County, Texas,* San Antonio, Texas: Published by Jim and Doris Menke, 1998.

Metzner, Henry, *A Brief History of the American Turnerbund,* Pittsburgh, Pennsylvania: National Executive Committee of the American Turnerbund, 1924.

Michener, James A., *Texas,* New York, Random House, 1985.

Moore, Frank, Ed., *Rebellion Record – A Diary of American Events,* Sixth Volume, New York, New York: Arno Press, 1982.

Moore, Ike, Arranged, *The Life and Diary of Reading W. Black: A History of Early Uvalde,* Uvalde, Texas: El Progreso Library, 1997.

Morgenthaler, Jefferson, *The German Settlement of the Texas Hill County,* Mockingbird Books, Boerne, Texas, 2007

Morgenthaler, Jefferson, *Promised Land: Solms, Castro, and Sam Houston's Colonization,* College Station, Texas, 2009.

Morrison, Michael D., Intro. *Fugitives from Justice: The Notebook of Texas Ranger Sergeant James B. Gillett,* Austin, Texas: State House Press, 1997.

Moursund, John Stribling, *Blanco County History,* Burnet, Texas: Nortex Press, 1979.

Moursund, John Stribling, *Blanco County Families For 100 Years,* Burnet, Texas: Nortex Press, 1981.

Mueller, Oswald, Translated, *Roemer's Texas 1845 to 1847 by Dr. Ferdinand Roemer,* Austin, Texas: Eakin Press, 1995.

Mundie, Jr., James A. et al, *Texas Burial Sites of Civil War Notables: A Biographical and Pictorial Field Guide,* Hillsboro, Texas: Hill College Press, 2002.

Murphy, Alexander and Douglas L. Johnson Editor, *Cultural Encounters with the Environment: Enduring and Evolving Geographic Themes*, New York, New York: Rowman and Littlefield Publishers, Inc., 2000.

Nagel, Charles, *A Boy's Civil War Story,* Philadelphia, Pennsylvania: Dorrance and Company, 1925.

Newcomb, William W., Jr., *Friedrich Richard Petri: German Artist on the Texas Frontier,* Austin, Texas: University of Texas Press, 1978.

North, Thomas, *Five Years In Texas: or What you Did Not Hear During the War,* Cincinnati, Ohio: Elm Street Printing Company, 1871.

Nunn, W. C., *Escape From Reconstruction,* Fort Worth, Texas: Texas Christian University Press, 1956.

Oaks, Stephen B., Ed., *RIP Ford's Texas by John Salmon Ford,* Austin, Texas: University of Texas Press, 1963.

Olmsted, Frederick Law, *A Journey Through Texas or, a Saddle Trip on the Southwestern Frontier*, Austin, Texas: University of Texas Press, 1978.

Orton, Brigadier General Richard H., *Records of California Men in the War Of The Rebellion—1861 To 1867,* Sacramento, California: California State Printing Service, 1890.

Pease, S. W., *They Came to San Antonio,* San Antonio, Texas: Published by S. W. Pease, c1975.

Penniger, Robert, *Fest Ausgabe fuenfzigjaehrigen Jubilaeum der Deutfchen Kolonie Friedrichsburg,* Fredericksburg, Texas: Herlag von Robert Penniger, 1896.

Perry, Garland, *Historic Images of Boerne, Texas*, San Antonio, Texas: Econ-o-Print, 1962.

Pickering, David and Judy Falls, *Brush Men and Vigilantes: Civil War Dissent in Texas,* College Station, Texas: Texas A and M University, 2000.

Pirtle, Caleb III and Michael F. Cusack, *Fort Clark; The Lonely Sentinel On Texas's Western Frontier,* Austin, Texas: Eakin Press, 1985.

Ponder, Jerry, *Fort Mason, Texas – Training Ground For Generals,* Mason, Texas: Ponder Books, 1997.

Ponder, Jerry, *Mason County's First Settlers,* Mason, Texas: Ponder Books, 1997.

Porter, Clyde H., *The Dresel Family,* San Antonio, Texas: Privately Published, 1952.

Pustay, John S., *Counter-Insurgency Warfare,* New York, New York: The Free Press of Glencoe and London: Collier Macmillan, 1965.

Ragsdale, Crystal Sasse, *The Golden Free Land: The Reminiscences and Letters of Women on an American Frontier,* Austin, Texas: Landmark Press, 1976.

Raines, C. W., Ed., *Six Decades in Texas or Memoirs of Francis Richard Lubbock, Governor of Texas in Wartime 186163: A Personal Experience in Business, War, and Politics,* Austin, Texas: Ben C. Jones and Company, 1900.

Randers-Pehrson, Justine Davis, *Adolf Douai, 1819 – 1888: The Turbulent Life of a German Forty-Eighter in the Homeland and in the United States,* New York: Peter Lang Publisher, 2000.

Ransleben, Guido E., *A Hundred Years of Comfort In Texas,* San Antonio, Texas: The Naylor Company, 1974.

Rapoport, Anatol, Ed., *Carl von Clausewitz On War*, Middlesex, England: Penguin Books 1982.

Reichstein, Andreas, *German Pioneers on the American Frontier: The Wagners in Texas and Illinois*, Denton, Texas, University of North Texas Press, 2001.

Rentschler, Thomas B., *Rifles and Blades of the German-American Militia and the Civil War: A Comprehensive Illustrated History of the Turners, a Unique German-American Gymnastic Society, and their Role in the Events Before and During the American Civil War*, Hamilton, Ohio: Blue Hills Press, 2003.

Rice, Edward E., *Mao's Way,* Berkeley, California: University of California Press, 1972.

Richardson, Rupert N., et al, *Texas: The Lone Star State,* Englewood Cliffs, New Jersey: Prentice-Hall, Inc., 1981.

Roberts, Colonel O. M. and Colonel J. J. Dickson, *Confederate Military History*, Volume XI, Texas and Florida, Secaucus, New Jersey: Blue and Gray Press, n. d. c1890.

Robinson, Robert R., Jr., *The Bremers and Their Kin in Germany and in Texas*, Austin, Texas: Nortex Press, 1986.

Rogers, Laura Wood, FLO: *A Biography of Frederick Law Olmsted*, Baltimore, Maryland: The John Hopkins University Press, 1973.
Ruggero, Ed., *Combat Jump: The Young Men Who Led the Assault into Fortress Europe, July 1943*, New York, New York, Harper Collins, 2003

Rybczynski, Witold, *A Clearing in The Distance: Frederick Law Olmsted and America in the 19th Century*, New York, New York: Simon and Schuster, 1999.

San Antonio Genealogical and Historical Society, *Index Wills and Inventories Of Bexar County, Texas 1742 – 1899*, San Antonio, Texas: San Antonio Genealogical and Historical Society.

Santleben, August, *A Texas Pioneer: Early Staging and Overland Freighting Days on the Frontiers of Texas and Mexico*, Castroville, Texas: Castro Colonies Heritage Association, 1994.

Sansom, John W., *Battle Of Nueces River In Kinney County, Texas August 10, 1862*, San Antonio, Texas: Published by John W. Sansom, 1905.

Scharf, Edwin E., *Frontier Freethinkers in the Texas Hill Country*, Comfort, Texas: Published by Edwin E. Scharf, c2001.

Schmidt, Charles Frank, Translated, *Texas in 1848 by Viktor Bracht*, Manchaca, Texas: German-Texan Heritage Society, 1931.

Schram, Stuart R., Translated, *Mao Tse-Tung Basic Tactics*, New York, New York: Frederick A. Praeger Publishers, 1966.

Schwartz, Stephan, *Twenty-Two Months A Prisoner Of War*, St. Louis, Missouri: A. F. Nelson Publishing Company, 1892, 1892.

Sealsfield, Charles, *The Cabin Book or National Characteristics*, Austin, Texas: Eakin Press, 1985.

Shefelman, Janice, *Sophie's War: The Journal of Anna Sophie Franzisky Guenther*, Austin, Texas, Eakin Press 2006.

Sibley, Marilyn McAdams, *Lone Stars and State Gazettes: Texas Newspapers Before the Civil War,* College Station, Texas: Texas A and M University Press, 1983.

Siemering, A., *Ein Verstehltes Leben*, San Antonio, Texas: Published by August Siemering, 1876.

Sifakis, Stewart, *Compendium Of The Confederate Armies: Texas,* New York, New York: Facts On File, Inc., 1995.

Simpson, Harold B., Ed., *Texas In The War 1861 – 1865, Compiled by Marcus J. Wright, Brigadier General, CSA,* Hillsboro, Texas: Hill Junior College Press, 1965.

Smiley, Jerome, Edited By, *History of Denver*, Denver, Colorado: The Denver Times—Sun Publishing Company, 1901.

Smith, David Paul, *Frontier Defense in the Civil War,* College Station, Texas: Texas A and M University Press, 1992.

Smith, Shirley, *Confederate Veterans of Kerr County, Texas,* Kerrville, Texas: Published by Adam R. Johnson Chapter No. 2498, United Daughters of The Confederacy, 1991.

Smith, Thomas C., *Here's Yer Mule – The Diary of Thomas C. Smith, 3rd Sergeant, Company G., Wood's Regiment, 32nd Texas Cavalry, CSA,* Waco, Texas: Little Texan Press, 1958.

Smith, Thomas T., *The U. Army and The Texas Frontier Economy 1845 – 1900,* College Station, Texas: Texas A and M Press, 1999.

Smith, Thomas Tyree, *Fort Inge: Sharps, Spurs, and Sabers On The Texas Frontier 1849 – 1869,* Austin, Texas: Eakin Press, 1993.

Smithwick, Noah, *The Evolution of a State or Recollections of Old Texas Days,* Austin, Texas: University of Texas Press, 1983.

Sowell, A. J., *Texas Indian Fighters*, Austin, Texas: State House Press, 1986.

Speer, John W., *A History of Blanco County,* Austin, Texas: The Pemberton Press, 1965.

Spellmann, Charles E., *The Texas House of Representatives: A Pictorial Roster 1846-1992,* Austin, Texas: Texas House of Representatives, 1992.

Spencer, John W. *Terrell's Texas Cavalry: Wild Horsemen of the Plains in the Civil War,* Austin, Texas: Eakin Press, 1982.

Stanley, D. S., *Personal Memories of Major General D. S. Stanley,* Cambridge, Massachusetts: Harvard University Press, c1917.

Stayer, Joseph R., *The Mainstream of Civilization,* New York, New York: Haracourt, Brace and World, Inc., 1969.

Stephens, Robert W., *Texas Ranger Indian War Pension,* Quanah, Texas: Nortex Press, 1975.

Stewart, Anne and Mike, *Comfort Women in Comfort History,* Comfort, Texas: Published by Anne and Mike Stewart, 1993.

Stewart, Anne and Mike, *Texas In The Civil War, Death On The Nueces, The Minna Stieler Stories,* Comfort, Texas: Published by Anne and Mike Stewart, 1997.

Stovall, Allan A., *Nueces Headwater Country: A Regional History,* San Antonio, Texas: The Naylor Company, 1959.

Stovall, Allan A., *Pioneer Days In The Breaks Of The Balcones: A Regional History,* Austin, Texas: Firm Foundation Publishing House, 1967.

Striegler, Selma Weyrich, Ed., *The Johan Frederick Gottlieb Striegler Family History: A Detailed Genealogy of the Descendants of Pioneer Citizens of Gillespie County, Texas,* Fredericksburg, Texas: Published by Selma Weyrich Striegler and Striegler Family Members, 1952.

Struve, Walter, *Germans and Texans: Commerce, Migration, and Culture in the Days of the Lone Star Republic,* Austin, Texas: University of Texas Press, 1996.

Sullivan, Jerry M., *Fort McKavett A Texas Frontier Post,* Austin, Texas: Texas Parks and Wildlife Department, 1981.

Sutherland, Daniel E., Ed., *Guerrillas, Unionists, and Violence on the Confederate Home Front,* Fayetteville, Arkansas: The University of Arkansas Press, 1999.

Thompson, Jerry, *Confederate General of the West: Henry Hopkins Sibley,* College Station, Texas: Texas A and M University Press, 1996.

Thompson, Jerry D., *Mexican Texans in the Union Army,* El Paso, Texas: University of Texas at El Paso Press, 1986.

Thompson, Jerry Don, *Vaqueros in Blue and Gray,* Austin, Texas: Presidial Press, 1976.

Thompson, Leroy, *The Counter Insurgency Manual: Tactics of the Anti-Guerrilla Professionals,* Mechanicsburg, Pennsylvania: Stackpole Books, 2002.

Tilley, Nannie M., *Federals on the Frontier: The Diary of Benjamin F. McIntyre 1862 – 1864,* Austin, Texas: University of Texas Press, 1963.

Tiling, Moritz, *History Of German Element in Texas From 1820 – 1850: And Historical Sketches of the German Texas Singers; League and Houston Turnverein From 1853 – 1912,* Houston, Texas: Published by Moritz Tiling, 1913.

Tolman, Arlene and Fred, *Cemeteries of Kerr County 1859 – 1976,* Kerrville, Texas: Kerrville Genealogical Society, 1980.

Tolzmann, Don Heinrich, Ed., *The German-American Forty-Eighters 1848 – 1998,* Indianapolis, Indiana: Indiana University Press, 1998.

Townsend, Mary Bobbitt, *Yankee Warhorse: A Biography of Major General Peter Osterhaus,* Columbia, Missouri: University of Missouri Press, 2010.

Trinquier, Roger, *Modern Warfare: A French View of Counter-insurgency,* New York, New York: Frederick A. Praeger Publishers, 1961.

Turner, Mary Lewis, *Julius Theodore Splittgerber (1819 – 1897), Volume One: His Life and Times,* San Antonio, Texas: Watercress Press, 2003.

Turner, Mary Lewis, *Julius Theodore Splittgerber (1819 –1897), Volume Two: His German Ancestors and American Descendants,* San Antonio, Texas: Watercress Press, 1997.

Tyler, Ron, et al., Ed., *The New Handbook of Texas,* Six Volumes, Austin, Texas: Texas State Historical Association, 1996.

Underwood, Rodman L., *Death on the Nueces: German Texans Treue der Union,* Austin, Texas: Eakin Press, 2000.

Van Wagenen, Michael Scott, *The Texas Republic and the Mormon Kingdom of God,* College Station, Texas: Texas A and M University Press, 2002.

Von Clausewitz, General Carl, *Principles Of War,* Harrisburg Pennsylvania: The Military Service Publishing Company, 1942.

Von Maszewski, W. M., *A Sojourn in Texas, 1846 – 47: Alwin II. Soergel's Texas Writings,* San Marcos, Texas: German-Texan Heritage Society, 1992.

Von Maszewski, Wolfram, Translated, *Voyage to North American 1844 – 45: Prince Carl of Solms's Texas Diary of People, Places, and Events,* Denton, Texas: University of North Texas Press, 2000.

Wallace, Ernest, *The Howling of the Coyotes: Reconstruction Efforts to Divide Texas,* College Station, Texas: Texas A and M University Press, 1979.

Waller, John L., *Colossal Hamilton of Texas: Andrew Jackson Hamilton Militant Unionist and Reconstruction Governor,* El Paso, Texas: The University of Texas at El Paso Press, 1968.

Watkins, Clara, *Kerr County Texas 1856 – 1976,* Kerrville, Texas: Hill Country Preservation Society, Inc., 1975.

Weaver, Bobby D., *Castro's Colony: Empresario Development in Texas 1842 – 1865,* College Station, Texas: Texas A and M University Press, 1985.

Webb, Walter P., Ed., *The Handbook of Texas, Volume I and II,* Austin, Texas: Texas State Historical Association, 1952.

Weber, Adolf Paul, *Die Deutsche Pioniers Zur Geschichtes des Deutschthums in Texas,* San Antonio, Texas: Published by Adolf Paul Weber, 1894.

Wight, Jermy Benton, *The Wild Ram of the Mountain: Lyman Wight,* Bedford, Wyoming: Afton Thrift Print, 1997.

Wiley, Bell Ervin, Ed., *Fourteen Hundred and 91 Days in the Confederate Army: A Journal Kept by W. W. Heartsill, Or Camp Life: Day-By-Day of the W. P. Lane Rangers,* Wilmington, North Carolina: Broadfoot Publishing Company, 1992.

Williams, R. H., *With the Border Ruffians: Memories of the Far West, 1852 – 1868,* Lincoln, Nebraska: University of Nebraska Press, 1982.

Wisseman, Dr. Charles L., Sr., Ed., *Fredericksburg, Texas ... The First Fifty Years: A Translation of Penniger's 50ᵗʰ Anniversary Edition,* Fredericksburg, Texas: Fredericksburg Publishing Company, Inc., 1971.

Witt, Gerald, *The History Of Eastern Kerr County,* Austin, Texas: Nortex Press, 1986.

Wittke, Carl, *Refugees of Revolution: The German Forty-Eighters in America,* Philadelphia, Pennsylvania: University of Pennsylvania Press, 1952.

Woolf, Henry Bosley, et al, eds., *Webster's New Collegiate Dictionary,* Springfield, Massachusetts: G and C. Merriam Company, 1979.

Zoeller Descendants, *A Family History of Hellwig Karl Ludwig Adolph Zoeller and Augusta Zoeller,* Boerne, Texas: Published by Zoeller Descendants.

Zucker, A. E. Ed., *The Forty-Eighters: Political Refugees of the German Revolution of 1848,* New York, New York: Columbia University Press, 1950.

Oral Accounts, Interviews and Pensentations

Briefing, Colonel Frank King, Commander Second Psychological Group, U. S. Army, Cleveland, Ohio, February 1980 – September 1983.

Off the Record Oral Interviews with several Confederate and Unionist family members who requested they not be identified as the source for information. The two most informative were July 16, 1997 and June 29, 2001. In each Off the Record Oral Interview the data was verified by other sources to the extent possible. In both the July 16, 1997 and June 29, 2001, interviews it was not possible to totally verify the data. The July 16, 1997, interview claimed one the men who executed the wounded insurgents was

Oscar Splittgerber. He had joined the Union Loyal League using the alias of Bargerum. The June 29, 2001, interview provided the names of the two Unionists who were captured and executed on the afternoon of August 9, 1862.

Oral Account by Clifton Stork, (Fredericksburg historian), April 1995.

Oral Interview with Dorothy Basse, (Luckenbach family member), May 1997.

Oral Interviews with Karl Biermann, (Itz family member) Member, November 1997 and December 1997.

Oral Interview with Kenneth Crenwelge (Gold family member), May 1997.

Oral Interview with L. J. Dean (Nueces Canyon Historian), March 1998.

Oral Interview with Amelia 'Mollie' Eckhardt Dennis (Duecker family member), May 1997.

Oral Interview with Clara Garber (Bock family member), May 1997.

Oral Interview with Louis Weiss Haufler, (Weiss family member), August 1996.

Oral Interviews with Ronald Hall, (Tegener family member), August 1996 and March 1997.

Oral Interview with Helen Hester (Hester family member), October 1996.

Oral Interview with Leo Itz, (Itz family member), August 1996.

Oral Interview with Joseph Kamlah, (Kammlah family member), April 1995.

Oral Interview with Peter Kleck, (Kleck family member), May 1997.

Oral Interview with Werner Klier, (Klier family member), April 1995.

Oral Interview with John Koepke (Scott family member), March 1997.

Oral Interviews with Gregory J. Krauter (Comfort historian), January 1995 until February 2003.

Oral Interview with Wayne Meier, (Zoeller family member), April 1995.

Oral Interview with Margie Morgan, (Zoeller family member), July 1997, Kerrville, Texas

Oral Interviews with Emilie Real Neill, (Real family member), August 1997 and October 1997.

Oral Interview with Dr. Ansgar Reiss, Regensburg University, Germany, Regensburg, Germany, October 6, 1998 at University of Dallas.

Oral Interview with David Paul Smith, (author and historian), April 1997.

Oral Interviews with Anne Stewart, (author and Comfort historian), June 1995 until February 2003.

Oral Interview with Clifton Stock, (Markwordt family member), October 1997.

Oral Interviews with Ester Bonnet Strange, (Bonnet family member), May 1997 and October 1997.

Oral Interview with Raymond Usener, (Usener family member), October 1997.

Oral Interview with Evelyn Woods, (Bock family member), May 1997.

Oral Presentation by T. R. Fehrenbach at *Nueces Encounter 1862: Battle or Massacre?* Fredericksburg, Texas March 22, 1997.

Oral Presentation by Roland Hall at *Nueces Encounter 1862: Battle or Massacre?* Fredericksburg, Texas March 22, 1997.

Oral Presentation by Rick Hamby, University of Texas of the Permian Basin, 'Texas, Germans, and the Battle of the Nueces River' at *1998 Texas State Historical Association Annual Meeting* Austin, Texas March 6, 1998.

Oral Presentation by Prince Johannes von Sachsen-Altenburg, Duke of Saxony, New Braunfels, Texas February 21, 2003, and Fredericksburg, Texas, April 5, 2003.

Oral Presentation by Mary Lewis Turner at *1997 German – Texan Heritage Society Convention* Kerrville, Texas September 7, 1997.

Oral Presentation by Raymond E. Usener at *Nueces Encounter 1862: Battle or Massacre?* Fredericksburg, Texas March 22, 1997.

Oral Presentation by Roberta Warren at *Nueces Encounter 1862: Battle or Massacre?* Fredericksburg, Texas March 22, 1997.

Telephone Interview with Roscoe Basse (Luckenbach family member), June 1997.

Telephone Interviews with Louis Leinweber, (Scott and Leinweber family member), January 1998 and February 1998.

Telephone Interview with Mrs. Temple [Wanda] Henderson (Henderson family member), April 1997.

Telephone Interview with Aleta Addicks Stahl, (Markwordt family member), October 1997.

Newspapers

Austin State Gazette, 1861-1863.

Arkansas Daily Gazette (Little Rock), 1869.

Arkansas Weekly Gazette (Little Rock), 1869.

Clarkesville Standard. 1861-1862

Comfort News, 1909, 1912, 1939, 1973.

Dallas Herald, 1860-1862.

Dallas Morning News, 1929, 1997.

Denver Republican, 1899, 1900.

Fredericksburg Radio Post, 1952.

Fredericksburg Standard-Radio Post, 2004.

Fredericksburg Standard-Radio Post, 2006.

Houston Chronicle, 1995.

Kerrville Mountain Sun, 1953.

Marble Falls Picayune, 1998.

New Braunfels Herald, 1963.

New Braunfels Zeitung, 1853-1862.

San Antonio Alamo Express, 1860-1861.

San Antonio Daily Express, 1869.

San Antonio Express, 1918, 1924, 1934, 1952.

San Antonio Express-News, 1996, 2003.

San Antonio Freir Presse fuer Texas, 1865, 1923.

San Antonio Herald, 1860-1862.

San Antonio Ledger and Texan, 1855, 1861-1862.

San Antonio Light, 1961-1962.

San Antonio Semi-Weekly News, 1861-1862.

San Antonio Weekly Herald, 1861-1862.

San Antonio Zeitung, 1854-1855.

Washington Times, 1996.

West Kerr Current, 2006.

Western American (Keosauqua, Iowa), 1852.

Periodicals

Alberthal, Vernel, "Bushwhackers In Them Thar Hills", *The Radio Post*, October 2, 1952.

Author not stated, "Battle of Nueces River", *A Twentieth Century History of Southwest Texas*, Volume 2, 1907.

Author not stated, "German Unionist In Texas", *Harper's Weekly*, January 20, 1866.

Author not stated, "Land Records Kinney County, Texas", *Branches and Acorns*, Volume III, No. 4, June 1988.

Baron, Frank, "German Republicans and Radicals in the Struggle for a Slave-Free Kansas" Charles F. Kob and August Bonde; *Yearbook of German-American Studies*, Volume 40, 2005.

Baron, Frank, "The Campaign against Slavery in Kansas" *Yearbook of German-American Studies Supplement Issue*, Volume 4, 2012

Barr, Alwyn, ed., "Records of the Confederate Military Commission in San Antonio July 2 – October 10, 1862", *Southwestern Historical Quarterly*, Volume LXX, July 1966; Volume LXX, October 1966; Volume LXX, April 1967; Volume LXXI, October 1967; Volume LXIII, July 1969; and Volume LXXIII, October 1969.

Baulch, Joe, "The Dogs of War Unleashed: The Devil Concealed in Men Unchained" *West Texas Historical Association Year Book,* Volume LXXIII, 1997.

Behr, Joyce, "Grapetown Cemetery Records", *The Gillespiean*, Volume 3, No. 2.

Biesele, R. L., "The Texas State Convention of Germans in 1854", *Southwestern Historical Quarterly*, Volume XXXIII, April 1930.

Boerne Genealogical Society, "List of Petitioners Against the Formation of Kendall County", *Keys To The Past*, Volume III, No. 2, April 1984.

Boerne Genealogical Society, "List of Petitioners For the Formation of Kendall County", *Keys To The Past,* Volume III, No. 2, April 1984.

Boerner, Bernhard Dr. Jur, "The Bonnet Family From Chambons In The Dauphine", An English Translation, *Deutfsches Geschlechterburh,* Volume 60, Part I, 1928.

Brockway, Michael, "Comfort in the Country", *Texas Highways,* Volume 42, No. 12, December 1995.

Clare, Mary, "Bloody Ground: The Incident on the Nueces", *Civil War,* Issue Number 70, October 1998.

Bibliography

Creech, Caren G., "Treue der Union", *Hill Country Vista Magazine,* Summer-Fall, 1996.

Crosby, David F., "The Wrong Side Of The River", *Civil War Times Illustrated,* Volume 36, Number 7, February 1998.

Croteau, Roger, "Historian tells of area settlers' breakaway plan", *The Journal,* Volume XXV, Number 1, Spring 2003.

Elliott, Claude, "Union Sentiment in Texas 1861—1865", *Southwest Historical Quarterly,* Volume L, April 1947.

Engelke, Louis B., "They Had Enough of Duff", *San Antonio Express Magazine,* date not known.

Frizzell, Robert W. "Killed by Rebels; A Civil War Massacre And Its Aftermath," *Missouri Historical Review,* Volume 71, Number 4, 1977.

Geue, C. W., "Die Lateinische Ansiedlung in Texas [The Latin Settlement in Texas] by August Siemering", '*Texana*', Volume V, Summer 1967.

Gold, Gerald R., "Gillespie County in the Civil War", *The Junior Historian.* Date not known.

Herbst, Walter, "Comfort In Texas", *The Junior Historian,* Volume VI, November 1945.

Heinen, Hubert, "German-Texan Attitudes toward the Civil War," *Yearbook of German American Studies,* Volume XX, 1985.

Hoffman, David R., ed., "A German-American Pioneer Remembers: August Hoffmann's Memoir", *Southwestern Historical Quarterly,* Volume CII, April 1999.

Hohenberger, Cynthia, "The Grapetown Legacy", *Texas Junior Historian,* Volume 26 #1, September 1965.

Humphrey, David C., "A Very Muddy and Conflicting View: The Civil War as Seen from Austin, Texas", *Southwestern Historical Quarterly,* Volume XCIV, January 1991.

Jordan, Terry, "A Religious Geography of the Hill County Germans of Texas", *Ethnicity on the Great Plains,* Frederick Lueble editor, Lincoln, Nebraska, University of Nebraska Press, 1980.

Kamphoefner, Walter D., "New Perspectives on Texas Germans and the Confederacy", *Southwestern Historical Quarterly,* Volume CII, April 1999.

Kelton, Elmer, "The Fleeing Sixty—A True Story", *Ranch Romances Magazine,* January 18, 1952.

Knopp, Kenn, "The Situation in Germany and in Texas 1840 – 1860; Julius Theodor Splittgerber (1819 – 1897)", *The Journal,* Volume XXVIII, Number 4, Winter 2006.

Knopp, Kenn, "Sam Houston and The Texas Hill County Germans", *The Journal,* Volume XXVIII, Number 2, Summer 2006.

Lack, Paul D. "Slavery and Vigilantism in Austin, 1840-1860" *Southwestern Historical Quarterly,* Volume LXXXV, No. 1, July 1981.

Lossing, Benson J., "Sufferings of Texan Loyalist", *Pictorial Field Book of The Civil War,* Volume II, 1997.

McDaniel, Ruel, "Massacre By White Men", *Frontier Times,* Volume 43, Number 5, August-September 1969.

McGowen, Stanley S., "Battle or Massacre? The Incident On The Nueces, August 10, 1862", *Southwestern Historical Quarterly,* Volume CIV, July 2000.

Miller, Gary L., "Historical Natural History: Insects and the Civil War", *American Entomologist,* Volume 43, Number 4, Winter 1997.

Moidenhauer, Roger, "Benton County Lutherans and the Battle of Cole Camp June 19, 1861", *Concordia Historical Institute Quarterly,* Number 61, Winter 1988.

Moore, Frank, ed. "Massacre Of The Germans In Texas", *Rebellion Record: A Diary of American Events,* Volume 6, 1977.

Muir, Andrew Forest, "San Houston and the Civil War", *Texana,* Volume VI, No 3, Fall 1966.

Nixon, Victor, Jr., "An Encounter With The Partisan Rangers", *The Junior Historian,* Volume XXIV, September 1963.

Olsson, Karen, "Rock the Boat: Giant Boulder Causes Discomfort in Hill County Town" *The Texas Observe,* Volume 18, Issue 12, 1999.

Roberts, Edwin A., Jr., "Mainstreams – The 'Last Gentlemen's War' Also Had Its My Lai", *The National Observer,* March 31, 1973.

Rogers, Laura Wood, "Frederick Law Olmsted and the Western Texas Free-Soil Movement", *The American Historical Review*, Volume LVI, No. 1, October 1950.

Rutherford, Phillip, "Defying The State of Texas", *Civil War Times Illustrated,* Volume 19, No. 1, April 1979.

Sansom, John W., "The German Citizens Were Loyal To The Union", *Hunter's Magazine,* Volume II, November 1911.

Sansom, John W., "The Bloody Battle of Nueces River", *K. Lamaity's Harpoon,* Volume Six, Number Eight, October 1, 1908.

Schutze, Albert, "Was a Survivor of the Nueces Battle", *Frontier Times,* Volume 2, October 1924.

Schweppe, Ida Altgelt, Mrs. "Ernst Herman Altgelt, Founder Of Comfort", *Hunter's Magazine,* November 5, 1936.

Selcer, Richard and William Paul Burrier, "What Really Happen On The Nueces River?", *North and South,* Issue #2, January 1998.

Shook, Robert W., "The Battle of the Nueces, August 10, 1862", *Southwestern Historical Quarterly*, Volume LXV, October 1961.

Siemering, August, "Die Lateinische Ansiedlung in Texas", *Der Deutsche Pionier,* Volume 10, 1874.

Smyrl, Frank H. "Texans in the Union Army, 1861 – 1865", *Southwester Historical Quarterly,* Volume LXV, October 1961.

Smyrl, Frank H. "Unionist in Texas, 1856—1861", *Southwest Historical Quarterly*, Volume LVIII, April 1965.

Stewart, Anne, "The Town With No Sunday Houses: Comfort, Texas", *The Journal,* Volume XXI, Number 3, Fall 1999.

Tausch, Egon Richard, "GOTT mit UNS: The Texas German Confederates", *Southern Partisan,* Fall Issue, 1985.

Trorpe, Helen, "Historical Friction", *Texas Monthly,* October 1997.

Wolff, Linda, "Along the Indianola Trails to New Braunfels and San Antonio", *The Journal,* Volume XXVIII, Number 2, Summer 2006.

Papers, Theses, And Dissertations

Biermann, Karl, *Bushwhackers and the Haengerbande – Texas Hill Country Germans and the Civil War,* San Antonio, Texas: Graduate Paper, History 5153, University of Texas at San Antonio, December 1997.

Curtis, Sara Kay, *A History of Gillespie County, Texas, 1846 – 1900,* Austin, Texas: M. A. Thesis, University of Texas, Austin, 1943.

Dykes-Hoffmann, Judith, *Treue Der Union: German Texan Women On The Civil War Homefront,* San Marcos, Texas: M. S. Thesis, Southwest Texas State University, 1996.

Felger, Robert Pattison, *Texas In The War For Southern Independence 1861 – 1865,* Austin, Texas: Ph. D. Dissertation, University of Texas, 1947.

Hall, Ada Maria, *The Texas Germans in State and National Politics, 1850 – 1865,* Austin, Texas: M. A. Thesis, University of Texas, 1938.

Heintzen, Frank W., *Fredericksburg, Texas During The Civil War And Reconstruction,* San Antonio, Texas: M. A. Thesis, St. Mary's University, 1944.

Johnson, Melvin C., *A New Perspective For The Antebellum And Civil War Texas German Community,* Nacogdoches, Texas: M. A. Thesis, Stephen F. Austin State University, 1993.

Weinheimer, Ophelia Nielsen, *The Early History of Gillespie County,* San Marcos, Texas: M. A. Thesis, Southwest Texas State Teachers College, 1952.

Unpublished Letters, Manuscripts, Journals, Diaries and Miscellanous

Author Not Stated, "Biographical Sketch of Jacob Luckenbach", an undated document.

Betzer, Roy J., "Early Fredericksburg and Fort Martin Scott", n. d. Copy in possession of author.

Brinkmann, Alex, "Memoirs Capt. John W. Sansom", Notes taken by Brinkman during interviews with John Sansom about 1918. Copy provided by Gregory J. Krauter, Comfort, Texas.

Comfort Heritage Foundation, Inc., Handout, n. d.

Committee of Safety of Medina Castroville, James Menke Collection, San Antonio, Texas, copy located in Castroville Public Library, Castroville, Texas.

Crenwelge, Kenneth O., "Kallenberg Family Data", Series of Kallenberg Family Information, n. d.

Diamond Jubilee Souvenir Book of Comfort, Texas, Commemorating 75th Anniversary, August 18, 1929.

Diary of Elise Tips Wuppermann, "Elise Tips Wuppermann Through Her First Years of Marriage, "September 1850-July 1860"

Dornwell, Anne, "The Sage of Captain Philip Braubach" An unpublished manuscript, n. d. c1986, copy provided by Gregory J. Krauter, Comfort, Texas.

Folder, Centennial Observance "Battle of the Nueces" held at the Union Monument Comfort, Texas on Sunday August 12, 1962, located at Institute of Texan Cultures, San Antonio, Texas.

Hall, Roland "Biographical Data on Frederick Tegener", Mesquite, Texas May 19, 1997.

Hamilton, Byrde Pearce, "The Old Dutch Battleground Of Kinney County", *Early History Of Uvalde and Surrounding Territory*, Section Two, El Progresso Memorial Library, Uvalde, Texas.

Harper, Rose Ann, "History of My Mother's Life", undated manuscript provided by John F. Koepke.

Heinen, Hubert, "Exhibit 3 – Genealogy – Stieler Family". Copy provided by Anne Stewart, Comfort, Texas.

Henderson, Wanda, "Henderson Cemetery Historical Marker Dedication", June 10, 1990.

Hopkins, D. P., Hopkins, Diary, *San Antonio, Express,* January 13, 1918.

Kleck, Peter G., "List of Luckenbach Buschwacher" n. d. (ca 1862).

Kleck, Peter G., "List of Luckenbach Rangers", n. d. (ca 1862).

Ledger, John W. Sansom's, located in Sansom's File at Daughters of Republic of Texas (DRT) Library at The Alamo, San Antonio, Texas.

Letter, Albrecht, Frank, Jr., March 12, 1995, Austin, Texas, contained in series of documents entitled "The Kusenberger Family", copy provided by Gregory J. Krauter, Comfort, Texas.

Letter, Albrecht, Frank, Jr., August 11, 1999, Austin, Texas to author.

Letter, Altgelt, Ernst, May 21, 1863 to John H. Thurmund, copy provided by Gregory J. Krauter, Comfort, Texas.

Letter, Carrigan, Catherine, August 1, 1991 to Gregory Krauter, copy provided by Gregory J. Krauter, Comfort, Texas.

Letter, Cramer, Ernest, October 30, 1862 to his parents. An English Translation, copy provided by Gregory J. Krauter, Comfort, Texas.

Letter, Degener, E. July 20, 1865 to Mr. Schwethelm Contained in *Heinrich Joseph Schwethelm His Life and Times,* Compiled by Comfort Heritage Foundation Archival Staff, Carolyn Lindemann Overstreet, Editor n.d.

Letter, Fehrenbach, T. R., March 26, 1997 San Antonio, Texas to author.

Letter, Fenwich, Robert and Penelope, February 5, 1998, Perth, Scotland, to author.

Letters, Gonsalves, Joanne M., Littleton, Colorado, December 2001 to March 2002 to author, containing numerous research material on James M. Duff.

Letter, Hoffmann, August, September 1, 1925 to his children, copy provided by Gregory J. Krauter, Comfort, Texas.

Letter, Kamphoefner, Walter D., Department of History Texas A and M University, College Station, Texas, February 26, 1995 to author.

Letters, Kemper, Patricia A., Golden, Colorado, July 1996 to August 1996 to author, containing numerous research material on James M. Duff.

Letters, Kleck, Peter, Fredericksburg, Texas to Paul Camfield, Gillespie County Historical Society, May 8, 1997 and June 3, 1997 with copies of rosters [lists] of 'Luckenbach Buschwhackers' and 'Luckenbach Rangers' found in billfold of Sylvester Kleck.

Letter, Kuechler, Jacob, undated (c1887) to Honorable James Newcomb, San Antonio, Texas, located in Jacob Kuechler's File, Institute of Texan Cultures, San Antonio, Texas.

Letter, Lawhon, John, Blanco County, January 7, 1862 to Governor Lubbock.

Letter, Montague, Charles, Bandera Justice of the Peace, Bandera, Texas, July 19, 1861 to Governor Clark.

Letter, Nowlin, J. C., Blanco County n. d. to Governor Lubbock.

Letter, Patton, B. F., Blanco County n. d. to Governor Lubbock.
Letter, Pilgrim, Michael, Textual Reference Division, National Archives, Washington, D. C. January 31, 1995, to author.

Letter, Robison, Neil, Curries Creek, December 10, 1861 to Governor Lubbock.

Letters, Rodriquez, Jesse, San Antonio, Texas to Donald Swanson 1997, Curator Emeritus, Old Guard House Museum, Fort Clark Springs, Texas.

Letter, Sansom, John W., August 14, 1907 to James T. DeShields, located in John W. Sansom's File, DRT Library, San Antonio, Texas.

Letter, Sansom, John W., August 18, 1907 to James T. DeShields, located in John W. Sansom's File, DRT Library San Antonio, Texas.

Letter, Schlickum, Julius, December 21, 1862 to his father-inlaw, copy provided by Anne Stewart, Comfort, Texas.

Letter, Schwethelm, Henry May 16, 1913 to his grandson Otto Schwethelm.

Letters, Simon, Ferdinand, from June 1862 to 1865 to his wife and other family members written while he was in prison in San Antonio. Letters are in possession of John Garnett, Brenham, Texas.

Letter, Tegener, Fritz, Austin, Texas to Herr August Duecker Gillespie County, Texas, August 23, 1875.

Miscellaneous Clippings, Articles, Notes, and other documents filed in 'Local History' Vertical File, Pioneer Memorial Library, Fredericksburg, Texas.

Notes containing family story of the Lich Family of Comfort, Texas, copy provided by Gregory J. Krauter, Comfort, Texas.

Sagebiel, Hatty L., "The Massacre", 1990, copy provided by Gregory J. Krauter, Comfort, Texas.

Schmidt, Eduard, English Translation of Address Commemorating The 50th Anniversary Of The Battle On The Nueces, August 1862, copy provided by Gregory J. Krauter, Comfort, Texas.

Schulz, Robert G., Jr., The Nueces Massacre, also known as the Battle of the Nueces, from a family history chapter entitled, "The Germans: *Geh Mit Ins Texas*", copy in possession of author.

Schweppe, F. W. "Bonnet Brothers", copy located in the Edith A. Gray Library, Boerne Public Library, Boerne, Texas, copy provided by Ester Bonnet Strange, Kerrville, Texas.

Siemering, August, "Texas, Her Past, Her Present, Her Future", Translated from *Texas Voewarts*, August 10 – October 12, 1894. Dresel File in San Antonio Public Library.

Stock, A. D., "Fritz Schellhase Journal" (c1912) dictated to A. D. Stock by Fritz Schellhase, copy provided by Anne Stewart, Comfort, Texas.

Travers, Douglas N. "Biographical Sketch of Colin D. McRae" an unpublished manuscript, 1997.

Usener, Raymond, "Jacob and Ludwig Usener Story" an unpublished article, 1997.

Whiteturkey, Lucille V., "The Carl Beselers of Welfare, Texas", an unpublished manuscript, ca 1995.

Zoeller Family Descendants, "A Family History of Hellwig Karl Ludwig Adolph Zoeller and Augusta Zoeller" an unpublished manuscript, n. d., copy provided by Margie Morgan, a Zoeller family descendant.

Index

Pehl, Mathias, 167, 448, 451, 518, 530, 531, 532, 546, 590, 616, 639, 668, 679, 684
Pettis, George H., 528, 549
Pfannensteil, August, 245
Pfannensteil, William, 245
Pfeiffer, Adolf, 536, 551
Pierce, Leonard, 228, 263, 293, 294, 310, 458, 515, 516, 527
Poetsch, Christian, 296, 639, 670
Poetsch, Joseph, 42, 248, 296, 448, 451, 530, 531, 532, 546, 590, 639, 669, 679, 684
Poetsch, Peter, 42
Preece, Richard L., 141, 292, 309
Preece, Wayne P., 292, 309
Preece, William M., 140, 292, 309
Preece, William Martin, 282, 307
Prescott, Aaron, 281, 647
Probst, Christian, 376, 419
Quintel, Adolph, 376, 419
Rabb, Thomas, 217, 222, 337, 572, 575
Rabb's Company, 219, 360, 574
Radeleff, Rudolph, 78, 149, 166, 172, 180, 181, 187, 188, 201, 245, 305, 323, 332, 339, 583, 613, 619, 639
Ragsdale, Thomas B., 235, 236, 265
Rahe, Frederick, 376, 419
Ransleben, Julius, 533
Rausch, Henry, 550, 590, 639, 670
Real, Adolph, 527, 548

Real, Casper, 369, 527, 548, 568, 607
Reed, Erastus, 284, 297, 307, 323
Reeh, Reinhardt, 558, 594, 639
Rees, Alonzo, 484, 589, 604, 607
Rhodius, Christian W., 71, 323, 643
Richardson, Samuel, 235, 265
Richarz, Henry J., 148, 173, 208, 220, 223, 236, 275, 287, 289, 621
Riordan, Thomas, 524, 525, 547
Robinson, Alfred, 360
Robinson, John Frank, 446, 475, 481, 502, 652, 672, 680
Rosenthal, Adolph, 356, 365, 376, 606, 643
Rosser, Benjamin Franklin, 441, 469, 481, 501, 503, 660, 673, 680
Ruebsamen, Adolph, 296, 430, 509, 512, 541, 543, 639, 668, 676, 682
Ruebsamen, Louis, 296, 509, 511, 542, 543, 639, 668, 676, 682
Ruetli League, 162, 163, 164
San Antonio Zeitung, 69, 71, 73, 77, 80, 264, 320, 722
San Patricio Battalion, 78, 81, 334, 360
Saner, Thomas, 206, 620
Sanger, H. C., 369, 643
Sansom, John W., 101, 151, 152, 153, 154, 155, 156, 157, 158, 159, 161, 165, 174, 175, 189,

Other books in this series byWm. Paul Burrier, Sr. are available by writing to W. Paul Burrier, P.O. Box 1096, Leakey, TX 78873. Price includes Texas state sales tax.

Germans in Texas during the Civil War	$30.00
Confederate Military Commission	$20.00
Nueces Source Documents	$45.00
Nueces Battle Myths Facts	$25.00